Lecture Notes in Electrical Engineering

Volume 190

For further volumes:
http://www.springer.com/series/7818

Society of Automotive Engineers of China
(SAE-China) · International Federation of
Automotive Engineering Societies (FISITA)
Editors

Proceedings of the FISITA 2012 World Automotive Congress

Volume 2: Advanced Internal Combustion Engines (II)

SAE CHINA

FISITA

Springer

Editors
SAE-China
Beijing
People's Republic of China

FISITA
London
UK

ISSN 1876-1100 ISSN 1876-1119 (electronic)
ISBN 978-3-662-51140-4 ISBN 978-3-642-33750-5 (eBook)
DOI 10.1007/978-3-642-33750-5
Springer Heidelberg New York Dordrecht London

Printed on acid-free paper

Springer is part of Springer Science+Business Media (www.springer.com)

Contents

Part IV After Treatment and Emission Control

Part V
Flow and Combustion Diagnosis

Simulation of EGR Stratification on Timing-Sequential Regionalized Diesel Combustion

Zhaojie Shen, Zhongchang Liu, Jing Tian, Kang Li and Jiangwei Liu

Abstract In this paper, the influences of EGR distribution on timing-sequential regionalized diesel combustion characters and emissions have been investigated numerically. Timing-sequential regionalized diesel combustion described in previous paper is realized by post injection and proper in-cylinder chemical atmosphere transport to keep post injection fuel injecting to high oxygen concentration region. Cases with EGR stratification have similar heat release rate character compare to uniform EGR case while lower soot emissions and identical NO emissions as local EGR rate equal to uniform case, similarly, lower NO emissions and identical soot emissions as EGR mass equal to uniform case. Axial and radical EGR stratification that defined artificially in cylinder under the condition of 1,650 r/min and 50 % load have been numerically analysed by commercial STAR-CD code. It is indicated that both axial and radical distribution of EGR stratification make NO and soot emissions decrease simultaneously, however, radical EGR distribution works better. NO emissions decrease while soot emissions keep constant under the case of EGR stratification with post injection using main injection timing advance and higher local EGR rate; it is also found that cyclic net work per cylinder is improved at some extent. Intake port shape and swirl rate also have been investigated to analyse in-cylinder gradient of EGR distribution to achieve EGR stratification. EGR gas distributes by gradient in cylinder as air and air-EGR mixture flow through two intake valve that connected to tangential and spiral intake port respectively.

F2012-A05-004

Z. Shen (✉) · Z. Liu · J. Tian
State Key Laboratory of Automotive Simulation and Control,
Jilin University, Changchun, China

K. Li · J. Liu
China FAW Group Corporation R&D Centre, Jilin, Changchun, China

SAE-China and FISITA (eds.), *Proceedings of the FISITA 2012 World Automotive Congress*, Lecture Notes in Electrical Engineering 190,
DOI: 10.1007/978-3-642-33750-5_1,

Keywords EGR stratification · Diesel emissions · Time-regionalized combustion · CFD simulation · Diesel combustion

1 Introduction

Diesel engine is facing a severe challenge due to the pressure from energy and environment protection even with high efficiency which is essential both in terms of fuel economy and the impact on global warming through CO_2 emissions [1] and what made this situation worse are the published research results from World Health Organization June 12, 2012 that exhaust from diesel engines can cause cancer [2]. As we all known, PM and NO_x emissions are the main culprits, in fact, diesel researchers have been working for searching the solutions to reduce the two kinds emissions from decades on. New advanced combustion conceptions like HCCI [3], LTC [4], PCCI [5], MK [6, 7], UNIBUSS [8], NADI [9], and so on have already been developed to meet more stringent emissions standards all over the world, thanks to flexible high pressure common rail injection system and high EGR rate. Cleaning combustion now is the only way to satisfy the ultra-low diesel emission standards for limitations from after-treatment system like difficulties of setup, limited efficiency, and high cost, even immature drive cycle control strategy and unsatisfactory fuel penalty, etc. PM and NO_x trade-off has yet been proven to be substantially alleviated by feasible technologies from the knowledge of Φ-T diagram [10]. EGR and flexible common rail injection system balance the generation conditions of NO_x and PM to obtain ultra-low engine-out emissions of both simultaneously in cylinder.

EGR effects in many ways on diesel combustion include positive effect on NO_x emission and negative effect on BSFC, PM and HC emissions [11–15], high EGR rate is generally used to decrease combustion temperature in advanced combustion technology such as LTC, PCCI, MK. Wagner [16] investigated the simultaneous reduction of nitrogen oxides (NO_x) and particulate matter (PM) in a modern light-duty diesel engine under high exhaust gas recirculation (EGR) levels by two approaches, one is using throttle to reach high EGR rate, and the other is combining the EGR and injection parameters. Unfortunately fuel penalty always accompanies with high EGR [16, 17]. The effect of cooled EGR temperature for various EGR rate on performance and emissions of a turbocharged DI heavy duty diesel engine is revealed that the decrease of EGR gas temperature has a positive effect on BSFC, soot (lower values) while it has only a small positive effect on NO [18, 19]. EGR effects on diesel combustion have been investigated by international engine researchers through experiments and CFD simulations.

Diesel spray vaporization is enhanced by flexible common rail injection system due to high injection pressure and multiple injection strategy. Post injection strategy play a more important role on soot emissions [20–23]. Mobasheri et al. [24] considered three parameters include EGR rate, the dwell of main injection and post-injection and the amount of injected fuel in each pulse to control diesel emissions. It has found that injecting adequate fuel in post-injection at an appropriate EGR allows significant soot reduction without a NO_x penalty rate. Bobba et al. [25]. issued that for LTC post-injection conditions, the absolute crank angle position of the post injection appears more important for soot reduction than the dwell between injections, The combination of EGR and post injection has been used widely and effectively to simultaneously decrease NO_x and PM emissions. However, the emission legislation is going to be more stringent, which is always accompanied by fuel penalty and increase of HC emissions. In-cylinder stratification is a possible method for simultaneously reducing NO_x and PM. In-cylinder stratification has mainly been studied in HCCI. Fuyuto et al have suggested in-cylinder EGR stratification concept that reduced smoke while NO_x was maintained at a medium speed/load level [26]. Choi et al. [27] evaluated the potential for combustion and emission control by in-cylinder EGR stratification and single injection strategy in a single cylinder engine experiment using a 2-step piston to improve the EGR stratification. Sandia National Laboratories Combustion Research Facility [28] provides PM and NO_x formation process by combination of CFD and laser diagnostics, it is found that soot and NO_x emissions generate at different combustion period of main injection as seen in Fig. 3 [left]. It is feasible to control NO_x and PM at different combustion period and different region. It is also testified by timing-sequential regionalized combustion [30].

In this paper, EGR stratification is considered based on timing-sequential regionalized combustion for two reasons. First, the use of high EGR rates by which the ultra-low NO_x engine-out emissions is obtained creates high need of EGR cooling requirement of intake gas, and adds considerably to engine cooling. Second, post injection together with proper in-cylinder air motion oxide soot emissions substantially obtained by previous studies seen in Fig. 2 [29, 30]. It has shown that oxygen concentration around the spray direction is very low at the end of main fuel injection, however as in-cylinder gas moved, when post injection start, oxygen concentration is made good around fuel spray which enhanced post injection fuel combustion to lower PM emissions. That is to say in-cylinder flow motion is of great consequence to PM oxidation by proper post injection. So it is enough to simply control NO generation at the main fuel combustion, complemented by proper post injection and in-cylinder gas flow motion to reach the balance of engine emissions and performance of fuel and power.

2 System Modelling and Model Validation

The computational mesh was created using STAR-CD code for a whole cylinder model, and calculated from 30° CA BTDC to 90° CA ATDC. Combustion model, turbulence model, spray model and emission model were chosen and computational model was validated by experimental results of CA6DL2-35R diesel engine

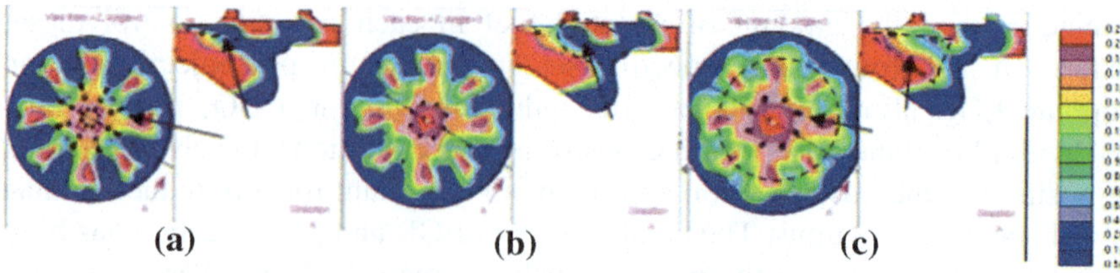

Fig. 1 Distribution of in-cylinder oxygen mass concentration from the end of main fuel injection (381° CA) to the start of post injection (385° CA)

and optical single cylinder engine in previous paper [29, 30]. Boundary conditions were set based on experimental results from CA6DL2-35R diesel engine (Fig. 1).

3 Analysis and Discussion

3.1 Comparison of Uniform EGR and EGR Stratification Cases

Comparisons of EGR stratification cases which achieved by means of setting in-cylinder EGR distribution using subprogram in STAR-CD code and uniform EGR case were discussed in this paper, and total injection fuel is kept constant for all simulation cases. The simulation model validation was conducted in previous paper [29, 30] which in good qualitative agreement with tested engine and optical single cylinder engine. In Fig. 2 (left), as EGR mass introduced in cylinder decrease in EGR stratification case with the same EGR rate of uniform EGR and less EGR region, heat release rate in case (b) gains a little, while soot mass decrease up to 40 %, NO equal to case (a) comparisons of case (a) and case (c) which keeps EGR mass introduced in cylinder consistent with that of case (a) are shown in Fig. 2 (right). The Peak of Heat release rate of case (c) and case (a) are equal. In fact, ignition delay of case (c) should longer than that of case (a) as EGR rate in case (c) is twice that in case (a) around combustion area, so main injection timing has been advanced to make up the negative effect of EGR. Better yet, NO emissions decrease 12 %, while soot equal to case (a). In Fig. 2, it is proved that EGR stratification has the potential to balance the trade-off between NO and soot emissions.

Combination of EGR stratification and double injection strategy based on timing-sequential regionalized diesel combustion [29] has been simulated. The comparison results showed in Fig. 3 (right) demonstrates that with post injection the peak of soot emissions is half of single injection strategy, yet obtained a bimodal in soot history curve which means that post injection should be adjusted with EGR stratification to gain the most optimal result which successively balance the trade-of between emissions and fuel economy. In Fig. 4, cyclic net work of simulation base points described in Table 1 have been calculated, it is indicated that EGR stratification strategy could improve engine cyclic net work which means

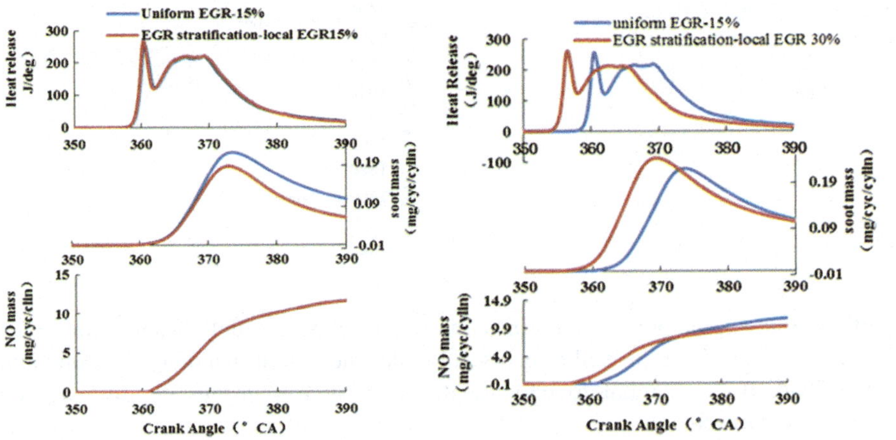

Fig. 2 Emissions and heat release comparisons between uniform EGR and EGR stratification cases with local EGR rate 15 % (*left*) and 30 % (*right*) respectively

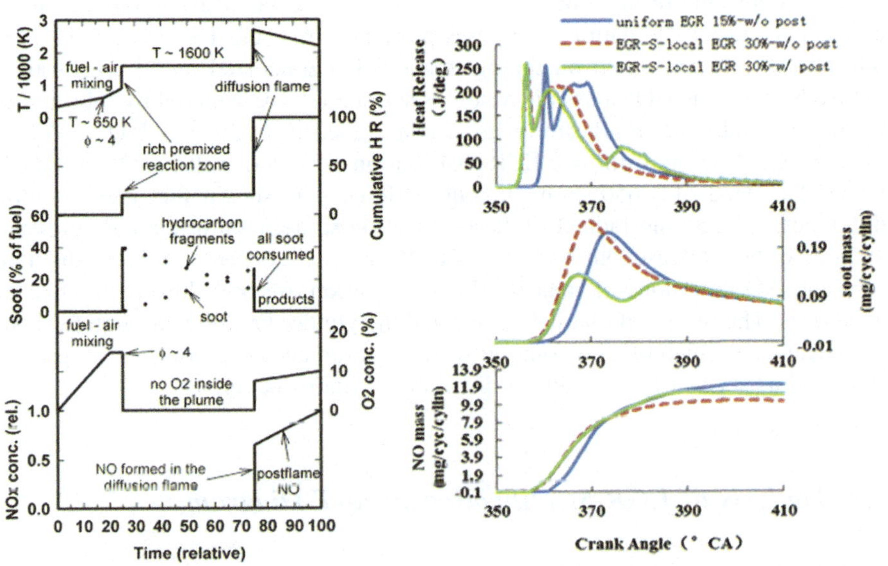

Fig. 3 Reaction speculation of emissions, temperature and heat release history during main injection combustion versus relative time which represent the start of fuel injection to the end of main injection combustion (*left* Flynn et al. [28]). Comparison s of uniform EGR and EGR Stratification with or without post injection cases on ROHR, NO and soot emissions (*right*)

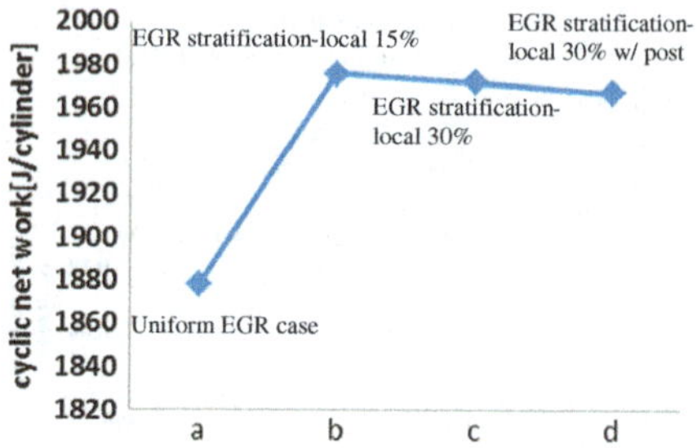

Fig. 4 Cyclic net work of four base simulation points [seen in Table 1] per cycle per cylinder

getting more engine power under the condition of constant fuel consumption. EGR stratification has been proved to be useful under the conditions of 1,650 r/min and 50 % load. It is essential to validate the strategy for various conditions even transient conditions.

3.2 Comparison Between Radical and Axial EGR Stratification Strategy

In this section, the results obtained for two kinds of EGR distributions in engine cylinder based on simulation base cases presented in Table 1 is considered. Post injection is selected based on timing-sequential regionalized diesel combustion. Figure 5 shows the CO_2 mass distributions set in STAR-CD model by subroutines for initial conditions. EGR stratification is designed from 30° CA BTDC. As in-cylinder gas flow and diffuse, EGR distribution persisted well to 8° CA BTDC under the speed of 1,650 r/min. In Fig. 6 (left), it is shown that there is little difference of axial and radical EGR stratification strategy on in-cylinder pressure and rate of heat release, but NO and soot emissions are different, it is notable that soot and NO emissions of radical EGR distribution are both lower than that of axial ones. The reason of that is fuel injected in cylinder first flow through fresh air in radical case which obtains some fresh air entrainment to release soot generation area, and as seen in Fig. 5, radical distribution retains better.

3.3 Analysis of EGR Stratification in *Φ*-T Diagram

Φ-T diagram is a useful method for analysing the way to balance the NO and soot emission trade-off. In Fig. 7 (right), NO emissions generate substantially during the combustion period for high combustion temperature and rich oxygen concentration. However, in Fig. 7 (left) as high EGR rate introduced to combustion area, NO generation condition is restrained during main combustion phase.

Table 1 Simulation base cases

	Uniform EGR (a)	EGR-S-radical[a] (b)	EGR-S-radical (c)	EGR-S-radical (d)
Operating case	B50	B50	B50	B50
Main injection timing	8° CA BTDC	8° CA BTDC	12° CA BTDC	12° CA BTDC
EGR rate	15 %	Local 15 %	Local 30 %	Local 30 %
Injection strategy	Single	Single	Single	Double[b]

[a] EGR stratification, distribution along radical direction
[b] With post injection 20 % of total fuel. In this paper post injection parameters are chosen based on timing-sequential regionalized diesel combustion [29]

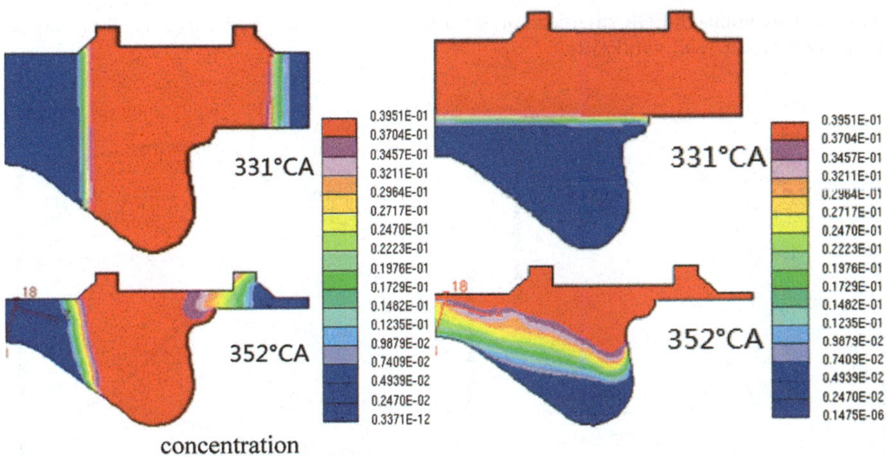

Fig. 5 Mass concentration distribution of CO_2 in cylinder at 352° CA radical EGR stratification (*left*) and axial EGR stratification (*right*)

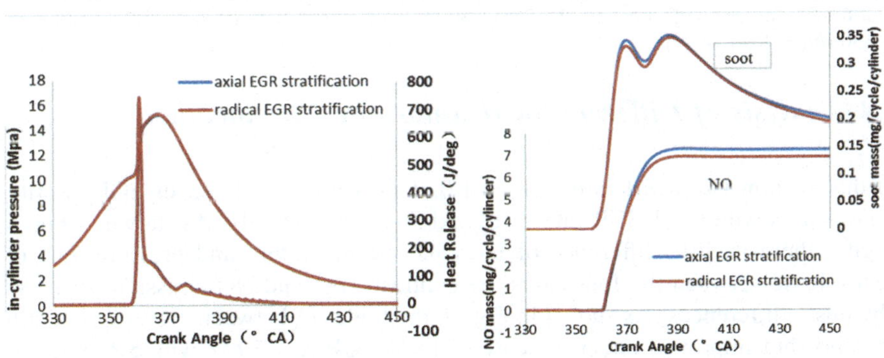

Fig. 6 Comparisons of axial and radical EGR stratification on engine emissions and heat release rate

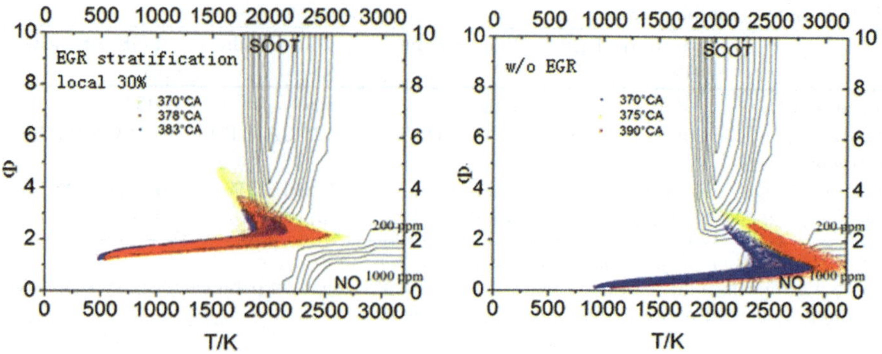

Fig. 7 Φ-T diagrams of EGR stratification (*left* 1,650 r/min, load of 50 %, local EGR rate 30 %) and uniform case (*right*, w/o EGR)

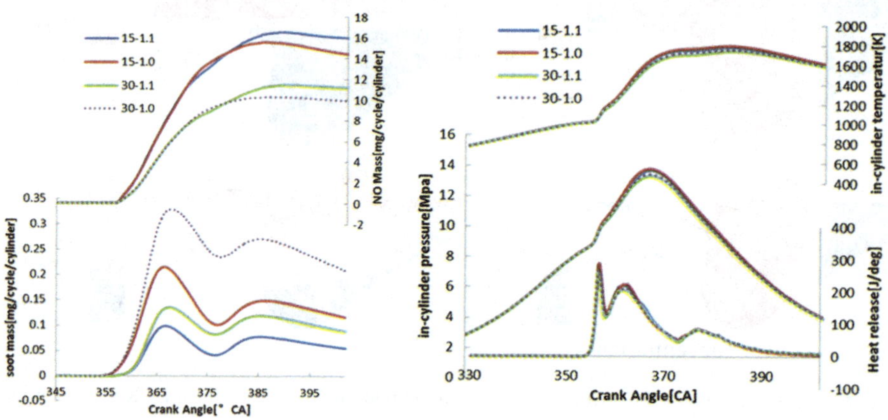

Fig. 8 Comparisons of diesel emissions and engine performance parameters between EGR stratification cases (with local EGR 15 and 30 %) with different swirl ratio (SR = 1.0, 1.1), (1,650 r/min, load 50 %)

3.4 Analysis of Different Swirl Ratio with Stratification

In this section two swirl ratio (SR = 1.0, 1.1) were considered in EGR stratification cases with local EGR rate of 15 and 30 % respectively. As shown in Fig. 8 (right), there is little difference of in-cylinder temperature and pressure, rate of heat release between the four cases, by contrast, soot and NO emissions displays obviously differences. As shown in Fig. 8, the trade-off between NO and soot still presents, but improved under the case of local EGR rate 30 % with SR = 1.1 as obtaining lower NO and soot emissions simultaneously. That is to say, swirl ratio affects diesel emissions more for higher EGR rate cases, and affects more on soot emissions.

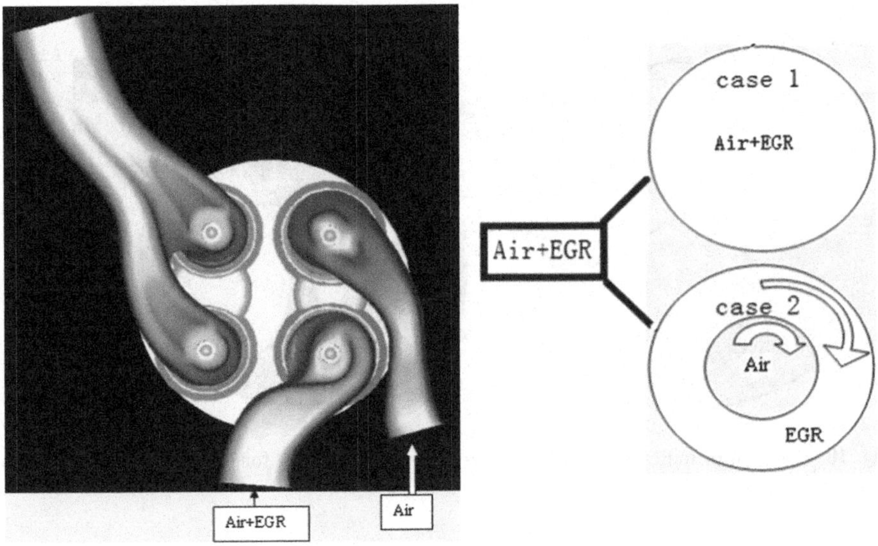

Fig. 9 Analysis of EGR stratification in four valve diesel engine, two intake valve connected to tangential and spiral intake port respectively

3.5 Flow Analysis of Radical EGR Stratification with Tangential and Spiral Intake Port

In this section, the formation of EGR stratification is discussed during intake process. Compared to setting EGR distribution by subroutines of software code, intake flow is considered to form in-cylinder EGR stratification. As two intake valve that connected to tangential and spiral intake port respectively, the distribution of EGR can be controlled by introducing different gas contents to the two ports with special shape to control gas motion. As combustion always happened around the piston bowl, EGR should distribute surrounding the bowl to decrease the combustion temperature, so spiral intake port is the way of EGR flow that can form a strong screw flow to keep EGR gas remaining around the piston bowl. The tangential intake port always connects to fresh air to keep no EGR around the cylinder wall. As shown in Fig. 9, spiral port is ventilated with air, and tangential one with air and EGR. In this section, two strategies of gas flow with spiral port are considered. One is ventilated with mixture of EGR and air; another is using an internal tube extended to valve ventilated with Air, EGR gas flow out of the pipe as seen in Fig. 9 (right, bottom). The simulation results of the two strategies are shown in Fig. 10. It is observed that the gradient of O_2 mass concentration is lower in first case than that of the second, but higher swirl intensity that forms a strong axial recirculation flow.

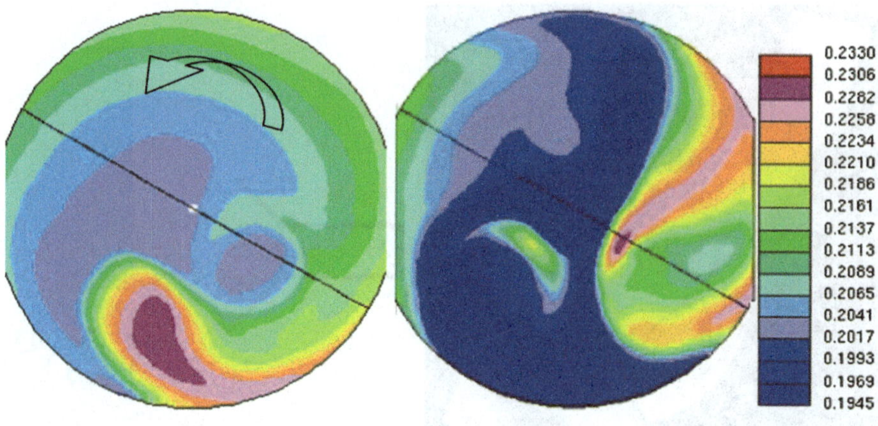

Fig. 10 Diagram of radical distribution of O_2 mass concentration for radical EGR stratification cases at 30° CA BTDC

4 Conclusion

In this paper, influences of EGR stratification on diesel combustion, emissions and heat release rate have been investigated by CFD code. EGR stratification mentioned above is achieved by subroutines in STAR-CD code. However, the method of controlling intake flow to form in-cylinder EGR stratification is also considered on four valve diesel engine by means of full engine simulation model calculated from intake valve open. EGR stratification has the potential to balance the trade-off between emissions and fuel consumption. The conclusions are summarized as follows:

As keeping local EGR rate of EGR stratification cases identical with the uniform EGR case, EGR mass introduced in cylinder decrease, soot emissions reduce substantially while NO emissions unchanged, cyclic net work per cylinder increased about 5.2 %. As keeping EGR mass introduced in cylinder of EGR stratification cases identical with the uniform EGR case, EGR rate is much higher, NO emissions decrease as combustion temperature decrease while soot emissions unchanged, cyclic net work per cylinder increased about 5 %. Diesel emissions and rate of heat release of EGR stratification distributed as radical and axial gradients in cylinder respectively have been compared. NO and soot Emissions decrease in both cases, and there is no difference on pressure and heat release rate. Yet radical EGR stratification works better. Φ-T diagram explains the reason of low NO emissions obtained by EGR stratification for higher concentration of EGR around combustion area during main combustion phase. Swirl ratio is investigated in this paper, the trade-off of NO and soot emissions still presents, but improved under the case of local EGR rate 30 % with SR = 1.1 as obtaining good NO and soot emissions simultaneously. Two intake strategies that introducing EGR to cylinder through spiral intake port has been compared, The gradient of O_2 mass concentration is lower for introducing the mixture of air and EGR to spiral intake port than that of introducing EGR and air separately by a pipe extended to intake valve in spiral port, but higher swirl intensity that forms a strong axial recirculation flow.

Acknowledgments This project is supported by the National Natural Science Foundation of China (50976046) and Graduate Innovation Fund of Jilin University (Project 20111053).

References

1. Blom D, Karlsson M, Ekholm K, Tunestål P, Johansson R (2008) HCCI Engine Modeling and Control Using Conservation Principles . SAE Technical Paper 2008-01-0789, Detroit, MI
2. Hellerman C (2012) WHO: Diesel exhaust can cause cancer. CNN June 12, 2012—Updated 1903 GMT (0303 HKT). http://edition.cnn.com/2012/06/12/health/diesel-fumes-cancer/index.html
3. Stanglmaier R, Roberts C (1999) Homogeneous charge compression ignition (HCCI): Benefits, compromises, and future engine applications, SAE Technical Paper 1999-01-3682, doi:10.4271/1999 01 3682
4. Neely G, Sasaki S, Huang Y, Leet J et al (2005) New diesel emission control strategy to meet US tier 2 emissions regulations, SAE Technical Paper 2005-01-1091, doi:10.4271/2005-01-1091
5. Simescu S, Ryan T, Neely G, Matheaus A et al (2002) Partial pre-mixed combustion with cooled and uncooled EGR in a heavy-duty diesel engine, SAE Technical Paper 2002-01-0963, doi:10.4271/2002-01-0963
6. Kimura S, Matsui Y, Kamihara T (1997) A new concept of combustion technology in small di diesel engines-4th report: the effects of fuel injection rates on Mk combustion, SAE Technical Paper 978224
7. Aiyoshizawa E, Muranaka S, Kawashima J, Kimura S (1998) Development of a new 4-valve/cylinder small di diesel engine-application of the Mk concept (report 2), SAE Technical Paper 988081
8. Hildingsson L, Persson H, Johansson B, Collin R et al (2005) Optical diagnostics of HCCI and UNIBUS using 2-D PLIF of OH and Formaldehyde, SAE Technical Paper 2005-01-0175, doi:10.4271/2005-01-0175
9. Walter B, Gatellier B (2002) Development of the high power NADI™ concept using dual mode diesel combustion to achieve Zero NOx and particulate emissions, SAE Technical Paper 2002-01-1744, doi:10.4271/2002-01-1744
10. Gan S, Ng HK, Pang KM (2011) Homogeneous charge compression ignition (hcci) combustion implementation and effects on pollutants in direct injection diesel engines. Appl Energy 88(3):559–567
11. Agarwala D, Singha SK, Agarwal AK (2011) Effect of exhaust gas recirculation (EGR) on performance, emissions, deposits and durability of a constant speed compression ignition engine. Appl Energy 88(8):2900–2907
12. Peng H, Cui Yi, Shi Lei, Deng K (2008) Effects of exhaust gas recirculation (EGR) on combustion and emissions during cold start of direct injection (DI) diesel engine. Energy 33(3):471–479
13. Hashizumea T, Miyamotoa T, Akagawaa H, Tsujimurab K (1999) Emission characteristics of a MULDIC combustion diesel engine: effects of EGR. JSAE Rev 20(3):428–430
14. Benajes J, Serrano JR, Molina S, Novella R (2009) Potential of atkinson cycle combined with EGR for pollutant control in a HD diesel engine. Energy Convers Manage 50(1):174–183

15. Shi L, Cui Y, Deng K, Peng H, Chen Y (2006) Study of low emission homogeneous charge compression ignition (HCCI) engine using combined internal and external exhaust gas recirculation (EGR). Energy 31(14):2665–2676
16. Wagner RM, Green JB, Dam TQ, Edwards DK, Storey JM (2003) Simultaneous low engine-out NOx and particulate matter with highly diluted diesel combustion. SAE Trans 112:432–439
17. Maiboom A, Tauzia X, Hétet JF (2008) Experimental study of various effects of exhaust gas recirculation (EGR) on combustion and emissions of an automotive direct injection diesel engine. Energy 33(1):22–34
18. Hountalasa DT, Mavropoulosa GC, Binder KB (2008) Effect of exhaust gas recirculation (EGR) temperature for various EGR rates on heavy duty DI diesel engine performance and emissions. Energy 33(2):272–283
19. Mehdi J, Tommaso L, Gianluca D'E, Bai X-S (2012) Effects of EGR on the structure and emissions of diesel combustion. In: Proceedings of the combustion institute available online July 2012
20. Hountalas D, Lamaris V, Pariotis E, Ofner H (2008) Parametric study based on a phenomenological model to investigate the effect of post fuel injection on HDDI Diesel engine performance and emissions-model validation using experimental data. SAE Technical Paper 2008-01-0641, doi:10.4271/2008-01-0641
21. Molina S, Desantes J, Garcia A, Pastor J (2010) A numerical investigation on combustion characteristics with the use of post injection in DI diesel engines. SAE Technical Paper 2010-01-1260, doi:10.4271/2010-01-1260
22. Desantes J, Arrègle J, López J, García A (2007) A comprehensive study of diesel combustion and emissions with post-injection. SAE Technical Paper 2007-01-0915, doi:10.4271/2007-01-0915
23. Payri F, Benajes J, Pastor J, Molina S (2002) Influence of the post-injection pattern on performance, soot and nox emissions in a HD diesel engine. SAE Technical Paper 2002-01-0502, doi:10.4271/2002-01-0502
24. Mobasheri R, Peng Z, Mirsalim SM (2012) Analysis the effect of advanced injection strategies on engine performance and pollutant emissions in a heavy duty DI-diesel engine by CFD modeling. Int J Heat Fluid Flow 33(1):59–69
25. Bobba M, Musculus M, Neel W (2010) Effect of post injections on in-cylinder and exhaust soot for low-temperature combustion in a heavy-duty diesel engine. SAE Int J Engines 3(1):496–516. doi: 10.4271/2010-01-0612
26. Fuyuto T, Nagata M, Hotta Y, Inagaki K, Nakakita K, Sakata I (2010) In-cylinder stratification of external exhaust gas recirculation for controlling diesel combustion. Int J Engine Res 11:1–15
27. Choi S, Park W, Lee S, Min K, Choi H (2011) Methods for in-cylinder EGR stratification and its effects on combustion and emission characteristics in a diesel engine. Energy 36(12):6948–6959
28. Flynn PF, Durrett RP, Hunter GL, zur Loye AO, Akinyemi OC, JE. Dec (1999) Diesel combustion: an integrated view combining laser diagnostics, chemical kinetics, and empirical validation. SAE Technical Paper 1999-01-0509
29. Liu ZC, Jin HY, Wang ZS, Shen ZJ, Li K (2010) Simulation of timing-sequential regionalized combustion of a diesel engine based on post injection strategy. J Combust Sci Technol. 16(1):23–29. doi:1006-8740(2010)01-0023-07
30. Shen Z, Liu Z, Yu L, Niu X (2011) Numeric investigation of EGR and post-injection diesel emissions. Transportation, Mechanical, and Electrical Engineering (TMEE), international conference on. TMEE 2011, vol 4, 1285–1288. doi:10.1109/TMEE.2011.6199440

Investigation on the Validity Region of Online Combustion and Torque Models of Gasoline Engines with Retarded Ignition

Fangwu Ma and Zheng Qu

Abstract The recently proposed online mean-value thermodynamic combustion model suffers from its error-prone CA50 prediction. In an effort to improve the accuracy of CA50 prediction in the case of ignition retard, this paper demonstrates the convergence of normalized IMEPH-CA50 predictions towards a unanimous characteristic curve indiscriminately for various gasoline engine selections, speeds and loads. For the first time, this paper reveals the physical principle of a characteristic curve via an ideal-heat-release model and thereby formulates a validity region of the IMEPH-CA50 predictions. The predicted values outside of the region will then be corrected by a surrogate of the characteristic curve. In this way, ECUs successfully identify invalid CA50 predictions online and modulate them towards the actual values. Large-scale experiments have shown the developed method improves the accuracy in CA50 prediction, while preserving the high accuracy of IMEPH prediction.

Keywords Torque model · Gasoline engine · Prediction accuracy · Validity region · Ignition retard

---

F2012-A05-007

F. Ma (✉) · Z. Qu
Geely Automobile Research Institute, Hangzhou, China
e-mail: mafw@rd.geely.com

SAE-China and FISITA (eds.), *Proceedings of the FISITA 2012 World Automotive Congress*, Lecture Notes in Electrical Engineering 190,
DOI: 10.1007/978-3-642-33750-5_2,

1 Introduction

Recent soaring gasoline price has placed a higher demand on gasoline engine control for the purpose of higher fuel efficiency. To optimize the control of air flow, modern vehicles replace the mechanical throttle with electronic throttle control (ETC.) for complete air path control [1, 2]. As a result, a valid torque model is required in order to project the desired driver torque to the desired air mass [3]. In the case of ignition angle (IGA) retard, the required air amount can vary vastly because torque magnitude can be altered by late ignition [4]. Current engine control units (ECU) makers have developed a simple torque model to represent the late IGA effects. The model uses a parabolic data-fitted curve to map a retarded ignition angle (IGA) to a normalized indicated-mean-effective-pressure-high (IMEPH) loss, which is named as the "data-fitted torque model". To further cut the cost, ECU makers have been practicing to use a unique model to support all engine fleets on the market. For years, this model has been considered sufficient in fulfilling the requirement of torque prediction accuracy for all gasoline engines. The data proof of this action, however, has been scarce in literature.

This paper demonstrates large IMEPH-IGA dataset over wide range of engine selections and evaluates the model performance with numerical proof. In Sect. 2.1, a normalized graph will render the torque data in the entire operation range and thereby reveal both the performance and limitation of the data-fitted torque model. The major limitation is found to be the large model error in large IGA retard region. In order to renovate the accuracy in the particular region, the authors recently proposed an online mean-value thermodynamic combustion model that generates cycle-to-cycle IMEPH and CA50 prediction [5, 6]. When applied to the case of ignition retard, the new model has shown more accuracy and robustness in torque prediction compared with that of a data-fitted model. This paper will further expound the amount of improvement in IMEPH precision with large-scale data proof. Meanwhile, this paper will also investigate into the quality of model's CA50 prediction and develop a method to explore its use in torque validity diagnosis.

With fluid energy dynamics fully monitored, the online combustion model can provide real-time prediction of combustion phasing, such as 50 % fuel burn moment (CA50). This feature turns out to be equally important as the torque prediction capability in today's engine control design, because CA50 has been widely used as the most representative index of combustion quality [7–9]. Some advanced control strategy relies heavily on the knowledge of CA50. One example is the self-calibration scheme that modulates the VVT set-points according to the real-time knowledge of torque and CA50 [8], in which way an engine can defy aging and maintain high efficiency throughout its life. There has been an effort to corporate the online combustion model with the self-calibration scheme. No significant success has been made due to more than 15 % averaged error in the predicted results by the combustion model (or other CA50 models) [5, 8]. This amount of error fails most advanced control strategies while their nominal benefit

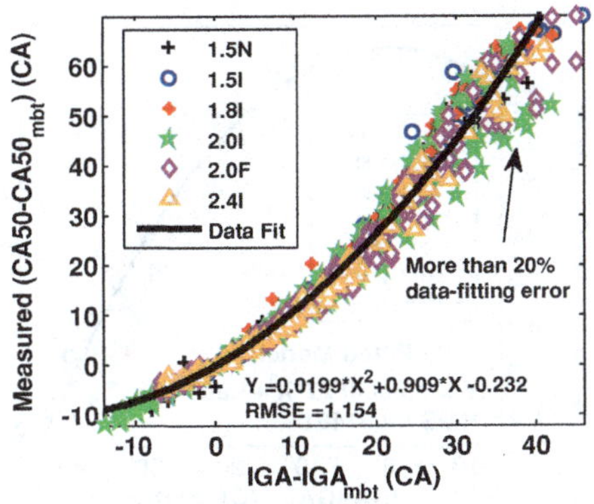

Fig. 1 If using IGA projection, the performance of CA50 prediction degrades in large IGA retard region

is commonly less than 8 %. As a result, there exists strong urge to improve the precision of online CA50 prediction.

Several efforts have been made to improve the CA50 prediction online. One is to carefully parameterize the coefficients of the combustion model according to different running conditions [5, 10]. Similar with the data-fitted torque model, this method can only be accurate on the test-bench and its performance will gradually degrade during long-time running. Another method is to corporate with a polynomial-fitted map that projects CA50 using ignition timing and combustion duration [8]. Figure 1 summarizes the performance of polynomial-fitted method through a plot of the IGA shift with respect to CA50 shift. It illustrates a fact that there exists more than 20 % fitting error in large ignition retard region. Such amount of error jeopardizes an optimization scheme using CA50 values.

This paper will propose a new method to improve the accuracy of CA50 prediction in large retard region which can be implemented in real-time operation. The new method employs a normalized IMEPH-CA50 characteristic curve to modulate the raw predictions of the combustion model. The engine IMEPH-CA50 characteristic curve has been found through experimental measurements on the test-bench and proven to be representative for wide engine selections and conditions [10]. In Sect. 2.2, experiments will show that the predicted IMEPH-CA50 results, which are calculated by the combustion model, also converge towards the characteristic curve with slightly larger excursion than the measured data do. Section 2.3 will reveal the analytical principle of the IMEPH-CA50 characteristic curve by investigating into the physics of a combustion cycle. Then this knowledge is used to formulate a "validity region" of IMEPH-CA50 prediction. Section 2.4 will integrate all model segments and propose a method to use the "validity region" as a criterion to distinguish and modulate the raw predictions. Section 3 will show the approach of the validation and the actual performance of our overall method.

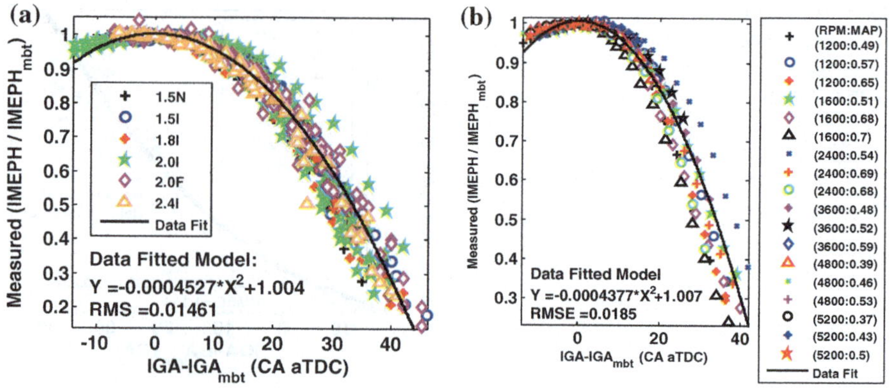

Fig. 2 The nomarlized IMEPH with respect to the ignition retarded from MBT showing the performance of the data-fitted torque model for **a** six different engines; **b** 2.0I engine of different engine speeds (RPM) and manifold-air-pressure (MAP) loads in (*bar*), with parabolic curve found by data-fitting technique

2 The Online Torque and Combustion Model

2.1 The Data-Fitted Torque Model: The Conventional Way

In the case of ignition retard from MBT timing, conventional ECUs of a light duty vehicle estimate gasoline engine torque, or $\widehat{IMEPH}$, using a MBT value and a scale function, as shown below in Eq. (1).

$$\widehat{\mathrm{IMEPH}} = \mathrm{IMEPH_{MBT}}(\mathrm{m_{air}}, \mathrm{N})\mathrm{h}(\Delta\theta) \tag{1}$$

where $IMEPH_{MBT}$ is the nominal MBT torque value which is almost proportional to the inducted air mass m_{air} and engine speed $N \cdot h(\Delta\theta)$, which is a scalar between 0 and 1, is the normalized torque loss due to ignition timing retarded from MBT, or $\Delta\theta$. A small amount of experiments have discovered that $h(\Delta\theta)$ is quite monotonic with little dependency on other engine variables [11]. ECU makers concluded such property and formulated a 1-D map between $\Delta\theta$ and h. Then the tabulated map is used in either torque estimation, or air mass calculation from a torque request. When applying this method to engine fleets, engineers employ slight calibration on h values. The performance of such practice is investigated in Fig. 2a. Although the data-fitted curve, which represents h, can cover the majority of data points, it still cannot explain wild data excursion (more than 18 % fitting error) in large IGA retard region. Figure 2b shows a single-engine data series for 2.0I engine and reveals that data series are isolated for different engine speeds and loads. The 18 % fitting error challenges the presumption that h is univariate. It urges a replacement of a new torque model which can bring more accuracy in large ignition retard region.

2.2 The Combustion Model: Predicting Engine Torque From Combustion

In additional to the low accuracy, there is another motivation to replace the data-fitted torque model: the current model is ignorant of detail combustion dynamics and thus vulnerable during aging. This phenomenon actually exemplifies a dilemma in engine modeling: the more a model relies on calibration (or the less on physical modeling), the less robustness it has when engine states change. One end of the dilemma is a full-calibration model, as illustrated by the data-fitted model in Eq. (1). The other end of the dilemma is exemplified by a multidimensional computational fluid dynamics (CFD) model which reproduces fluid molecular motions and pressure thermodynamics completely [12]. The CFD model calculates IMEPH via definition Eq. (2) using the predicted in-cylinder pressure P. The CFD model, which is physics-based and therefore employs almost no calibration, turns out to be more robust in life-time torque prediction. However, it is computational expensive and obviously infeasible online.

$$IMEPH = \frac{\int PdV}{V_d} \tag{2}$$

To introduce more physical content in a torque model while still balancing its efficiency, the authors have proposed an online mean-value thermodynamic combustion model [5, 6]. The online combustion model takes in the measurement of cycle initial states and uses a mean-value thermodynamic equation to predict the spatially-averaged fluid states thereafter. On the platform that supports the data-fitted model, the new model only requires a few more slow sensors and thus has remarkable cost-effectiveness on light-duty vehicles. The following equation describes the governing thermodynamics of this method.

$$m_c c_v dT(k) + P(k)dV(k) = dQ_g(k) - dQ_{crbl}(P(k)) - dQ_{ht}(T(k), P(k)) \tag{3}$$

Equation (3) follows the first law of thermodynamics. It treats the chamber as a variant control volume with chemical heat generation dQ_g, mass exchange into the crevice volume dQ_{crbl}, and heat transfer to the ambient dQ_{ht}. The net heat gain by the fluid then is calculated and interpreted in the form of fluid states change, as shown by the left hand side of the equation. In order to calculate the fluid state dynamics, all three terms on the right-hand side need to be modeled. Following the fundamental works by previous researchers, all the energy gain and loss can be calculated using equations in (4). All equations are modified versions of the original ones [9] with improvements in efficiency in large ignition retard region.

$$\begin{aligned} dQ_g(k) &= \eta_c m_f Q_{LHV} dx_b(k) \\ dQ_{ht}(T(k), P(k)) &= h_w(P(k)) A_s(k)(T(k) - T_{wall}) \\ dQ_{crbl}(P(k)) &= \left(h'(k) - u(k)\right) dm_{crbl}(P(k)) \end{aligned} \tag{4}$$

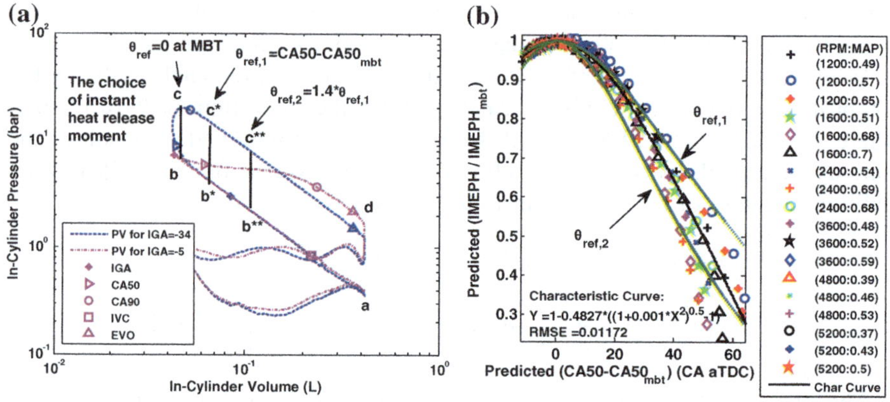

Fig. 4 **a** The schematics of the instant heat release process on the P–V diagram of two typical combustions, one for a MBT ignition timing and the other for a retarded ignition; **b** the two different choices of heat release moment determine the boundaries that enclose IMEPH-CA50 result series, forming a validity region

prediction validity, i.e. the unanimous characteristic curve in the normalized IMEPH-CA50 plot. The scheme then is to use accurate IMEPH prediction to cross-correct the CA50 prediction so that the characteristic curve can be reproduced in the prediction. While strict convergence onto the curve is over-demanding, the authors decided to develop a relaxed version of criterion by computing the up-and-down boundaries which enclose the characteristic curve, forming a "validity region" for prediction. Although manual placement of the boundaries can be performed given sufficient valid data, yet development of the boundary from the combustion physics is preferred for two reasons. First, to understand the physical mechanism behind such boundaries; second, to make sure the boundaries exclude invalid data that are physically impossible. ECUs also want to know an analytic solution of such boundaries in order to calculate them quickly online. To do so, the authors first examine the PV-diagram of a combustion in the case of IGA retard, as shown in Fig. 4a.

Figure 4a illustrates the P–V diagrams of two combustion cycles of different ignition timing. The blue curve for IGA at -34 CA aTDC shows that the areas "a-b-c-d" enclosed by the compression and expansion stroke traces are almost a parallelogram. The IGA therefore maximizes the effective work area "a-b-c-d" and thus maximizes the engine torque given a specific amount of fuel energy. By definition this ignition timing is very close to the MBT timing. On the other hand, the case of IGA at -5 CA aTDC illustrates the negative effect of ignition retard. The retarded IGA ignites the mixture during expansion and raises the fluid pressure in a much slower rate than that in the MBT case. It therefore truncates the effective work area as shown by the enclosed area "a-b-d" in red.

The round curve blue "b-c" or red "b-d" is a thermodynamic process governed by the interaction between combustion heat release, fluid temperature gain and work output to the piston head. The mean-value combustion model, shown previously in Eqs. (3–6), describes the spatially-averaged fluid behaviour during this period in an efficient way. In order to provide fast estimation of boundaries, further simplifications on the model and the computations are required. One good approximation is to assume that instantaneous heat release (or equivalently, the in-cylinder pressure rise) occurs sharply at the CA of θ_{ref} [8, 10]. Additionally, a polytropic compression and expansion process is assumed because of the high linearity of the log P–V curve in this period. With these two assumptions, the effective work area is then approximated as a parallelogram area "a-b-c-d" in MBT ignition case. By definition shown in Eq. (2), the estimated $IMEPH^*$ can be computed using Eq. (7).

$$IMEPH^* = \left(\frac{P_b V_b^{\gamma_c}}{1-\gamma_c}\right)\left(V_b^{1-\gamma_c} - V_{BDC}^{1-\gamma_c}\right) - \left(\frac{P_c V_c^{\gamma_e}}{1-\gamma_e}\right)\left(V_c^{1-\gamma_e} - V_{BDC}^{1-\gamma_e}\right) \quad (7)$$

Polytropic coefficient of expansion and compression are found to be close within an error of 0.01. Thus they are treated as identical, i.e., $\gamma^c = \gamma^e \equiv \gamma$. Since instantaneous heat release is assumed, the cylinder volume before and after heat release is equal at the crank angle θ_{ref}, i.e., $V_b = V_c \triangleq V_{ref} \triangleq V(\theta_{ref})$. And using the equivalent polytropic conversion $PV^\gamma = TV^{\gamma-1}$, Eq. (7) is reformed as a function of combustion heat release $\Delta T \triangleq T_c - T_b$ and cylinder volume at the heat release moment V_{ref}, as shown in Eq. (8):

$$IMEPH^* = \left(\frac{\Delta T}{1-\gamma}\right)\left(1 - \left(\frac{V_{ref}}{V_{BDC}}\right)^{\gamma-1}\right) \quad (8)$$

In Eq. (8), $IMEPH^*$ reaches maximum as V_{ref} reaches its minimum at V_{TDC}. In MBT case, V_{ref} is close to V_{TDC} as shown by Fig. 4a and thus $V_{ref} = V_{TDC}$ is assigned. In ignition retard case, θ_{ref} will be relocated to θ_{b^*} and the corresponding approximated effective work area will be the parallelogram area enclosed by "a-b*-c*-d". If a specific amount of fuel is combusted and constant thermal efficiency is assumed, the overall heat release amount ΔT is identical for both ignition cases. Thus the $\overline{IMEPH^*}$, which is the $IMEPH^*$ normalized by $IMEPH_{mbt}$, can be calculated using Eq. (9). Note V_{ref} is the cylinder volume at the moment θ_{ref}, and as a result, $\overline{IMEPH^*}$ is a function of θ_{ref} only.

$$\overline{IMEPH^*}(\theta_{ref}) = \left[1 - \left(\frac{V(\theta_{ref})}{V_{BDC}}\right)^{\gamma-1}\right] / \left[1 - \left(\frac{V_{TDC}}{V_{BDC}}\right)^{\gamma-1}\right] \quad (9)$$

The choice of θ_{ref} is at user's discretion and is important. It solely determines the how representative the performance boundary is. Geometrically, Fig. 4a shows that at MBT ignition timing, the θ_{ref} can be placed at 0 CAD approximately. For the case of ignition retard, θ_{ref} can be placed at ($CA50$-$CA50_{mbt}$) so that the parallelogram area "a-b*-c*-d" can be approximately equal to the area enclosed by the red curve "a-b-d". Thus the authors choose $\theta_{ref,1} = CA50 - CA50_{mbt}$ and plug into Eq. (9). Figure 4b shows that generally $\overline{IMEPH^*}(\theta_{ref,1})$ overestimates the torque-loss. The approximate IMEPH is quite accurate in small retard region and yet loses its accuracy gradually as CA50 is further retarded. The reason lies on the fact that in small retard region, the combustion duration is short and the instant-heat-release assumption holds. In the large ignition retarded area, the combustion duration, as exemplified by CA50 in Fig. 1, grows drastically and alters the overall shape of "a-b-c-d". In this case, the parallelogram area "a-b*-c*-d", as determined by the assumption, will be larger than the actual area and the difference keeps growing. Figure 4b shows $\overline{IMEPH^*}(\theta_{ref,1})$ sets the up-bound of IMEP-CA50 data series. While the placement of $\theta_{ref,1} = CA50 - CA50_{mbt}$ seems aggressive, a more conservative choice is introduced: $\theta_{ref,2} = 1.4(CA50 - CA50_{mbt})$ as illustrated in Fig. 4a. The model results with $\theta_{ref,2}$ generally under-estimate the torque value as shown in Fig. 4b. It is hypothetical that no combustion has a smaller effective work area than "a-b**-c**-d". The two curves, $\theta_{ref,1}$ and $\theta_{ref,2}$, encloses the characteristic curve and almost all predicted values of IMEPH-CA50 with a span range of 8 $\sim$ 10 %.

2.4 "Trinitiy" Correction Module

The up-bound of the validity region is a value from the ideal combustion case which is impossible to be surpassed. The low-bound is a performance calibre which is not supposed to be breached for a good engine. The "envelope" then becomes a physical principle of reciprocating machines. During online operation, after the $(CA50 - CA50_{mbt})$ is predicted by the online combustion model, the limit of IMPEH will be calculated via Eq. (9). The validity of IMEPH-CA50 prediction can then be checked weather it is within the boundaries. If it is not, ECU will trust IMEPH predictions over CA50 predictions and use the former to cross-correct the latter. The correction criterion is to follow a unanimous characteristic IMEP-CA50 curve. While true values of the characteristic curve is unknown online, the curve can be surrogated by a polynomial curve that is fitted over cascaded raw predictions during online operation, as demonstrated in Fig. 3b. The overall structure, named as "TRINITY" module, is illustrated by a flowchart below in Fig. 5.

3 Experiment Validation

In order to validate our model and method, steady-state test runs on various engine selections were conducted with our model running simultaneously with the measurements. The engines used for these experiments are Geely's SI VVT port-fuel-injection engines with different compression ratios and different VVT configurations. To further diverse the operation conditions, the engines were tested with

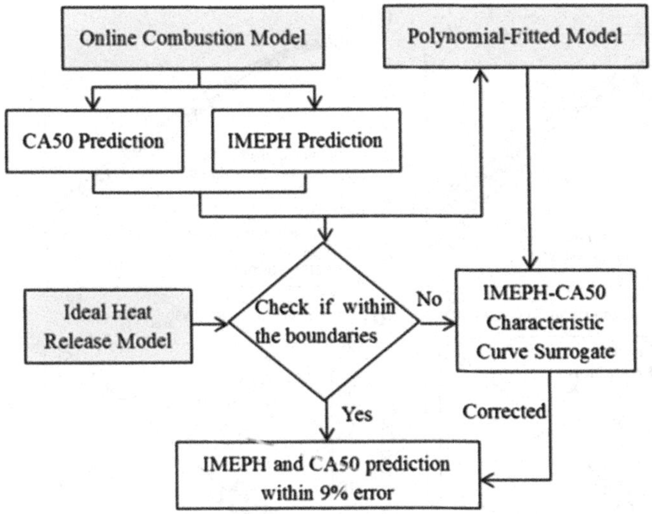

Fig. 5 The "TRINITY" correction module to guarantee the accuracy of CA50 and IMEPH prediction

Table 1 The specifications of experiment engines

Engine number	1.5 N	1.5I	1.8I	2.0I	2.0F	2.4I
Cylinder number	4	4	4	4	4	4
Displacement (L)	1.5	1.5	1.8	2.0	2.0	2.4
Compression ratio	10.3	10.3	10.0	10.0	9.7	10.0
Bore dia. (mm)	77.8	77.8	79.0	85.0	85.0	88.7
Stroke (mm)	78.8	78.8	91.5	88.0	88.0	96.2
Connect rod (mm)	153.3	153.3	129.4	138.6	138.6	130.5
VVT type	Dual[a]	Int[b]	Int	Int	Dual	Int
IVO (CA bTDC)	5.0	5.0	16.0	20.5	20.5	9.8
IVC (CA aBDC)	65.0	65.0	70.5	74.5	74.5	74.2
EVO (CA bBDC)	57.0	57.0	55.5	53.0	53.0	57.8
EVC (CA aTDC)	5.0	5.0	17.0	32.5	30.0	10.2

[a] Dual independent VVT. [b] Intake VVT only

different valve lift timings or durations. All engine's specification are shown in Table 1.

For each test run at an engine speed and load, we first located the MBT point by sweeping the ignition timing. Then we shifted ignition timing from MBT at the step of 2 CAD and conducted steady-state measurements and model calculation for each step. The typical speed selections were 1200, 2000, 2800, 3600, 4000, 4800 and 5200 RPM. The typical load selections were 0.5, 0.6 and 0.7 bar of MAP for each engine speed. All tests were performed with AFR at stoichiometry. All the sensors locations, resolutions, measurements were following the previous experiment setup

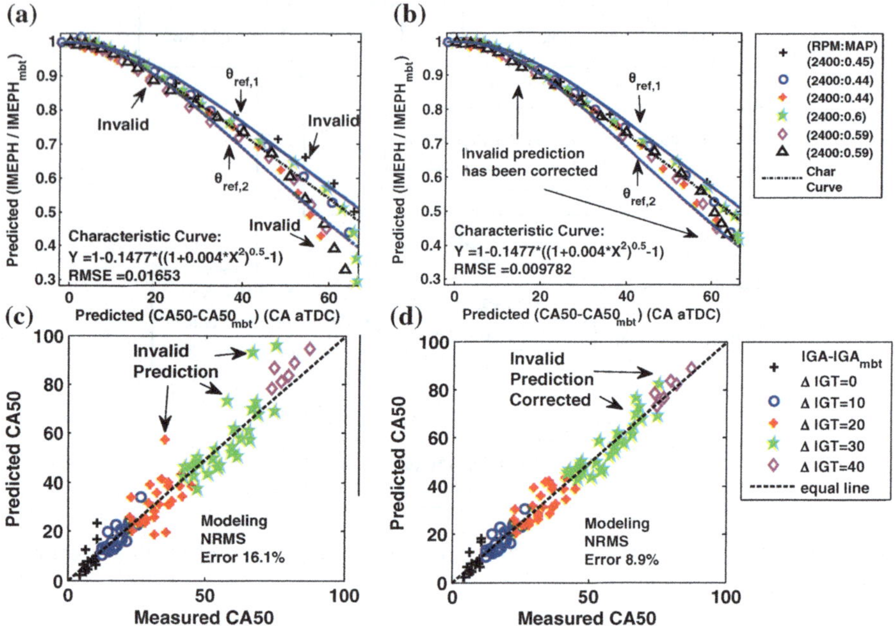

Fig. 6 The "TRINITY" module improved CA50 prediction from 16.1 to 8.9 % when ignition is retarded from MBT timing: **a** and **c** the module first identifies the invalid CA50 prediction on the original prediction; **b** and **d** the invalid IMEPH-CA50 results have been corrected and converged towards the engine characteristic curve and the predicted CA50 has converged towards the actual values

described in [5]. For reference, the test bench measured in-cylinder pressure using a piezoelectric transducer and convert to CA50 and IMEPH readings. To validate our model results, we coded our model in dSPACE rapid hardware prototype and ran the algorithm in real-time. Three torque models were computed parallel and their results are captured and coordinated by the TRINITY module. To compare with the time-averaged measurements, all predictions were averaged using a time frame of 15.6 ms.

First we plotted the normalized IMEPH-IGA data on a single graph for all engines, as shown in Fig. 2a. We then parabola-fitted all data points and reveals 0.01462 RMSE of the data-fitting. We also retrieved the results predicted by the online combustion model and plotted them on a normalized IMEPH-CA50 figure, as shown in Fig. 3a. Compared with Figs. 2a, 3a shows more convergence of data points with 0.0118 RMSE, 20 % less than that of Fig. 2a. Noted in IMEPH-CA50 plot, the CA50 is predicted by our online combustion model and is error-prone. The "TRINITY" correction module is applied on all engines in real time. Figure 6 below show the mechanism of the correction on CA50 predictions as exemplified on 1.5 N engine.

Figure 6a shows that the original IMEPH-CA50 prediction curve has excursions from validity region and they are successfully recognized as "invalid". Figure 6b shows that after being corrected by "TRINITY" module, the invalid IMEPH-CA50 result points have been shifted inside of the validity region. Correspondently, the corrected CA50 prediction are much more consistent with the actually measured values and the prediction error drops from 16.1 to 8.9 %, as shown by Fig. 6c, d. The benefit of accuracy improvement mainly comes from the convergence of the wide excursions in large ignition retard region. These excursions are the data points with large cycle-to-cycle variation and were previously difficult to predict using the online combustion model.

4 Conclusions

This paper has shown that with the fusion of three torque models, a "hybrid" method to promote the accuracy of IMEPH and CA50 prediction can be developed. The current data-fitted torque model has the simplest IMEPH prediction procedure and therefore suffers accuracy loss in large ignition retard region. The online combustion model renovates the accuracy in that particular region despite of the excursions from the engine characteristic IMEPH-CA50 curve. The inconsistency between the predicted results and the characteristic curve is attributed to error-prone CA50 prediction. In this paper, the authors successfully reconcile the inconsistency by modulating the invalid CA50 predictions according to a surrogate of the characteristic curve. The online identification of the invalid prediction has been proven effective if using a criterion developed based on an ideal-heat-release model. The criterion sets up-and-down torque boundaries of a combustion process and therefore reveals the physical principle of the characteristic curve. Large-scale experiments have validated the effect of our method in CA50 prediction improvement. Further research should employ delicate cycle-to-cycle analysis and reveal the statistical performance of the proposed method.

References

1. Huber W, Lieberoth-Leden B, Maisch W, Reppich A (1991) Electronic throttle control. Automot Eng Int, 99(6):15–18
2. Jankovic M, Frischmuth F, Stefanopoulou AG, Cook JA (1998) Torque management of engines with variable cam timing. IEEE Control Syst Mag 18(5):34–42
3. Stefanopoulou AG, Cook JA, Freudenberg JS, Grizzle JW (1998) Control-oriented model of a dual equal variable cam timing spark ignition engine. ASME J Dyn Syst Meas Contr 120:257–266
4. Hallgren B, Heywood J (2003) Effects of substantial spark retard on SI engine combustion and hydrocarbon emissions, SAE Technical Paper 2003-01-3237. doi:10.4271/2003-01-3237

5. Qu Z, Ma M, Zhao F (2012) An online crank-angle-resolved mean-value combustion model of gasoline engines including effects of cycle initial states, SAE Technical Paper 2012-01-0129. doi:10.4271/2012-01-0129
6. Qu Z, Ma M, Zhao F (2012) Estimation and analysis of crank-angle-resolved gas exchange process of spark-ignition engines, SAE Technical Paper 2012-01-0835. doi:10.4271/2012-01-0835
7. Rausen DJ, Stefanopoulou AG, Kang J-M, Eng JA, Kuo T-W (2005) A mean-value model for control of homogeneous charge compression ignition (HCCI) engines. J Dyn Syst Meas Control Trans ASME 127(3):355–362
8. Lee D, Jiang Li, Yilmaz H, Stefanopoulou AG (2010) Preliminary results on optimal variable valve timing and spark timing control via extremum seeking. In: 5th IFAC symposium on mechatronic systems
9. Heywood JB (1988) Internal combustion engine fundamentals. McGraw Hill, New York
10. Jiang L, Vanier J, Yilmaz H, Stefanopoulou AG (2009) Parameterization and simulation for a turbocharged spark ignition direct injection engine with variable valve timing. SAE Paper 2009-01-0680
11. Ayala FA, Gerty MD, Heywood JB (2006) Effects of combustion phasing, relative air-fuel ratio, compression ratio, and load on SI engine efficiency, SAE paper 2006-01-0229, Transactions, 115, J Engines, Section 3
12. Kong SC, Marriott CD, Reitz RD (2001) Modeling and experiments of HCCI engine combustion using detailed chemical kinetics with multidimensional CFD, SAE Technical Paper 2001-01-1026. doi:10.4271/2001-01-1026
13. Woschni G (1967) Universally applicable equation for the instantaneous heat transfer coefficient in the internal combustion engine, SAE Technical Paper 670931. doi:10.4271/670931

Development of Real Time Inlet Air Model of Diesel Engine Based on 'V' Cycle Mode

Chao Ma, Yong Hang, Xiaowu Gong and Fu Wang

Abstract In recent years, more attentions are paid to environment protection and energy crisis in China so demands are being put on improving fuel economy and low emissions. On the other hand, many people in the automotive community never give up looking for innovative methods to meet the demands. Model-based control and diagnosis technology is very potential and now has been used in many aspects. Particularly, Model-based methods are very popular in development of Engine control unit and some relevant developing tools also arise such as MATLAB/Simulink, dSPACE. A standard development flow called 'V' cycle is refined. In order to improve control efficiency and extend the diagnostic methods of fuel injection and after treatment equipments, It needs to get the accurate amount of fresh air charged into diesel cylinder which greatly influences engine combustion process. According to the popular "V" development mode, we build a real time model of fresh air based on Simulink tool and realize the off-line simulation, rapid prototype validation, data scaled, code auto generation, and engine bench test. Real engine test results indicate that the model can satisfy the real time and accuracy requirements completely. This chapter describes the creation and validation of a control-oriented diesel inlet model and doesn't consider the effect of exhausted air which will no doubt increase the complexity and decrease the real time efficiency, but can extend the model used scope greatly. During construction of the model, We not only consider the dynamic parameters that influence the amount of fresh air such as volumetric efficiency, temperature and pressure of the inlet air, but also consider the difference between static test standards and dynamic running states.

F2012-A05-010

C. Ma (✉) · Y. Hang · X. Gong · F. Wang
FAW Co., Ltd R&D Center WFIERI, Wuxi, China
e-mail: mc@wfieri.com

SAE-China and FISITA (eds.), *Proceedings of the FISITA 2012 World Automotive Congress*, Lecture Notes in Electrical Engineering 190,
DOI: 10.1007/978-3-642-33750-5_3, © Springer-Verlag Berlin Heidelberg 2013

Keywords Inlet air model · 'V' cycle mode · Real time · Data scaled · Simulation and test

1 Introduction

Cylinder charge is an important parameter which has heavy influence on engine power and emissions. By adjusting the amount of fuel injected into cylinder, we can get accurate control result. The amount of fuel is easily determined especially using electronic control fuel injection system such as common rail injection system, but it is necessary to fix a sensor to get the accurate amount of fresh air which increases the cost. Develop of a real time model of diesel cylinder charge executed on diesel controller is more convenient and economic than a sensor [1]. In the past decades various kinds of inlet model of diesel engine were developed by intelligent engineers and scientists but these models are usually used in specific area and involve complex math equations which can hardly be realized in microchip. Mathematical models of air inlet are broadly classified as either non-thermodynamic or thermodynamic models [2]. Thermodynamic models are derived from thermodynamic principles and can be classified into three groups according to their complexity as follows: the quasi-steady model (QSM), filling and emptying model (FEM), and gas dynamic model(GDM). The characteristics of these modeling techniques can be summarized in Table 1 [3]. Non-thermodynamic model, which is also called MAP model, is mainly obtained through test data. This kind of model looks like black block, only has inputs and outputs, and doesn't contain detail information about researched object, but the non-thermodynamic model is easily realized in control unit and the calculation speed is very high.

In recent years, the focus in automotive industry is placed on model-based control algorithm development and the benefits from using a real time capable component model is more and more clear. Software-in-the-loop, Rapid control prototyping (RCP), so-called 'V' develop mode is now widely used by many OEMs. In this study, a dynamic nonlinear model of cylinder charge of a common rail direct injection diesel engine with a turbocharger (showed as Fig. 1) is proposed and validated through engine tests. The development process of this model is according to the 'V' cycle: build the model based on MATLAB/Simulink firstly and get primary results through off-line simulation, obtain embedded codes of the model using auto code-generation tool provided by Math Works corporation, integrate the resulting codes into the whole control program of diesel engine and test the model on the test bench, calibrate necessary data for all MAPs, finally compare the off-line simulation results and the test bench results and give a conclusion.

Table 1 Comparison of various modeling methods

Item	QSM	FEM	GDM
Mathematical complexity	Simple	Complex	Complex
Computing accuracy	Low	High	High
Reliance on empirical data	High	Low	Low
Calculation speed	Fast	Slow	Slow

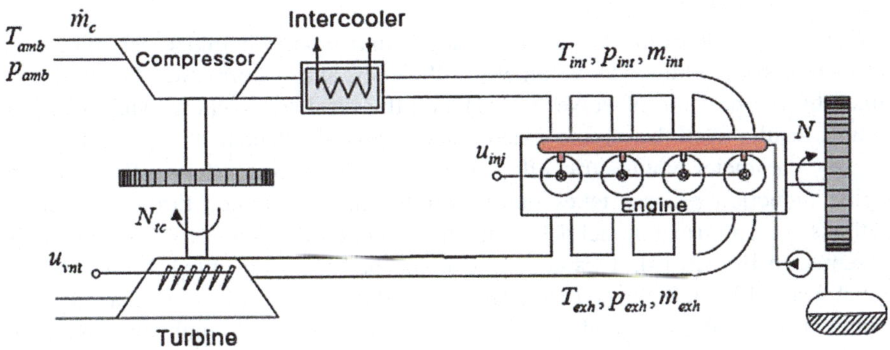

Fig. 1 Schematic of common rail direct injection diesel engine with turbocharger

2 Model Description

The cylinder charge model is to accomplish some tasks, for example, to approximate air amount which cannot be measured by calculating them from the available sensor data, to calculate physically meaningful substitute signals when relevant sensors are in malfunction state, to provide dynamically reliable information which is not impaired by the response limit of the sensor or by gas travel delay in the system, to provide correction for sensor signals. We try to compute the amount of fresh air charged into engine cylinders at minimum computing cost, because complex empirical or thermodynamic equations can hardly satisfy real time demand so not proposed, on the other hand, though simple algebraic equations or data MAPs don't involve complex computation, computing accuracy is not enough, so we compromise between the two methods. Whenever possible, the model is described by analytic equations rather than maps. In doing so, the true behaviour of the system is abstracted to the extent that the control unit can calculate the model in real time.

Figure 1 shows the structure of the physical system. The identifiers of the physical variables used correspond to the formula symbols in the equations. The process in intake manifold is assumed to be adiabatic because the temperature change in the intake manifold is very slow so the heat transfer in the intake manifold is negligible. The estimated intake manifold air volume per minute which is used to compute the cylinder air charge is expressed as Eq. (1).

$$V_{\text{int}} = \eta \cdot \lambda_{VGT} \cdot \frac{n}{2} \cdot V_{eng} \tag{1}$$

where

V_{int} estimated air volume, m^3

η engine volumetric efficiency,

n engine speed, r/min

V_{eng} engine displacement, m^3

The amount of fresh air volume charged into engine cylinder per minute is calculated from product of volumetric efficiency and a correction factor determined by the position of valve of VGT and the ideal gas volume. The ideal gas volume results from the engine speed n and engine displacement V_{eng}. In Eq. (1), the key parameter volumetric efficiency η which measures how efficiently an engine induction process takes place overall can be obtained through different methods. An empirical model describing the volumetric efficiency for the engine is presented in Eq. (2) [4]. This equation should be regression fitted, because there are at least 12 coefficients that have to be determined, of course, a large of computation is not avoided which is not consistent with embedded system. Additionally, empirical determination of the volumetric efficiency has limited accuracy [5].

$$\begin{aligned} \eta = {} & a_0 + a_1 \cdot n + a_2 \cdot n^2 + a_3 \cdot p_{\text{int}} + a_4 \cdot n \cdot p_{\text{int}} + a_5 \cdot T_{\text{int}} + a_6 \cdot n \cdot T_{\text{int}} \\ & + a_7 \cdot p_{exh} + a_8 \cdot n \cdot p_{exh} + a_9 \cdot \frac{p_{\text{int}}}{\sqrt{T_{\text{int}}}} + a_{10} \cdot \frac{p_{exh}}{p_{\text{int}}} + a_{11} \cdot f + a_{12} \cdot f^2 \end{aligned} \tag{2}$$

where:

a_x coefficient of terms, x = 0:12

n engine speed, r/s

p_{int} pressure of intake manifold

T_{int} temperature of intake manifold

p_{exh} pressure of exhausted manifold

In this study, we consider two main parameters that influence the volumetric efficiency greatly the engine speed and the engine load which is presented by injection fuel amount per cycle or efficient torque of diesel engine. At every steady engine speed, change engine load step by step and record the true air mass m_{real} that charged into cylinder and the temperature and pressure of intake manifold. We neglect the pressure and temperature difference between the inlet entrance and the engine cylinder, that means the pressure and the temperature equal that in the engine cylinder. Volumetric efficiency of current operating condition is calculated from Eq. (3). Under the intake manifold circumstance, air mass that charged into engine cylinder ideally m_{refe} is calculated from ideal gas state Eq. (4). The result MAP of volumetric efficiency is presented in Fig. 2. Axis x presents engine speed from 700 to 2500 rpm. Axis y presents engine efficient torque from 0 to 1500 N.m.

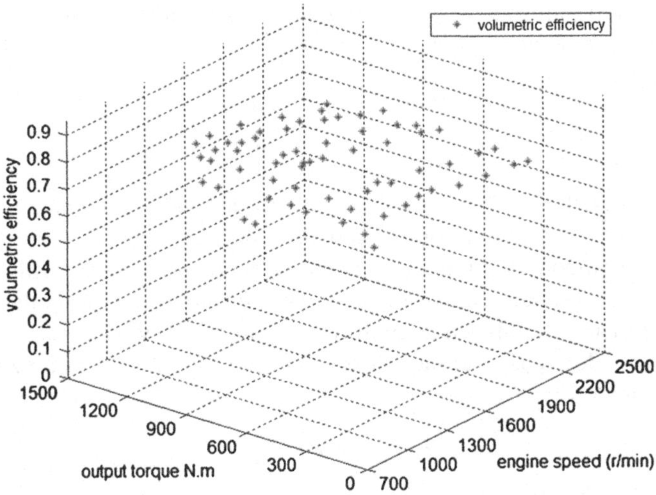

Fig. 2 Volumetric efficiency MAP of a 6 cylinder diesel engine with turbocharger

It should point out that all the volumetric efficiency data is obtained at engine steady state and control signal of VGT is fixed at maximum value which makes VGT work in its maximum capability. On other untested steady states, linear lookup algorithm is used to get the volumetric efficiency. During transient process, volumetric efficiency value is approached by a PT1 filter algorithm to model the transition process.

$$\eta = \frac{m_{real}}{m_{refe}} \tag{3}$$

$$m_{refe} = \frac{n}{2} \cdot V_{eng} \cdot \frac{p_{\text{int}}}{R \cdot T_{\text{int}}} \tag{4}$$

Considering Eqs. (1) (3) and (4), the true amount of fresh air mass charged into engine cylinder is expressed as Eq. (5)

$$m_{real} = \frac{n}{2} \cdot V_{eng} \cdot \frac{p_{\text{int}}}{R \cdot T_{\text{int}}} \cdot \eta \tag{5}$$

In addition to the both parameters engine speed and engine load that mainly determine the volumetric efficiency, there are many other parameters that have influence on it. Especially, the temperature and the pressure of intake manifold should be considered. While measuring the true amount of fresh air, the temperature and pressure of intake manifold should also be recorded formatting maps of reference temperature and reference pressure. These reference parameters express the circumstance of volumetric efficiency on tested state. As a result of the referenced parameters, two other correction factors λ_{temp} and λ_{press} that compensate a dependency of the difference between the referenced parameter values and the true values are used to improve the computing accuracy of the fresh air amount. These two parameters are expressed as Eqs. (6) and (7).

$$\lambda_{temp} = f(T_{\text{int}} - T_{ref}) \tag{6}$$

$$\lambda_{press} = f(P_{\text{int}} - P_{ref}) \tag{7}$$

When the turbocharger is electronically controlled, different amount of fresh air will be absorbed at different control signal of turbocharger. A correction factor λ_{VGT} is used to embody the influence of the turbocharger control. This correction factor can be got through a curve mapped data. In order to express the gas travel delay characteristic, the objective control position of the VGT actuator should be filtered by a PT1 low-pass and the output value is as input of the curve map. λ_{VGT} is expressed using Eq. (8)

$$\lambda_{VGT} = f(PT1[r]) \tag{8}$$

Where r: the objective position of control actuator of the VGT

Combined with Eqs. (5), (6), (7), (8), the final amount of fresh air absorbed into diesel cylinder is got expressed as Eq. (9).

$$m_{final} = \frac{n}{2} \cdot V_{eng} \cdot \frac{p_{\text{int}}}{R \cdot T_{\text{int}}} \cdot \eta \cdot \lambda_{temp} \cdot \lambda_{press} \cdot \lambda_{VGT} \tag{9}$$

3 Model Realization and Integration

Equation (9) is realized with simulink blocks based on MATLAB/SIMULINK. The whole computing diagram is as flowing Fig. 3 which includes four maps and two curves. They are basic volumetric efficiency map on steady state, correction map for VGT control signal, reference temperature map for intake manifold, reference pressure map for intake manifold, two correction curves for real temperature and pressure. In addition to these maps and curves, some parameters are also needed which are presented by constants such as engine displacement, fresh air constant R and some other unit conversion coefficients. Here, we assume the gas as ideal gas, so gas constant equals 287 J/(kg.K). It is necessary to point out that in order to keep all variables consistent, all standard units should be used.

This computing diagram contains five inputs including average speed of diesel, amount of fuel injection per cycle per cylinder, true temperature and pressure of intake manifold, the filtered objective position of control actuator of VGT and three outputs that are volume and mass of fresh air absorbed per second and correction factor product of pressure and temperature. Here, more attention should be given the input diesel engine speed which must be filtered, that means the speed value should be approached by mean algorithm or other useful algorithm because

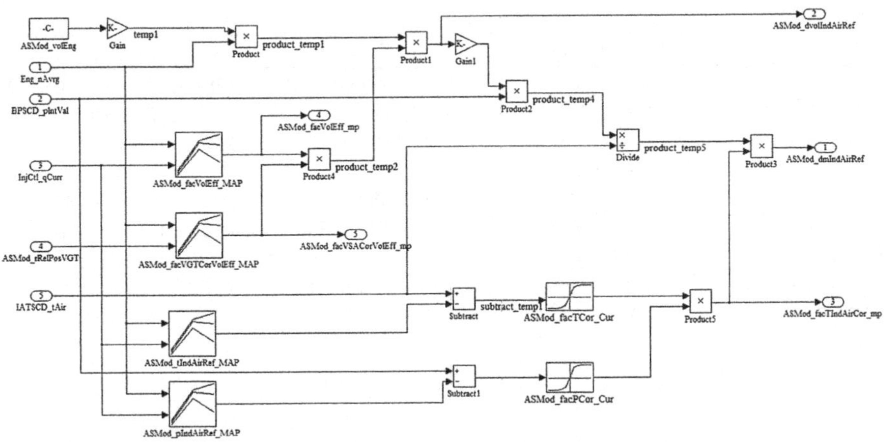

Fig. 3 Computing diagram of fresh air charged into cylinder based on SIMULINK

of the gas travel delay characteristic. On the contrary, if the transient value is used in this diagram, the output volume and mass of fresh air might acutely vibrate.

The computing diagram realized is just the first step because all the parameters and signals are double precision and multiplication and division algorithm are executed on PC platform. The air predicted model is intend to be executed in real time and our target platform is microchip which doesn't support double precision arithmetic now. Moreover, it generally has limited hardware source which is not enough for complex algorithm. From the analysis above, we get the conclusion that signals and parameters must be scaled into integer and some number operation and lookup map algorithm should be redesigned. Fortunately, a tool called real time workshop embedded coder supplies a way to help us accomplish these complex tasks. Firstly, considering the variable precision and range, we make a series of rules according to which the physical variables are converted into integers. Now, all data type of parameters and signals become fixed point supported by a majority of microchip. Though precision of algorithm loses a bit, the speed of computation improves greatly. As Table 2 presents, all the key variables conversion range and formula are listed. Before making conversion from double precision to fixed-point precision, it usually need to do off-offline simulation to determine a suitable integer range which mainly depends on presentation precision that means different integer makes different least significant bit (LSB).

Figure 4 shows the comparison of calculation result using fixed-point variables and using double ones on 65 different operation conditions. In up part, fixed-point calculation results are indicated in symbols of asterisk and dashed line. Double precision calculation results are expressed in symbols of solid circle. Most fixed-point values are a little larger than double point values. The down part of Fig. 4 which shows the difference percent between the two results also proves the conclusion because all the difference percent values are positive. Difference percent is calculated according to the flowing formula:

Table 2 Rules of physical variables conversion

Item	Integer range	Physical range	Conversion formula
Speed	[0, 25600]	[0, 10000]	(Physical value/100)*256
Fuel	[0, 61440]	[0, 600]	(Physical value/2.5)*256
Coefficients	[-25600, 25600]	[-100, 100]	Physical value/256
Volume	[0, 64000]	[0, 10000]	(Physical value/40)*256
Mass	[0, 40000]	[0, 10000]	Physical value *4
Temperature	[-1040, 960]	[-1000, 1000]	Physical value −40
Pressure	[0, 25600]	[0, 1000]	(Physical value/10)*256
Intermediate value	[0, 4.2906e + 009]	[0, 4190000]	Physical value *2^10

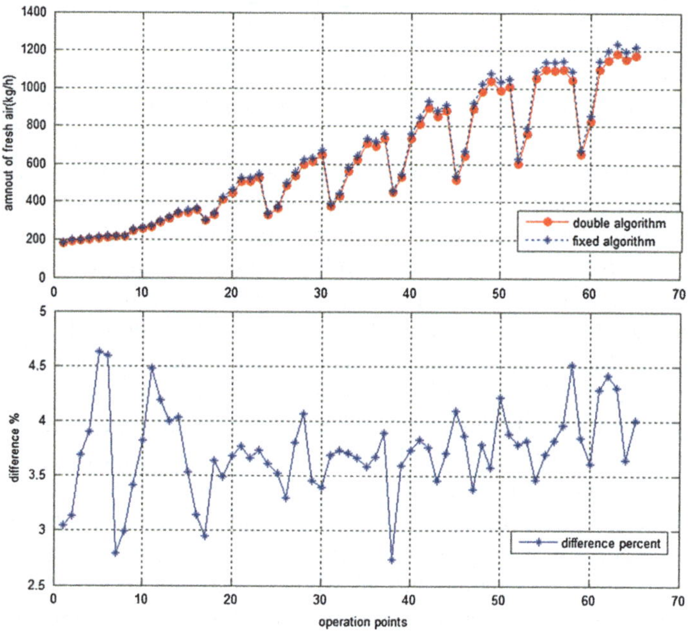

Fig. 4 Calculation result comparison between fixed-point algorithm and double algorithm

$$\lambda = \frac{V_{fixed} - V_{double}}{V_{double}} * 100\% \tag{10}$$

The maximum of difference percent is below 5 % which is accepted in estimating the amount of fresh air per second. At the cost of the precision lost, the calculation speed improves greatly and a real time model becomes possible.

When the offline simulation and validation is accomplished, the real time fresh air model could be generated into embedded codes that have to be integrated into whole control program from which the inputs of the model are supplied. The code generation process is automated after a series of correct configuration with the help of real time workshop embedded coder tool. All requirements from embedded

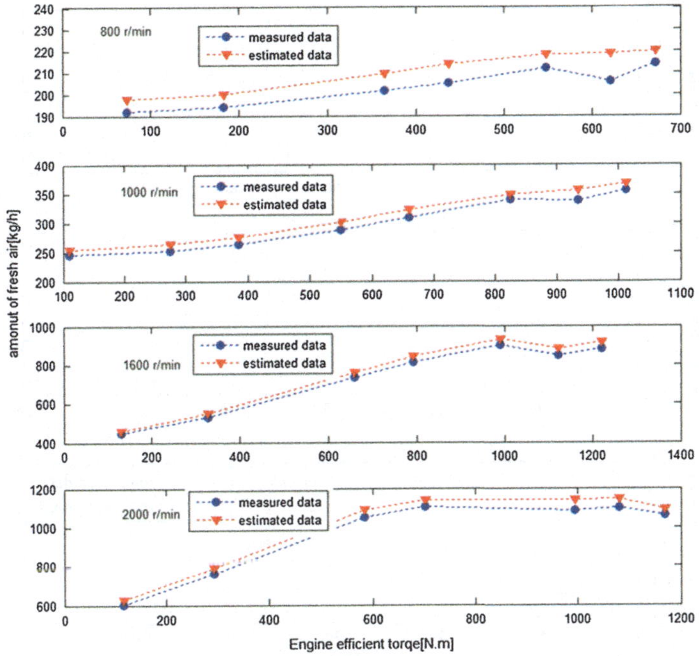

Fig. 5 Validation of fresh air model charged into cylinder on steady state at constant speed

system such as the fixed step request, data separated from program, stack optimization and so on must be met or the whole program might not be compiled or executed correctly. Process integration refers to how to schedule the real time model that means the model is executed periodically or by a trigger. Due to the cyclic characteristic of diesel engine, this real time model is scheduled periodically and the parameter of period also has influence on transient process model because output in transient process is filtered by a PT1 algorithm whose exponent is division result of scheduled period and delay constant.

4 Model Validation

For the validation of the inlet air model, 8.6 L, an inline 6 cylinder common rail direct injection diesel engine with exhausted turbocharger and an eddy current dynamometer are used. Signals of diesel engine speed, brake torque, temperature and pressure of intake manifold, air flow rate are measured [6].

In order to confirm the accuracy of the air inlet model, the measured data and estimated data on 60 operating conditions are compared. The speed range is from 800 to 2100 rpm with step of 200 rpm, and the load range, is from 10 to 95 % of the most load at each speed. Figure 5 shows part of results on comparison between

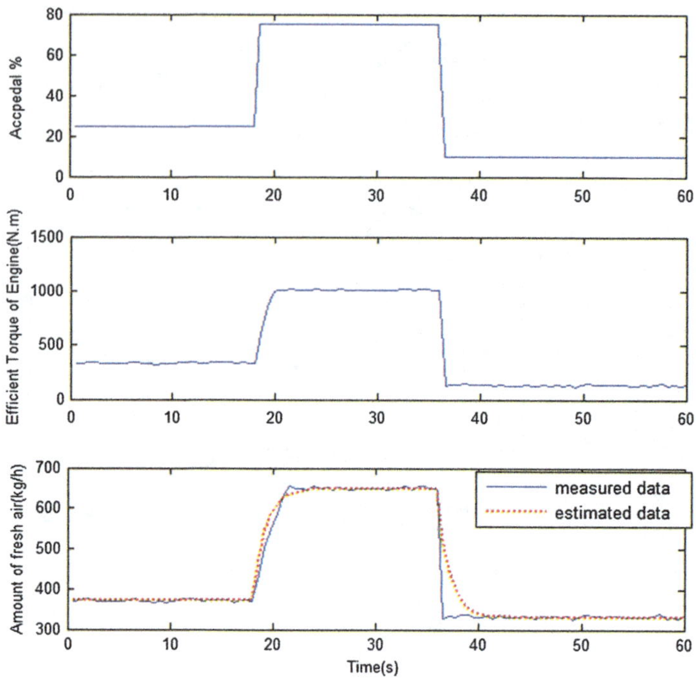

Fig. 6 Validation of fresh air model charged into cylinder on transient state at constant speed

measured data expressed in blue colour and estimated data expressed in red colour on steady state. At some constant speed, adjust load of engine and then record the measured values. The comparison indicates they are in good agreement with each other. All difference is below 5 % in all operating conditions.

For validation of the model in transient state, the dynamometer is also operated on a constant speed 1,300 rpm. Adjust the accelerator pedal from 25 to 75 % acutely, keep it for a while and then change it back to 10 %. Observe and record relevant parameter values, As Fig. 6 shows, the brake torque's response is similar to that of accelerator pedal and a delay is observed in air mass flow rate which is due to the gas delay and turbocharger inertia. From the comparison of measured data and estimated data, it is obvious that real time model's computing accuracy also satisfies the demand in transient process only slight errors exist.

5 Conclusions

According the 'V' cycle development flow, a nonlinear model of fresh air absorbed into the cylinder of diesel is developed. The model is realized and tested in embedded control system and successfully satisfy control and state predicted demand. Some conclusions are derived as follows:

Using of modern development flow 'V' cycle in model-based control saves time and cost greatly and can be used more widely.

Control and state predicted model embedded into microchip should not be complex because of the software and hardware resource limit. So some empirical models must be simplified using data maps or other techniques though some computing accuracy is lost, the computing speed improves greatly.

The developed real time inlet air model is validated to an accuracy of 5 % by comparison of estimated data and measured data obtained from steady operating conditions.

In transient process, the estimated results are good agreement with experimental data and the model satisfies the real time and accuracy demands completely.

References

1. Grimes MR, Verdejo JR, Bogden DM (2005) Development and usage of a virtual mass air flow sensor. SAE paper 2005-1-0074
2. Grondin O, Stobart R, Chafouk H, Maquet J (2004) Modeling the compression ignition engine for control: review and future trends. SAE paper 2004-01-0423
3. Horlock JH, Winterbone DE (1986) The thermodynamics and gas dynamics of internal combustion engines vol 2. Clarendon Press, Oxford
4. Lee B, Guezennec YG, Rizzoni G, Trombley D (2003) Development and validation of control-oriented model of a spark-ignition direct injection engine. In: Proceedings of 2003 ASME IMECE, DSCD
5. Hendirics E, Chevelier A, Jensen A, Sorenson SC (1996) Modeling of the intake manifold filling dynamics. SAE paper 960037
6. Namhoon C, Sunwoo K, Myoungho S (2005) Nonlinear dynamic model of turbocharged diesel engine. SAE paper 2005-01-0017

CFD Simulation and Optical Engine Diagnostics of Mixture Formation Processes in DI Gasoline Engine with Flexible Valvetrain

Yi Zheng, Po-I Lee, Atsushi Matsumoto, Xingbin Xie and Ming-Chia Lai

Abstract The interactions of multi-hole direct injection (DI) gasoline sprays with the charge motion are investigated using computational fluid dynamics simulation and high-speed imaging of sprays inside engines. Advanced flexible valve-train, with valve-deactivation and variable valve-lift, produces very dynamic charge flow motions, with varying tumble and swirl ratios. The resultant turbulent flow interact with off-axis multiple-hole DI injections, has important implications for the engine mixing and resultant combustion performance. The effects of injection timing on the bulk flow motion and fuel–air mixing in an optical accessible engine, in terms of tumble and swirl ratios, turbulence, and fuel wall film behaviors are first discussed for the conventional baseline engine geometry. The early- and late-variable intake valve closing events are then tested in a metal engine. The effects of different valve lifts and valve deactivation on the mixing and combustion are then discussed. Using integral analyses of the simulation results, the mechanisms in reducing fuel consumption and emissions in a variable valve-actuation engine, fuelled by side-mounted multi-hole DI injectors are illustrated. The implications to the engine mixing and the resultant combustion in a metal engine are demonstrated.

Keywords Direct injection · Gasoline engine · Flexible valve · CFD · Optical engine

F2012-A05-011

Y. Zheng (✉) · P.-I. Lee · A. Matsumoto · X. Xie · M.-C. Lai
Wayne State University, Detroit, MI, USA
e-mail: mclai@wayne.edu

SAE-China and FISITA (eds.), *Proceedings of the FISITA 2012 World Automotive Congress*, Lecture Notes in Electrical Engineering 190,
DOI: 10.1007/978-3-642-33750-5_4, © Springer-Verlag Berlin Heidelberg 2013

1 Introduction

Advanced valvetrain coupled with Direct Injection (DI) provides an opportunity to simultaneously reduce fuel consumption and emissions [1]. Because of their robustness and cost performance, multi-hole nozzles are currently being adopted as the gasoline DI fuel injector of choice, mostly in the side-mount configuration. In addition, ethanol and ethanol-gasoline blends are being used in the down-sized, down-speed and variable-valve-train engine architecture, because of their synergy in improving the turbo-charged DI gasoline performance. There has been much research in the literature carried out with interactions of DI gasoline sprays and the in-cylinder flow fields [2–10], but very little study on the side-mounted multi-hole nozzle with the interaction of charge motions, which is the focus of the current study. Multi-dimensional computational fluid dynamics (CFD) offers a promising alternative to experiments for its capability to offer much more detailed information on in-cylinder mixture formation. Numerical method has been used to analyze the injector nozzle flow, near-field primary spray evolution, and wall impingement of gasoline and diesel engine, and compare with experimental observation [11–15]. The charge motion and its interaction with spray were studied computationally and experimentally, and showed that the mixture preparation prior to combustion is important for predicting combustion characteristics and emissions [16–18]. The ignition process and combustion regimes of gasoline DI engine were investigated by using CFD method [19]. These research works also show that the significant influence of turbulence modelling on in-cylinder flow predicting and Reynolds-averaged Navier–Stokes (RANS) turbulence method is probably still the best compromise between reliability and computationally expensive. In this chapter, the spray characteristics and simulations multi-hole DI injectors without charge motion are first discussed. Then interactions with charge motions inside optical engine are presented. Finally the implications to the engine combustion and emission performance are summarized.

2 In-Cylinder Spray/Mixing Characterizations

Mie scattering, back-lighting, and Schlieren visualization have been widely adopted for spray visualization [1, 4–8, 20–23]. Compared to Diesel sprays, the DISI sprays have smaller enclosed cone angle, more volatile fuels, shorter aspect-ratio nozzle holes and consequently more unstable vortices. As a result, more interactions among the various spray plumes are expected. Therefore, in addition to the embedded grid control used to reduce the grid size at sensitive areas such as the injector tip, Adaptive Mesh Refinement (AMR) is also used to automatically enhance mesh resolution in critical areas of spray development within the spray or combustion chamber. The Kelvin–Helmholtz and Rayleigh–Taylor (KH–RT) breakup model, the No Time Counter (NTC) collision model, and the renormalization group (RNG)

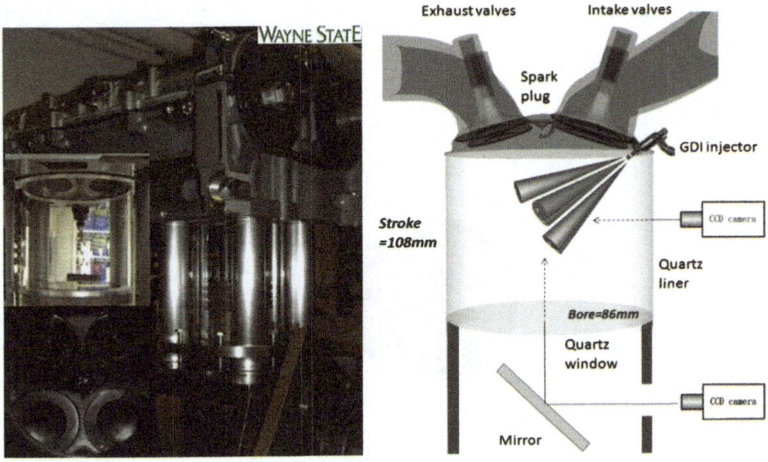

Fig. 1 Schematic and photograph of the optical engine

k-ϵ model are used, in conjunction with the RANS solver. The uncertainty in the liquid fractions of the multi-phase jets and the interactions among different spray plumes, as well as and the potential of flash boiling [5, 15, 20–23] make it difficult to simulate the dynamic spray plumes using RANS simulation, although reasonable agreement in spray plume penetration could be achieved. The spray characteristics of the two injectors A and B used in this paper has been well characterized in spray chambers and validated using CFD previously [14, 15].

An optical accessible engine (OAE) which utilizes the same cylinder head, injector, cams, and shares the same 86 mm bore as a production 2 l engine, is used for the spray mixing investigation and compared with CFD simulations. The optical engine has quartz liner and Bowditch quartz to provide side- and bottom-view spray visualizations inside the cylinder (Fig. 1). A Phantom 7.1 CMOS high-speed digital camera is used for the imaging with solid lighting provided by continuous projector light source for both the Mie and Schlieren imaging techniques. Optical engine piston has a flat top while the piston head of metal engine has a bowl and valve recesses as shown in Fig. 1. The number of cells is in the range of 75,000 and 250,000 depending on the piston position, and the average cell size is 0.125 mm near the spark plug, 0.5 mm for the each nozzle direction, 1 mm for the mesh refined automatically, and 2 mm for the other area (Fig. 2). The cam profiles and phasing range are illustrated in Fig. 3.

2.1 *Effects of Injection Timing for Baseline Case*

Figure 4 shows the comparison of spray visualization and the simulated spray particle clouds, color-coded with drop size down to 1 micron. The simulated spray

Fig. 2 Geometry and mesh strategy

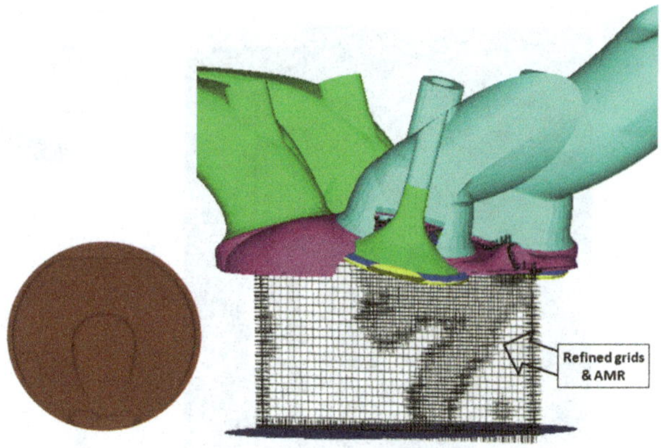

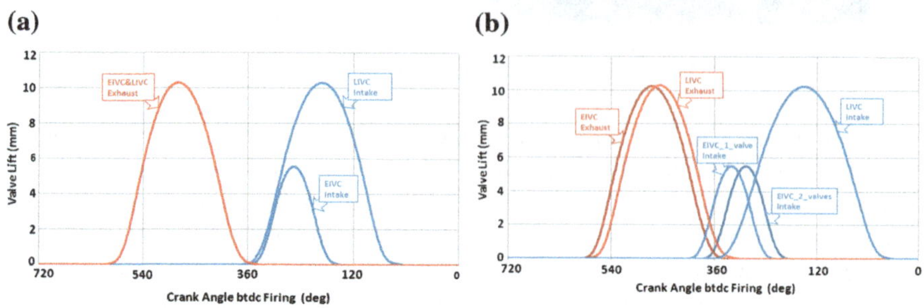

Fig. 3 Cam profiles. Black profiles: exhaust cam; *red* high-lift intake cam (HLC), *blue* low-lift intake cam; (LLC). **a** Optical engine. **b** Metal engine

distributions cover a larger area than the Mie scattering visualization, which may have problem resolving smaller drops. In general, the agreement is good for this case.

The air flow at the intake valve is very strong and is the dominant turbulence generation mechanism for ICE engines. The interactions with multi-hole spray is very complicated both spatially and temporally, especially for the deactivated cases, and will have a significant influence on the spay impingement and mixing. The charge motion is a combination of both swirl and tumble motions which is very dynamic and would require 3-D postprocessing or CFD to resolve the mixture formation processes and interactions. However, the interactions of the fuel injections and the charge motions are bettered illustrated by the integral analyses. The predicted effects of injection timing on the tumble and swirl ratios of the charge motion and the total turbulent kinetic energy are shown in Fig. 5. The swirl is mostly negligible for the two inlet port geometry used in this engine. Depending on the injection timing and phasing of the tumble dynamics, the fuel injection could inhibit or enhance the tumble motion. Injecting too early actually decreases the tumble ratios; later injections at 120° and 180° crank angle degree (CAD) after the intake TDC, the tumble flow was enhanced by the fuel injection. In these cases,

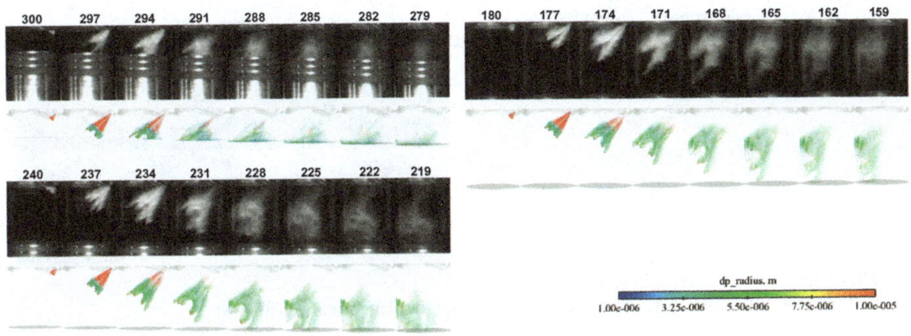

Fig. 4 Comparison between CFD and OAE results of the in-cylinder spray processes SOI = 120° aTDC, 1000 rpm, injector B, 100 MPa, 16 mg (0.88 ms duration)

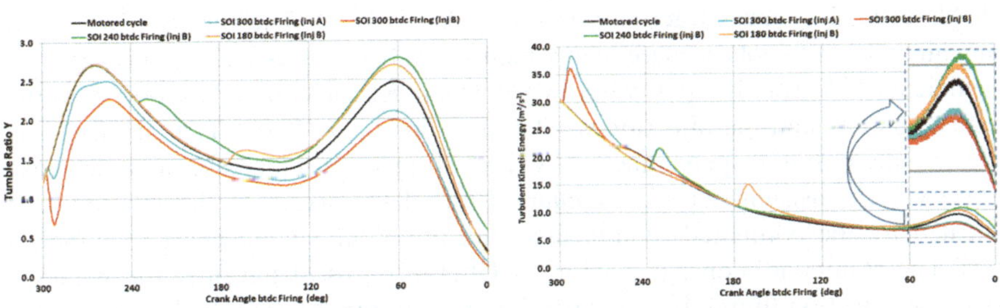

Fig. 5 The effects of SOI on the Tumble and Swirl ratios (*left*) and integrated turbulent kinetic energy (*right*), injector B

enhanced turbulent kinetic energy (TKE) due to the eddy breakdown of the tumble motion is also observed at the end of compression stroke, where faster flame computed by CFD of a few injection timings is shown in Fig. 5. With the early start of injection (SOI) at 60°, a large amount of liquid impinges on the surface due to the relatively shorter injector-piston distance. For this case, the side- and bottom-view of the wetted footprints on the piston at 90° aTDC are also shown in Fig. 6. At this timing, the maximum wall film amount for Injector B is almost twice as much as Injector A, in spite of less liner wetting because of its spray targeting. This piston wetting can be avoided by retarding the injection timing. If the injection timing is set to be later than 120°, the maximum film mass is predicted to reduce by 62 %. Since the spray injected at 120° and later was not found to touch the piston, the film mass could be located on the side wall.

A 4-cylinder 2 l metal engine which has the same configuration of cylinder head, injector, cams, and bore size as the optical engine, was used to carry out the combustion and emission tests [24]. This present chapter focuses only on the low-speed and naturally aspiration conditions, primarily because using Early Intake Valve Closing (EIVC) at higher speeds with one valve inactive limits the peak load of the engine. The engine has a compression ratio of 11.9, to enable flex-fuel (E0 ~ E85) operation.

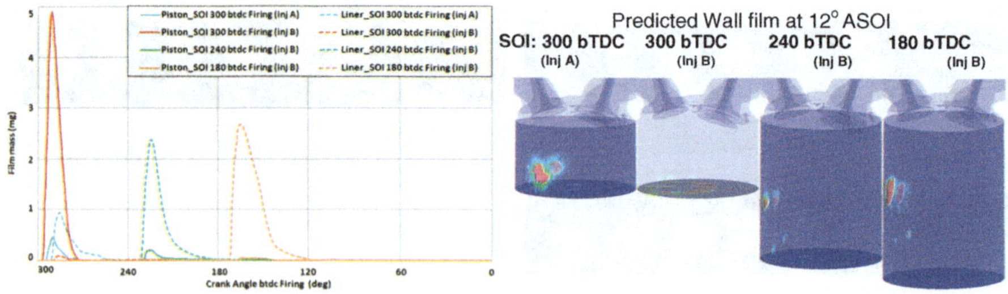

Fig. 6 Piston wall wetting footprints and total liquid film mass evolution

A typical metal engine test results in Fig. 7 show the effects of start of injection (SOI) timing on soot (Filtered Soot Number or FSN, measured with an AVL 415 smoke meter), coefficient of variability (%COV) of BMEP, and the burning speed in CAD (10–75 % mass fraction burnt, or CAD10 ~ 75). The injection widow is constrained by the soot measurement which strongly depends on wall wetting, and Coefficients of Variability (COV) of IMEP which depends on the uniformity of the mixture. As a result, the injection window is around 120–200° aTDC. During this window the burning speed and COV are observed to increase and decrease respectively, consistent with the enhanced turbulent mixing of the DI sprays as indicated by the computed tumble flow and turbulent kinetic energy characteristics. The results are consistent with the CFD in terms of the injection window, where the injection timing is constrained by wall impingement which is correlated to soot (FSN), and incomplete mixing which is correlated to combustion instability (COV). Injection too early or too late result into wall impingement; similarly injection too early or too late usually results into insufficient mixing due to the interaction of spray injection with the tumble flow. Higher turbulence enhances mixing before ignition and enhances combustion after ignition; therefore, there is optimal injection timing for faster mixing and combustion speed within the injection window.

2.2 *Effect of Valve Actuation for Low-Speed Low-Load Case*

Naturally aspirated DISI engines usually has poor charge motion at low speeds with EIVC, and high particulate levels when operating at high loads on gasoline and intermediate ethanol blends. Use of valve deactivation has been shown to improve stability and reduce particulate emissions at low loads and very low (2 ~ 3 mm) lift valves [7, 8]. In order to evaluate the effects of flexible valvetrain, the engine is modified with a dual-independent cam-phaser (DICP). The 2-Step variable valve lift system was installed to provide compression ratio management using Late- and Early- Intake Valve Closing (LIVC and EIVC). These strategies allowed change of effective CR and load control by use of cam phasing.

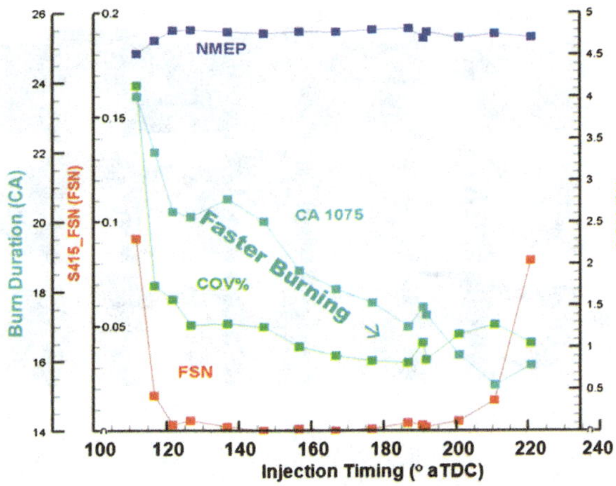

Fig. 7 Effect of injection timing on combustion: 1500 rpm, 4.7 BMEP, low-lift EIVC, injector B, E85

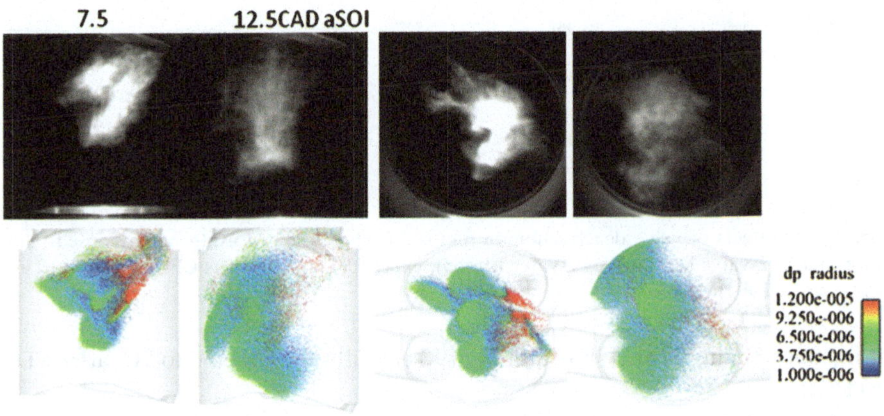

Fig. 8 CFD validation with injection at 270CAD with LIVC deactivation configuration, 1000 rpm, Inj. A

Simulation was also carried out to compare and validate with experimental result for valve deactivation. For example, Fig. 8 shows images at 7.5 and 12.5CAD aSOI with injection at 270CAD with LIVC deactivation configuration. The simulation results show the distribution of the particles with radius information in color. Even though CFD sprays overshoot for some extent, agreement of overall spray shape is in fair agreements for more asymmetric fuel distributions. Because of the higher lift, there is spray impingement on the intake valve which should be avoided. High lift is also not necessary at low load cases.

The effects on valve deactivation can be visualized by false coloring the spray images and superimpose them. For example, Fig. 9 shows the effects of valve-deactivation on the in-cylinder sprays, with the two-valve case colored red, and the one-valve case, green; therefore the overlapped spray regime is yellow. The effect

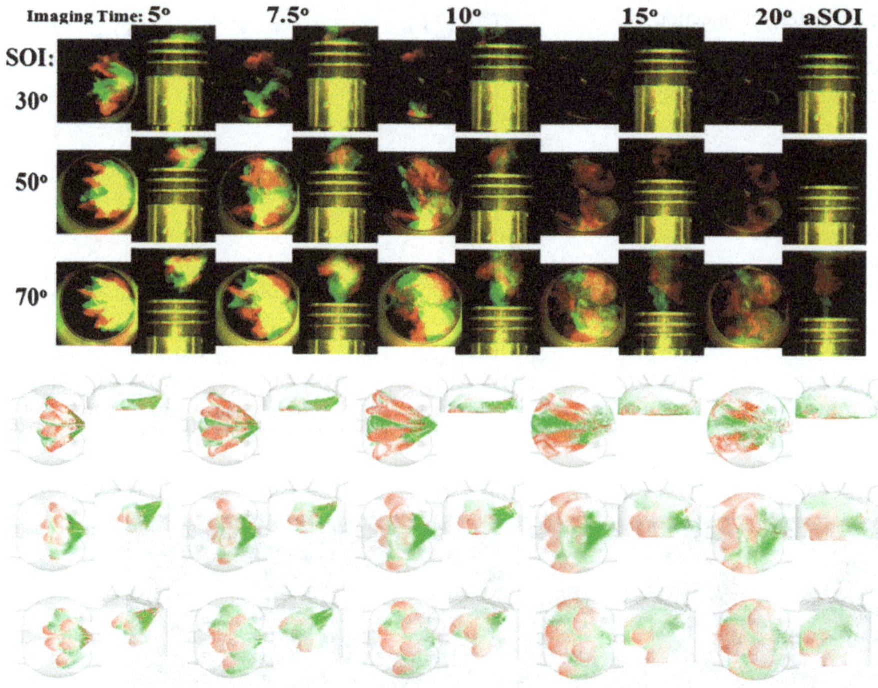

Fig. 9 The effects of valve deactivation on fuel sprays as visualized using superimposed false-color image. 2-valve (*red*) versus 1-valve (*green*); experiment (*top*) versus CFD (*bottom*); EIVC, 1000 rpm, 1 bar BMEP

of injection timing is also shown in Fig. 9 vertically from the 30 to 70° after TDC (aTDC).

Injection too early at 30° aTDC caused wall impingements and incomplete mixing, as a result, the engine produced high soot and COV. For this engine test, cam timing optimization was carried out at each operation point to maximize engine efficiency for both the conventional 2 valve and the 1 valve (valve deactivation) modes. The cam timing optimization was conducted independently for the different strategies to allow comparison at an optimal performance point. During testing, two fundamental issues at low speed operations were identified with the conventional 2-valve configuration and that valve deactivation was able to address. At low loads with EIVC, combustion stability was poor under un-throttled conditions. To improve combustion stability and efficiency the engine had to run under throttle with minimal valve overlap to limit residual.

Another simple analysis of the OAE spray images can be carried out by calculating the fractions of the spray cloud area as viewed through the extended transparent piston. There is no significant difference in the vaporization rate for too early injection as piston impingement dominates the spray behavior, but as the injection is retarded, the interference with the piston decreases and interaction with stronger turbulent charge motion increases. As evident by the smaller liquid spray

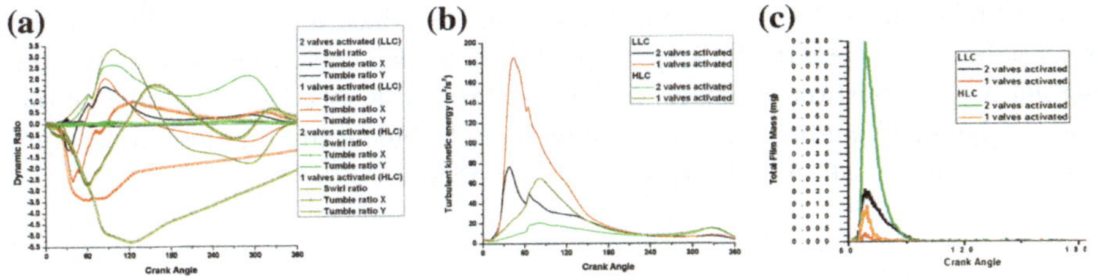

Fig. 10 Computed integral analysis: SOI = 60° aTDC, 1000 rpm, 1 bar BMEP equivalent. **a** Charge motion dynamic ratios. **b** Turbulent kinetic energy. **c** Liquid film mass

clouds, the spray mixing and vaporization is improved significantly, and the benefits of valve deactivation is demonstrated by lower soot and gaseous emissions, as well as better mixing (lower COV), consistent with engine data.

The optical engine data also show that valve deactivation deflects the spray and entrain it into the higher velocity intake flow. Visualization of sprays with valve deactivation seems to show faster impingement on the piston surface for early SOI at conditions that soot was also observed for the fired engine. The increased turbulent mixing due to the stronger bulk motion increases the rate of vaporization as the green spray of the deactivated-valve case is shown to disappear faster than the red baseline case. The acceptable injection timing window is limited in the engine data by soot (>0.1 FSN) for early timings, and poor combustion stability (>1.5 %COV) for later timings. Valve deactivation is shown to increase the injection window by stronger turbulent mixing due to the reduced intake opening and therefore higher intake velocity.

The computed integral analysis of the bulk flow dynamic ratios, which consists of a swirl ratio and two (X and Y) tumble ratios, are plotted in Fig. 10. The baseline 2-valve charge motions were dominated by the in-plane (or Y direction) counter-clockwise tumble motion, similar to Fig. 5, with the HLC case greater than LLC one. When one valve was deactivated, both the swirl and the out-of-plane (or X) tumble motions significantly appear, with the tumble changing directions or signs during the mixture preparation cycle, sometimes more than once. Even though HLC produces overall stronger and more coherent charge motion, the dynamic ratios of LLC develops faster at the beginning, due to the faster inlet air velocity coming through the reduced inlet valve open area. Consequently, the mass-averaged turbulent kinetic energy (TKE) during the mixture formation period is much higher; more than double that of the LLC cases. Similarly, the TKE of the deactivated cases more than double that of the two-valve cases. However, the higher TKE dissipates quickly and does not sustain well into the flame initiation phase, but fall below that of the two-valve cases. The enhancement of TKE by the fuel injection depends on the phasing of valve opening and is shown to favor the HLC cases for the engine and injector configurations. However, this TKE enhancement advantages is countered by the stronger bulker motion which sends more fuel sprays toward the combustion chamber walls. The deactivated 1-valve cases have not only stronger turbulent mixing but stronger 3-D rotational flows, which produce an insular effect on the spray plumes and keep the bulk spray plumes from the walls. Figure 18 shows the cumulative combustion chamber wall film mass. The mixing CFD calculations confirm the metal engine data which show more turbulent mixing, less liquid film, more stable combustion and wider injection window.

2.3 *Effect of Variable Valve Actuation for Medium and High-Load Case*

Higher valve lift is usually beneficially at higher load due to better breathing. The use of valve deactivation at this condition allowed a wide degree of cam phasing with good stability. Un-throttled conditions could be achieved, although optimal efficiency was under slightly throttled conditions. This allowed a significant reduction in pumping losses, for example at 2000 RPM, 2 Bar BMEP case, resulting from reduced throttling from 53 kPa MAP with 2 valves to 95 kPa MAP with 1 valve deactivated. The lower pumping work resulted in a reduction of fuel consumption. Another problem identified was the undesirably high soot emissions at low-speed high-load conditions with the LIVC strategy. Even though combustion stability was improved with wider timing authority, the soot could not be reduced by retarding injection timing. A typical operation point is at 1500 rpm 8 Bar BMEP, for the LIVC timing sweep for the 2 valve configuration. A review of the soot and the hydrocarbons emissions shows increased levels with injection retard. This indicates that fuel impinging on the combustion chamber wall is not adequately vaporized and mixed prior to combustion, which may be in the mode of pool fire. Operating the engine with one valve deactivated increases swirl and tumble and provides a significant advantage in soot emissions. Injection Timing is still limited at early timings by impingement on the piston but later timing allows a significant reduction in soot and hydrocarbons. CO emission, which is indicative of the level of uniformity of the combustible mixture, tracks the trends of COV and BSFC quite closely.

In the CFD simulation, four cases pertaining to operation of the EIVC and LIVC at 2000 rpm 2Bar BMEP, and 1500 rpm 8Bar BMEP, to illustrate the interactions of multi-hole DI sprays with VVA charge motions. The SOI is fixed at 60° aTDC for simplicity. The computed flow fields are highly dynamic and three-dimensional as shown by the instantaneous streamlines which is color-coded by the velocity magnitude in Fig. 11 for these four cases.

It is clearly visible that valve deactivation promotes more vigorous charge motion through the intake stroke, and direct the high speed flow advantageously to the center of the cylinder for mixing although larger pumping loss may be the trade-off. The simulated spray pattern and droplet distribution are shown for the EIVC 2000 RPM, 2 Bar BMEP cases are presented in Fig. 12 for side view and Fig. 13 for bottom view respectively, with (a) and without (b) valve deactivation. Injection timing is at 60° aTDC. Valve deactivation shows behavior similar to the

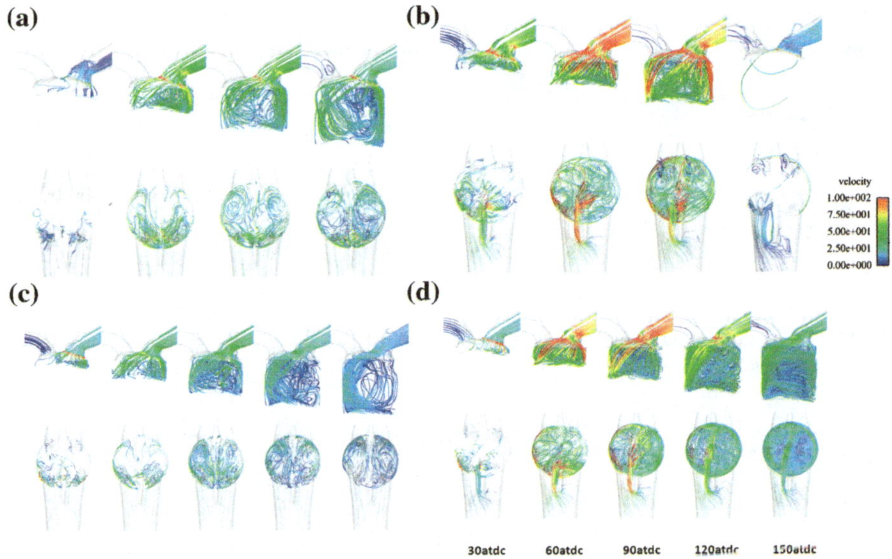

Fig. 11 The streamlines of EIVC cases at 2000 rpm: **a** 2-valve versus **b** 1-valve; LIVC cases at 1500 rpm: **c** 2-valve **d** versus 1-valve

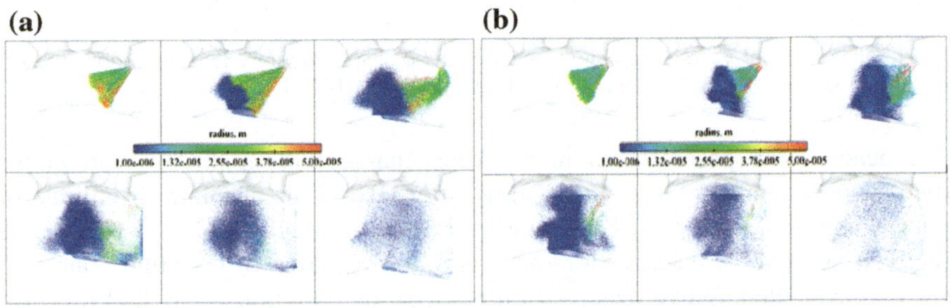

Fig. 12 The side view of fuel droplet distributions of EIVC cases at 2000 rpm, 2 Bar BMEP. **a** 2-valve versus **b** 1-valve

OAE visualization where the spray clouds are entrained toward the active intake valve into the inlet jet. Vaporization of the spray is faster with valve deactivation. The in-cylinder dynamic ratios were analyzed and the results are compared with and without valve deactivation in Fig. 13. The results show the swirl and two components of tumble around the x and y axis of Fig. 14. When both valves are active the flow is tumble dominated, initially with a reverse tumble followed by a forward tumble around the y-axis. When a valve is deactivated swirl and a cross tumble around the x-axis become significant and the initial reverse tumble is reduced. The location of the SOI timing is shown for reference.

The dynamic ratios for the LIVC condition are presented in Fig. 14. The results with 2 valves show the tumble dominated condition similar to EIVC. The use of

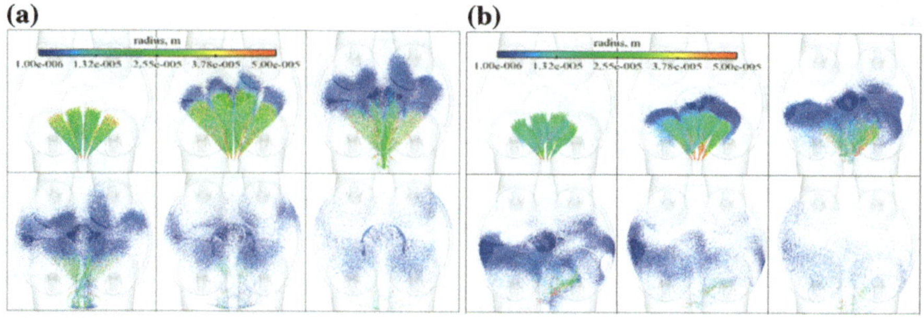

Fig. 13 The bottom view of fuel droplet distributions of EIVC cases at 2000 rpm, 2 Bar BMEP. **a** 2-valve versus **b** 1-valve

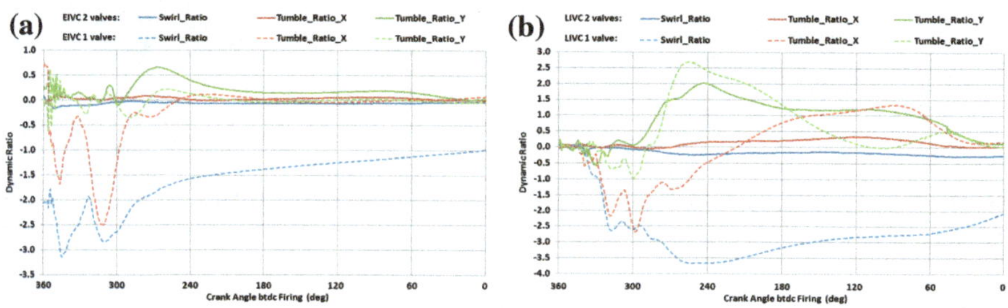

Fig. 14 The effects of valve deactivations on computed charge motions dynamic ratios. **a** EIVC case: 2000 rpm 2 bar BMEP. **b** LIVC case: 1500 rpm 8 Bar BMEP

valve deactivation promotes more vigorous charge motion through the intake stroke since the valve closes significantly after BDC. The absolute magnitude of the swirl and tumble increases and swirl persists up through TDC. To evaluate the effectiveness of valve deactivation on mixing the variation of in-cylinder air–fuel ratio was evaluated. Valve deactivation reduces the variation of air fuel ratio for both strategies. The LIVC has more variation earlier in the injection stroke as wall films and rich regions are mixed in at the higher load condition. At the time appropriate for ignition and combustion the more vigorous charge motion from the LIVC strategy has provided a more uniform mixture.

References

1. Zhao F-Q, Harrington DL, Lai M-C (2002) Automotive gasoline direct injection engines. Society of Automotive Engineers, Pittsburgh, R-315
2. Li JW et al (2001) Liquid fuel impingement on in-cylinder surfaces as a source of hydrocarbon emissions from direct injection gasoline engines. J Eng Gas Turbines Power 123:659–668

3. Ra Y, Reitz RD (2002) A model for droplet vaporization for use in gasoline and HCCI engine applications. In: Proceedings of ICEC'02, ASME spring technical conference, Rockford, Ill., 14–17 April 2002
4. Joh M, Huh KY, Yoo J-H, Lai M-C (2002) Numerical prediction and validation of fuel spray behavior in a gasoline direct-injection engine. J Fuels Lubr 110(4):2387–2404
5. Schmitz I, Ipp W, Leipertz A (2002) Flash boiling effects on the development of gasoline direct injection sprays. SAE 2002-01-2661
6. Horie K, Kohda Y, Ogawa K, Goto H (2004) Development of a high fuel economy and low emission four-valve direct injection engine with a center-injection system. SAE 2004-01-2941
7. Stansfield P et al (2007) Un-throttled engine operation using variable valve actuation: the impact on the flow field, mixing and combustion. SAE 2007-01-1414
8. Patel R et al (2008) Comparison between un-throttled, single and two-valve induction strategies utilizing direct gasoline injection: emissions, heat-release and fuel consumption analysis. SAE 2008-01-1626
9. Xue Q, Kong S-C (2008) Simulation of direct injection gasoline sprays using multi-level dynamic mesh refinement. IMECE
10. Wang Z et al (2007) Study of multimode combustion system with gasoline direct injection. J Eng Gas Turbines Power 129(4):1079
11. Lai M-C et al (2011) Characterization of internal flow and spray of multihole DI gasoline spray using X-ray imaging and CFD. SAE 2011-01-1881
12. Befrui B, Corbinelli G, Spiekermann P, Shost M, Lai M-C (2012) Large eddy simulation of GDi single-hole flow and near-field spray. SAE Int J Fuels Lubr 5(2):620–636
13. Battistoni M et al (2012) Coupled simulation of nozzle flow and spray formation using diesel and biodiesel for ci engine applications. SAE 2012-01-1267
14. Zheng Y, Xie X, Lai M-C, VanDerWege B (2012) Measurement and simulation of DI spray impingements and film characteristics. ICLASS, 2012-1253
15. Matsumoto A, Zheng Y, Xie X, Lai M-C et al (2011) Spray characterization of ethanol gasoline blends and comparison to a CFD model for a gasoline direct injector. Int J Engines 3(1):402–425
16. Gao J, Trujillo M, Deshpande S (2011) Numerical simulation of hollow-cone sprays interacting with uniform crossflow for gasoline direct injection engines. SAE Int J Engines 4(2):2207–2221
17. Renberg U, Westin F, Angstrom H, Fuchs L (2010) Study of junctions in 1-D & 3-D simulation for steady and unsteady flow. SAE 2010-01-1050
18. Dempsey A et al (2012) Comparison of quantitative in-cylinder equivalence ratio measurements with CFD predictions for a LD low temperature combustion diesel engine. SAE Int J Engine 5(2):162–184
19. Yang XF et al (2012) Ignition and combustion simulations of spray-guided SIDI engine using arrhenius combustion with spark-energy deposition model. SAE 2012-01-0147
20. Yoo J, Zhao F-Q, Liu Y, Lai M-C, Lee K-S (1997) Characterization of direct-injection gasoline sprays under different ambient and fuel injection conditions. Int J Fluid Mech Res 24:480–487
21. Park J, Im K-S, Kim H, Lai M-C (2004) Characteristics of wall impingement at elevated temperature conditions for GDI spray. Int J Automot Technol 5(3):155–164
22. Marriott C, Wiles W, Gwidt J, Parrish S (2008) Development of a naturally aspirated spark ignition direct-injection flex-fuel engine. SAE 2008-01-0319
23. Rimmer J et al (2009) The influence of single and multiple injection strategies on in-cylinder flow and combustion within a DISI engine. SAE 2009-01-0660
24. Moore W et al (2011) Charge motion benefits of valve deactivation to reduce fuel consumption and emissions in an SIDI, VVA engine. SAE 2011-01-1221

Multidimensional CFD Simulation of a Diesel Engine Combustion: A Comparison of Combustion Models

Arif Budiyanto, Bambang Sugiarto and Bagus Anang

Abstract The objective of this study is to simulate combustiuon process and pollutant formation in the combustion chamber of a DI diesel engine. The modelled results were validated by comparing predictions against corresponding experimental data for a diesel engine. The predicted and measured in-cylinder pressure and emission data were in good agreement. Computational fluid dynamics (CFD) is able to significantly reduce the number of experimental test and measurement and lower the development time and costs. Some parameter which are needed for CFD calculation must be achieved experimentally such as turbulence time scale constant. The CFD simulations demonstrated good agreement to the measured data. The results show that, applying appropriate constant of each combustion model including eddy break up model (Ebu), caracteristic timescale model (Ctm) and extended coherent flamelet model (Ecfm) causes the computational result to be in agreement with experimental results. Furthermore the result show that the nearest prediction in comparasion with experimental result is by applying the Ecfm model.

Keywords Diesel engine · CFD · Combustion models

F2012-A05-012
Received 26 June 2012; accepted 8 September 2011

A. Budiyanto · B. Sugiarto
Internal Combustion Engine Laboratory Mechanical Engineering,
University of Indonesia, Depok, Indonesia

B. Anang (✉)
Internal Combustion Engine Laboratory, BTMP- BPPT- Indonesia, Jakarta, Indonesia
e-mail: muhammad.arif11@ui.ac.id

SAE-China and FISITA (eds.), *Proceedings of the FISITA 2012 World Automotive Congress*, Lecture Notes in Electrical Engineering 190,
DOI: 10.1007/978-3-642-33750-5_5,

Nomenclature

u	Velocity
p	Pressure
T	Temperature
C	Species concentration
$\dot{m}$	Mass flow rate
$\dot{Y}$	Mass fraction
r_f	Stoichiometric coefficient
n	Engine speed
h_v	Maximum valve lift
Greek symbols	
$\hat{\rho}$	Density
μ	Viscosity
ϕ	Equivalence ratio
$\dot{\omega}$	Combustion reaction rate
ε	Turbulent Dissipation rate
Abbreviations	
CFD	Computational fluids dynamics
TKE	Turbulence kinetic energy
Ebu	Eddy break-up model
Ctm	Caracteristic timescale model
Ecfm	Extended coherent flamelet model

1 Introduction

Diesel engine development is currently facing many challengges. On the one hand, there is a need to lower emissions in order to improve air quality, while, on the other, despite the diesel engine's good fuel economy compared to many other power sources, there is continous competition to further reduce fuel consumption. In addition, new fuels, novel combustion techniques and combustion conditions have been developed to lower emissions and obtain better engine efficiency. All these advancements are great challenges for the engine manufacturers [1].

The improvement of science and technology the optimization of combustion process for reduction of exhaust emissions and fuel consumption is still to be attractive objects for researchers. Numerous experiments and computations have been performed by large number of researchers to improve combustion and control exhaust pollutants.

Modeling an internal combustion engine is a complex problem. This is due to numerous phneomena that each one related to the other. The combustion models plays an important role in the modeling of internal combustion.

Many combustion simulations of dual fuel engines have used various model to analyze the combustion.

Pirouzpanah et al. [2] was developed a model to simulate combustion characteristics at part loads using a quasi—dimensional multi zone combustion model. In another research, a wor was done, conducted to investigate the combustion characteristics at part loads using a single zone combustion model with detailed chemical kinetics for combustion of natural gas fuel [3].

Ghasemi et al. [4] simulate combustion process, plolutant formation and flow field in the combustion chamber of diesel gas engine using Eddy breakup model. Adrian et al. was simulate diesel engine using Mixing Controlled Combustion model was chosen for the prediction of the combustion characteristics. This model considers the effects of the premixed and diffusion controlled combustion processes [5].

Mansour et al. developed a computer program to model the combustion process in dual fuel engines. A detailed chemical reaction mechanism of natural gas and were used to predict the main combustion characteristics. This semi-empirical dual fuel engine model uses the Wiebe's function combustion model for different speeds and air fuel ratios [6].

Scarcelli using coherent flamelet model to predict effect of improvement the lean burn operation of natural gas/air mixtures [7]. The 3-zone extended coherent flame model for computing premixed/diffusion combustion was simulated by Colin et al. [8].

The objective of this work is multidimensional simulation of diesel dual fuel engine using AVL FIRE CFD tool. The characteristic and comparation of combustion models are investigated; eddy breakup model (Ebu) and extended coherent flamelet model (Ecfm). First the simulation is performed in complete diesel condition and then dual fuel combustion process is simulated. Results are validated by means of experiments performed on the Ricardo Hydra DI diesel engine.

2 Experimental and Numerical Setup

Experimental set up and engine specifications and operating condition based on Ricardo Hydra DI diesel engine is used in this study. Specifications and operating conditions of the engine are shown in Table 1. A schematic diagram of system is shown in Fig. 1. In this study exhaust gas emission such as NOx (by signal analyser) and soot (by a AVL smoke meter) were measured.

Computational grid generation based on the geometry description, a set of computational meshes covering 360° crank angle was created. The 3D mesh of the modeled engine is shown in Fig. 2 containing cylindrical Hydra type piston bowls. 90° sector mesh was used in this study considering that the diesel injector has four nozzle holes. This mesh resolution has been found to provide adequtely independent grid result. The multi-block structure of grid, containing spray and injector blocks is shown in Fig. 3.

Table 1 Engine specification

Subjects	Description
Model	Ricardo hydra
Engine type	DI diesel
Number of cylinder	1
Number of valve	2
Swept volume (lit)	0.45
Bore (mm)	80.26
Stroke (mm)	88.9
Compression ratio	19:1
Combustion chamber shape	Bowl in piston
Injector operation pressure	250
Valve timing	IVO 8 BTDC, IVC 42 ATDC, EVO 60BBDC, EVC 12 ATDC

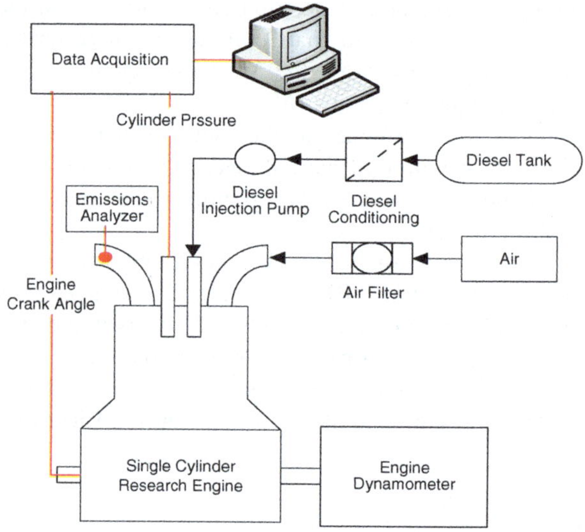

Fig. 1 Schematic of experimental setup

After modeling of the combustion chamber, the initial and boundary conditions should be define. After creating the moving mesh, the condition of the engine at the end of the induction stroke is considered as the initial condition. The initial operating conditions of the engine are shown in Table 2 and boundary conditions are show in Fig. 4.

Thera are two parameter in the engine namely turbulence kinetic energy and turbulence leght scale that are need to be defined. The turbulence kinetic equation is as follows:

$$TKE = \frac{3}{2} u'^2 \tag{1}$$

where, u' is defined as follows:

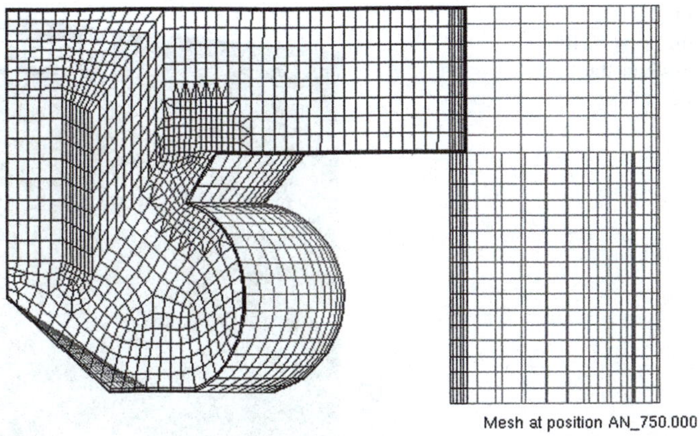

Fig. 2 3-Dimensional grid of the modeled engine showing Hydra type piston bowls

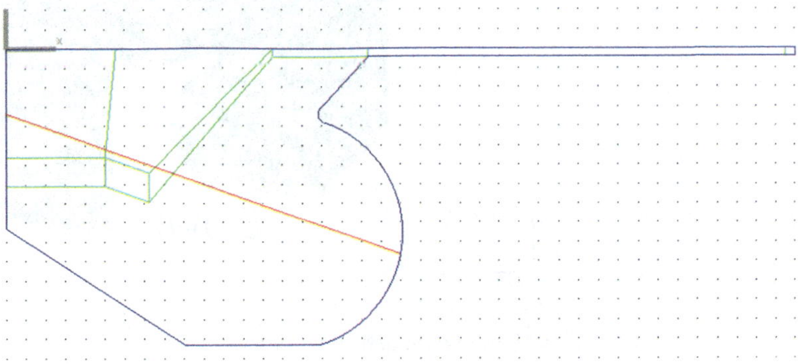

Fig. 3 Multi-block structure of the grid containing spray and ijector blocks

Table 2 Initial operationg condition of the engine

Parameter	Unit	Diesel	DDF 90 %
Start angle	Deg	560	560
End angle	Deg	840	840
Cylinder head temp.	K	570.15	570.15
Cylinder liner temp.	K	470.15	470.15
Piston Surface temp.	K	570.15	570.15
Cylinder pressure	Pa	101324	101120
Inlet gas temperature	K	305.15	298.05
Engine Speed	rpm	1200	1200
Turbulence kinetic energy	m^2/s^2	0.968	0.968
Turbulence lenght scale	m^2/s^3	0.00400	0.00400
Mass of solar	kg	1.793E−05	1.197E−06
Mass of CNG	kg	0	0.24520
Mass of air	kg	0.00049	0.00037

Fig. 4 Boundary conditions definitions of the volume control when the piston is located at *bottom dead center*

$$u' = 0.7x\,(mean\,piston\,velocity) \tag{2}$$

where, mean piston velocity is defined as follows:

$$mean\,piston\,velocity = \frac{2\,x\,n\,x\,S}{60} \tag{3}$$

where, n is the engine speed equals to 1200 rpm and S is the stroke of the engine that equals to 88.9 mm. And turbulence lenght scale is defined as:

$$TLS = \frac{h_v}{2} \tag{4}$$

Where, $h_{v,}$ is maximum valve lift that for engine.

Computational fluids dynamics (CFD) simulation based on basic equation are presented for the following dynamic and thermodynamic properties. Mass (Equation of continuity)

$$\frac{\partial}{\partial t}\rho(x,t) + \frac{\partial}{\partial x_i}(\rho(x,t)v_i(x,t)) = 0 \tag{5}$$

Momentum equation (Navier–Stokes)

$$\rho\left(\frac{\partial}{\partial t}+v_j\frac{\partial}{\partial x_j}\right)v_j-\frac{\partial}{\partial x_j}\left(\tau_{ij}\left[\frac{\partial v_k}{\partial x_l}\right]\right)=-\frac{\partial p}{\partial x_i}+f_i \tag{6}$$

Energy (1st Law of Thermodynamics)

$$\rho\left(\frac{\partial}{\partial t}+u_j\frac{\partial}{\partial x_j}\right)u-\frac{\partial}{\partial x_j}\left(\lambda\frac{\partial T}{\partial x_i}+\rho D_{(k)}\sum_{(k)}h_{(k)}\frac{\partial c_{(k)}}{\partial x_i}\right)=-p\frac{\partial v_i}{\partial x_i}+\tau_{ij}\frac{\partial v_i}{\partial x_i} \tag{7}$$

Concentration of species Equation

$$\rho\left(\frac{\partial}{\partial t}+v_j\frac{\partial}{\partial x_j}\right)Z-\frac{\partial}{\partial x_i}\left(D\rho\frac{\partial}{\partial x_j}Z\right)=0 \tag{8}$$

Eddy Breakup Model (Ebu). The Ebu model is described by Magnussen and Hjertager [9]. This model assumes that in premixed turbulent flames, the reactants (fuel and oxygen) are contained in the same eddies and are separated from eddies containing hot combustion products. The chemical reactions usually have time scales that are very short compared to the characteristics of the turbulent transport processes. Thus, it can be assumed that the rate of combustion is determined by the rate of intermixing on a molecular scale of the eddies containing reactants and those containing hot products, in other words by the rate of dissipation of these eddies. The attractive feature of this model is that it does not call for predictions of fluctuations of reacting species. The mean reaction rate can thus be written in accordance with Magnussen and Hjertager:

$$\overline{\rho\dot{r}_{\mathrm{fu}}}=\frac{\mathrm{C_{fu}}}{\tau_{\mathrm{R}}}\bar{\rho}\min\left(\bar{y}_{\mathrm{fu}},\frac{\bar{y}_{\mathrm{Ox}}}{\mathrm{S}},\frac{\mathrm{C_{Pr}}\cdot\bar{y}_{\mathrm{Pr}}}{1+\mathrm{S}}\right) \tag{9}$$

The first two terms of the "minimum value of" operator "min(…)" simply determine whether fuel or oxygen is present in limiting quantity, and the third term is a reaction probability which ensures that the flame is not spread in the absence of hot products. Cfu and CPr are empirical coefficients and τR is the turbulent mixing time scale for reaction. The value of the empirical coefficient Cfu has been shown to depend on turbulence and fuel parameters (Borghi; Cant and Bray). Hence, Cfu requires adjustment with respect to the experimental combustion data for the case under investigation (for engines, the global rate of fuel mass fraction burnt) [10].

Characteristic Timescale Model. In diesel engines a significant part of combustion is thought to be mixing-controlled. Hence, interactions between turbulence and chemical reactions have to be considered. The model described in this section combines a laminar and a turbulent time scale to an overall reaction rate [10]. The time rate of change of a species m due to this time scale can be written as follows:

$$\frac{dY_m}{dt} = -\frac{Y_m - Y_m^*}{\tau_c} \tag{10}$$

where Ym is the mass fraction of the species m and Ym* is the local instantaneous thermodynamic equilibrium value of the mass fraction. τc is the characteristic time for the achievement of such equilibrium. This whole approach is conceptually consistent with the model of Magnussen. The initiation of combustion relies on laminar chemistry. Turbulence starts to have an influence after combustion events have already been observed. The combustion will be dominated by turbulent mixing effects in regions of $\tau l \ll \tau t$. The laminar time scale is not negligible in regions near the injector where high velocities cause a very small turbulent time scale. Auto ignition is calculated by the Shell Model which is integrated in the specific model description. The ignition model is applied where ever T < 1,000 [K].

Coherent Flame Model. A turbulent premixed combustion regime can be specified using different properties such as chemical time scale, integral length scale and turbulence intensity. Due to the assumption that in many combustion devices the chemical time scales (e.g. reciprocating internal engines) are much smaller in comparison to the turbulent ones, an additional combustion concept can be applied: the Coherent Flame Model or CFM. The CFM is applicable to both premixed and non-premixed conditions on the basis of a laminar flamelet concept, whose velocity SL and thickness δL are mean values, integrated along the flame front, only dependent on the pressure, the temperature and the richness in fresh gases. Such a model is attractive since a decoupled treatment of chemistry and turbulence is considered. All flamelet models assume that reaction takes place within relatively thin layers that separate the fresh unburned gas from the fully burnt gas. Using this assumption the mean turbulent reaction rate is computed as the product of the flame surface density Σ and the laminar burning velocity SL via:

$$\overline{\rho \dot{r}_{fu}} = -\omega_L \Sigma \tag{11}$$

with L ω as the mean laminar fuel consumption rate per unit surface along the flame front. Where fufr is the partial density of the fresh gas that is define as

$$\omega_L = \rho_{fu,fr} S_L \quad \text{with} \quad \rho_{fu,fr} = \rho_{fr} Y_{fu,fr} \tag{12}$$

In this equation ρfu, fr is the partial fuel density of the fresh gas, ρfr the density of the fresh gas and yfu, fr is the fuel mass fraction in the fresh gas. When combustion starts new terms are computed, source terms and two quantities in order to use equation Σ and SL Currently, three different CFM's are available which are described in increasing complexity in the following chapters. First the standard CFM is described, than the MCFM for application under very fuel rich or lean conditions and finally the ECFM which is coupled to the spray module in order to describe DI-SI engine combustion phenomena.

The ECFM (E stands for extended) has been mainly developed in order to describe combustion in DI-SI engines. This model is fully coupled to the spray model and enables stratified combustion modeling including EGR effects and NO

Table 3 Case set up

Case	1	2	3
Operating condition	Diesel	Diesel	Diesel
Combustion model	EBU	CTM	ECFM
Turbulent time scale	0.001	0.001	0.001

formation. The model relies on a conditional unburned/burnt description of the thermochemical properties of the gas [10].

3 Results and Discussion

Table 3 shows the simulations condition which performed on the mentioned engine in this research.

Comparison of mean cylinder pressure for all combustion models (Case_1, Case_2, Case_3) and experiments id demonstrated in Fig. 5. As can be seen, the agreement between three results is very good.

The combustion processes in diesel engines running divided into four stages. They are, starts from period of ignition delay, and next is premixed (rapid pressure rise) combustion, continued by controlled or difussion combustion, and the ended by late combustion [11]. Point at start of ignition delay that is the start of fuel injection and end of the ignition delay that is a start of combustion. Figure 6 exhibits diesel combustion behavior containing premixed and diffusion stages of combustion. During the premixed combustion period of each model has a different shape, ha shows that in this period is very sensitive to the approach to the calculation by each model. Can be seen in the graphs that Ecfm model has the form that most closely with the experimental data, this suggests that the velocity and thickness of flame this suggests that the velocity and thickness of flame the flame is agreeable for calculating for premixed combustion.

Variations of mean cylinder temperature with crank angle for case 1 is shown in Fig. 6. Comparing this diagram with temperature and pressure diagrams shows high rates of temperature and pressure rise during the premixed combustion period. After diffusion combustion the pressure decrease slowly accordanceby piston angle.

Heat release rate diagram for case 1 is presented Fig. 7. Ecfm model has a higher form than others. Whereas Ctm form that take more models have low compared to other models. It shows the time scale given in this model is less influencial for computing the diesel combustion.The heat release diagram of diesel fuel operation shows an apparent negative heat release prior to the main start of combustion. This is due to the cooling effect of the injected liquid fuel. The pilot fuel is then flattened the heat release rate within this region [12].

The Nox emissions has been show in Fig. 8. The Zeldovich mechanism has been used together with the characteristic time combustion model. It was observed

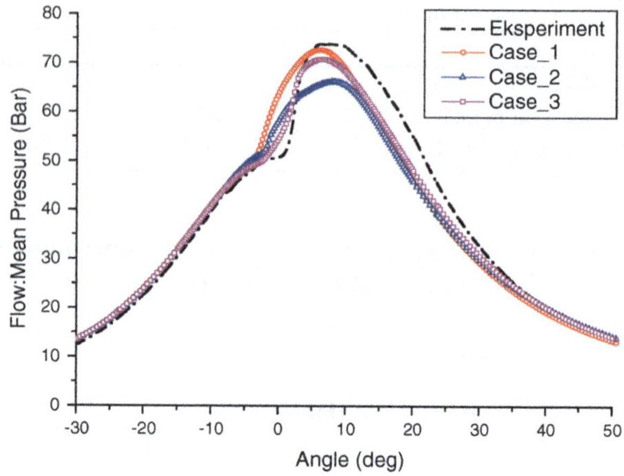

Fig. 5 Comparison of mean cylinder pressure for all combustion models and Experiment

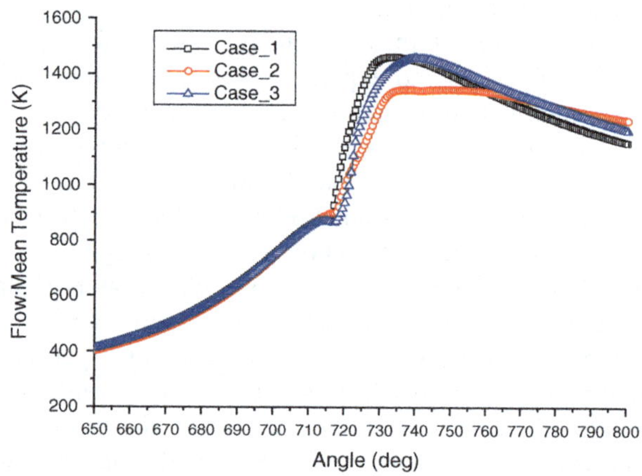

Fig. 6 Variation of mean cylinder temperature with crank angle for all case

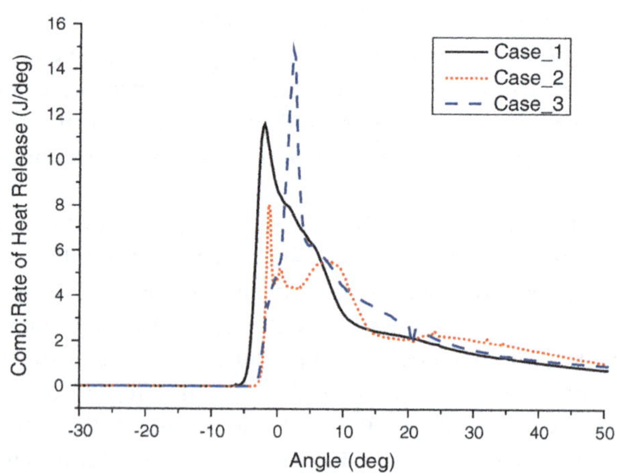

Fig. 7 Heat release rate diagram for case 1

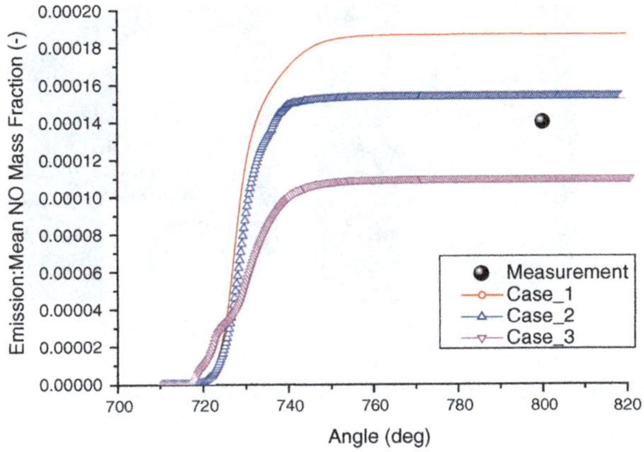

Fig. 8 Comparasion of NOx precdiction for model in Case_1, Case_2 and Experiment

that the default NOx model for Ebu and Ctm parameter in AVL Fire Code gave NOx results that were too high compared to measurement, but model Ecfm gave NOx result that were too low compared to measurement.

NOx chemistry is slow compared to the main chemistry and sonsequently, may not be predicted well with fully mixing limited approaches. This issue is studied in Fig. 8 where the measured and computed NOx results are shown in the Hydra engine the Ebu model together with extended Zeldovich mechanism, Ecfm model together with extended Zeldovich mechanism, or Ctm combustion model together with the standard Zeldovich mechanism. Quantitatively, the Ctm model plus the extended Zeldovich mechanism gives better agreement with measurement. The trend predictive capabilitis are not so evident, in conclusion, the Ctm combustion model plus the extended Zeldovich mechanism for NOx gives better agreement with the experimental results, especially in quantitative terms, but the trend predictive capabilities were only modestly better (Fig. 9).

Soot foormation and Oxidation chemistry is a complex issue involving thousands of reacting and hundreds of species, and consequently, simplified solution must be used within CFD modeling. In this case, semi-empirical models, wich offer low computational cost but still retain predictives capabilities, have been used. The soot results obtained from CFD modeling are dependent also on the accuracy of the other models used. For example, it was shown above that soot formation tendency depends very much on the local fuel–air ratio and is thus strongly affected by the accuracy of the flow field (turbulence) and fuel spray simulation [13]. Despite these challenge, soot models have been succesfully implemented, developed, and used in this research.

The Hiroyasu-Magnussen (HM) soot formation and oxidation model was implemented into AVL FIRE code and compared with the engine measured. It was observed that the predicted soot emission may be sensitive to the turbulence model used, as show in Fig. 10, and that chemistry may effect the soot results in an unexpected way. The maximum soot concentration during combustion was generally higher with the eight-step mechanism than with the two-step approach and

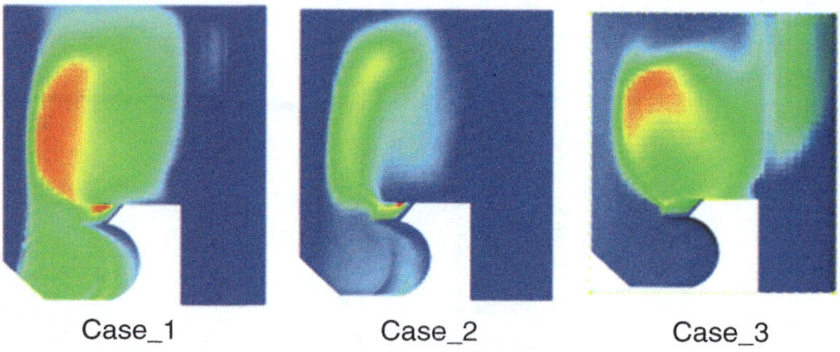

Fig. 9 Comparison of NOx formation contours for all case in 780 CA degree

Fig. 10 Comparasion of soot for model in case_1 and case_2

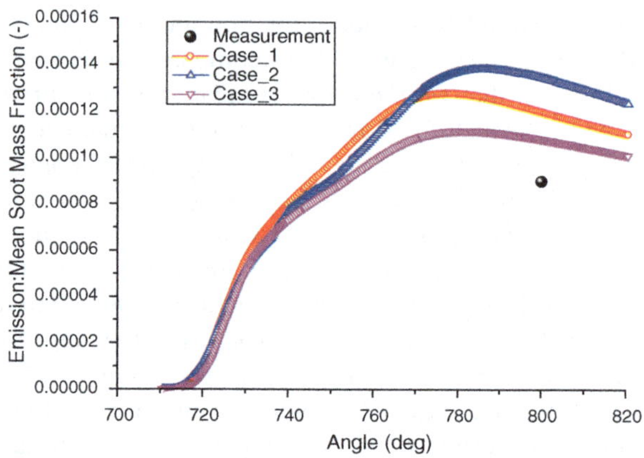

the final soot amount was generally lower with eight-step model. These trends were closer to the experimental observation and, as such, there may provide some proof that the maximum soot concentration during combustion is under predicted by the traditional two-step models [14]. in conclusion, the Ecfm combustion model plus the Hiroyasu-Magnussen mechanism for Soot formation gives better agreement with the experimental results, especially in quantitative terms, but the trend predictive capabilities were only modestly better (Fig. 11).

The Eddy break-up model (Ebu) and the Extended coherent flame model (ECfm) combustion model have been compared with the different turbulent length scale. The predicted and measurement pressure curve are seen in Fig. 12. It is seen that, on average, there are has difference between the range of turbulent lenght scale constant which applied.

As described by Magnussen and Hjertager, the Ebu model that assumes the reactants and the hot combustion products are contained in separated eddies. The chemical reactions usually have time scales that are very short in comparison to the characteristics of the turbulent transport processes. Thus, it can be assumed that

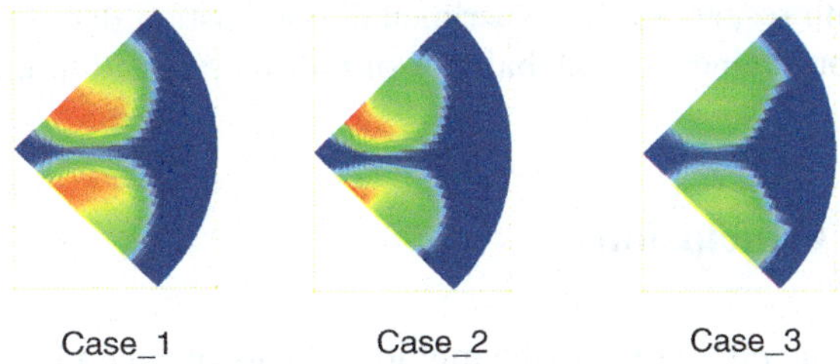

Fig. 11 Comparison of NOx formation contours for all case in 780 CA degree

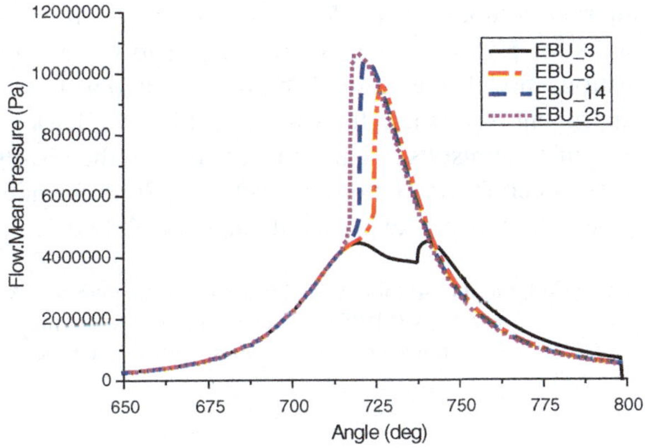

Fig. 12 Comparison between the model turburlent lenght scale constants in Eddy break up model

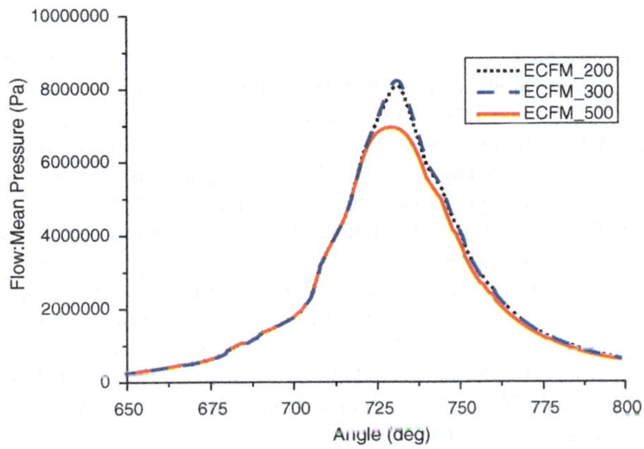

Fig. 13 Comparison between the model turburlent lenght scale constants in Ecfm model

the rate of combustion is determined by the rate of intermixing on a molecular scale of eddies containing reactants and those containing hot products.

The useful feature of this model is that it does not call for predictions of fluctuations of reacting species. The mean reaction rate can thus be written in accordance with Magnussen and Hjertager. The value of the empirical coefficient Cfu depends on turbulence and fuel parameters. Hence, Cfu requires adjustment with respect to the experimental combustion data for the case under investigation (for engines, the global rate of fuel mass fraction burnt) (Fig. 13).

4 Conclusion

In the present study the multidi-dimensional combustion modeling was carried out for diesel engine combustion. Also flow field simulation was perfromed for three different combustion models in diesel mode. Results were validated via experiments for Hydra DI engine. According to the results of pressure for three combustion model the Ctm combustion model has the least difference to the experiment results and the Ctm model has the least variation in comparison to the experimental results. It can be inferred from the results that CFD simulations led to more accurate prediction of NO and Soot formation. There have been good agreement between experiments and the CFD calculations.

Acknowledgments Authors would like to express thanks to Mechanical Engineering University of Indonesia and BTMP-BPPT (Agency for the Assessment and Application of Technology) Indonesia for their financial support and laboratory facilities in the research project.

References

1. Moshaberi R, Fotrosy Y, Jalalifar S (2009) Modeling of spark ignition engine combustion: a computaional and experimental study of combustion process effects on NOx emissions. Asian J Appl Sci, ISSN 1996-3343, Malaysia
2. Pirouzpanah V, Khoshbakhti Saray R (2006) A predictive for the model combustion process in dual fuel engine at part loads using a quasi dimensional multi zone model and detailed chemical kinetics mechanism. IJE Trans B: Appl 19(1)
3. Pirouzpanah V, Saray KR (2007) Comparasion of thermal and radical effect of EGR gases on combustion process in dual fuel engine at part loads. Energy Convers Manage 48:1909–1918
4. Ghasemi A, Djavareskhian MH (2010) Investigation of the effect of natural gas equivalence ratio and piston bowl flow field on combustion and pollutant formation of a DI dual fuel engine. J Appl Sci, ISSN 1812-5654, Asian Network for Sciencific Information
5. Britas A, Chiriac R (2011) U.P.B. Sci Bull, Series D, 73(4), ISSN 1454-2358
6. Mansour C et al (2001) Gas-diesel (dual-fuel) modeling in diesel engine environment. Int J Therm Sci 40:409–424
7. Scarcelli R (2008) Lean burn operation for natural gas/air mixtures: "the dual-fuel engines". Phd Thesis Universita' Degli Studi Di Roma 'Tor Vergata
8. Colin O, Benkenida A (2004) The 3-zones extended coherent flame model (ECFM3Z) for computing premixed/diffusion combustion. Oil Gas Sci Technol—Rev IFP 59(6):593–609, Copyright © 2004, Institut français du pétrole
9. Magnussen BF, Hjertager BH (1977) On mathematical modeling of turbulent combustion with special emphasis on soot formation and combustion. Symp (Int) Combust 16:719–729
10. AVL FIRE combustion module.revised E 17-May-2004 CFD Solver v8.3—combustion 08.0205.0699. Copyright © 2004, AVL

11. Sahoo a BB, Sahoo b N, Saha b UK (2008) Effect of engine parameters and type of gaseous fuel on the performance of dual-fuel gas diesel engines—a critical review, 1364-0321/$—see front matter 2008 Elsevier Ltd. All rights reserved. doi:10.1016/j.rser.2008.08.003
12. Nwafor OMI (2003) Combustion characteristics of dual fuel diesel engine using pilot injection ignition. Inst Eng (India) J Apr 2003 84:22–25
13. Kaario O, Larmi M, Tanner FX (2002) Relating integral leght scale to to turbulrnt scale and comparing k-e and RNG k-e turbulence models in diesel combustion simulation. SAE 202 Trans J Engines, pp. 1886–1900, SAE 2002-01-1117
14. Kaario OT, Larmi M, Tanner FX (2002) Comparing single-step and multi-step chemistry using the laminar and turbulence characteristic time combustion model in two diesel engine. SAE Paper 2002-01-1749, Reno, 6.5.-9.5.2002

Part VI
Engine Design and Simulation

Dual Fuel CNG-Diesel Heavy Duty Truck Engines with Optimum Speed Power Turbine

Alberto Boretti

Abstract The turbocharged direct injection lean burn Diesel engine is the most efficient engine now in production for transport applications with full load brake engine thermal efficiencies up to 40 to 45 % and reduced penalties in brake engine thermal efficiencies reducing the load by the quantity of fuel injected. The major downfalls of this engine are the carbon dioxide emissions, the depletion of fossil fuels using fossil Diesel, the energy security issues of using foreign fossil fuels in general, and finally the difficulty to meet future emission standards for soot, smoke, nitrogen oxides, carbon oxide and unburned hydrocarbons for the intrinsically "dirty" combustion of the fuel injected in liquid state and the lack of maturity the lean after treatment system. CNG is an alternative fuel with a better carbon to hydrogen ratio therefore permitting reduced carbon dioxide emissions. It is injected in gaseous form for a much cleaner combustion almost cancelling some of the emissions (even if unfortunately not all of them) of the Diesel and it permits a much better energy security within Australia. The paper discusses the best options currently available to convert Diesel engine platforms to CNG, with particular emphasis to the use of these CNG engines within Australia where the refuelling network is scarce. This option is determined in the dual fuel operation with a double injector design that couples a second CNG injector to the Diesel injector. This configuration permits the operation Diesel only and the operation Diesel pilot and CNG main depending on the availability of refuelling stations where the vehicle operates. Results of engine performance simulations are performed for a straight six cylinder 13 litres truck engine with a novel power turbine

F2012-A06-003

A. Boretti (✉)
Missouri University of Science and Technology, USA

A. Boretti
University of Ballarat, Australia

SAE-China and FISITA (eds.), *Proceedings of the FISITA 2012 World Automotive Congress*, Lecture Notes in Electrical Engineering 190,
DOI: 10.1007/978-3-642-33750-5_6, © Springer-Verlag Berlin Heidelberg 2013

connected to the crankshaft through a constant variable transmission that may be by-passed when non helpful to increase the fuel economy of the vehicle or when damaging the performances of the after treatment system. The Diesel operation permits full load efficiencies of 45 % thanks to the power turbine arrangement. The Diesel-CNG operation permits slightly reduced full load efficiencies working at different air-to-fuel ratios with about same torque output as desired in the vehicle control.

Keywords Compressed natural gas · Power turbine · Internal combustion engines

Abbreviations

BDC	Bottom dead centre
CNG	Compressed natural gas
CVT	Continuously variable transmission
EGR	Exhaust gas recirculation
EVO	Exhaust valve opening
EVC	Exhaust valve closure
HDT	Heavy duty truck
HPDI	High pressure direct injection
IVO	Intake valve opening
IVC	Intake valve closure
LHV	Lower heating value
LNG	Liquefied natural gas
TC	Turbocharged
TDC	Top dead centre

1 CNG as a Fuel Substitute for Diesel

Compressed natural gas (CNG) is a fossil as well as a renewable fuel substitute for traditional gasoline (petrol) or Diesel fuels. Renewable natural gas, also known as sustainable natural gas, is a biogas obtained from biomass that may possibly supplement if not replacing the fossil natural gas in the future. Natural gas is one of the world's most abundant sources of primary energy, and has been in use as a transportation fuel for more than 60 years. Natural gas is non-toxic. It is much lighter than air and diffuses quickly if released, unlike diesel fuel that pools on the ground. When it comes to protecting public health and the environment, natural gas is the cleanest fuel that is widely available today. Natural gas is composed primarily of methane, and burns more cleanly than diesel fuel or gasoline because it has more hydrogen and less carbon. Although its combustion also produces carbon dioxide CO_2 at the tail pipe, it is a more environmentally clean alternative

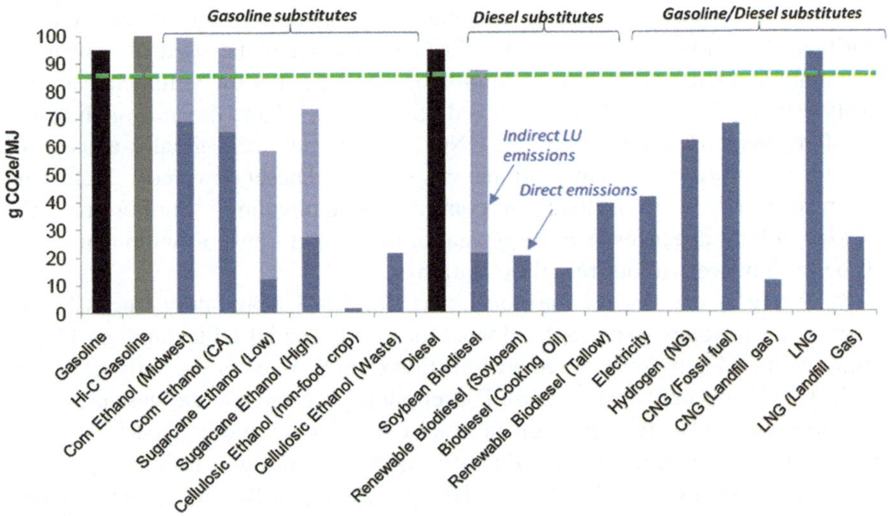

Fig. 1 CARB fuel cycle values of CO_2-e per MJ of fuel energy for diesel, gasoline/petrol and some renewable and non renewable alternative fuels [1, 2]. CNG fossil is rated about 70 g CO_2-e/MJ vs. the 95 gCO_2-e/MJ of gasoline and diesel. CNG from landfill is rated about 10 gCO_2-e/MJ and LNG from landfill is rated about 25 gCO_2-e/M. Results of life cycle analyses of transportation fuels vary from country to country and from author to author depending on the many factor affecting the result and the future scenarios that are predicted

to those fuels for the much better C to H ratio permitting a lower value of CO_2 per MJ of fuel energy. CNG is much safer than other fuels in the event of a spill because it is lighter than air and disperses quickly when released. When produced from landfills or waste water, the fuel life cycle carbon dioxide CO_2-e per MJ of fuel energy may be one order of magnitude lower than with fossil gasoline and Diesel. Figure 1 presents the CARB fuel cycle values of CO_2-e per MJ of fuel energy for Diesel, gasoline/petrol and some renewable and non-renewable alternative fuels (picture from [1], with data presented in [2]). The combustion of the CNG delivered in gas phase is much cleaner than the combustion of the Diesel delivered as a mixture of liquid hydrocarbons. And this reduces the formation of pollutants, smoke and soot first of all. Finally, virtually all of Australia's natural gas supply is continentally sourced, making natural gas less vulnerable to foreign supply disruption and price volatility.

CNG is made by compressing natural gas which is mainly composed of methane [CH_4] to less than 1 % of the volume it occupies at standard atmospheric pressure. It is stored and distributed in hard cylindrical or spherical containers at a pressure of 200–250 bar. CNG is widely used in traditional spark ignition gasoline internal combustion engines that have been converted into bi-fuel vehicles (gasoline/CNG). CNG may be used as a replacement of gasoline in a spark ignition engine. However, it is as a replacement of Diesel in a compression ignition engine that it may work better thanks to the better top fuel conversion efficiencies and part load efficiency penalty changing the load. In response to fuel prices, energy security and environmental concerns, CNG is starting to be used also in commercial applications. The major downfalls of CNG are the low volumetric energy density estimated to be 40 % of LNG and 25 % of Diesel and the unavailability of a capillary network for distribution. CNG vehicles require a greater amount of space for fuel storage than conventional gasoline and Diesel powered vehicles and this ends in a loss of payload for commercial applications. Furthermore, the coverage of long distances as those typical of the freight transport within Australia will be prohibitive without refueling stations.

The Westport engines set the benchmark for CNG heavy-duty trucks [3, 4]. Westport has developed two technologies for CNG. The HPDI technology [3–13] uses a dual fuel High Pressure Direct Injection (HPDI) injector for both the Diesel and the CNG fuels delivering a small Diesel pilot spray and a larger gas spray. The diesel averages only about 5 % of the total energy input and is used to start the main combustion of high-pressure directly-injected natural gas. Efficiencies of the converted engine are Diesel-like, with BMEP values that have been increased up to 24 bar. The CNG-DI technology [5] uses a Compressed Natural Gas Direct Injection (CNG-DI) injector for only the CNG. Ignition is then forced with either a hot surface or a spark. In the HPDI technology, basically the main CNG injection occurs within an environment where the diffusion controlled combustion of the CNG is made possible by the temperature and pressures created by the previous pilot Diesel injection. In the CNG-DI technology, the start of combustion is more complicated requiring a continuously operated glow plug to bring the temperatures up to the CNG auto ignition or even using a spark discharge to initiate combustion changing the Diesel-like operation to gasoline-like. The CNG-DI technology certainly requires much more effort to work properly than the HPDI technology.

In the present paper, the adoption of two injectors, one for the Diesel and one for the CNG, is considered to deliver the amount of fuel needed in the two possible mode of operation, Diesel only and Diesel pilot and CNG main. In the dual fuel operation, more complicated strategies of simultaneous Diesel and CNG injections are possible but not considered just for sake of simplicity. Figure 2 presents this Diesel igniting dual fuel design. The main chamber LNG or CNG injector introduces the most part of the fuel. The Diesel injection may occur prior, simultaneously or after the LNG or CNG injection.

The engine considered as the baseline is a Diesel engine having a power turbine. Internal combustion engines have a maximum theoretical fuel conversion efficiency that is similar to that of fuel cells and considerably higher than the mid-40 % peak values seen today [14, 15]. The primary limiting factors to approaching these theoretical limits of conversion efficiency start with the high irreversibility in traditional premixed or diffusion flames, but include heat losses during combustion and expansion, untapped exhaust energy, and mechanical friction. One area where there is large potential for improvements in ICE efficiency is losses from the exhaust gases [14, 15].

Exhaust losses are being addressed by analysis and development of compound compression and expansion cycles achieved by variable valve timing, variable

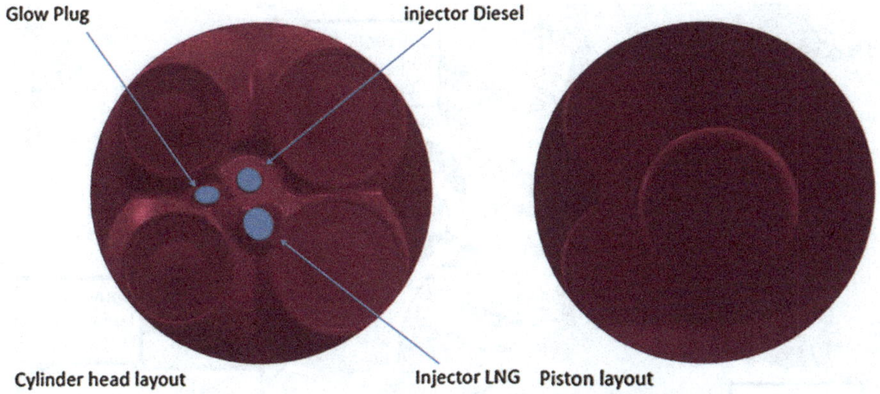

Fig. 2 The main chamber CNG injector introduces the most part of the fuel. The diesel injection may occur prior, simultaneously or after the CNG injection

stroke use of turbine expanders, regenerative heat recovery, and application of thermoelectric generators. Employing such cycles and devices, it has been claimed to have the potential to increase engine efficiency by 10 % [14, 15]. Of all these technologies, the more mature is the use of two turbines in series; one waste gated driving the upstream compressor, and one to producing additional power to the driveline, historically known as turbo compound.

2 Optimum Speed Power Turbine

The use of turbo compounding goes back quite a long way. The concept was originally used back in the late 1940s on aircraft engines [16]. Its promise of low fuel consumption was soon overtaken by the rapid development of the turboprop engine. For automotive diesel engines, the introduction of a power turbine downstream of the turbocharger generates more work by re-using the exhaust gases from the conventional turbocharger [17, 18]. The work generated by the power turbine is then fed back into the engine crankshaft via an advanced transmission. A gear is fitted to the power turbine shaft. To assist the power turbine, the turbine for the conventional turbocharger is designed for a reduced expansion ratio. This small turbocharger gives another system advantage to the turbo compound engine, providing better transient response and higher boost pressure for improved low speed torque. The behavior of the two turbines in series—turbocharger and power turbine—offers a dynamic response across the engine speed and air flow range. However, it requires optimum matching of the turbines.

Figure 3 presents one of the most successful applications of turbo compound technology to a truck engine (from [17]). The turbo compound includes a downstream axial flow power turbine plus the speed reduction gears, the fluid coupling and the final gear reduction to crankshaft to supplement crankshaft power. In the

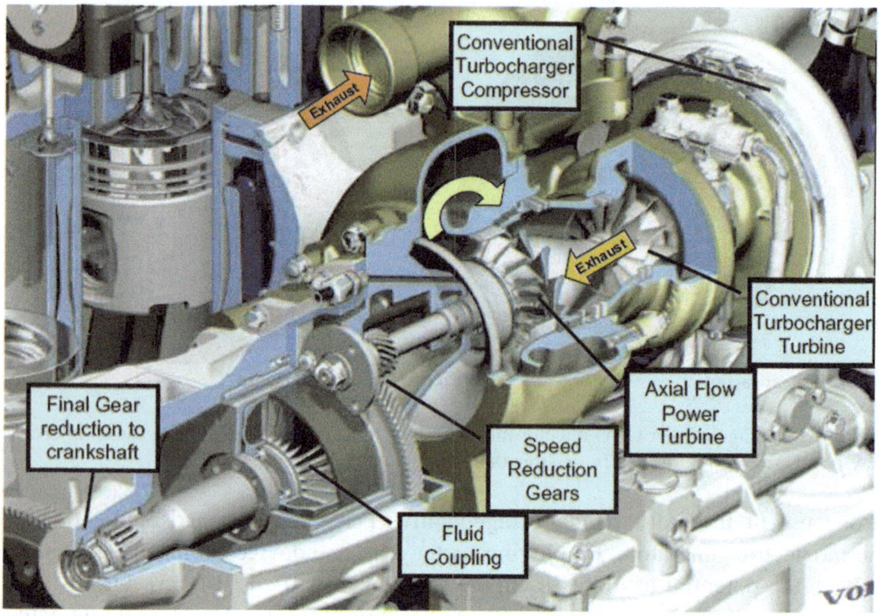

Fig. 3 Traditional turbo compound design (from [17])

mechanical transmission made up of a gear train and a hydraulic coupling, the turbine is running with maximum speeds of up to 70,000 rpm. The gear ratio from the exhaust turbine to the turbo compound intermediate shaft is around 6:1, and the gear ratio from the intermediate shaft to the crankshaft is around 5:1 for an overall gear ratio of about 30:1. Torsional vibrations caused from the internal combustion engine process would be increased by the overall gear ratio exceeding 30:1 and could possibly destroy the turbine. To reduce torsional vibrations the turbo compound intermediate shaft is equipped with a hydrodynamic coupling with a slip inside the coupling that is normally around 2 %.

Despite few truck manufacturers' claim of better efficiency and increased power output as a result of the additional power turbine, this solution has not encountered so far the favor of the vast majority of truck and the totality of car manufacturers. Because the amount of energy available downstream of the turbocharger turbine is small and the complexity to add a second power turbine geared to the crankshaft is high, turbo compound is expected in principle to provide limited advantages in a limited area of operation of the engine at high costs in terms of design, control, packaging and weight. Engine manufacturers that have had or will have Turbo compound Engines include Volvo, Iveco (off-highway) and Scania. Detroit Diesel, Cummins, CAT, Mercedes, and International have also considered the technology.

In the present paper, the traditional gear train coupling of the power turbine is replaced by a constant variable transmission (CVT). A by-pass of the power turbine is also included. This permits to narrow the range of speeds where the

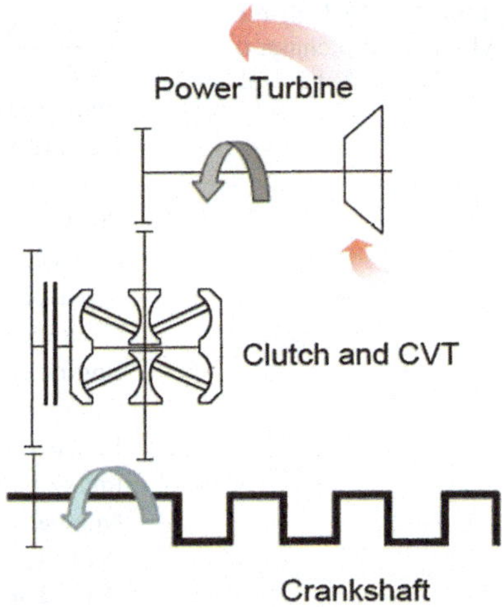

Fig. 4 Sketch of the novel power turbine layout. With reference to the arrangement previously proposed with just a gear in between the crankshaft and the power turbine shaft, this design now include a clutch and a continuously variable transmission to disconnect the power turbine when bypassed by the exhaust gases and to operated the power turbine at the most favorable speed

turbine operates producing more power than the power loss for back pressures while permitting temperatures to the downstream after treatment system high enough. A sketch of the novel design is proposed in Fig. 4.

3 Engine Analysed

Engine performance simulations have been performed for a 12.8 L in-line six cylinder turbo charged directly injected Diesel engine. The engine geometry (baseline) is presented in Table 1. IVO is the intake valve opening, IVC the intake valve closure, EVO the exhaust valve opening and EVC the exhaust valve closure. IVO of 314° means the intake valves open 46° before TDC. IVC of 602° means the intake valves close 62° after BDC. EVO of 100° means the exhaust valves open 80° before BDC. EVC of 400° means the exhaust valves close 40° after TDC. A power turbine downstream of the turbocharger is used to recover a percentage of the thermal energy that would normally be lost through the engine's exhaust.

Engine simulations have then been performed without a power turbine. A by-pass of the power turbine is also included. This permits to operate the power turbine when the turbine produces more power than the power loss for the back pressures while permitting temperatures to the downstream after treatment system high enough. The CVT permits operation of the power turbine at the more efficient speeds. The engine has a target performance of 2,600 Nm torque 1,000–1,450 rpm, and of about 400 kW of power 1,450–1,900 rpm, with BSFC values around 190 g/kWh, corresponding to brake efficiencies of 44 %.

The engine is compliant with EURO-5 emission standards. The additional cooling of the exhaust gases through the power turbine may in principle reduce the effectiveness of the exhaust after treatment systems. In principle, this may require

Table 1 12.8 L in-line six cylinder turbo charged directly injected diesel engine

Displacement per cylinder [l]	2.13
Number of cylinders	6
Engine layout	I-6
Compression ratio	18
Bore (mm)	131
Stroke (mm)	158
Connecting rod length (mm)	300
Wrist pin offset (mm)	0
Clearance volume (l)	0.125
Engine type	C.I.
Number of intake valve per cylinder	2
Intake valve diameter (mm)	46.3
Intake valve maximum lift (mm)	12.1
Number of exhaust valve per cylinder	2
IVO (deg)	314 (−46)
IVC (deg)	602 (+62)
Exhaust valve diameter (mm)	41.5
Exhaust valve maximum lift (mm)	13.4
EVO (deg)	100 (−80)
EVC (deg)	400 (+40)

more active regenerations for particulate filter, or less use of cooled EGR. The additional cooling may certainly reduce the time when NOx systems are effective (LNA, SCR, or LNC). However, using the power turbine for high loads and speeds only, this is not an issue, cause temperatures for these loads and speeds will be otherwise much higher than those of medium to low loads and speeds.

4 Dual Fuel Diesel-Like Combustion in Engine Performance Codes

The direct injection of gas and Diesel within the cylinder is not a novelty. Since the early 1980s, several researches have been made on the subject [6–13]. A small amount of Diesel fuel was injected before a main injection of natural gas to start the combustion of the engine. With this technology, Diesel engines were able to run on natural gas without compromises in performance and fuel efficiency.

The major advantage of the Direct Injection Diesel engine is the bulk, lean, diffusion combustion in a high compression ratio highly boosted engine with reduced heat losses to the liner and head walls. The engines using the stoichiometric, premixed combustion controlled by a spark discharge suffer the disadvantage of the higher heat losses to the liner and head walls. The combustion is wall initiated, and the flame has to reach all the walls for burning all the homogeneously distributed fuel. Furthermore, the compression ratio is reduced to avoid the abnormal combustion phenomena. Furthermore, in the lean burning Diesel the load may be controlled by the quantity of fuel injected, while the stoichiometric gasoline requires throttling the intake to vary the load and this bring a further advantage of the Diesel able to perform with smaller penalties from already higher efficiencies reducing the load.

In case of dual fuel operation of a Diesel engine with CNG as the second fuel, it seems therefore logical to inject both the Diesel and the CNG fuel within the cylinder to achieve the same as Diesel efficiencies both full load and part load, with the load being controlled by the quantity of fuel injected.

The single dual fuel injector has the advantage to be easily incorporated into an existing Diesel engine with minimal or no modifications to the engine cylinder head, while the two injectors' option requires a new design of the cylinder head only possible on novel engines of large bores. Late-cycle, high-pressure direct injection ensures diffusion type combustion for the CNG and therefore retains the high power, torque, and efficiency of the Diesel engine with possibly the advantages of the higher power densities.

Not considered here, flexible operation of the two injectors may also permit different combustion strategies, not only Diesel-like but also gasoline-like or alternative low temperatures as the homogeneous charge compression ignition modes. Homogeneous charge compression ignition modes may be achieved directly injecting the CNG early during compression, Diesel-like operation may be achieved preceding a late CNG main injection with a Diesel pre injection, gasoline-like operation may be achieved injecting the Diesel after the CNG, and mixed Diesel-like and gasoline-like operation may be finally obtained performing simultaneously the two injections. Finally, operating the glow plug continuously may permit auto ignition of the CNG without the Diesel injection.

The CNG injector has to manage the high volumetric flow rates required for the CNG fuel much less dense than Diesel and the carbonizing and thermal fluctuations that can result from the reduced cooling capacity of a gaseous fuel. The major disadvantage of the CNG-Diesel fuel injector is the complex fuel routing for two fuels within one injector. This injector provides a small diesel pilot spray and a much larger CNG gas spray. The Diesel averages only about 5 % of the total energy input and is used to start the main combustion of high-pressure directly-injected CNG gas and the reduced pollutants of the CNG.

Figure 5 shows one latest generation common-rail multiple injection strategy (top), first generation common-rail injection strategy (middle) and a typical heat release rate profile vs. crank angle for this latter injection profile (bottom). Diesel particulate filter regenerations obviously do not apply to steady state map points.

For the purpose of fuel conversion efficiency computations only, the complex injection of Fig. 5 on top may be replaced by a single equivalent injection event. Obviously, soot, smoke, NOx, HC and CO emissions are not included in the prediction. Engine performance models are not able to represent complicated injection and combustion phenomena as those of Fig. 5 on top, but they can tackle reasonably well single injection or pilot and main combustion events. Therefore, having measured pressure traces for the Different map points with the Diesel, the heat release analysis may provide the equivalent injection and combustion model

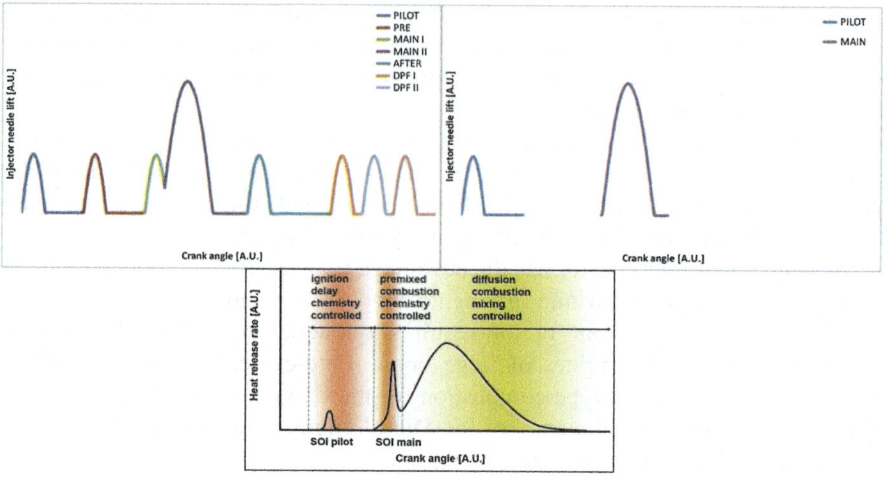

Fig. 5 Latest generation common-rail multiple injection strategy (*top left*), first generation common-rail injection strategy (*top right*) and modelled heat release rate profile vs. crank angle with diesel pilot and main Diesel or gas injections (*bottom*). This heat release profile is the one adopted in the simulations described below where a Diesel Wiebe function is used to describe the combustion evolution

parameters for all the different (BMEP, speed) map points. Then, after tuning on the start of injection and the injection duration, the engine model may provide maps of brake specific fuel consumption vs. BMEP and speed very close to the experimental ones. Even when the accuracy of the combustion model parameters reduces, tuning on the start of injection and the injection duration may still produce acceptable results.

In what follows, the assumption made is that with Diesel only and Diesel pilot and CNG main, the only thing that changes are the fuel properties, but the combustion parameters and the start and duration of injection set for the Diesel are also valid for the Diesel and the CNG working at same speed and percentage of maximum load. The fuel properties are therefore the only parameters changed in the engine model.

The Diesel-like combustion (M1) is modeled with the GT-POWER DI Wiebe model [19]. This model imposes the burn rate using a three-term Wiebe function (the superposition of three normal Wiebe curves). These Wiebe curves approximate the "typical" shape of a DI compression ignition burn rate. The purpose of using three functions is to make it possible to model pre-ignition (the large initial spike) and larger tail. The injection profile does not influence the burn rate except if, at any instant, the specified cumulative combustion exceeds the specified injected fuel fraction. The inputs of the Wiebe equations are *SOI* = Start of Injection, *ID* = Ignition Delay, *DP* = Premix Duration, *DM* = Main Duration, *DT* = Tail Duration, *FP* = Premix Fraction, *FT* = Tail Fraction, *EP* = Premix Exponent, *EM* = Main Exponent, *ET* = Tail Exponent and CE = Fraction of Fuel Burned (also known as "Combustion Efficiency"). The calculated constants of these equations are FM = Main Fraction, WCP = Wiebe Premix Constant, WCM = Wiebe Main Constant and WCT = Wiebe Tail Constant given by equation below.

$$F_M = (1 - F_P - F_T) \tag{1}$$

$$WC_P = \left[\frac{D_P}{2.302^{1/(E_P+1)} - 0.1051^{1/(E_P+1)}}\right]^{-(E_p+1)} \tag{2}$$

$$WC_M = \left[\frac{D_M}{2.302^{1/(E_M+1)} - 0.1051^{1/(E_M+1)}}\right]^{-(E_M+1)} \tag{3}$$

$$WC_T = \left[\frac{D_T}{2.302^{1/(E_T+1)} - 0.1051^{1/(E_T+1)}}\right]^{-(E_T+1)} \tag{4}$$

With θ = Instantaneous Crank Angle, the burn rate is finally given by the equation below:

$$\begin{aligned} Combustion\ (\theta) =& (CE)(F_P)\left[1 - e^{-(WCp)(\theta - SOI - ID)^{(E_P+1)}}\right] \\ &+ (CE)(F_M)\left[1 - e^{-(WCM)(\theta - SOI - ID)^{(E_M+1)}}\right] \\ &+ (CE)(F_T)\left[1 - e^{-(WC_T)(\theta - SOI - ID)^{(E_T+1)}}\right] \end{aligned} \tag{5}$$

The cumulative burn rate is calculated, normalized to 1.0. Combustion starts at 0.0 (0.0 % burned) and progresses to 1.0 (100 % burned).

The values of the model inputs may be obtained using the excel worksheet accompanying the software from a recorded pressure trace [19]. With pressure traces recorded at different loads and speeds, then the model inputs can be made a tabular function of speed and load [19] for all the inputs. Tables of combustion parameters may be transferred in between similar engines.

The Westport HPDI concept is very well established. It has been demonstrated in a huge number of heavy duty truck Diesel engine conversions to run dual fuel Diesel and LNG both engine dynamometer and road vehicle tested [3–7, 20]. After roughly 5 % of the total fuel energy is introduced with the Diesel burning as usual, the pressures and temperatures and the availability of radicals of the partially burned combustion gases makes the combustion of the LNG introduced next a diffusion controlled more than a kinetically controlled process. It has been proved numerically with coupled fluid dynamic and detailed chemical kinetic simulations that as soon as the in-cylinder conditions have been made suitable for the diffusion combustion of a low Cetane number gas, either hydrogen, methane or propane, then the gas burns diffusion controlled once injected. This is the rationale behind the use of the Diesel Wiebe equation for representing the heat release rate with fuels of different Cetane number [20–22]. Their injection does not occur in air, but in hot partially burned combustion products.

For the operation gasoline-like (M3), the Diesel post injection occurs after the main chamber CNG is injected. The main injection combustion is modeled with the GT-POWER SI Wiebe model [19]. This model can be used with any type of injection, but if the fuel is injected directly into the cylinder, the start of injection must precede the start of combustion so that there is fuel in the cylinder to burn when combustion starts. At any instant, the specified cumulative burned fuel fraction must not exceed the specified injected fuel fraction. The inputs of the Wiebe equations are: AA = Anchor Angle; D = Duration; E = Wiebe Exponent; CE = Fraction of Fuel Burned (also known as "Combustion Efficiency"); BM = Burned Fuel Percentage at Anchor Angle; BS = Burned Fuel Percentage at Duration Start; BE = Burned Fuel Percentage at Duration End. The calculated constants of these equations are Burned Midpoint Constant BMC, Burned Start Constant BSC, Burned End Constant BEC, Wiebe Constant WC and Start of Combustion SOC given by equation below with θ the Instantaneous Crank Angle, BMC = -ln(1-BM), BSC = -ln(1-BS), BEC = -ln(1-BE):

$$WC = \left[\frac{D}{BEC^{1/(E+1)} - BSC^{1/(E+1)}}\right]^{-(E+1)} \tag{6}$$

$$SOC = AA - \frac{(D)\cdot(BMC)^{1/(E+1)}}{BEC^{1/(E+1)} - BSC^{1/(E+1)}} \tag{7}$$

$$Combustion\ (\theta) = (CE)\cdot\left[1 - e^{-(WC)(\theta - SOC)^{(E+1)}}\right] \tag{8}$$

The cumulative burn rate is calculated, normalized to 1.0. Combustion starts at 0.0 (0.0 % burned) and progresses to 1.0 (100 % burned). The Wiebe constants may be matched to an apparent burn rate that has been calculated from measured cylinder pressure [13]. With pressure traces recorded at different loads and speeds, then the model inputs can be made a tabular function of speed and load [13] for all the inputs.

In mixed Diesel/gasoline modes of operation (M2), the Diesel injection occurs in the main chamber after only part of the main chamber CNG is injected. The engine then operates mixed Diesel-like/gasoline-like.

Homogeneous Change Compression Ignition operation (M4) is obviously also theoretically possible without any igniting Diesel. Being HCCI quite difficult to be controlled and limited to a very narrow window of operating conditions, an HCCI assisted operation is made possible by the post injection of the Diesel approaching top dead centre in M3.

The novel designs permit complicated strategies coupling premixed and diffusion combustions within the main chamber, but unfortunately these strategies cannot be modeled with the available Wiebe function approach. An experimental campaign is certainly the best option available to address all the issues and fully explore the capabilities of the systems.

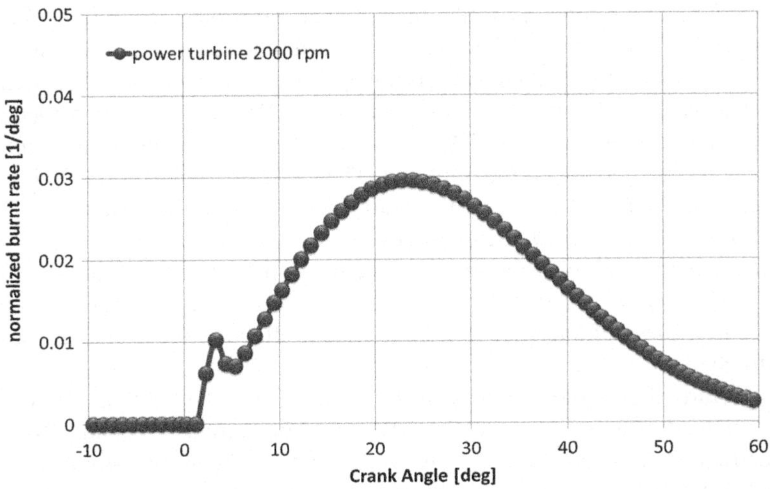

Fig. 6 Normalized burnt rate measured on the baseline diesel engine working with $\lambda = 1.45$ at 2000 rpm

5 Results of Non Predictive Combustion Models

Engine performance simulations have been performed with the GT-POWER code [19]. GT-POWER is the industry leading engine simulation tool used by every major engine manufacturer worldwide. Engine simulation, sometimes referred to as cycle simulation, is a mainstream activity conducted by all engine manufacturers well documented in the literature. Within this activity, GT-POWER is used to perform a variety of both steady state and transient analyses. Engine simulations usually provide accurate descriptions of the full load torque curve and the brake specific fuel consumption map. Conversely, these simulations are not able to accurately predict the brake specific fuel consumption emissions maps.

The Diesel fuel is considered a liquid hydrocarbon of composition C = 13.5 and H = 23.6, fuel heating value at 298.15 K of 43.25 MJ/kg and stoichiometric air-to-fuel ratio of 14.33. The CNG fuel is considered methane gas of composition C = 1 H = 4, fuel heating value at 298.15 K of 50 MJ/kg and stoichiometric air-to-fuel ratio of 17.12.

Simulations have been performed for the Diesel mode of operation (**M1**) full and part load and for the gasoline/HCCI mode of operation (**M3–4**) at part load only. The mixed Diesel/gasoline-like modes of operation, possibly the most interesting option to explore, have not been investigated because of the limit of the simulation tool when dealing with different fuels. Figure 6 presents the normalized burnt rate working with the equivalent air-to-fuel-ratio $\lambda = 1.45$ at 2000 rpm. This burnt rate is the one measured on the baseline Diesel engine. This same burnt rate is used for the same λ and speed with all the CNG options.

Figures 7 and 8 presents the brake mean effective pressure and brake efficiency operating with Diesel pilot and CNG main. The reference case is the baseline Diesel only operation where the CNG injector is not operational and only the Diesel injector is used. The air-to-fuel equivalence ratios are $\lambda = 1.35$ and $\lambda = 1.45$ respectively. The main fuel account roughly 95 % of the total fuel energy input, the remaining fuel energy being supplied by the pilot fuel. The mode of combustion is **M1**, i.e. the pilot fuel is injected first and then the main fuel enters the in-cylinder and burns almost diffusion controlled. The further development of the concept to be performed experimentally will certainly require the injection of part of the CNG also before the igniting fuel injection or even a larger amount of igniting fuel being injected. Consequently, these efficiencies and brake mean effective pressures are possibly over estimated.

This picture shows the opportunity to achieve even better than the baseline Diesel full load efficiencies with the M1 mode of combustion and CNG main chamber fuel. Results are more in line with what has been measured in engines modified following the HPDI concept and operated dual fuel Diesel-LNG.

Figure 9 presents the brake efficiency vs. engine speed operating at 20, 15, 10, 5, 2.5 and 1 bar/0.1 MPa mean effective pressure (left to right, top to bottom). Mode of operation is Diesel-like M1 and gasoline/HHCI-like M3-4 for the intermediate loads where the λ values are below the lean burn limit while permitting a peak in-cylinder pressure below a maximum value. The reference case is the Diesel only operation where the CNG injector is not operational and only the Diesel injector is used in the Diesel igniting design. This picture shows the opportunity to achieve even better than the baseline Diesel part load efficiencies not only with the M1 mode of combustion, but also with the M3–4 mode in selected intermediate loads where the air-to-fuel equivalence ratio λ is high enough but the peak pressure is still well below the 180 bar/18 MPa considered the maximum acceptable.

Results of Figs. 7 to 9 serve as a reference for the capabilities of the novel design, the more consolidated Diesel-like, having the support of the past experience done by Westport with the HPDI, and the less developed gasoline-like, still far from a production stage with the concept limited so far to just a few experiments and simulations. Again, more complicated injection strategies for the igniting Diesel and the CNG better modulating premixed and diffusion combustions may possibly deliver even better results than those presented. Also worth of mention is that the system feed with the standard Diesel fuel retains the same efficiency.

6 SRM Detailed Chemistry Simulations

The SRM (Stochastic Reactor Model) suite [23] is a sophisticated engineering tool for carrying out computations of fuels, combustion and emissions formation in an IC engine context. The software has five major features that distinguish it from

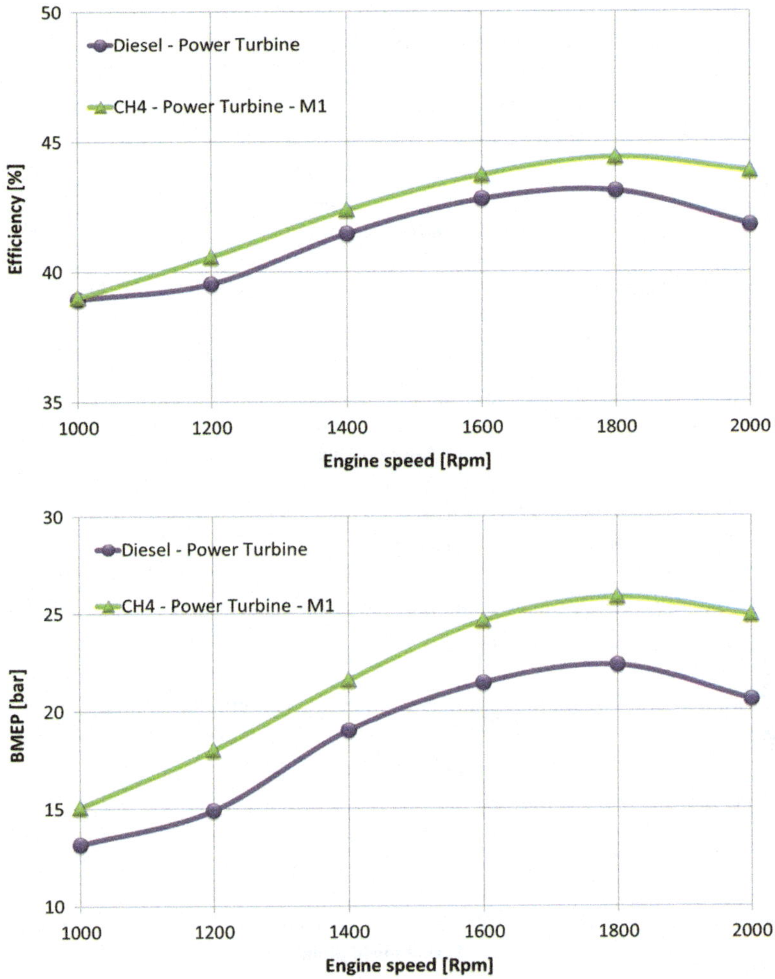

Fig. 7 Brake efficiency and mean effective pressure vs. engine speed operating with CNG and diesel with $\lambda = 1.45$. Mode of operation is diesel-like M1

conventional 1 or 3D fluid dynamics approaches. It can solve for in-cylinder in-homogeneities (in contrast to 1D codes), enabling to deal with composition/thermal stratification. On account of its unique approach, the code is faster than 3D CFD, enabling to carry out model parameter and design optimizations. It includes detailed combustion chemistry, enabling to solve for fuels, combustion characteristics, emissions and engine performance. The gaseous emissions of NOx, CO and HCs are computed based on their formation starting as a fuel molecule, during combustion and expansion. Soot emissions can also be computed using a detailed soot model, which solves for soot particle dynamics providing information such as particle size distribution including composition and morphology of soot

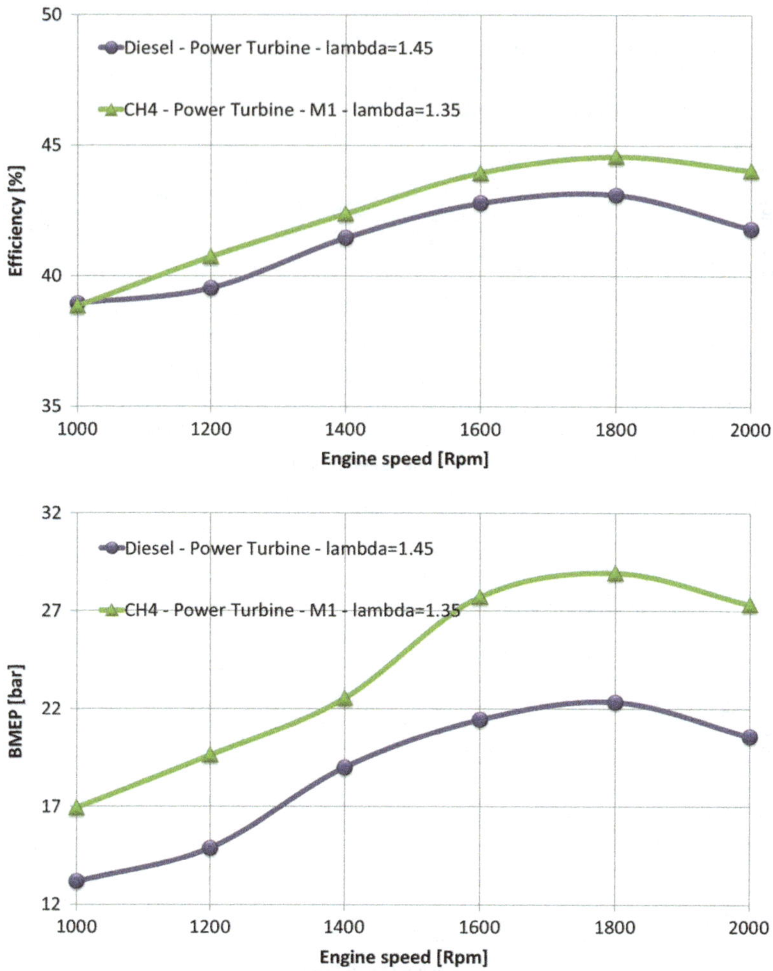

Fig. 8 Brake efficiency and mean effective pressure vs. engine speed operating with CNG $\lambda = 1.35$ and diesel with $\lambda = 1.45$. Mode of operation is diesel-like M1

aggregates. SRM is either and both a 1 or 3D code. The code combines the features from both approaches, the speed of 1D and mostly of the predictive capabilities of 3D, but with the addition of detailed combustion chemistry.

The stochastic term means combustion is modeled as a process involving a sequence of observations each of which is considered as a random sample from a probability distribution. Stochastic fluctuations are inherent to any internal combustion engine operation and need to be accounted in engine models. Each element of the probability distribution is termed a stochastic particle. In a way a stochastic particle represents the air–fuel parcel. Depending on the application, between 50 and 500 stochastic particles are typically required to properly sample in-cylinder processes. IN the simulations performed here after, 250 parcels are used. The SRM

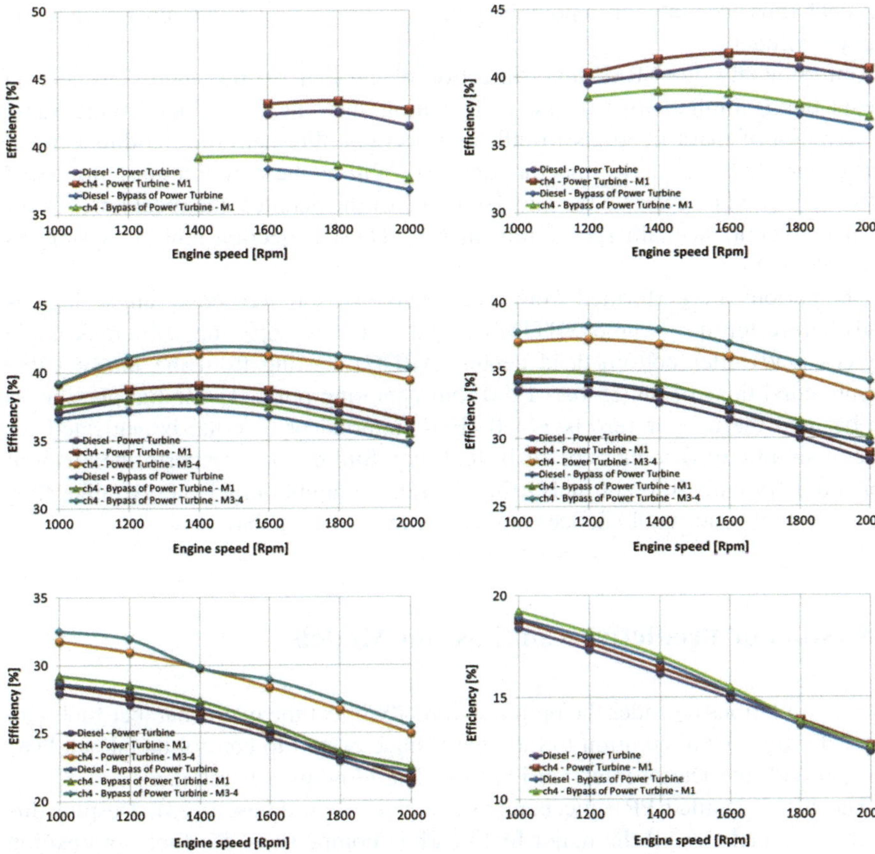

Fig. 9 Brake efficiency vs. engine speed operating at 20, 15, 10, 5, 2.5 and 1 bar/0.1 MPa mean effective pressure (*left to right*, *top to bottom*). Mode of operation is diesel-like M1 and gasoline/HHCI-like M3-4 for the intermediate loads where lambda values are below the lean burn limit while permitting a peak in-cylinder pressure below a maximum value

Suite uses a probability density function (PDF) based approach that simplifies many of the most computationally expensive fluid-dynamic processes whilst maintaining much of the predictive capability of the 3D CFD codes. This is achieved by down sampling the required number of grid points associated with CFD from tens-of-thousands down to a few hundred stochastic particles which best represent the in-cylinder mixture composition states. The method retains most of the spatial information required to achieve a predictive model without incurring the computational overhead. The ensemble of the stochastic particles approximates the distribution of the in-cylinder properties such as the composition of chemical species and the temperature. The model is then solved to account for the influence of the processes such as fuel injection, combustion kinetics, turbulent mixing, piston motion and convective heat loss on the multi-dimensional PDF. The approach thus accounts for in homogeneities in composition and temperature in the engine cylinder.

Unfortunately, the SRM direct injection of the fuel permits multiple injection events but is limited to a single composition fuel, i.e. to a fuel having same prescription of mass fractions for all the injections. Furthermore the direct injection occurs from a single injector location. Within these limitations, the SRM software is used to improve the understanding of the detailed chemical kinetics for the unconventional multi fuel, multi injector, Diesel injection ignition system as described below.

Simulations are performed from intake valve closure 120 crank angle degrees (cad) before top dead center (BTDC) to exhaust valve opening 100 crank angle degrees (cad) after bottom dead center (ABDC) starting from the intake valve closure conditions of the 25 bar, 1900 rpm operating point. The initial pressure is 2.6 bar and the temperature is 344 K. EGR internal or external is neglected for sake of simplicity. The mass of air is 4620 mg. Simulations are performed lean of stoichiometry with different fuels using the same amount of fuel energy rather than the same fuel-to-air equivalence ratio to compare their behaviors.

7 Results of Predictive Combustion Models

These simulations consider the option to directly inject the main chamber fuel with a specific injector in addition to the Diesel injector and to control the about TDC ignition with the Diesel injection ignition. The glow plug is retained.

The SRM detailed PRF mechanism chemistry is now used [23]. Despite this mechanism includes all the major hydrocarbon components, the fuel composition is limited to a mixture of N-C_7H_{16} and I-C_8H_{18} surrogate of Diesel and gasoline. The SRM detailed PRF mechanism [23] includes 208 different species. However, simulations with H_2, CH_4, C_3H_8, and NH3 are not possible because the fuel composition may only be I-C_8H_{18} or N-C_7H_{16}.

The simulations will be limited to the reference N-C_7H_{16} pilot plus main injection combustion, plus two conditions where the top dead center ignition is obtained auto igniting the I-C_8H_{18} injected early during compression (homogeneous charge compression ignition mode). In a first case, no further injection of I-C_8H_{18} is made during expansion. This is basically a particular gasoline-like mode. In a second case, the early compression injection of I-C_8H_{18} is followed by a second injection of I-C_8H_{18} during the expansion, as in the mixed Diesel/gasoline-like mode. The Diesel-like operation with pilot N-C_7H_{16} and then I-C_8H_{18} is not investigated because of the multi fuel issue. Similarly, the support of the auto ignition process of I-C_8H_{18} injecting N-C_7H_{16} at top dead center is not analyzed.

Results are presented in Fig. 10, with the same fuel energy input of the previous paragraph in case of I-C_8H_{18} of 16a and N-C_7H_{16} of 16b. The figure presents in-cylinder pressure, temperature, heat release rate and cumulative heat release rate. The fuel input of I-C_8H_{18} of 3c is one half of the total of 16a or 16b. The

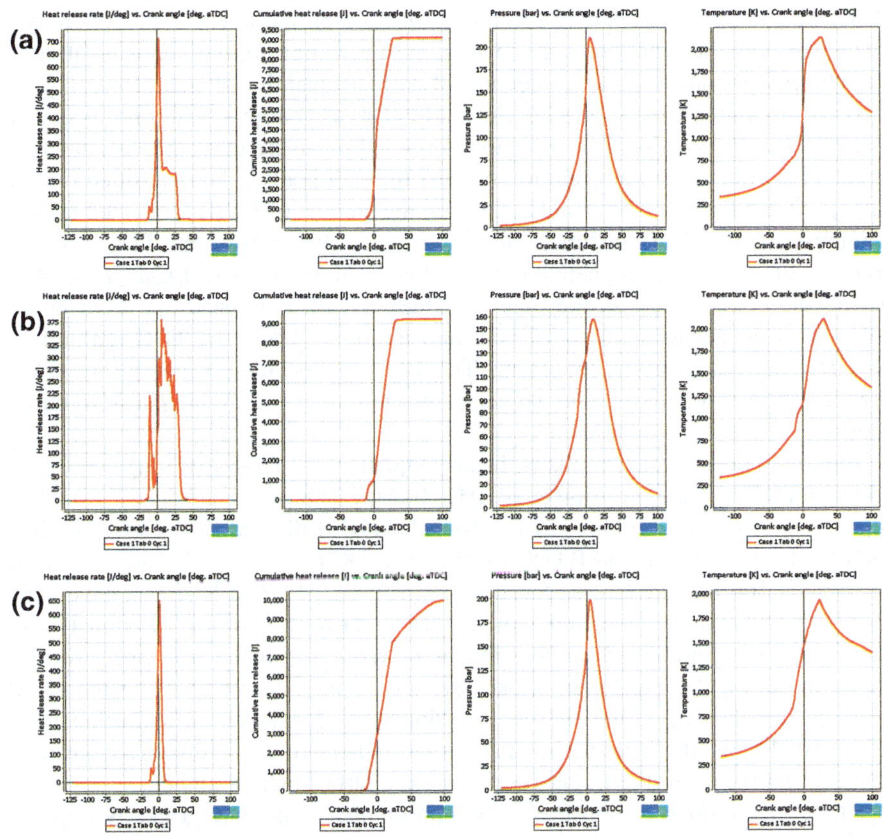

Fig. 10 Top to bottom heat release rate, cumulative heat release rate, in-cylinder pressure and temperature with 100 + 100 mg of I-C8H18 (**a**), 20 + 180 mg of N-C7H16 (**b**), and 100 mg of I-C8H18 (**c**)

simulations are performed for same pressure, temperature and composition at intake valve closure of the prior paragraph. The injection of 100 mg of liquid I-C_8H_{18} occurs −100 – −75 cad BTDC. Under these specific circumstances, no Diesel ignition is needed to start combustion, because the early injected I-C_8H_{18} auto ignites before TDC, with heat release becoming significant about TDC. Then a second injection of 100 mg of liquid I-C_8H_{18} occurs from TDC to 25 cad ATDC for the results of Fig. 10a, while no more fuel is injected in the case of Fig. 10c. Auto-ignition of I-C_8H_{18} is quite difficult to be controlled even in simulations; therefore a pilot Diesel injection is always practically needed to make the TDC ignition stable and repeatable. Even in the simulations, small changes bring to much earlier auto ignition or not significant auto ignition.

Injection of the 200 mg of liquid I-C8H18 earlier during the compression stroke does not make any sense, because it would only produce huge compression work and compression pressure. With the 100 + 100 mg splitting, part of the fuel burns

HCCI-like at TDC, and part of the fuel burns Diesel-like after TDC. This maximizes the pressure work for a given fuel energy within normal mechanical constraints. AS previously noted, injection of a small quantity of liquid N-C_7H_{16} is theoretically not needed at TDC in this particular case but practically necessary to make robust the process. This injection is needed to ensure that combustion starts about TDC. Being the auto-ignition process quite difficult to be controlled over a full speed and load map by simply adjusting the amount of I-C_8H_{18} injected during compression, the TDC injection of N-C_7H_{16} is what makes the TDC start of combustion stable and repeatable at any load and speed. The I-C_8H_{18} injected during the compression stroke may be adjusted with the speed and load up to values that do not incur in significant auto ignitions. Then, the TDC ignition of N-C_7H_{16} is what starts combustion. The I-C_8H_{18} injected earlier during compression burns premixed, the I-C_8H_{18} injected after during expansion burns diffusion.

Figure 10 presents the results with early injection of I-C_8H_{18} 100 mg $-100 - -75$ cad BTDC and late injection of I-C_8H_{18} 100 mg TDC to 25 cad ATDC (a), pilot injection of N-C_7H_{16} 20 mg $-25 - -10$ cad BTDC followed by a main injection of N-C_7H_{16} 180 mg -5 cad BTDC to 25 cad BTDC (b) and finally early injection of I-C_8H_{18} 100 mg $-100 - -75$ cad BTDC (c). I-C_8H_{18} and N-C_7H_{16} have very close lower heating value and stoichiometric air-to-fuel ratios. Obviously the almost isochoric premixed combustion of I-C_8H_{18} permit the best pressure distribution during the expansion stroke (positive work) vs. pretty much the same pressure distribution during the compression work (negative work) during non-firing conditions thus producing the larger indicated mean effective pressure (IMEP) and therefore the larger output and the larger fuel efficiency.

These simulations demonstrated the opportunity to produce a nearly optimal pressure build up within limited maximum pressure constraints modulating the premixed and diffusion combustion events. Worth of mention is that the very high in cylinder temperatures for this mild lean of stoichiometry operation is an opportunity to further increase the fuel conversion efficiency by injecting water and expanding steam as shown in [13, 14].

To properly study all the possible combustion modes with better details, obviously a fully 3D CFD+ chemical kinetic simulation is needed for the two phase, multi fuels, reacting flows but more than that a detailed experimental campaign is needed for the further engine development as well as the refinement of the models.

8 Discussion and Conclusions

Up to 20–25 % of the fuel energy in a modern heavy duty diesel is exhausted, and by adding a power turbine in the exhaust flow, theoretically up to 20 % of this exhaust energy can be recovered. In reality, only a minor part of this fuel energy can be practically recovered, first of all because of the added exhaust back pressure also increasing the pumping losses. Further limits to the use of a power turbine arise from the after treatment requiring temperatures above threshold values to operate efficiently. The increasing use of cooled EGR further limits the perspectives of this technique. Crank train coupling and power turbine add weight, complexity in design, control and service and costs, with a definitively negative trade-off in light load applications.

The additional cooling of the exhaust gases reduces the effectiveness of exhaust after treatment systems, requiring more active regenerations for particulate filter and reducing the time NOx control systems are effective. This is not an issue for the proposed application where a by-pass system avoids the cooling for expansion in the power turbine except than when useful and without implications on the after treatment. The use of external cooled EGR is negative with a turbo compound because the exhaust energy decreases due to energy extracted into cooling system reducing the energy available to turbo charger and power turbines. Space requirements of turbo compound further constrain packaging of exhaust gas recirculation and turbochargers.

Simulations performed for a 12.8 L in-line six cylinder turbo charged directly injected Diesel engine have shown the opportunity to gain up to 2 percentage points in efficiency at high loads and speeds where the wasted energy is relatively large. At low engine loads the converted energy available downstream of the turbine is not enough to compensate for back pressure losses and the efficiency deteriorates. The turbo compound has therefore the advantage of improved fuel conversion efficiency in applications where high engine loads are dominating, but negative impact at light load. Also worth of mention the increase in maximum power output about maximum speed, where otherwise the power output sharply reduces. These improvements do not impact on the efficiency of the after treatment as well as on the use of cooled EGR to control the emissions, because the power turbine is by-passed when needed.

The use of a constant variable transmission to replace a gear ratio in between the power turbine and the crankshaft has the advantage of permitting operation of the power turbine within a narrow range of speeds decoupled from the speed of the crankshaft. The better efficiency of the power turbine translates in a better recovery of the exhaust energy. The idea of using a CVT and a by-pass is relatively novel and the model will certainly need further refinements. Prototyping of the solution will certainly help considerably in understanding the limits of the technology only partially addressed in this paper.

Dual fuel CNG and Diesel internal combustion engines may operate more efficiently and with increased power density adopting direct injection with high pressure and fast actuating injectors. The key area of concern for these engines is the further development of injectors able to independently inject the Diesel and the CNG fuels with high flow rates and speeds of actuation.

An equivalent single injection event may replace the multiple event injection of the latest Diesel-only engines for brake engine thermal efficiency simulations. The model set up for Diesel-only injection and combustion may then be used as a first approximation of the pilot Diesel and main CNG injection and combustion. Same combustion model parameters and time of start of injection and end of injection are used with Diesel only or with Diesel pilot and CNG main injections for same speed and percentage of load. Providing the volumetric flow rate is fast enough for the CNG injection, the differences in terms of combustion evolution should then be minimal.

The dual fuel CNG-Diesel internal combustion engine has top brake engine thermal efficiency approaching 45 % as in the original Diesel-only engine and similarly reduced penalties in efficiency reducing the load. However, the novel dual fuel engine permits larger brake mean effective pressures thanks to the much closer to stoichiometric operation with CNG permitted by the injection of a gas rather than a mixture of liquid hydrocarbons and the then much easier mixing with air and combustion.

The opportunity to run the engine almost stoichiometric full load follows the availability not only of an extremely fast and high flow rate dual fuel injector, but also the adoption of a water injector to control the temperature of gases within the cylinder and to the turbine at higher loads.

The major advantages of the dual fuel CNG Diesel operation are the reduced CO_2 emissions, both tailpipe and fuel life cycle analysis, for the better C–H ratio of the CNG fuel and the lower CO_2 emissions when producing and distributing the fuel, the reduced smoke and particulate emissions thanks to the gaseous fuel, and finally the better energy security. Running lower lambda because of the better mixture formation and combustion evolution properties of the CNG, the CNG engine may also deliver more power at high speeds. The dual fuel operation also permits acceptable driving ranges also in cases of a minimal refueling network for CNG, with CNG being the fuel used around capital cities and the Diesel being the fuel used elsewhere.

The injection system is the major area where to focus the research on these engines. A further developed dual fuel injector or the use of two off-the-shelf fuel specific injectors are enablers of novel modes of operation possibly delivering even better fuel efficiencies than the Diesel-like operation when preceding the CNG injection with a Diesel pre injection. These other modes may include homogeneous charge compression ignition directly injecting the CNG early during the compression stroke; gasoline-like operation injecting the Diesel after the CNG; and finally mixed Diesel-like and gasoline-like combustions performing simultaneously the two injections. Furthermore, operating the glow plug continuously may also permit the auto ignition of the CNG without any Diesel injection.

This paper presents simulation trends and the case to improve turbo compounding. The baseline engine model was delivering same performances of the production target engine in terms of power and brake specific fuel consumption. The model for the engine with a power turbine and for the engine operating Diesel-CNG are derived from this model using the same consolidated process where a single design variable is changed at the time and the results are guessed by simulations. The claims about the CVT follow a very solid logic, because it is very well known that a power turbine speed dictated by the crank shaft through a fixed gear ratio is much less efficient that the best speed permitted. The fact that turbo compounding improves efficiency is rather well known. The turbo compounding does not improve the power density, because the power density is actually larger without, and this is another interesting finding. It is not known the amount possible with an optimum speed power turbine as the one enabled here by the addition of the clutch and the CVT, and this is shown in the paper.

References

1. Yeh S (2011) LCA for fuels regulation, workshop LCA of GHG emissions for transport fuels: issues and implications for unconventional fuel sources, Calgary, Canada, Sept 2010.www.arb.ca.gov/fuels/lcfs/121409lcfs_lutables.pdf, 15 Dec 2011
2. Wesport HD, Westport HD engine advantages, www.westport-hd.com/advantages/engine, 15 Dec 2011
3. Wesport HD, Technology, www.westport-hd.com/technology, 15 Dec 2011
4. Westport, Fuel injectors, www.westport.com/is/core-technologies/fuel-injectors, 15 Dec 2011
5. Hodgins KB, Ouellette P, Hung P, Hill PG (1996) Directly injected natural gas fueling of diesel engines, SAE P. 961671, SAE, Warrendale 1996
6. Mtui PL, Hill PG (1996) Ignition delay and combustion duration with natural gas fueling of diesel engines. SAE P. 961933, SAE, Warrendale 1996
7. Dumitrescu S, Hill PG, Li G, Ouellette P (2000) Effects of injection changes on efficiency and emissions of a diesel engine fueled by direct injection of natural gas, SAE P. 2000-01-1805, SAE, Warrendale 2000
8. Kamel M, Lyford-Pike E, Frailey M, Bolin M, Clark N, Nine R, Wayne S (2002) An emission and performance comparison of the Natural Gas Cummins Westport Inc. C-gas plus versus diesel in heavy-duty trucks, SAE P. 2002-01-2737, SAE, Warrendale 2002
9. Harrington J, Munshi S, Nedelcu C, Ouellette P, Thompson J, Whitfield S (2002) Direct injection of natural gas in a heavy-duty diesel engine, SAE P. 2002-01-1630, SAE, Warrendale 2002
10. McTaggart-Cowan GP, Bushe WK, Rogak SN, Hill PG, Munshi SR (2003) Injection parameter effects on a direct injected, pilot ignited, heavy duty natural gas engine with EGR, SAE P. 2003-01-3089, SAE, Warrendale 2003
11. McTaggart-Cowan GP, Jones HL, Rogak SN, Bushe WK, Hill PG, Munshi SR (2006) Direct-injected hydrogen-methane mixtures in a heavy-duty compression ignition engine, SAE P. 2006-01-0653, SAE, Warrendale 2006
12. Brown S, Rogak S Munshi S (2009) A new fuel injector prototype for heavy-duty engines has been developed to use direct-injection natural gas with small amounts of entrained diesel as an ignition promoter. SAE P. 2009-01-1954, SAE, Warrendale 2009
13. Department of energy, energy efficiency and renewable energy, vehicle technology program, "FY 2008 progress report for advanced combustion engine technologies". www1.eere.energy.gov/vehiclesandfuels/pdfs/program/2008_adv_combustion_engine.pdf, 15 Dec 2011
14. Department of energy, energy efficiency and renewable energy, vehicle technology program "FY 2009 progress report for advanced combustion engine research and development". www1.eere.energy.gov/vehiclesandfuels/pdfs/program/2009_adv_combustion_engine.pdf, 15 Dec 2011
15. Gerdan D, Wetzler JM (1948) Allison V-1710 compounded engine, SAE P. 480199, SAE, Warrendale 1948
16. Greszler A (2011) Diesel turbo-compound technology, ICCT/NESCCAF workshop improving the fuel economy of heavy-duty fleets II, San Diego, Feb 2008. www.nescaum.org/documents/improving-the-fuel-economy-of-heavy-duty-fleets-1/greszler_volvo_session3.pdf, 15 Dec 2011

17. Wagner R et al (2010) Investigating potential light-duty efficiency improvements through simulation of turbo-compounding and waste-heat recovery systems, SAE P. 2010-01-2209, SAE, Warrendale 2010
18. Gamma Technologies GTI (2011) Engine performance simulation", www.gtisoft.com/applications/a_Engine_Performance.php, 15 Dec 15 2011
19. Westport (2011) Direct injection natural gas demonstration project, www.liquegas.com/WPT-AHD_HPDI_Fuel_System_MED1.pdf, 15 Dec 2011
20. Boretti A (2011) Advantages of the direct injection of both diesel and hydrogen in dual fuel H2ICE. Int J Hydrogen Energy 36(15):9312–9317
21. Boretti A (2011) Advances in hydrogen compression ignition internal combustion engines. Int J Hydrogen Energy 36(19):12601–12606
22. Boretti A (2011) Diesel-like and hcci-like operation of a truck engine converted to hydrogen. Int J Hydrogen Energy 36(23):15382–15391
23. CMCL Innovations (2012), SRM suite. www.cmclinnovations.com/html/srmsuite.html Internet. Cited 26 April 2012

The Effect of Intake Port Shape on Gasoline Engine Combustion in Cylinder

Xiaodong Chen and Zhangsong Zhan

Abstract The effect of intake port shape on gasoline engine combustion performance was studied by CFD technology. The correlation of CFD simulation and experiment data is good, the all results shows the combustion performance of intake port with high tumble is better than that of intake port with low tumble, the high tumble in cylinder transfers into turbulence kinetic energy as the piston close to top dead center during compression stoke, it is good for accelerating the process of combustion in cylinder and improving the performance of combustion.

Keywords Intake port · CFD · Combustion · Tumble · Gasoline engine

1 Introduction

The flow characteristic of intake port has a substantial influence on the combustion process in IC engines, and directly decides the engine performance and emission. The intake port shape of gasoline engine has a significant effect on tumble motion in cylinder, the tumble can accelerate fuel and air to form mixture. The high tumble in cylinder transfers into turbulence kinetic energy as the piston close to top dead center during compression stoke, that is good for accelerating the process of combustion in cylinder and increasing the flame spread speed and improving the engine performance. Intake port development is a very important component of the

F2012-A06-006

X. Chen (✉) · Z. Zhan
Changan Auto Global R and D Center of Changan Automobile Co Ltd, Chongqing, China
e-mail: sunqiemail@163.com

SAE-China and FISITA (eds.), *Proceedings of the FISITA 2012 World Automotive Congress*, Lecture Notes in Electrical Engineering 190,
DOI: 10.1007/978-3-642-33750-5_7,

combustion system design because the charge motion in cylinder is generated during the intake stroke and it is mostly dependent on the intake port design [1]. The intake port with reasonable tumble level which was optimized by experiment and CFD has been one of the key technologies for gasoline engine development. Currently the steady CFD and test rig are used to investigate the flow capacity and tumble characteristic of intake port of gasoline engine at different valve lift, this is a very convenient way to optimizing the flow pattern, especially the CFD plays a important role in this optimization process, but the steady method is difficult for associating these parameters of intake port with engine combustion performance. It is known that the experimental characterization of the near top dead center flow field in cylinder is not practical and cost effective in an engine development. Instead, CFD is more convenient, more economical, and more versatile to study the in cylinder flow physics if its accuracy is validated with experimental results [2]. This paper presents a transient CFD computational study of charge motion and combustion process for medium and high tumble intake port in order to evaluate the effect of different intake port shape on gasoline engine combustion in cylinder. Comparison of the CFD results with the available experimental results is indicated and showed a good agreement.

2 CFD Analysis of Charge Motion and Combustion in Cylinder

The geometry model including port, valve, liner, combustion chamber, piston, spark plug (Fig. 1). A short straight pipe was added to the inlet and outlet of port for better stability and convergence. The sketch of two intake ports with different tumble is shown in Fig. 2, the MT is the symbol of medium tumble, the HT is the symbol of high tumble. The comparison of flow capacity and tumble level for two intake ports is illustrated in Fig. 3, the MT intake port with higher flow capacity and lower tumble level relative to HT intake port. The natural aspirated engine which was studied has the following specifications in Table 1.

2.1 CFD Model and Methodology

The geometry of the flow domain is made up of triangular surface. The triangulated surface has formed a completely closed solid, without holes, gaps or overlaps. Surfaces need to be grouped into boundaries representing each engine component (port, valves, etc.). The prepared model was imported to ES-ICE to build the moving mesh. The four strokes of exhaust, intake, compression, combustion were simulated by STAR-CD, the computation is started at the opening of exhaust valve. GT-POWER 1-D engine simulation was used in order to get the pressure and temperature at boundary or zone to define the 3-D CFD boundary

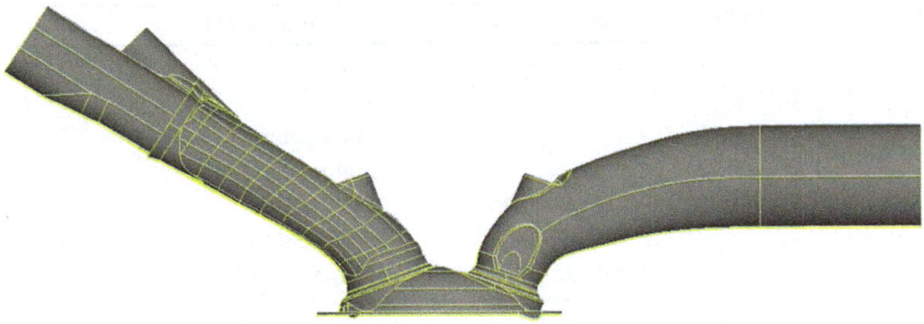

Fig. 1 The geometry model

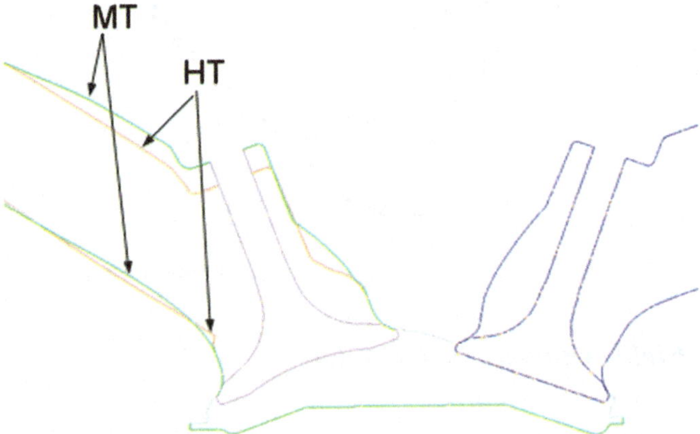

Fig. 2 The sketch of the two intake ports

Fig. 3 Performance comparison of two intake ports

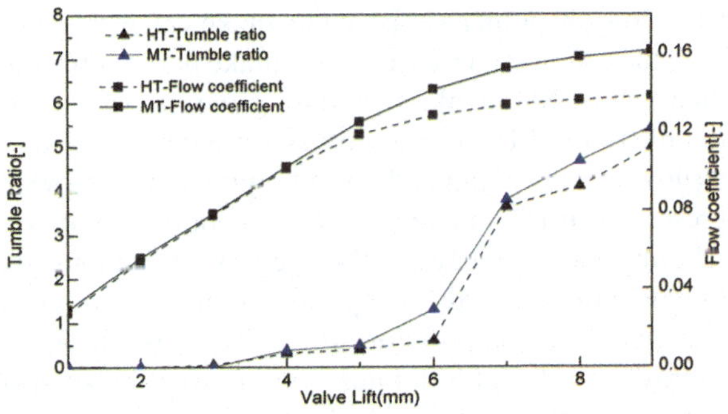

condition and initial condition of the run. The combustion model is ECFM-3Z, the heat transfer model in cylinder is Angel Berger. The fuel spray and atomization is not considered in this model, a assumption is that the fuel and air has formed the homogeneous mixture at the inlet of intake port.

Table 1 Load and engine specification

Load	6000 rpm (WOT)	
Engine parameters	Bore	76 mm
	Stroke	71.6 mm
	Connecting rod length	140.7 mm
	Compression ratio	10.2
Valve timing	IVO:366CA	IVC:608CA
	EVO:128CA	EVC:378CA
Ignition time	698CA(MT)	700.4CA(HT)

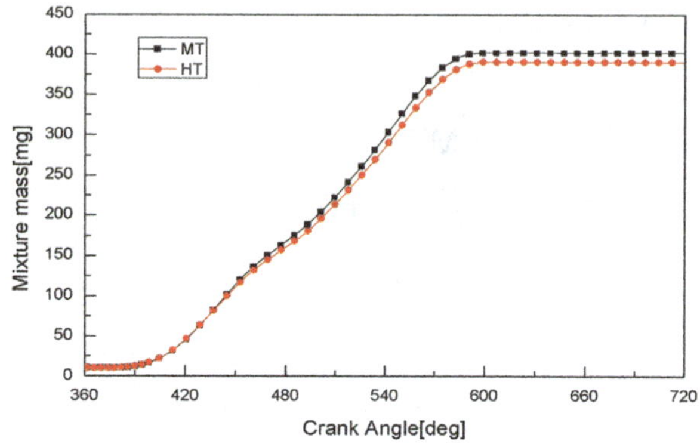

Fig. 4 mixture mass in cylinder

2.2 Flow Field Analysis in Cylinder

The results analysis is mainly focus on the air mass, tumble ratio, turbulence kinetic energy and flow field in cylinder. Fig. 4 illustrate the charge mass variations during the intake and compression. It is obvious that the mass enters to cylinder with MT intake port is higher than that of HT intake port at the same boundary condition. No back flow phenomena exist at the intake valve closing and the mass remains constant until the end of compression. HT intake port tumble ratio in cylinder is always higher than that of MT intake port during the intake and compression (Fig. 6), the peak tumble ratio of HT intake port is more than twice the MT peak. This indicates that airflow is well organized and the momentum provided by intake port is also well preserved in HT intake port, higher tumble ratio is very helpful for formation of mixture and accelerating the process of combustion in cylinder. Macroscopical tumble motion is crushed by moving piston and transferred into turbulence kinetic energy during compression stroke. There is a little difference of turbulence kinetic energy for HT and MT intake port during intake stroke, but big difference of turbulence kinetic energy at the location of spark plug appears during the compression stroke (Figs. 5 and 7), the third peak of turbulent kinetic energy does not exist in cylinder with MT intake port, but appears in cylinder with HT intake port, it is seen that the turbulent kinetic energy in cylinder with HT intake port is more than double that of the MT intake port during the period of spark time, higher turbulence kinetic energy is good for stable ignition and increasing the flame spread speed.

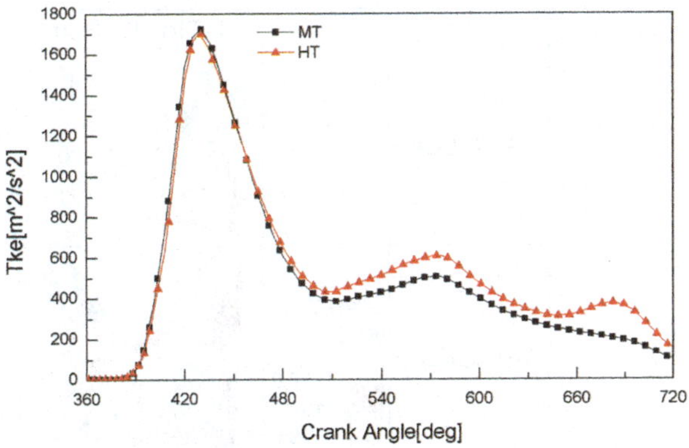

Fig. 5 Turbulence kinetic energy in cylinder

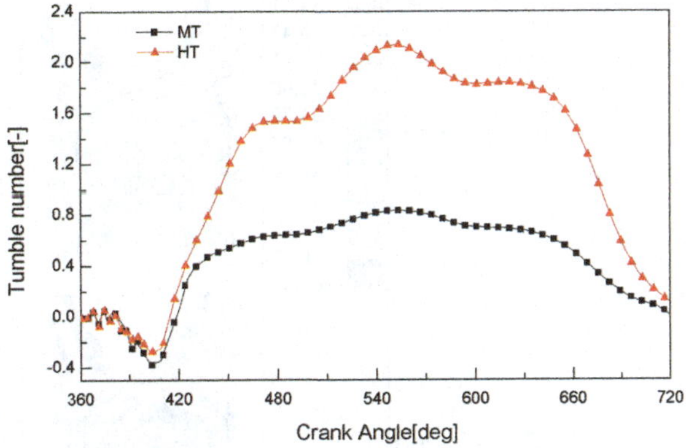

Fig. 6 Tumble ratio in cylinder

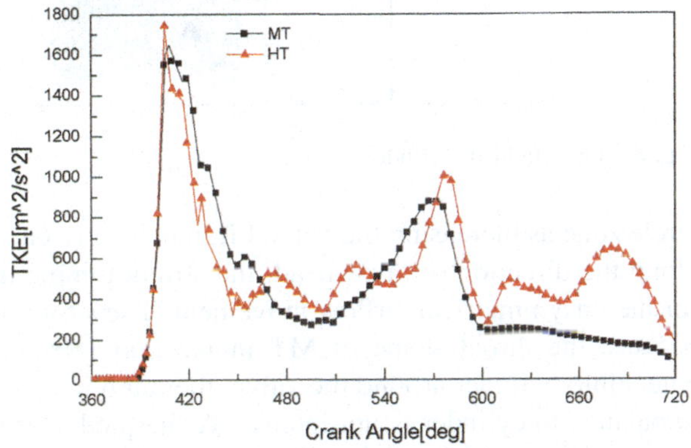

Fig. 7 Turbulence kinetic energy at spark plug

At low valve lift, the flow pattern in cylinder and in intake port of the two intake ports is very similar (Fig. 8), the airflow along intake port wall and no flow separation occurs, it is found that the tumble motion in cylinder is weak from the vector plot owing to airflow symmetrically enters to cylinder around the intake valve. The HT intake port appears strong flow separation at the sharp edge in the

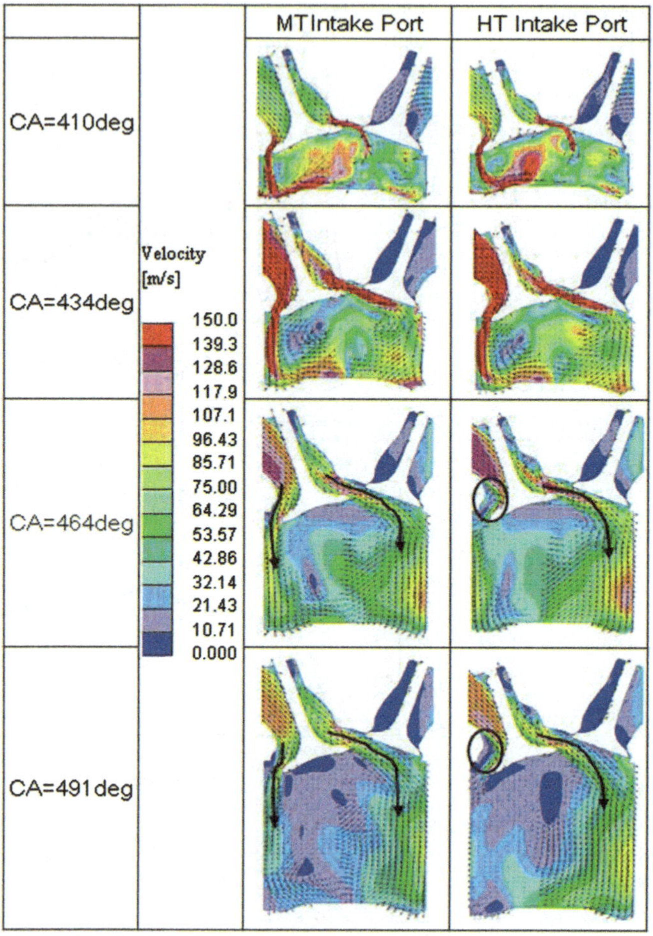

Fig. 8 Flow field in cylinder

circle zone as increasing the valve lift, much more charge mass enters into cylinder along the direction of arrowhead, the strong tumble motion in cylinder is formed for the unsymmetrical airflow movement. The strong flow separation doesn't exist because the throat shape of MT intake port is smooth, so airflow can equably enters into cylinder around the valve, the tumble motion is weak due to the airflow imbalance in cylinder is not intense. A sharp edge at the circle zone is adopted to increase the tumble ratio in cylinder (Fig. 8), the dimension of sharp edge can be adjusted to satisfy the tumble demand of natural aspirated engine and turbo charge and GDI engine.

The Fig. 9 shows the flow field plot along the direction of crankshaft, the difference of flow pattern for the HT and MT intake port is very obvious, the flow structure in cylinder is dissimilar. The airflow of MT intake port moves from

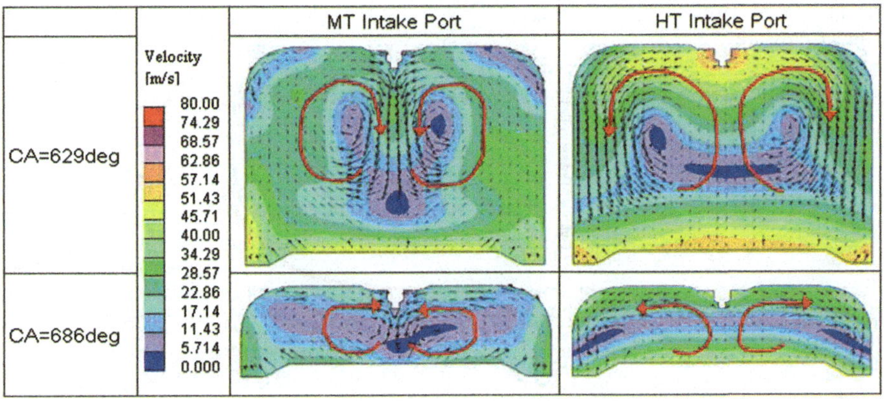

Fig. 9 Flow field in cylinder along the crankshaft

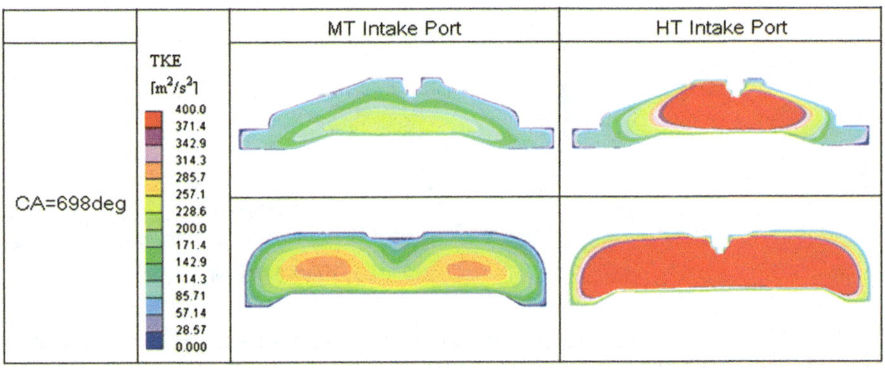

Fig. 10 Turbulence kinetic energy at spark time

exterior to interior in cylinder. The flow structure of HT intake port is opposite to MT intake port, the airflow moves from interior to exterior, this flow structure is helpful to transport flame surface to circumference quickly and accelerate the process of combustion in cylinder. The turbulence kinetic energy in cylinder of HT intake port is much higher than that of MT intake port at spark time because the HT intake port can produce the strong tumble motion during intake stroke and more turbulence kinetic energy was generated during compression stroke (Fig. 10).

2.3 Analysis of Combustion in Cylinder

The spark time of HT intake port delay 2.4 CA than that of MT intake port, but the peak pressure in cylinder advance 5.1 CA relative to MT intake port, see Fig. 11.

Fig. 11 Comparison of cylinder pressure

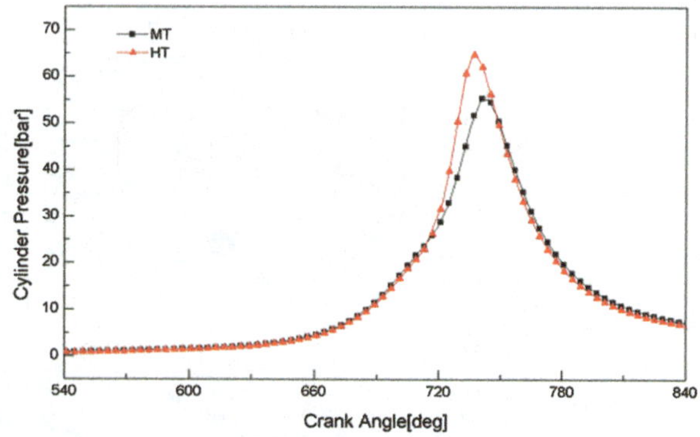

Fig. 12 Comparison of heat release

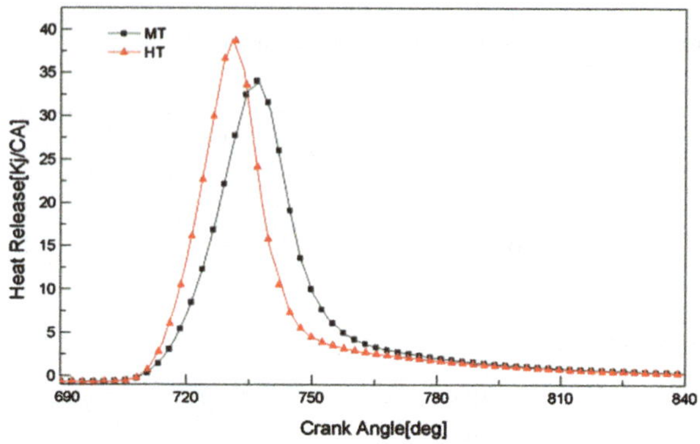

The Fig. 12 shows the heat release of combustion system with HT intake port is higher than that of MT intake port, the HT intake port make the combustion system possess good performance. The combustion duration of 50 % burned fuel of HT intake port is shorter 7.7 CA than that of MT intake port, the combustion duration from 10 to 90 % burned fuel of HT intake port is shorter 3.8 CA than that of MT intake port. The combustion speed of HT intake port is faster than that of MT intake port from the C8H18 concentration distribution variation, and the combustion symmetry of HT intake port is better. The unburned mixture of combustion system with MT intake port occurs at the intake side of cylinder, this is bad for increasing the efficiency and quality of combustion Fig. 13.

3 Comparison of CFD and Experiment

The CFD and test cylinder pressure curves of two intake ports show good correlation (Figs. 14 and 15), the peak pressure error is within 1 % between CFD and test, and the CFD and test crank angle of peak pressure is nearly same, see Table 2. The CFD and test can get the same results after the CFD model has been validated

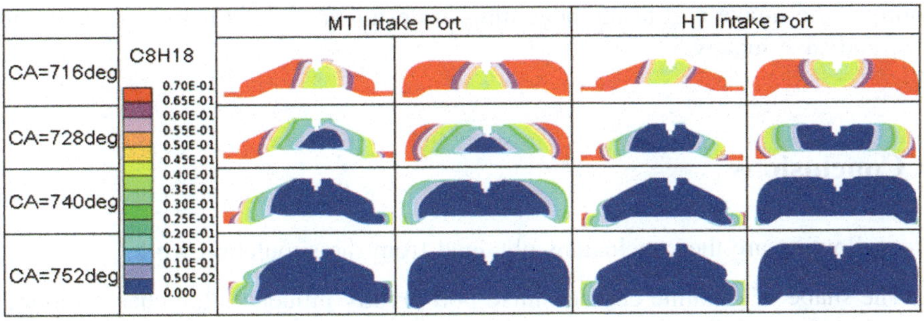

Fig. 13 Comparison of C8H18 concentration

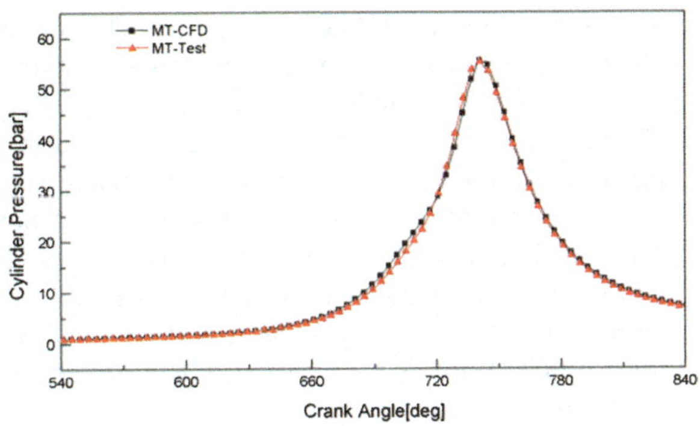

Fig. 14 Cylinder pressure comparison of MT

and the advanced CFD model has been selected. The CFD is a useful tool to

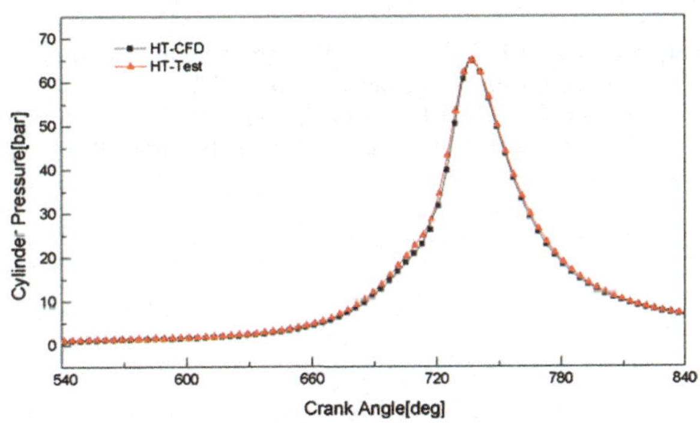

Fig. 15 Cylinder pressure comparison of HT

Table 2 Comparison of peak pressure

	CFD	Test
MT intake port	55.64bar@742.1CA	55.28bar@740CA
HT intake port	64.89bar@737CA	65.02bar@737CA

optimize the intake port at the beginning of design stage, and help to shorten the time and save money.

4 Conclusions

The following are the conclusions obtained from the simulation work.

1. The shape of gasoline engine intake port greatly influences the charge motion and mixture formation and combustion in cylinder. The key dimension of intake port with small modification will change the performance of engine. The developments of intake port need to consider the charge motion and tumble level, and that the intake port must match with chamber and piston well.
2. The charge motion and combustion in cylinder can be accurately analyzed by CFD, CFD simulation can provide abundant flow and combustion information to help engineer to understand the essential difference of different intake port. The CFD results shows the strong tumble motion in cylinder can accelerate the process of combustion and increase the flame speed. The modern engine with high tumble motion in cylinder will be a tendency.
3. The correlation of CFD simulation and experiment data is good, the accuracy of CFD is good enough to fulfilling the demand of engineering. A lot of shape optimization work of intake port and combustion chamber and piston can be done by CFD during the stage of CAD design.

References

1. lyer C and Yi J (2009) 3D CFD upfront optimization of the in-cylinder flow of the 3.5L V6 ecoboost engine, SAE paper 2009-01-1492
2. Qi Y, Liu H, Midkiff K and Puzinauskas P (2010) A feasible CFD methodology for gasoline intake flow optimization in a HEV application—Part 2: prediction and optimization, SAE paper 2010-01-2238

Cycle-Resolved Computations of Stratified-Charge Turbulent Combustion in Direct Injection Engines

Tomoaki Kubota, Nobuhiro Shinmura and Ken Naitoh

Abstract Cyclic variations of stratified-charge turbulent combustion in direct injection gasoline engine can be simulated over ten continuous cycles based on the multi-level formation for the compressible Navier–Stokes equation and also a spray model. Computational results are compared with experimental data. Then, a factor generating the cyclic variations is revealed, which leads to an effective way to control instability of combustion at very lean burning conditions.

Keywords Numerical analysis · Mixture formation · Cyclic variations of combustion · Stratified-charge engine · Very lean burning

1 Introduction

Electric battery is expected as a future power train system because of no emissions. However, there are still some problems such as short cruising distance and high price of battery. Moreover, the society depending only on electricity should be avoided in order to maintain our economical activities. Thus, improvement of fuel consumption rate of engines is also necessary for reduction of CO_2 emission. One of the engine concepts for achieving the purpose is extremely lean burning engines [1]. To achieve this concept, stable combustion should be realized for a wide range of operating

F2012-A06-007

T. Kubota (✉) · N. Shinmura · K. Naitoh
Faculty of Science and Engineering, Waseda University, Shinijuku, Japan
e-mail: k-naito@waseda.jp

SAE-China and FISITA (eds.), *Proceedings of the FISITA 2012 World Automotive Congress*, Lecture Notes in Electrical Engineering 190,
DOI: 10.1007/978-3-642-33750-5_8,

conditions at air–fuel ratios over 40.0. Numerical analysis is also important to attach this target because only experimental researches are time-consuming. Thus, we have built some simulation models of engines.

The ensemble-averaged models based on the Reynolds-averaging Navier–Stokes (RANS) equation [2] are world-widely used, which cannot predict cyclic variations. Thus, in this report, we propose the simulation model based on the Large Eddy Simulation (LES) [3–6], which can realize the cycle-resolved computations of the fuel mixing and turbulent combustion processes in stratified-charge direct injection engines.

Engine specifications, operating conditions and configuration of our simulation model are based on the experiment [7]. And, computational results are compared with experimental data. Then, a factor generating the cyclic variations is revealed, which leads to an effective way to control instability of combustion at very lean burning conditions.

2 Governing Equation and Numerical Method

Unsteady compressible Navier–Stokes equation in three-dimensional space is used as the governing equation for analyzing the compressible turbulence and combustion processes. The governing equation is transformed to the more complex equation system based on the multi-level formulation [3, 4].

A mesoscopic averaging window, smaller than that for the continuum assumption, yields a stochastic compressible Navier–Stokes equation in a non-conservative form in the case where specific heat and viscosity coefficient have constant values, written as the following governing equation,

$$F \equiv \begin{bmatrix} f_1 \\ f_2 \\ f_3 \end{bmatrix} = \begin{bmatrix} \dfrac{\partial \bar{u}_i}{\partial x_i} + \dfrac{1}{\bar{\rho}} \dfrac{D}{Dt} \bar{\rho} \\ \dfrac{D\bar{u}_i}{Dt} + \dfrac{\partial \bar{p}}{\partial x_i} - \dfrac{1}{\mathrm{Re}} \dfrac{\partial^2 \bar{u}_i}{\partial x_j^2} \\ \dfrac{D\bar{T}}{Dt} - \dfrac{D\bar{p}}{Dt} - \dfrac{1}{Pe} \dfrac{\partial^2 \bar{T}}{\partial x_i^2} \end{bmatrix} = \begin{bmatrix} \varepsilon_1 \\ \varepsilon_2 \\ \varepsilon_3 \end{bmatrix} \tag{1}$$

where u_i, p, ρ, $\varepsilon(i = 1 - 3)$, t, Re, T and Pe denote the dimensionless quantities of velocity components in the i-direction, pressure, density, random forces, time, Reynolds number, temperature and Pelclet number, respectively.

Equation (1) is transformed to Eq. (2), which is a multi-level formulation [4].

$$F \equiv \begin{bmatrix} f_1 \\ f_2 \\ f_3 \\ f_4 \end{bmatrix} = \begin{bmatrix} \hat{D} + \dfrac{1}{\bar{\rho}} \dfrac{D}{Dt} \bar{\rho} \\ \dfrac{D\bar{u}_i}{Dt} + \dfrac{\partial \bar{p}}{\partial x_i} - \dfrac{1}{\mathrm{Re}} \dfrac{\partial^2 \bar{u}_i}{\partial x_j^2} \\ \dfrac{\partial^2 \bar{p}}{\partial x_i^2} + \bar{\rho} \dfrac{\partial}{\partial t} \hat{D} + \Phi \\ \dfrac{D\bar{T}}{Dt} - \dfrac{D\bar{p}}{Dt} - \dfrac{1}{Pe} \dfrac{\partial^2 \bar{T}}{\partial x_i^2} \end{bmatrix} = \begin{bmatrix} \varepsilon_1 \\ \varepsilon_2 \\ \varepsilon_3 \\ \varepsilon_4 \end{bmatrix} \tag{2}$$

where $\hat{D}(= \partial \bar{u}_i / \partial x_i)$ denotes the divergence of velocity and also the equation for f_3 is the spatial derivative of that of f_2, while the third term on the left-hand side

$$\Phi = \frac{\partial}{\partial x_i} \left[\bar{u}_j \frac{\partial \bar{u}_i}{\partial x_j} - \frac{1}{\mathrm{Re}} \frac{\partial^2 \bar{u}_i}{\partial x_i^2} \right] \tag{3}$$

denotes the spatial derivative of the convection and viscosity terms in f_2. Four variables of $\hat{D}$, $\bar{u}_i$, $\bar{p}$, and $\bar{T}$ can be solved by the four equations in Eq. (2). The second term on the left-hand side of the energy conservation law in Eq. (2) can approximately be eliminated for low Mach number conditions.

Next, let us consider ε_1 in Eq. (2). Stronger inlet disturbances result in a smaller characteristic scale, because they lead to more inhomogeneous velocity distributions at the starting point of instability inside the laminar boundary layer. Thus, the relation

$$\sum_{n=1}^{N} \varepsilon_1 / N = C_e \delta \tag{4}$$

is used with the inlet disturbance δ, an arbitrary constant C_e and the number of grid points N. In this report, C_e is assumed to be 1.0 [8, 9].

Stochastic terms ε_2, ε_3 and ε_4 in Eq. (2) should also be added to the momentum conservation law, the Poisson equation, and the energy conservation law averaged in the mesoscopic averaging window. However, in this study, ε_2, ε_3 and ε_4 are set to be 0.0, because their role is similar to small variations of Reynolds and Peclet numbers in space and time, resulting in less influence on the transition point than in the case of the mass conservation law. The stochastic terms ε_2, ε_3 and ε_4 have relatively weak influence on the transition phenomenon, because those in momentum and energy conservation laws can also be dissipative, while mass with ε_1 cannot essentially be dissipative. There are two ways to calculate the random force term of ε_1: One which involves the use of a random number generator and the other which is a special numerical disturbance close to the actual random number.

Then, an approach based on the G-equation proposed by Williams [3, 4] is included to track the flame front accurately. And we use a model for heat loss from engine wall [2]. The discrete droplet model [5] is used for computing fuel spray. The Monte Carlo method is also employed in order to reduce computational time. Moreover, the present spray model includes that for time-dependent parcel radius [5].

A third-order upwind scheme is employed for convection terms in the Navier–Stokes equation, while the Euler backward scheme is used for temporal developments. The convection term in the G-equation is discretized by a flux-corrected finite difference method. The other terms are discretized by second-order central difference. Numerical algorithm is extended from that of the MAC and SIMPLE methods [2–4].

The present formulation based on the multi-level formulation and the numerical method can compute the unsteady turbulent flow and transition turbulence in pipes for a wide range of Mach numbers from zero to supersonic conditions, which can simulate the four strokes of intake, compression, expansion and exhaust processes including combustion and shock waves [3, 4].

3 Calculation Conditions of Stratified-Charge Direct Injection Engine

Figure 1 shows the engine configuration for simulating the fuel mixing process and turbulent combustion in the direct injection gasoline engine, which has a pent-roof and piston bowl. Tumble control valves (TCV) are set for the intake ports to make tumble vortex. Exhaust ports are also included for computing the exhaust process. Then, a swirl type of injector of fuel (Sauter mean particle size of 20 microns) is displaced between two intake ports [6]. Engine specifications and operating conditions are shown in Tables 1 and 2. Engine speed and air–fuel ratio are set to be 1,600 rpm and 40.0–60.0, respectively. Ignition timing is fixed at 35° BTDC. The number of grid points is approximately 5,900,000. Homogeneous and orthogonal grid system is used.

The continuous cycles of four strokes including exhaust process are computed, while only the blow down stage just after opening the exhaust valves is eliminated because of a very short period. Although our numerical method can simulate the blow down process in a supersonic regime [3, 4, 8], the very short period of blow down is eliminated in the present paper because we assume that the pressure inside the cylinder suddenly becomes atmospheric just after opening the exhaust valves. Turbulent flow, spray motion, vaporization, and combustion process in the four-stroke stages are computed during ten cycles.

4 Combustion Instability Generated by Cyclic Variations of Air Flows

Cyclic variations of combustion in direct injection gasoline engines, combustion instabilities, come from some main factors such as cyclic variations of air flows and spray motions. This section shows the influence of cyclic variations of

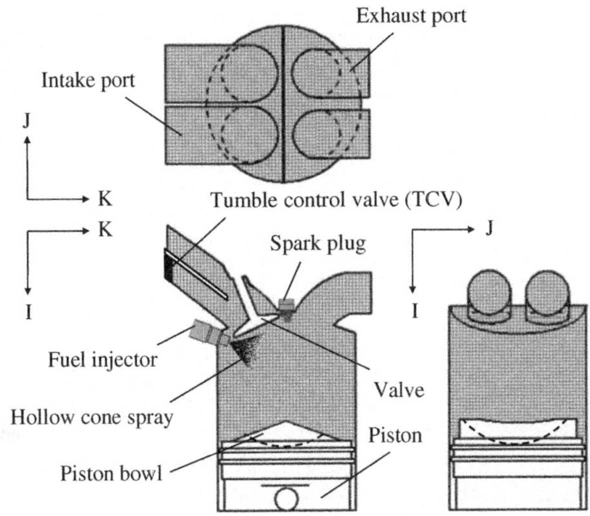

Fig. 1 Configuration of stratified-charge direct injection SI engine

Table 1 Engine specifications of stratified-charge direct injection engine

Bore mm	85.6
Stroke mm	81.8
Compression ratio	10.5
Combustion chamber	Pent-roof

Table 2 Operating conditions of stratified-charge direct injection engine

Engine speed rpm	1,600
Air–fuel ratio	40.0, 50.0, 60.0
Charging efficiency %	75
Fuel	Gasoline (Isooctane)
Grid points (i × j × k)	267 × 139 × 160
Tumble control valve (TCV)	Close
Injection end timing degBTDC	57
Injection angle deg	10
Injection quantity mg	10.6, 8.50, 7.08
Spray angle deg	80
Spray geometry	Hollow cone
Average droplet diameter μm	20
Ignition timing degBTDC	35

"Air flows" on combustion instabilities, while other variation factors such as cyclic variations of spray motions are fixed for each cycle of engine. Figures 2, 3 and 4 show time histories of pressure and heat release rate at A/F = 40.0, 50.0 and 60.0. Table 3 shows cyclic variations rate (CV rate) of mean effective pressures at each air–fuel ratio.

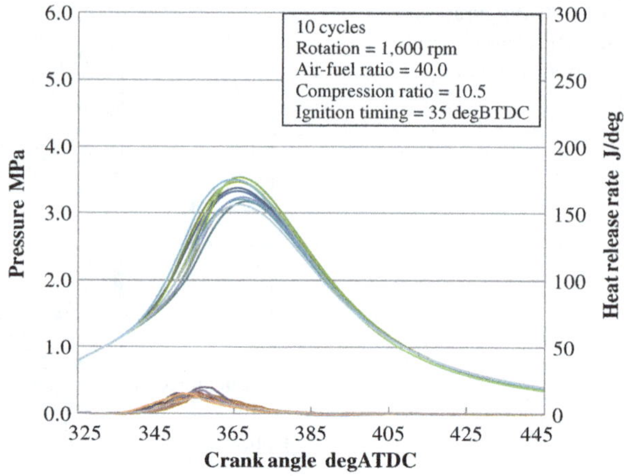

Fig. 2 Time histories of pressure and heat release rate. (Stratified-charge with A/F = 40.0, cyclic variations of combustion generated by the cyclic variations of air flows)

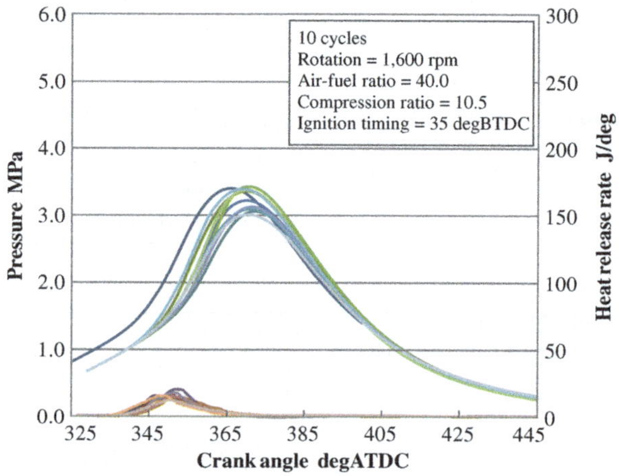

Fig. 3 Time histories of pressure and heat release rate. (Stratified-charge with A/F = 50.0, cyclic variations of combustion generated by the cyclic variations of air flows)

The present model simulated the flow field with random disturbance given at the edges of intake ports during the suction process by using the nonlinear stochastic Navier–Stokes equation with chemical reaction [8–11]. The computations are done at the inlet disturbance condition of 4.10 % on average. The random inlet disturbance varies cycle by cycle. (The random disturbance is put only at the intake port. There is no random disturbance inside the combustion chamber at the start of intake process. The flow at the end of the exhaust process of the previous cycle is the initial condition for the next cycle).

That is to say, one of factors generating the cyclic variations of combustion is those of air flow field, which is related to the random inlet disturbance of the intake pipe in the inlet stroke. In addition, the flow field of the residual air in the cylinder which varies cycle by cycle could be considered as the factor, because there is no flow at the initial condition of the first cycle, whereas the later cycles have turbulent flows, and also because the nonlinearity in the Navier–Stokes equations, i.e., chaotic effect, works.

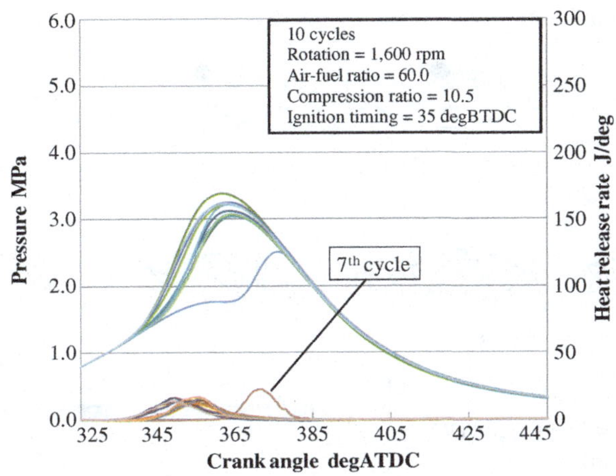

Fig. 4 Time histories of pressure and heat release rate. (Stratified-charge with A/F = 60.0, cyclic variations of combustion generated by the cyclic variations of air flows)

Table 3 Cyclic variations rate of mean effective pressures. (Stratified-charge with each air–fuel ratio, cyclic variations of combustion generated by the cyclic variations of air flows)

A/F	40.0	50.0	60.0
CV rate %	3.11	3.88	4.65

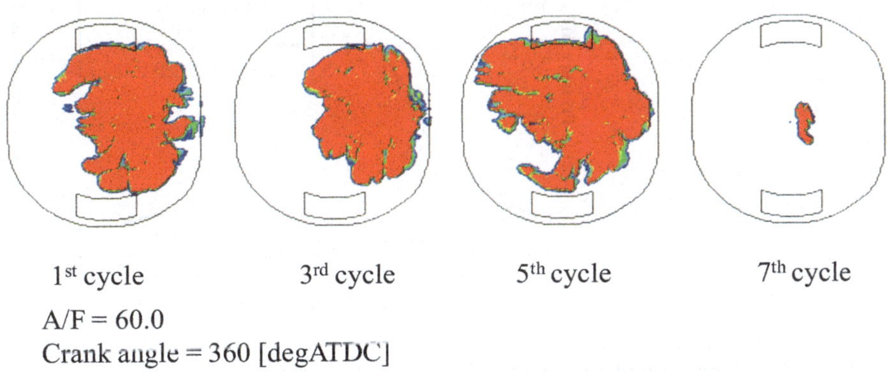

Fig. 5 Cyclic variations of combustion area. (Stratified-charge with A/F = 60.0, cyclic variations of combustion generated by the cyclic variations of air flows)

Figures 2, 3 and 4 show that time histories of pressure and heat release vary cycle by cycle. That is to say, we can see that our simulation model can simulate cyclic variations. From Table 3, we can look at the qualitative tendency that cyclic variations of combustion is stronger with increasing air–fuel ratio.

Figure 4 shows unstable combustion at one cycle (7th cycle) because of very lean burning condition. We must examine whether or not this instability level of

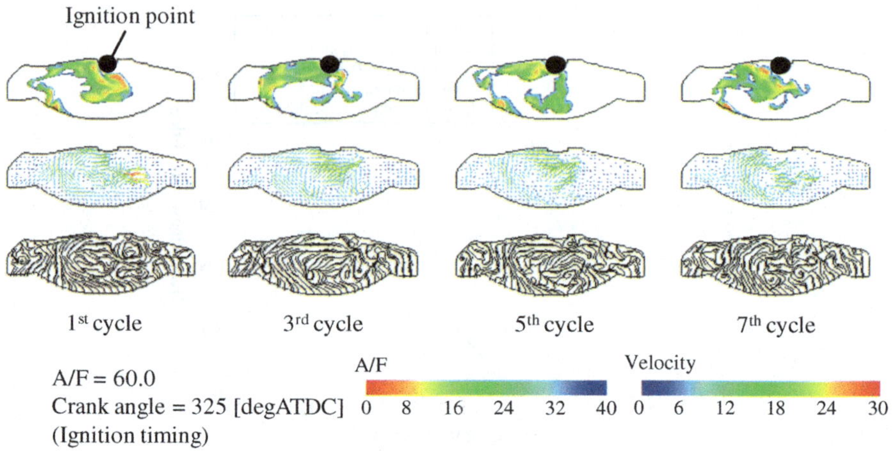

Fig. 6 Temporal evolutions of fuel vapor distribution, air velocity, and streamline. (Stratified-charge with A/F = 60.0, cyclic variations of combustion generated by the cyclic variations of air flows)

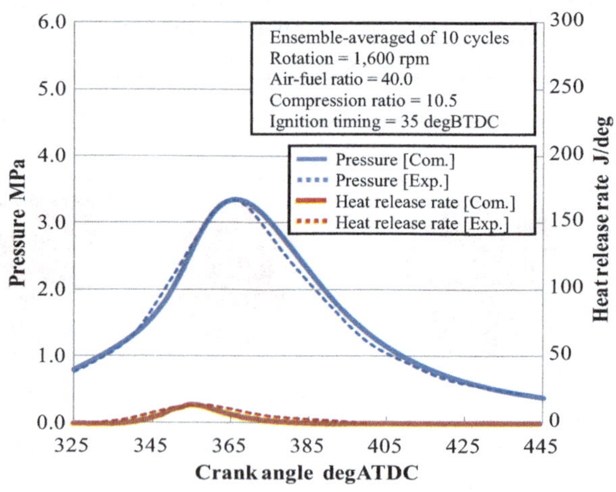

Fig. 7 Time histories of pressure and heat release rate computed and corresponding experimental data

combustion computed and the temporal pattern of heat release rate are correct by comparing computational results with experimental data.

Figure 5 shows cyclic variations of combustion area, and Fig. 6 shows fuel vapor distribution, air velocity and streamline at each cycle (A/F = 60.0). From Fig. 5, we can also see unstable combustion as slow flame propagation at one cycle (7th cycle). Usually, fuel vapor is carried close to the spark plug by air flows. However, in the cycle of unstable combustion, we can see that fuel vapor is not carried because of relatively weak stream flows at Fig. 6.

5 Comparison Between Computational Results and the Corresponding Experimental Data

Time history of pressure and heat release rate computed for a stratified-charge direct injection engine are compared with the corresponding ensemble-averaged experimental data taken by Teraji and Tsuda [7] at Fig. 7. The computational results are ensemble-averaged during ten continuous cycles.

The computational results agree fairly well with the corresponding experimental data. Thus, the present analysis shows the possibility that our simulation model is applicable for analyzing cyclic variations.

6 Conclusion

In direct injection gasoline engines, cyclic variations of combustion come from some main factors such as those of air flows and spray motions. In this paper, we examined the influence of cyclic variations of "Air flows" on combustion instabilities, while other variation factors are fixed for each cycle of engine.

One of factors generating the cyclic variations of combustion is those of air flow field, which is related to the inlet disturbance of the intake pipe during the inlet stroke. The present model simulates the flow field with random inlet disturbance by using the nonlinear stochastic Navier–Stokes equation with chemical reaction. On the basis of the stochastic Navier–stokes equation, we developed the model in order to predict stratified-charge combustion in direct injection gasoline engines. Then, we computed the flow and combustion in continuous cycles. Computational results averaged during continuous cycles agree fairly well with the corresponding experimental data.

Our simulation model has a potential for computing some effects which cyclic variations of air flows gives on combustion instabilities as cyclic variations rate (CV rate). However, level of cyclic variations computed should be compared with the corresponding experimental data.

In near future, we will find an effective way to control instabilities of combustion at very lean burning conditions.

Acknowledgments The authors express sincere thanks for their help on using the super computer of Osaka University, Cybermedia center. Then, the authors express sincere thanks to Dr. Atsushi Teraji of Nissan Motor Co. for permission of reuse of the experimental data.

References

1. Takagi Y, Itoh T, Muranaka S, Iiyama A, Iwakiri Y, Urushibara T, Naitoh K (1998) Simultaneous attainment of low fuel consumption, high output power and low exhaust emissions in direct injection SI engines. SAE Paper, (980149)
2. Amsden AA, O'Rourke PJ, Butler TD (1989) KIVA-II: a computer program for chemically reactive flows and sprays. Los Alamos National Laboratory report, (LA-11560-MS)
3. Naitoh K, Takagi Y, Kuwahara K (1993) Cycle-resolved computation of compressible turbulence premixed flame in an engine. Comput Fluids 22(4/5):623–648
4. Naitoh K, Kuwahara K (1992) Large eddy simulation and direct simulation of compressible turbulence and combusting flows in engines based on the BI-SCALES method. Fluid Dyn Res 10:299–325, [Also in AIAA paper, 2011, 2011–2316]
5. Naitoh K, Takagi Y (1996) Synthesized spheroid particle (SSP) method for calculating spray phenomena in direct-injection SI engines. SAE Paper, (962017)
6. Suruga Y, Shirai Y, Naitoh K (2009) Large eddy simulation for predicting combustion-stability and fuel mixing process in SI engine. Proceedings of 2009 JSAE annual congress (Spring), (20095061)
7. Teraji A, Tsuda T (2004) Development of a three dimensional combustion simulation for spark ignition engines. NISSAN Technical report, 55:22–26
8. Naitoh K, et al (2010) Stochastic determinism. Proceedings of 6th international symposium on turbulence and shear flow phenomena (TSFP6), [Also in Proceedings of ETMM8]
9. Shinmura N, Morinaga K, Kiyohara T, Naitoh K (2011) Cycle-resolved computations of stratified-charge and homogeneous-charge turbulent combustion in direct injection engines. SAE Paper, (JSAE 20119038, SAE 2011-01-1891)
10. Naitoh K, Nakagawa Y, Shimiya H (2008) Stochatic determinism approach for simulating the transition points in internal flows with various inlet disturbances. Proceedings of ICCFD5. Comput Fluid Dyn 2008, Springer, New York
11. Naitoh K, Shimiya H (2011) Stochastic determinism capturing the transition point from laminar flow to turbulence. Jpn J Ind Appl Math

Research on Torque-Angle Tightening of High Strength Bolt in Internal Combustion Engine

Wenfeng Zhan, Jian Wu, Fake Shao and Chuhua Huang

Abstract Torque-angle tightening is widely used in the internal combustion engine. Actually, Torque-angle tightening is an indirect method of length measurement tightening, used to make the bolt's plastic elongation rate right after tightening. Usually, the bolt's plastic elongation after tightening could only be measured by experiments. We researched on how to calculate the plastic elongation of the bolt after tightening when both elastic and plastic elongation occur, considering the tolerance of the torque and angle. What is the most important point but never be mentioned in most of the research.

Keywords Bolt · Torque-angle tightening · Elastic · Plastic · Elongation

1 Introduction

More and more high performance bolts are used in the ICE as the higher and higher requirement about lightweight and performance. In order to make full use of the bolt's material, we must control the preload force, which depended on the tighten method. There are many tighten methods for bolt which can be classified to three by principle as below in Table 1 [1].

Torque angle tightening method, which can decrease the scatter of the bolt's preload force, increase the usage factor of the material and also get a high

F2012-A06-016

W. Zhan (✉) · J. Wu · F. Shao · C. Huang
Guangzhou automobile group co., LTD automotive engineering institute, Mainland, China
e-mail: zhanwenfeng@gaei.cn

SAE-China and FISITA (eds.), *Proceedings of the FISITA 2012 World Automotive Congress*, Lecture Notes in Electrical Engineering 190,
DOI: 10.1007/978-3-642-33750-5_9,

Table 1 Three methods of bolt tightening

Tightening method	Principle	Scatter ratio (%)	Material usage (%)	Tightening tools	Cost	Example
1 Torque controlled tightening	Control the bolt's force by torque according to the relation between bolt force and torque	±25	40–60	Simple wrench	Cheap	Ordinary bolt joint
2 Torque-angel controlled tightening	Tighten the bolt using torque and angle, which can control the plastic elongation of the bolt. The bolt force only depends on the bolt's material	±15	90	Simple torque wrench protractor	Cheap	Important bolt joint
3 Yield controlled tightening	Detect the deformation of the bolt during tightening in order to control the bolt in the yield point after tightening. The bolt's force only depends on material	±3–5	95	Complicated machine to detect the bolt's deformation	Expensive	Rarely used

performance and lightweight, is widely used in high performance bolts in ICE such as cylinder head, main bearing cap, conrod and so on.

2 Design Method of Torque Angle Tightening

2.1 Choose Bolt

Choose a bolt according to the requirements of the joint. The bolt's ensure load force should be equal or higher than the necessary preload force of the joint. Usually, GB/T3098.1-2000 can be used for reference [2]. VDI2230 can be used if the bolt's quality is good enough [3], in which the ensure force is 15 % higher than the same type bolt in GB/T3098.1-2000.

2.2 Torque Tightening

At first, Torque tightening as a preloaded makes sure the interfaces are completely closed. The torque T including two parts, one is frictional torque in thread T_1 and the other is frictional torque between bolt head and boss or wash T_2.

$$T = T_1 + T_2 = F' \tan(\varphi + \rho_V)\frac{d_2}{2} + \frac{F'\mu}{3} \times \frac{D_W^3 - D_0^3}{D_W^2 - D_0^2} = KF'd$$

$$K = \tan(\varphi + \rho_V)\frac{d_2}{2d} + \frac{\mu}{3d} \times \frac{D_W^3 - D_0^3}{D_W^2 - D_0^2}$$

d Bolt diameter = outside diameter of thread (nominal diameter), mm;
F' Preload force, N;
d_2 Pitch diameter of the bolt thread, mm;
D_w Outer diameter of the bolt head contact face, mm;
D_0 Inner diameter of the bolt head contact face, mm;
Φ Helix angle of the bolt thread, angle of the substitutional deformation cone, mm;
ρ_v Angle of friction for μ_v, $\rho_v = \arctan\ \mu_v$;
μ_v Coefficient of friction in the thread;
μ Coefficient of friction in the head contact face;
K Tightening factor.

The relation between tightening torque and preload force depend on two frictional coefficient μ_v and μ. Frictional coefficient is a value by test, which effected by roughness, lubrication situation, humidity and so on. Because the frictional

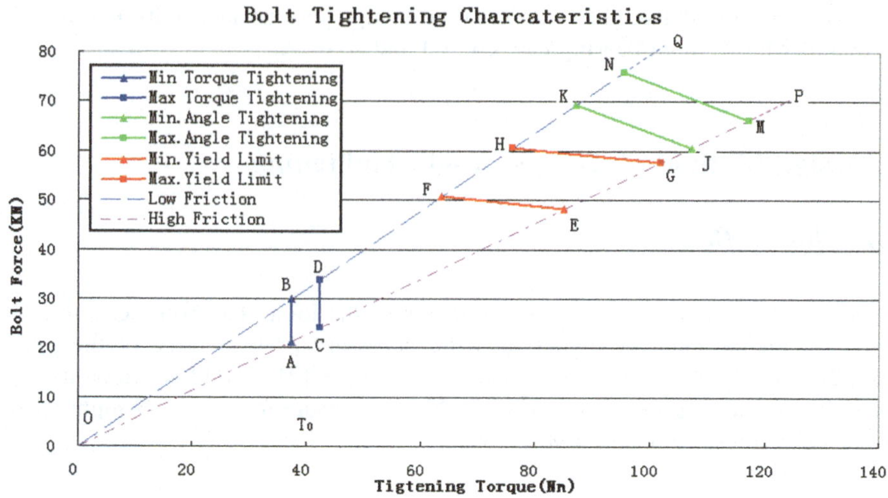

Fig. 1 Bolt tightening characteristics

coefficient will change according to the environment, we can set a range by give a minimum and a maximum considering the environment. Using torque as coordinate X, bolt preload force as coordinate Y, we can draw two lines OQ and line OP based on max and min coefficient respectively, which show the relation between tightening torque and preload force shown in Fig. 1.

Using 40 % of the total clamped force used when choose bolt, we can find the initial tightening torque T_0, Line AB and line CD can be drawn considering the tolerance ΔT based on the line OP and line OQ. The quadrangle ABDC means the bolt's preload force distribution area after the torque tightening T_0 at first step.

2.3 Calculate the Preload Force at Yield Limit

Yield limit of the bolt's material is $\delta_{0.2}$. The maximum possible preload (load at yield point) of the bolt is influenced by these simultaneously acting tensional and torsional stresses δ and τ. The resulting total stress is reduced by means of the deformation energy theory to an equivalent uniaxial stress state as below.

$$\delta_{xd4} = \sqrt{\delta^2 + 3\tau^2} \leq [\delta] = \delta_{0.2}$$

So that the bolt's maximum possible preload F'_m will be

$$F'_m = A_S \cdot \frac{\delta_{0.2}}{\sqrt{1 + 3\left[\frac{3d_2}{2d_S}\tan(\varphi + \rho_V)\right]^2}}$$

F'_m Bolt's maximum possible preload, N;
A_s Area at stress cross section, mm^2;
d_s Diameter at stress cross section A_s, mm.

Yield limit of the bolt's material $\delta_{0.2}$ has a tolerance also. Considering the maximum and minimum $\delta_{0.2}$, we can draw line EF and line GH at the maximum and minimum frictional coefficient line. As shown in Fig. 1, quadrangle EFHG means the distribution area while the bolt was tightened to the yield point.

2.4 Angle Tightening

With the assumption that the bolt always work at elastic area (below the yield point), the bolt's elongation after angle tightening θ can be calculated as below according to the thread's characteristics [4].

$$\frac{\theta}{360} \times P = \lambda_L + \lambda_F = \frac{F''}{C_L} + \frac{F''}{C_F}$$

P Thread pitch, mm;
θ Tightening angle, deg;
λ_L Elongation of the bolt, mm;
λ_F Compression of the clamped parts, mm;
C_L Stiffness of the bolt, N/mm. It can be calculated by the equations from handbook of mechanical design or VDI2230, which has already considered the stiffness of bolt head and engaged thread.
C_F Stiffness of the clamped parts, N/mm. It can be calculated by the equations from handbook of mechanical design or VDI2230, which has already considered the cones and sleeve of the clamped parts.

Considering the tolerance $\Delta\theta$ at angle tightening, we can draw line JK and line MN as shown in Fig. 1. Quadrangle JKNM means the distribution area after angle tightening base on the assumption bolt was tightened below the yield at angle tightening.

2.5 Calculate the Bolt's Elongation

Angle tightening is base on the assumption bolt was tightened below the yield limit. Obvious, the bolt has fully overstepped the yield limit point after angle tightening comparing the quadrangle JKNM and EFHG in Fig. 1. It means our assumption was wrong. The bolt was tightened to the plastic area (overstep the yield limit point), which is the result we like. We must adjust the tightening angle until the bolt was fully tightened into the plastic area, which is the actual purpose of angle tightening.

Make sure the bolt was fully tightened into plastic area by angle tightening, but too much overstep the yield limit point will increase the risk of bolt's fracture. We need to design a proper plastic elongation for the angle tightening in the next step. From the figure, we can find the equation to calculate the bolt's plastic elongation as below:

$$\varepsilon_p = \frac{F' - F'_m}{C_L \cdot L_p} \times 100\%$$

ε_p Bolt's plastic elongation, %;
F' Bolt's preload force under the assumption that bolt was tightened below the yield limit;
L_P Length for plastic elongation at bolt, the length of the minimum diameter d_0 the relevant smallest cross section of the bolt, mm

In the case of necked-down bolts (reduced-shank bolts), the length of d_0 is to be used; otherwise the length of free load thread will be used.

Usually, ε_p can be designed to approximate 0.5 %, and 0.2 % $\leq \varepsilon_p \leq$ 1 % is the normal range for high performance bolt from the material theory. Actually, the value of ε_p should consider about the factors such as material's elongation ratio, load situation and reused times. For example, designing a bolt which can be reused for three times, we must make sure the bolt's material can work well after three times plastic elongation.

3 Test Validation

We calculated all the important bolts in engine. Here we used conrod bolt to test for example.

The bolt is M8*1, 12.9 grade and 20 N m torque and 90° angle. All the calculation was integrated in a program in excel. The tightening result by calculation was shown in Fig. 2.

Considering the tolerance torque 3 NM, angle 2.5° and frictional coefficient from 0.08 to 0.14, bolt clamped force distributed between 37.6 and 47.5 KN and the plastic deformation from 0.2 to 0.48 %. The bolt clamped force, which only depended on the yield limit of bolt's material, achieved the maximum force in the material's limit, because it was tightened beyond the yield point.

After the calculation, we test the bolt's plastic elongation after tightening. 10 bolts were measured in this test. Four were tested for one time and others were tested for three times considering the requirement of conrod bolts. Tighten and loosen and then measured the length of the bolts. All the results were shown in

GAEI Bolt Calculator V1.6 RESULTS FOR PRINT					
Engine Name	1.8T				
Bolt Connection	Conrod				
Bolt Dimension			Size(mm)		Pitch(mm)
			8.00		1.00
			Min.	Mean	Max.
Bolt Yield Strength	δs	MPa	1100.00		1230.00
Bolt Force	F'	KN	37.62		47.46
Fricition Coeff. on Bolt Head	μk	-	0.08	0.11	0.14
Fricition Coeff. in Thread	μg	-	0.08	0.11	0.14
inner/Outer Dia. of Bolt Head Contact	dw	mm	9.20		14.20
Pressure under Bolt Head	δh	MPa	409		516
Inner/Outer Dia. of Washer Contact	d	mm	0.00		0.00
Pressure under Washer	δw	MPa	#DIV/0!		#DIV/0!
Torque & Angle Tightening					
Step 1: Bolt Tightening Torque	T	Nm	17.50	20.00	22.50
Step 2: Bolt Tightening Angle	θ	deg	87.00	90.00	93.00
Bolt Tensional Stiffness	CL	N/mm	1.83E+05		
Boss stiffness (cylindrical sections)	CF	N/mm	7.65E+05		
Total Clamped Length	-	mm	33.00		
Length for PLASTIC Deformation	-	mm	21.00		
Relative Plastic Deformation	-	%	0.2036%		0.4777%

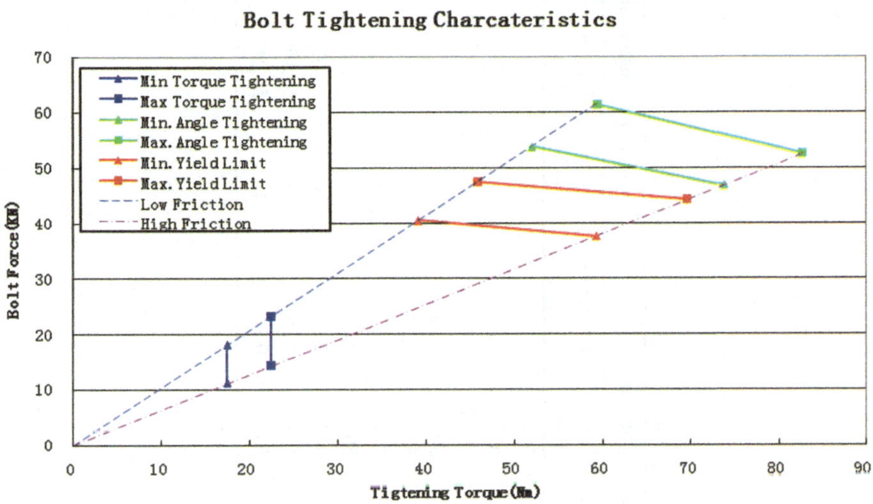

Fig. 2 Result of the bolt tightening characteristics

Table 2 Test results of the elongation

No	Original length (mm)	First time tightening			Second time tightening			Third time tightening			Over all
		Torque (NM)	Length (mm)	Elongation ratio (%)	Torque (NM)	Length (mm)	Elongation ratio (%)	Torque (NM)	Length (mm)	Elongation ratio (%)	Elongation ratio (%)
1	56.02	59	56.08	0.29	–	–	–	–	–	–	–
2	56.11	58	56.21	0.48	–	–	–	–	–	–	–
3	55.92	58	56.02	0.48	–	–	–	–	–	–	–
4	56.01	57	56.07	0.29	–	–	–	–	–	–	–
5	56.09	51	56.13	0.19	52	56.18	0.24	52	56.23	0.24	0.67
6	56.01	56	56.04	0.14	58	56.1	0.29	61	56.15	0.24	0.67
7	56.08	44	56.11	0.14	44	56.16	0.24	48	56.19	0.14	0.52
8	55.97	56	56.03	0.29	56	56.07	0.19	53	56.12	0.24	0.71
9	56.02	64	56.08	0.29	61	56.19	0.52	59	56.25	0.29	1.10
10	56.05	45	56.09	0.19	46	56.21	0.57	52	56.25	0.19	0.95

Table 2. The relative elongation distributed from 0.14 to 0.57 % after tightening, which was roughly coincident with calculation result. The small difference mainly because the friction coefficient 0.08–0.14 was given by experience.

4 Conclusion

Torque angle tightening make the high performance bolt works at the yield limit point, whose preload doesn't influenced by environments but only depends on material. Furthermore, the preload scatter was reduced from 50 % using torque tightening to 15 %, so that the usage ratio of the bolt's material will increase. The bolt joint will get the benefit of lightweight using this method.

Torque angle tightening is an indirect method of length measurement, since the linear deformation of the bolt over the pitch of the thread is (theoretically) directly proportional to the angle of rotation covered. We find out how to calculate the plastic elongation of the bolt after tightening when both elastic and plastic elongation occurs, which can be only tested by experiments after prototype.

References

1. Daxian C Handbook of mechanical design, 5th edn. Vol 2
2. Systematic calculation of high duty bolted joints, joints with one cylindrical bolt VDI2230 part 1, 2003
3. Zhang Q (2003) Research of tightening technology of cylinder head bolt, part 2, automotive technology
4. Li W, Zhuang M (2005) Research on the application of angle tighten at yield area. In: Automotive powertrain conference, 2005

Computational Study of Soot Entrainment via Thermophoretic Deposition and Crevice Flow in a Diesel Engine

Shin Mei Tan, Hoon Kiat Ng and Suyin Gan

Abstract Soot deposition on cylinder liner and the entrainment into engine oil have been correlated to oil starvation and damage of the engine piston. In this computational work, different injection strategies are implemented to investigate their respective effects on spatial evolution of combustion soot and the soot entrainment process in a light-duty diesel engine, taking into account the thermophoretic soot deposition on the cylinder liner as well as entrance of soot into the crevice region. Numerical computation of diesel combustion is undertaken by means of linking a plug-in chemistry solver namely, CHEMKIN-CFD into ANSYS FLUENT 12, a commercial Computational Fluid Dynamics (CFD) software. A chemical reaction mechanism of *n*-heptane combined with soot precursor formation mechanism is employed as the diesel surrogate fuel model. Here, turbulence-chemistry interaction is represented by the Eddy-Dissipation Concept (EDC) model. The inclusion of top land volume in the computation mesh and implementation of crevice model in the CFD solver allow inspections of soot entrainment. Soot mass in the top land volume obtained from the simulation is used to represent soot mass transport into crevice via blowby. By assuming thermophoresis as the primary mechanism of soot deposition on the liner, the mass of soot deposited is obtained through calculation of thermophoretic deposition

F2012-A06-017

S. M. Tan (✉) · H. K. Ng
Department of Mechanical, Materials and Manufacturing Engineering,
The University of Nottingham Malaysia Campus, Jalan Broga,
43500 Semenyih, Selangor, Malaysia
e-mail: keyx9tsm@nottingham.edu.my

S. Gan
Department of Chemical and Environmental Engineering,
The University of Nottingham Malaysia Campus, Jalan Broga,
43500 Semenyih, Selangor, Malaysia

SAE-China and FISITA (eds.), *Proceedings of the FISITA 2012 World Automotive Congress*, Lecture Notes in Electrical Engineering 190,
DOI: 10.1007/978-3-642-33750-5_10,

velocity. During the closed cycle combustion process, the mass of soot deposited on liner via thermophoresis is more dominant than those entrained into crevice region through blowby. Variation of start of injection (SOI) does not have considerable effect on the amount of soot entrainment. On the other hand, the amount of soot entrainment into engine oil is not favourable when the split-main injection is employed with late SOI, although the total soot formed during the combustion is reduced. While thermophoresis mechanism relies on the temperature gradient at the wall, the soot concentration near the wall and its temporal evolution are deduced to have a more significant effect on the deposited soot mass. Variation of the injection strategy is found to affect the soot entrainment process by influencing the in-cylinder gas motion, the location of combustion and the evolution of soot cloud. This simulation study provides an insight to the effects of injection parameters on key in-cylinder, which can assist in engine design for minimising soot deposition in the future.

Keywords Computational fluid dynamics · Diesel combustion · Thermophoresis · Soot entrainment · Crevice flows

1 Introduction

The regulations of diesel emissions have prompted concentrated efforts in recent years in improving engine designs, but certain modifications have shown to have detrimental effects on other pollutant species and level of contaminant in engine oil. It is well known that high soot concentration in engine oil can clog filters and pyrolise into hard carbon deposit, which eventually lead to oil starvation and damage to the engine. Additionally, the viscosity of engine oil is also raised, resulting in higher friction and invariably an increase in engine wear rate. During normal diesel engine operations, high soot concentrations are produced within the combustion chamber, and often exceed tenfold the exhaust concentrations. Within few hundred microns from the chamber wall, the local temperature is much lower and the soot particles in this region may escape from oxidation [1].The existence of this temperature gradient gives rise to the thermophoretic process where the particles move towards the region of lower temperature and deposit on the chamber wall. Thermophoresis is found to be the major contributor of the soot deposition on the in-cylinder surfaces [2]. These soot particles are eventually scraped off by moving piston and entrain into the engine oil through the crevice during exhaust blowdown [3]. Another pathway of soot entrainment into the engine oil is via crevice blowby, where the bulk mass of combustion product is transferred into the crankcase region through the crevice gap between the piston and the cylinder liner.

Table 1 Test engine and injector specifications

Displacement per cylinder (L)	0.5
Compression ratio	18.2
Bore x Stroke (cm)	8.6 × 8.6
Piston bowl volume (cm^3)	19.29
Connecting rod length (cm)	16.0
Number of injector holes	6
Nozzle diameter (mm)	0.149

In cylinder soot formation is one of the most complicated processes occurring inside a diesel engine, and is strongly dependent on the evolution of the combusting fuel spray. In order to investigate the phenomena of soot deposition and entrainment, it is crucial to evaluate the effects of the operating strategies on spatial and temporal evolution of soot. Nevertheless, in situ measurements to elucidate these processes may not be economically feasible to be set up. As such, high fidelity CFD models are used for numerical study to allow visualisation into the complex processes of combustion, emissions and transport processes. The objective of this work is to computationally study the effects of varying injection parameters on the soot entrainment process into engine oil in a light duty diesel engine, taking into account the thermophoretic deposition on the liner and the soot entrance into the crevice region. The injection strategies include single and split-main injection with different SOI, which have been developed by various researchers to extend the operating load range and improve emissions in diesel engine [4, 5].

2 Engine Geometries and Operating Conditions

In this study, the simulation is conducted based on a test engine with the specifications shown in Table 1. This simulation focuses on the closed part of the engine cycle, with fixed engine speed of 1600 rev/min. The initial conditions are set to match the experimental setup, which are pressure of 1 atm and temperature at 313 K. The fuel injection scheme comprises a 0.52 mg/cycle pilot injection, followed by main fuel injection of 25 mg/cycle at 25 crank angle degree (CAD) after the pilot injection. Two single injection cases and two split-main injection cases of different SOI are selected for illustration of the observed effects on soot entrainment. The SOI are set at -6 CAD after top dead centre (ATDC) for early injection case and +2 CAD ATDC for late injection. The split-main injections have dwell period of 0 CAD between the two main injections pulses, and is denoted as the close-coupled injection. The limits of each of the injection variable for this study are selected based on the permissible engine operating limits and the capabilities of the experimental setup. A ratio of 60/40 is selected for the first

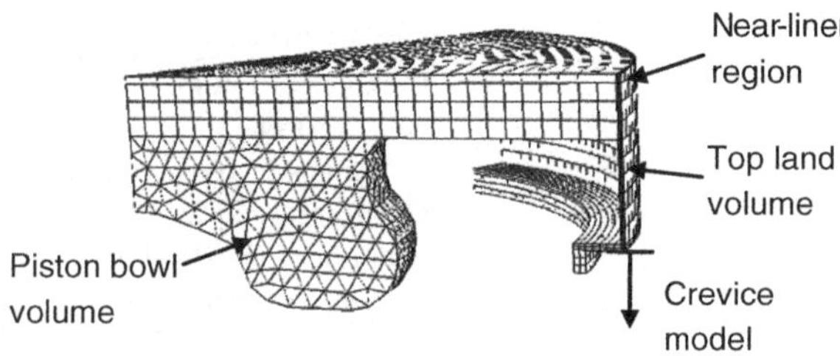

Fig. 1 Computational domain of the combustion chamber

injection pulse fuel ratio to the second pulse in the close-coupled injection as it has been found to give superior reduction in the exhaust soot emission.

3 Numerical Models

Numerical computation is conducted using ANSYS FLUENT 12, a commercial CFD software with the association of CHEMKIN-CFD as the chemistry solver. As shown in Fig. 1, the computational mesh of 60° sector of the combustion chamber is utilised in this study due to the symmetry imposed by the six equally spaced injector nozzle holes. The mesh is created with cell size of 1.5 mm. Grid independent result was achieved with this setting and further refinement in the resolution was found to give insignificant improvement in the predicted results [6]. An additional grid is introduced to account for the volume above the top ring which includes the top land and the top groove, to represent the crevice region and allow inspections of the soot entrainment process. Finer cell size of 0.11 mm thickness is utilised in this region as well as the layers adjacent to the liner. Below the top ring, a zero-dimensional ring-flow crevice model in ANSYS FLUENT is activated to simulate the crevice flow effect. Incorporation of crevice model has been found to affect the axial flow velocity near the liner and the piston head [7], which has a crucial effect on species transport and subsequently the soot entrainment.

The fuel spray breakup is modelled by the Kelvin–Helmholtz/Rayleigh–Taylor (KH-RT) breakup model with the value of breakup time constant, B_1 set as 10 [8]. The wall-jet model is applied as the discrete phase boundary condition for the combustion chamber surfaces. Turbulence is solved using the Renormalisation Group (RNG) k-ε turbulence model, and the Standard Wall Function is implemented for the near wall treatment. A reduced mechanism of n-heptane, the Nottingham Diesel Surrogate (NDS) [9], consisting of 44 species and 109 reactions is employed in this study to calculate the chemical kinetics of the diesel fuel oxidation and the formation of soot precursor species. The reduced mechanism has been validated under 48 shock tube conditions, in which the ignition delay (ID) timings are reproduced [9].The chemistry of the n-heptane is integrated with CFD solver through CHEMKIN-CFD. The Eddy-Dissipation Concept (EDC) model is implemented to incorporate the finite-rate chemistry with turbulent mixing, where the chemical reactions are assumed to occur in fine scales of small turbulent structure. The Moss-Brookes soot model and the Fenimore-Jones oxidation model in ANSYS FLUENT are implemented to predict soot formation and oxidation processes, respectively. In this model, the nucleation of soot precursor, coagulation and surface growth are the subjects to the solution of soot particle concentration and soot mass fraction. For a balance between small chemistry size and improved representation of soot formation process, the soot precursor species for soot inception is set as benzene ring, while acetylene is exploited as the surface growth species [10]. As proposed by [11], the mass density of soot and the mass of incipient soot particle are set at 2000 kg/m^3 and 1200 kg/k mol, respectively for application with higher hydrocarbon fuels. The model constants are set at the values given by [12].

3.1 Soot Deposition Mechanism

Thermophoretic deposition occurs when small particles suspended in gas experience a force in the direction opposite of the temperature gradient. In this study, the soot deposition rate at the cylinder liner is obtained by using the equation derived by [13], where the force in the y direction, F_{ty} is calculated by

$$F_{ty} = \frac{12\pi\mu v R_s C_s\left[\frac{k_g}{k_{soot}} + C_t Kn\right]}{(1 + 3C_m kn)\left[1 + 2\frac{k_g}{k_{soot}} + 2C_t Kn\right]}\left(\frac{1}{T}\right)\frac{dT}{dy} \tag{1}$$

where μ is the gas dynamic viscosity, and v is the kinematics viscosity. R is the particle radius, T is the temperature of the gas, while k_g and k_{soot} are the thermal conductivity of the gas and soot particle, respectively. k_{soot} is assumed constant at 0.26 W/m.K [14], and R_s is assumed to be a constant of 50 nm [15]. The values of the constants C_s, C_t and C_m are 1.17, 2.18 and 1.14, respectively. The Knudsen number is

$$Kn = \frac{\lambda}{R_s} \tag{2}$$

and λ is the gas mean free path, defined as

$$\lambda = \frac{2\mu}{\rho}\sqrt{\frac{\pi}{8R_u T}} \tag{3}$$

To determine the thermophoretic velocity, the thermal force is equated to the drag force of the particle [16], which is given by

$$F_{ty} = \frac{6\pi\mu V R_s}{C_c} \tag{4}$$

where V is the thermophoretic velocity. The Cunningham slips correction, C_c

$$C_c = 1 + Kn\left(A + Be^{-C/Kn}\right) \tag{5}$$

where the constants A, B and C have the values of 1.2, 0.41 and 0.88, respectively. Substituting Eq. 4 into Eq. 1 gives the thermophoretic velocity of the particle

$$V_{ty} = \frac{2C_s C_C v\left[\frac{k_g}{k_{soot}} + C_t Kn\right]}{(1 + 3C_m kn)\left[1 + 2\frac{k_g}{k_{soot}} + 2C_t Kn\right]}\left(\frac{1}{T}\right)\frac{dt}{dy} \tag{6}$$

and the equation can be written as

$$V_{ty} = v\mathbf{K_t}\left(\frac{1}{T}\right)\frac{dt}{dy} \tag{7}$$

where K_t is the thermophoretic constant.

The mass of soot deposited is determined by calculating the swept volume of soot from the thermophoretic deposition velocity, multiplied with the computed soot concentration at the gas phase region adjacent to the liner [16]. The volume of the swept area is the product of the element surface area and the distance travelled by the particle. It is assumed that all the soot contained in the swept volume is deposited on the wall. Several experimental studies on soot deposition [2, 17, 18] found a non-linear relationship between the soot deposition rate and the engine run time, where the deposition rate is high at the initial cycles and decreases as the layer of deposited soot increase in thickness. As more soot mass is deposited, a higher portion of the deposited soot can be removed from the wall by blow-off, thus the deposition rate of one cycle is able to give an estimation of the gross rate of soot deposition.

4 Validation

Results of the simulated test cases are first validated against experimental data, which include the pressure trace, heat released rate (HRR) and tailpipe soot emission as shown in Fig. 2. Comparisons between the experimental and simulated results show that the over predictions in the ID of the main combustion event are maintained within 1.5 CAD. The percentage error of the peak pressure is less than 3.8 % with the offset in the peak pressure location of less than 1 CAD. Similar HRR curves are also observed between the simulation and experimental data for all cases, and the locations of the peak HRR are matched with a maximum offset of 1 CAD. The predicted and experimental exhaust soot density are normalised to the same order of magnitude, by setting the +2 CAD SOI single injection case which produces the highest soot value as the baseline. This provides a clear comparison of the trend of soot emission when injection variable is varied. The simulated soot emissions are found to correlate well with the experimental results. Generally for

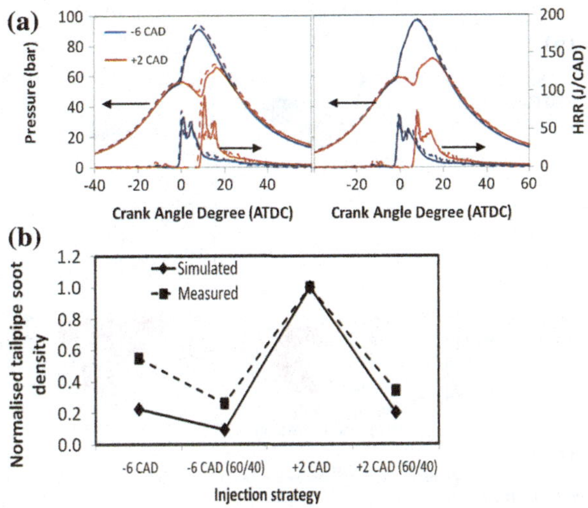

Fig. 2 **a** Experimental (*dotted lines*) and simulated (*solid lines*) pressure and HRR profiles for single injection cases (*left*) and close-coupled injection cases (*right*), and **b** normalised tailpipe soot density

all the test cases, the simulated results of the pressure, HRR and the soot emissions formation agree well with the obtained experimental data. This shows that the selected models have been appropriately calibrated for the simulation of this combustion system, and the results are exploitable for the subsequent investigation of the in cylinder effects.

5 Results and Discussions

Discussion on the effects of implementing different injection strategies on soot transfer is presented here using the illustrations of soot concentration distribution in the combustion chamber. For ease of visualisation, the images of predicted soot concentration are shown in Fig. 3a as the side view projection of the 3-D soot iso surfaces, where the soot cloud layers are featured as semi-transparency. For each case, the soot images are obtained at +20 and +40 CAD after start of injection (ASI) to permit the visualisation of soot propagation.

The analysis also comprises the thermophoretic soot deposition rate on the liner and soot mass in the top land volume. Results associated to the soot entrainment process are shown in Fig. 4. The mass of soot entering the near-liner region is obtained by monitoring the total soot mass in the volume next to the cylinder liner with thickness of 0.335 mm, which is equivalent to the crevice gap. It is then coupled with the temperature gradient on the liner surface to evaluate the effects on the thermophoretic soot deposition rate on the liner. This serves to reflect the potential amount of soot entrained into engine oil, as the soot deposited on the liner can be transferred into the crankcase due to the scraping action by the piston ring. The simulated mass of soot in the top land volume is employed to represent

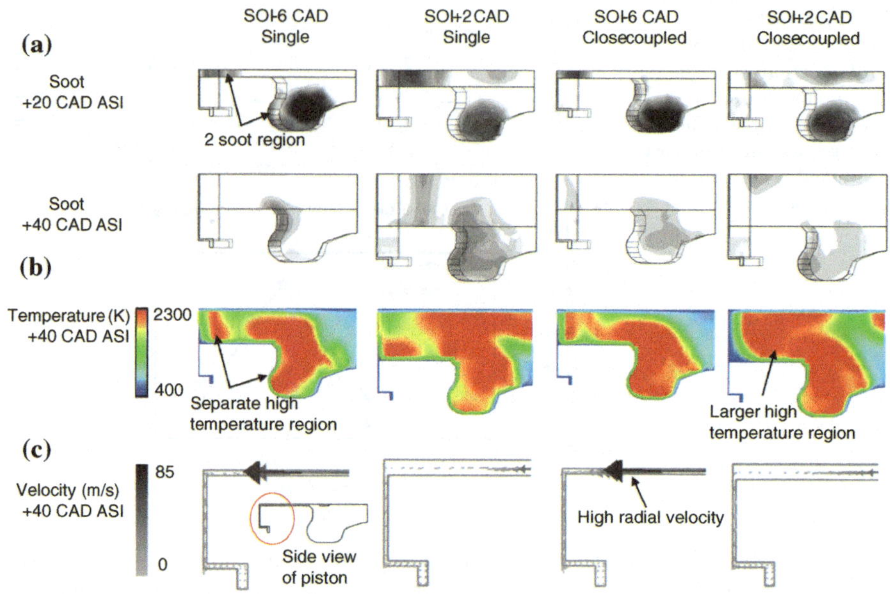

Fig. 3 **a** Spatial evolution of soot, **b** temperature contour and **c** velocity vector at the cross sectional plane along the spray axis for all test cases

the level of soot entrance into the crevice region via blowby process. This is because soot which accumulates in the crevice region can be transported into engine oil through the bulk gas motion of the crevice flow. The soot mass in crevice at exhaust valve opening (EVO) and the total deposited soot mass on the liner are then summed up to represent the total soot entrainment into engine oil. Although the discussions presented here are based on the results of the selected test cases, the key features drawn on the observed trends are applicable to other simulated cases which are not shown here for simplicity.

5.1 Single Injection with Varying SOI

Effects of varying the SOI on the amount of soot entering the near-liner region are evaluated here. From the graph of soot mass near liner in Fig. 4a, the soot entry is divided into two phases, which are immediately after ignition and at the later stage of expansion stroke. In addition, based on the visual examination of soot distribution in the combustion chamber at +20 CAD ATDC in Fig. 3a, it is observed that the soot formed occupies two regions, which are inside the piston bowl and near the liner. After the drop of the mass of first soot entry due to soot oxidation, soot which was produced inside the piston bowl has travel outwards and eventually approaches the liner, thus resulting in the subsequent increase of the soot mass

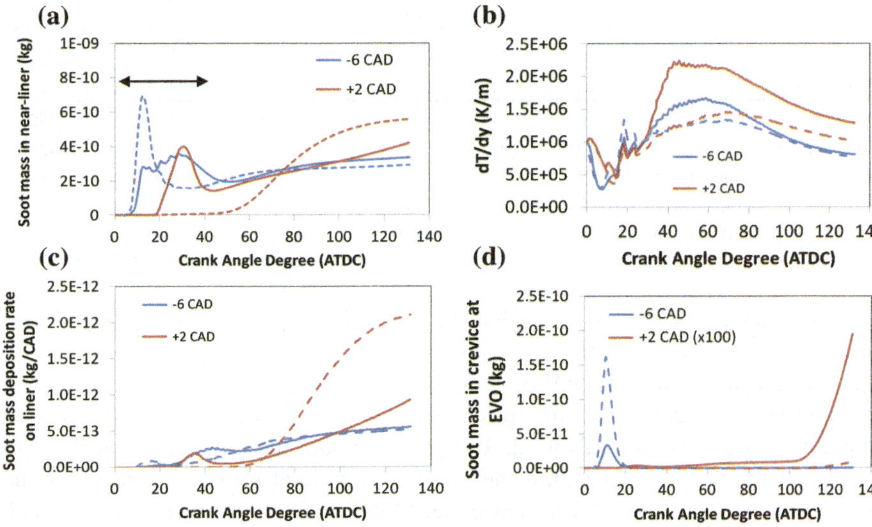

Fig. 4 **a** Soot mass in near-liner region, **b** temperature gradient at the liner, dT/dy, **c** soot mass deposition rate on the liner, and **d** soot mass in crevice for all test cases

near the liner. The explanation to the first soot entry is that after top dead centre (TDC), the space in the expanding swept volume is replaced by the charge from the piston bowl, thus accelerating the product of combustion towards the liner at an early stage. When the SOI timing is at -6 CAD, soot entry into near-liner region occurs earlier at +5 CAD ATDC. The first phase of soot entry in the late injection case is found to last for a shorter duration, which is from +20 CAD to +40 CAD ATDC. Both cases demonstrate similar amount of peak soot mass in the near-liner region, although in the late injection case, lower gas velocity towards the liner is found at +10 CAD ASI, which is the onset of soot entry. This can be observed in the velocity vector in Fig. 3c. Due to the delayed injection, a portion of the fuel escapes from the piston bowl due to the piston downward motion and is delivered into the swept volume, resulting in fuel combustion near the liner. This is verified by the high temperature region near the liner from the temperature contour in Fig. 3b. Based on the observation on spatial evolution of soot, the late injection case demonstrates the higher soot concentration in the cylinder, thus an increase is seen in the near-liner soot mass towards the EVO. From the graph of temperature gradient on the liner surface in Fig. 4b, +2 CAD SOI case demonstrates the highest value. Regardless the difference in the temperature gradient, the soot mass deposition rate on the liner demonstrates identical curve to the near-liner soot mass, where the value in the late injection case exceeds that of the early injection case at approximately +100 CAD ATDC. The lower deposition rate on the liner at earlier stage of the late injection case is compensated by the escalated rate towards the end. This results in similar amount of total deposited soot mass in both early

Table 2 Results of soot entrainment for all test cases

Injection strategy	Single		Close-coupled	
SOI (CAD ATDC)	−6	+2	−6	+2
Total soot mass ($\times 10^{-8}$ kg)	5.35	23.60	2.24	4.82
Deposited soot mass on liner ($\times 10^{-11}$ kg)	3.97	3.73	3.79	8.91
Soot mass in crevice ($\times 10^{-13}$ kg)	5.12	19.55	4.91	0.90
Total soot mass entrained ($\times 10^{-11}$ kg)	4.02	3.93	3.84	8.92
Percentage of total soot entrained (%)	0.08	0.02	0.17	0.19

and late injection cases as recorded in Table 2, which are $3.97 \times 10^{-11}$and 3.73×10^{-11} kg, respectively.

The effect of varying the SOI on the soot mass in crevice does not demonstrate similar trend to the deposited soot mass on liner, where retarding the SOI increases the crevice soot as seen in Table 2. The graph of soot mass in crevice in Fig. 4d shows that the soot entrance into the crevice is divided into two phases for the early injection case, as similar to that in near-liner region. However, after the peak at +10 CAD ATDC, nearly all the soot which enters the crevice during the first phase is removed. The reduction in the soot mass is resulted from the compressed gas in the crevice region being expelled, as the piston moves downwards during the expansion stroke. The second phase of soot mass entrance into crevice occurs near EVO. Contradicting to the first soot entrance event, retarding the SOI produces the highest amount of crevice soot mass at EVO, which is approximately 4 times the level in -6 CAD SOI cases. The soot mass in crevice is found to correlate to the soot concentration in the combustion chamber.

Total soot entrainment is obtained by summing up the total deposited soot mass on the liner and the soot mass in crevice. Observing from Table 2, although the −6 CAD SOI gives rise to a higher soot mass deposition on the liner, the higher crevice soot mass in the late injection case resulted in the amount of total soot entrainment which is only 2 % lower than that in the early injection case.

5.2 Close-Coupled Injection with Varying SOI

In this section, the close-coupled injection cases with fuel mass ratio of 60/40 are compared against the single injection case for both SOI timings at −6 and +2 CAD. Implementing the close-coupled injection strategy has reduced the total soot mass for both cases, as the separation of the main injection improves air entrainment into the fuel spray, and the soot formation region at the spray tip is not replenished when the injection is paused.

For the late injection case, the first phase of soot entry into the near-liner region is not observed in the close-coupled injection case, but the soot mass is found to increase more significantly at the later stage. By employing close-coupled injection, the spray penetration length is reduced due to the pause in the injection which causes momentum losses of the spray particles. Thus the injected fuel resides in the piston bowl, instead of escaping to the swept volume as occurred in the single injection case. From the temperature contour in Fig. 3b, the separate high temperature region near the liner seen in the single injection case is not replicated in the close-coupled injection case, but the combustion region in the piston bowl is found to be extended further towards the liner. This is due to the turbulence induced by the separation in the fuel delivery which improves the air–fuel mixing and increases the volume of combustible mixture. It can be seen from Fig. 3 that at +40 CAD ASI, the soot cloud dwells at a closer proximity to the liner although smaller soot volume is found in the close-coupled injection case. Figure 4b shows that the temperature gradient at the liner drops in the close-coupled injection case, which is due to the absence of the separate higher temperature region. However, the large soot mass in the vicinity of the liner promotes higher deposition rate, resulting in the highest amount of total soot deposited among all the cases, which is 8.9×10^{-11} kg as recorded in Table 2.

In the early injection case, the effect of the injection strategy on the high temperature region is similar to that in the late injection case, as depicted in Fig. 3b. Although the separate high temperature region is not prominent, high peak of soot mass near the liner is seen at +10 CAD ATDC, followed by a rapid drop. This is resulted by the higher radial velocity in the squish region, as seen in the velocity vector in Fig. 3c. The difference in the velocity is attributed to the higher in-cylinder pressure shown in Fig. 2a, which creates a large pressure gradient between the combustion chamber and the crevice region. As opposed to the late injection case, implementing close-coupled injection at SOI of -6 CAD does not increase the soot mass in the near-liner region towards the end of expansion stroke. This is attributed to the high soot oxidation rate and its total soot mass which is the lowest among all the case as given in Fig. 1b, thus results in the soot cloud to decrease rapidly in size, limiting the amount of soot that reaches the liner. From Table 2, the total deposited soot mass on the liner for both single and close-coupled injection cases at −6 CAD SOI are 3.97×10^{-11} and 3.79×10^{-11} kg, respectively. Therefore, variation of the injection strategy at early SOI does not impose large effect on the soot deposition.

From Table 2, employment of close-coupled injection with late SOI is found to decrease the soot mass in crevice as corresponding to the total soot mass. On the other hand, only a reduction of 4 % is seen in the early injection case, due to a certain amount of soot that is trapped in the crevice after the first phase of entrance. Similar to the soot mass in near-liner region, higher peak mass of the first entry of soot into the crevice is observed for close-coupled injection case in Fig. 4d.

The total mass soot entrainment for the close-coupled injection cases is similar to the soot deposited on the liner due to low soot mass in crevice. Comparing the percentage of total soot mass being entrained into the engine oil from Table 2, both the close-coupled injection cases have the higher values, which are approximately 0.2 %. Although there is high percentage of soot entrained, implementing the close-coupled injection at advanced SOI does not increase the total soot mass entrained due to the lower overall soot mass in the combustion chamber.

6 Conclusion

A computational investigation on the effect of different injection strategies on the soot entrainment into engine oil in a light-duty diesel engine is conducted. In all the test cases, soot deposited on the liner is found to occupy a larger fraction of the total soot entrained and is 20–1,000 times higher than the mass of soot in the crevice. For single injection cases, delaying the SOI does not affect the mass of soot entrainment significantly, but higher amount of crevice soot is found in the late injection case. Although employment of the close-coupled injection reduces the total soot mass, the turbulence induced increases the volume of combustible mixture, extending the soot cloud towards the liner. The high concentration of soot near the liner increases the deposition rate considerably. This effect is not apparent in the early SOI case, due to decreased amount of total soot. On the other hand, the difference of the temperature gradient between the test cases does not have significant effect on the soot entrainment as compared to the soot concentration near the liner. As a conclusion, the evaluation of the soot entrainment process in this study has demonstrates the interdependence between the spatial and temporal evolution of soot as well as the development of the combustion on the outcome. For more detailed prediction in further studies, wall thickness of the cylinder liner can be included in modelling the boundary condition to improve the heat transfer prediction and to investigate the associated effect on the soot deposition.

References

1. Eastwood P (2008) Particulate Emissions from Vehicles. Wiley, Chichester
2. Suhre BR, Foster DE (1992) In cylinder soot deposition rates due to thermophoresis in a direct injection diesel engine . SAE Paper 921629
3. Agarwal D, Singh SK, Agarwal AK (2011) Effect of exhaust gas recirculation (egr) on performance, emissions, deposits and durability of a constant speed compression ignition engine. Appl Energy 88(8):2900–2907
4. Gan S, Ng HK, Pang KM (2011) Homogeneous charge compression ignition (hcci) combustion: implementation and effects on pollutants in direct injection diesel engines. Appl Energy 88(3):559–567
5. Suh HK (2011) Investigations of multiple injection strategies for the improvement of combustion and exhaust emissions characteristics in a low compressions ratio (CR) engine. Appl Energy 88(12):5013–5019
6. Pang KM, Ng HK, Gan S (2012) Investigation of fuel injection pattern on soot formation and oxidation processes in a light-duty diesel engine using integrated CFD-reduced chemistry. Fuel 96:404–418

7. Rakopoulos CD, Kosmadakis GM, Dimaratos AM, Pariotis EG (2011) Investigating the effect of crevice flow on internal combustion engines using a new simple crevice model implemented in a CFD code. Appl Energy 88(1):111–126
8. Beale JC, Reitz RD (1999) Modeling spray atomization with the Kelvin-Helmholtz/Rayleigh-Taylor hybrid model. Atomization Spray 9:623–650
9. Pang KM, Ng HK, Gan S (2011) Development of an integrated reduced fuel oxidation and soot precursor formation mechanism for CFD simulations of diesel combustion. Fuel 90(9):2902–2914
10. Pang KM, Ng HK, Gan S (2012) Simulation of temporal and spatial soot evolution in an automotive diesel engine using the moss-brookes soot model. Energy Convers Manage 58:171–184
11. Hall RJ, Smooke MD, Colket MB (1997) Physical and chemical aspects of combustion. Gordon and Breach, Amsterdam
12. Moss JB, Perera SAL, Stewart CD, Makida M (2003) radiation heat transfer in gas turbine combustors . In: Proceedings of 16th international symposium on air breathing engines, Cleveland
13. Talbot L, Cheng RK, Schefer RW, Willis DR (1980) Thermophoresis in a heated boundary layer. J Fluid Mech 101(4):737–758
14. Tree DR, Foster DE (1994) Optical soot particle size and number density measurements in a direct injection diesel engine. Combust Sci Tech 95:313–331
15. Han Z, Reitz RD (1997) A temperature wall function formulation for variable-density turbulent flows with application to engine convective heat transfer modelling. Int J Heat Mass Transfer 40(3):613–625
16. Wiedenhoefer JF, Reitz RD (2002) A multidimensional radiation model for diesel engine simulation with comparison to experiment. J Num Heat Transfer 125:27–38
17. Tree DR, Svensson KI (2007) Soot processes in compression ignition engines. Prog Energy Combust Sci 33:272–309
18. Kittelson DB, Ambs JL, Hadjkacem H (1990) Particulate emissions from diesel engines: influence of in-cylinder surface. SAE technical paper No. 900645, international congress and exposition detroit michigan

Experiment and Numerical Analysis of Temperature Field of Cylinder Head Based on a GW4D20 Diesel Engine

Baoxin Zhao, Dingwei Gao, Jingqian Shen, Zheng Zhao, Hao Guan, Gang Liu and Ying Guan

Abstract *Research and/or Engineering Questions/Objective*: The thermal stress field need to be evaluated for the problem of superheat of cylinder head. In this paper, the temperature field of the cylinder head of GW4D20 diesel engine was analyzed by experiment and simulation. *Methodology*: By the coupled finite element method (FEM), the heat transfer simulation was carried out for the combustion chamber. The method coupled together the flow of circulating coolant, the convective heat transfer of gas in-cylinder, and the heat conduction of main parts of diesel engine, such as cylinder block, cylinder head, cylinder gasket, valve seat, valve guide and so on, which were regarded as a coupled unit. The same 3D Fluid–Solid Coupled simulation analysis was also made for an improved scheme. Finally, the results of the simulation would be compared with the experimental results based on hardness plug method. *Results*: Temperature field of head was calculated by 3D Fluid–Solid Coupled simulation analysis exactly. The problem of superheat of GW4D20 diesel engine cylinder head was solved finally with the change of increasing baffle for each cylinder. *Limitations of this study*: In addition to the study on head temperature field, it also needed to consider the effect of other assembly parts and thermal mechanical fatigue of the cylinder head. The lack of experimental data limited the role of FEA in cylinder head design. *What does the paper offer that is new in the field in comparison to other works of the author*: The nucleate boiling was applied to research how to influence the heat transfer coefficient (HTC) of coolant side. Wall temperature of coolant jacket which was obtained by FEA in each case was made as real wall boundary of further

F2012-A06-018

B. Zhao (✉) · D. Gao · J. Shen · Z. Zhao · H. Guan · G. Liu · Y. Guan
Engine R&D Centre, Great Wall Motors Company Limited, Baoding, China
e-mail: engine_cae@gwm.cn

SAE-China and FISITA (eds.), *Proceedings of the FISITA 2012 World Automotive Congress*, Lecture Notes in Electrical Engineering 190,
DOI: 10.1007/978-3-642-33750-5_11,

computational fluid dynamics (CFD) calculation and more actual heat boundary of coolant side were mapped. Finally, real Fluid–Solid Coupled simulation analysis was realized and the temperature field of cylinder head was further assessed. *Conclusion*: The improvement scheme with increasing baffle is more feasible than the original one. And, the simulation results by the Fluid–Solid Coupled method showed a good agreement with the experimental results on an engine test bench.

Keywords Temperature field · Diesel engine · Fluid–Solid Coupled · Hardness plug method · Nucleate boiling

1 Introduction

With the increase in the peak and brake mean effective pressure (BMEP), the cylinder head has become one of the most critical parts in engine durability assessment [1]. Aluminium was used more and more for cylinder head, and the main difficulty is to justify the temperature field reliability based on few experiments. This relationship between simulated reliability and experimental reliability becomes more difficult to establish [2].

By the coupled finite element method (FEM), the heat transfer was simulated based on a combustion chamber. For the method, several elements, such as cylinder block, cylinder head, cylinder gasket, etc., were regarded as a coupled unit. The method will be validated by FE analysis and controlled experiments. And the steady-state analysis of cylinder head of GW4D20 diesel engine will be used to indicate the accuracy and practicality.

2 Main Section

This paper mainly describes the simulation of the temperature field which is a part of the cylinder head thermal stress simulation process. The deployment of the flowchart (Fig. 1) needs to be addressed in order to obtain reliable and accurate results:

- Steady-state heat transfer
- Metallic materials-tensile testing
- FE analysis and temperature field experiment

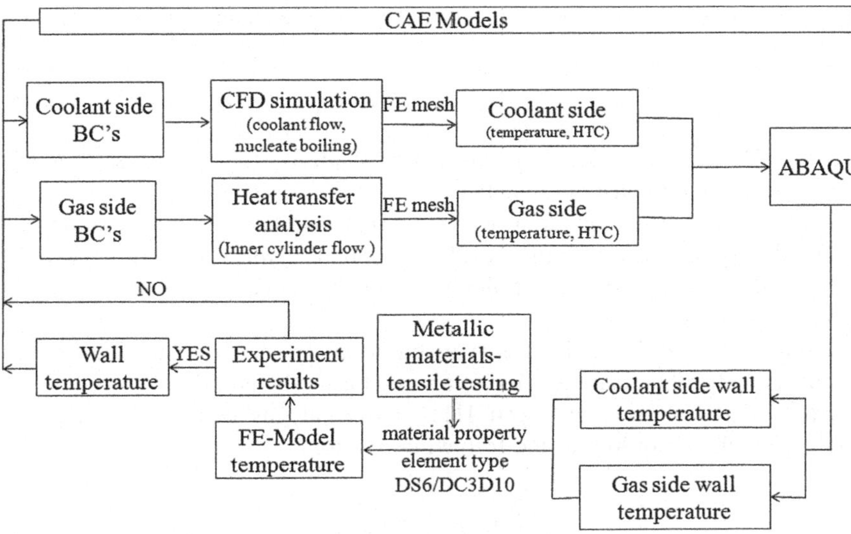

Fig. 1 Calculating flowchart of fluid–solid coupled

2.1 Heat Transfer Theory

2.1.1 Coolant Side Heat Transfer

The complexity of the coolant jacket geometry results from the four-valve per cylinder design and the requirements for the installation of the fuel injector. This makes it difficult to realize optimum cooling conditions throughout the cylinder head [1]. The model is implemented in this paper and it is based on the "Bubble departure lift-off" (BDL) model [3]. The wall heat flux according to the BDL formulation is,

$$q_w = h_{fc}(T_w - T_b) + h_{nb}(T_w - T_s)S_{sub}S_{flow}, \tag{1}$$

where h_{fc} is the forced convection HTC of coolant side, T_w is the coolant side wall temperature, T_b is the fluid temperature, h_{nb} is the nucleate boiling HTC, T_s is the saturation temperature of the fluid, S_{sub} and S_{flow} are the subcooling and flow induced suppression factors [4], respectively. The nucleate boiling HTC implementation is according to the formulation by Forster and Zuber [5].

2.1.2 Gas Side Heat Transfer

Heat transfer is mainly caused by forced convection and the temperature gradient of the working gas to combustion chamber walls. Newton's Law of Cooling is generally used to describe heat transfer. The heat exchange process can be described as follow,

$$q_g = h_g\left(T_g - T_{wg}\right), \tag{2}$$

where h_g is HTC of gas side, T_{wg} is combustion side wall temperature, T_g is gas temperature in the combustion chamber.

For calculating the spatial variations of the HTC and local gas temperature, 3D CFD simulation of one full engine cycle is applied in this paper. But the CFD analysis still indicates several difficulties which include efficient and rapid meshing for moving meshes [1].

The time-averaged data, which is available at the boundary of moving meshes of combustion chamber, is transferred to a static FE mesh and then averaged on the time axes to obtain cycle averaged HTC and local gas temperature [1]. The averaging is done according to the following relation,

$$\begin{aligned} \bar{h}(x) &= \tfrac{1}{720}\int_0^{720} h(\phi,x)d\phi \tfrac{n!}{r!(n-r)!} \\ \bar{T}_g(x) &= \tfrac{1}{\bar{h}(x)720}\int_0^{720} h(\phi,x)T_g(\phi,x)d\phi, \end{aligned} \tag{3}$$

where $\bar{h}$ and $\overline{T_g}$ are the locally time-averaged HTC and the effective gas temperature. Furthermore, h and T_g are the time dependent heat transfer and local gas temperature on the FE mesh.

2.1.3 Steady-State Heat Transfer Process

The calculation of HTC and temperature can be obtained during a steady-state. The heat flux in the solid reads according to the formulation,

$$q_s = \frac{\lambda}{\delta}\left(T_{wg} - T_w\right), \tag{4}$$

where λ is thermal conductivity of the cylinder wall, δ is the wall thickness, T_{wg} is combustion side wall temperature, T_w is coolant side wall temperature.

The steady-state heat transfer process is shown in Fig. 2. For FEM, the sequences normally are: generating only tetrahedrons in solid part of CFD-model, and then at every time step, transferring solid part temperature results to structure elements, executing thermal conduct analysis for getting mid node temperature loads and the temperature distribution within the solid.

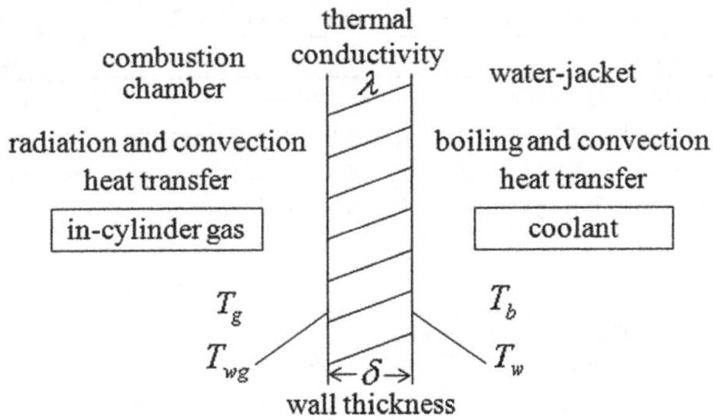

Fig. 2 Heat transfer process between coolant jacket and combustion chamber

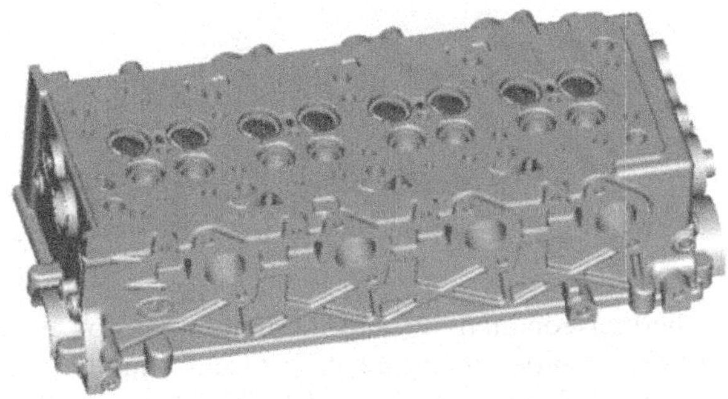

Fig. 3 GW4D20 cylinder head

2.2 Engine Description and Metallic Materials-Tensile Testing

2.2.1 Engine Description

A turbo charged in-line four-cylinder diesel engine is studied in this paper. The cylinder head is made of aluminium silicon alloy (Fig. 3). The combustion chamber is of ω type with four valves per cylinder and a central injector. All the eight exhaust valves and eight intake valves are arranged in-line respectively. Table 1 summarizes the main parameters of GW4D20 diesel engine.

2.2.2 Metallic Materials-Tensile Testing

Material characteristics of aluminium alloy cylinder head (ZL101A) can be obtained by metallic materials-tensile testing. Samples and cylinder head use the same heat treatment in T6 conditions: Solution treatment at a temperature of 535 °C (for a period of 8 h), quenched in water (temperature ranging from 70 to 80 °C, less than 15 s) and heat aged treatment at a temperature ranging from 170 to 180 °C (for a period of 6 h).

Table 1 Main technical parameters of GW4D20 diesel engine

Engine	Four-cylinder, forced water-cooling, GW4D20		
Bore × stroke(mm)	83.1 × 92	Compression ratio	16.7:1
Maximum BMEP(kPa)	1951@1800~2800 rpm	Maximum power(kW)	110@4000 rpm
Maximum torque($N \cdot m$)	310@1800~2800 rpm	Displacement(L)	1.996
Fuel injection system	Common rail	Combustion chamber	ω type

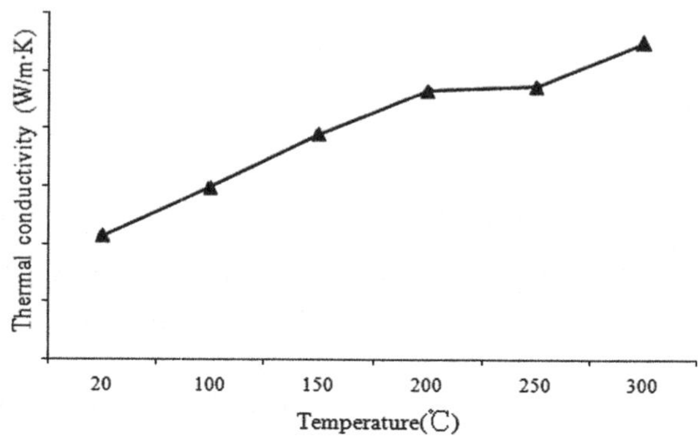

Fig. 4 Thermal conductivity of the test-bars

Thermal conductivity and thermal expansion coefficient can be measured by test-bars in order to establish a "true" simulation. Figures 4 and 5 show the thermal conductivity and thermal expansion coefficient of the material.

2.3 Temperature Field Experiment and FE Analysis

Due to its precision in temperature testing and easy installation on the fire deck face of cylinder head, hardness plug method was implemented to measure temperature field in this paper.

2.3.1 Hardness Plug Method

Hardness plug method utilizes the characteristic of the reducing of hardness with increasing temperature. This decrease in hardness of metal is called "age softening". The phenomenon occurs if the equilibrium phase diagram reveals partial solid solubility of alloying element at higher temperature than lower temperature [6]. To obtain accurate results from the temperature field experiment, hardness measurements shall be done under the same ambient temperature (room temperature).

Ten electron micro hardness plugs are equipped on the fire deck face per cylinder. Figure 6 shows the measurement system in the testing engine.

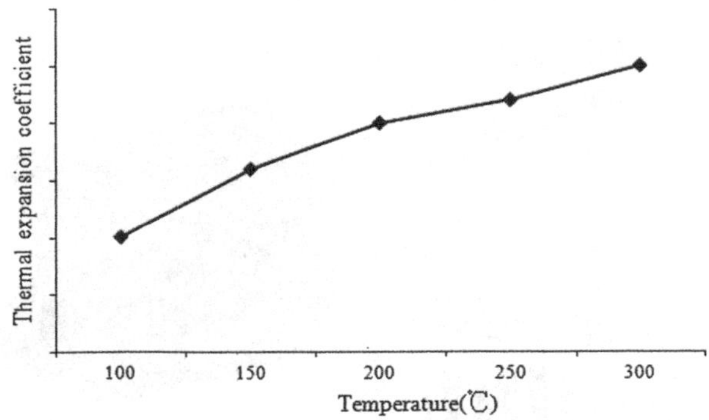

Fig. 5 Thermal expansion coefficient of the test-bars

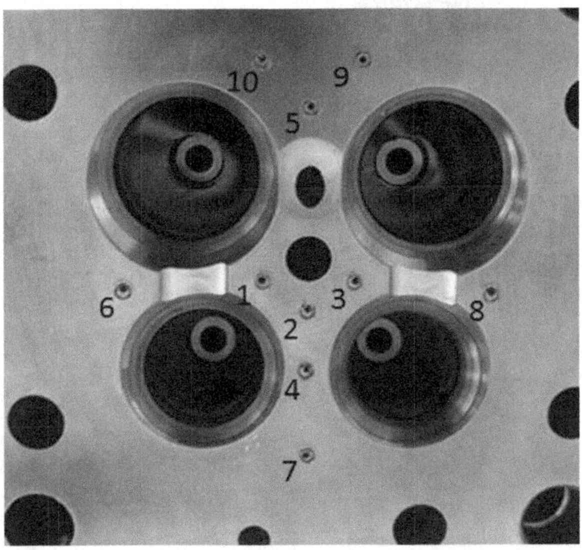

Fig. 6 Locations of the hardness plugs on the fire deck face

2.3.2 Description of Temperature Field Experiment

The experiment was conducted for 2 h after attaining peak power and the testing data was noted per 10 min. Hardness would be measured after the cylinder head dropped to room temperature.

2.3.3 FE Analysis and Simulation Results

To obtain the temperature field of FE-Model, the surface temperature method [7] is applied. The FE-model used to analyze the temperature field consists of about 1,043,402 parabolic tetrahedron elements (Fig. 7).

The wall temperature and forced convection HTC of gas side and coolant side were calculated from CFD simulation results (Figs. 8 and 9). The areas of the

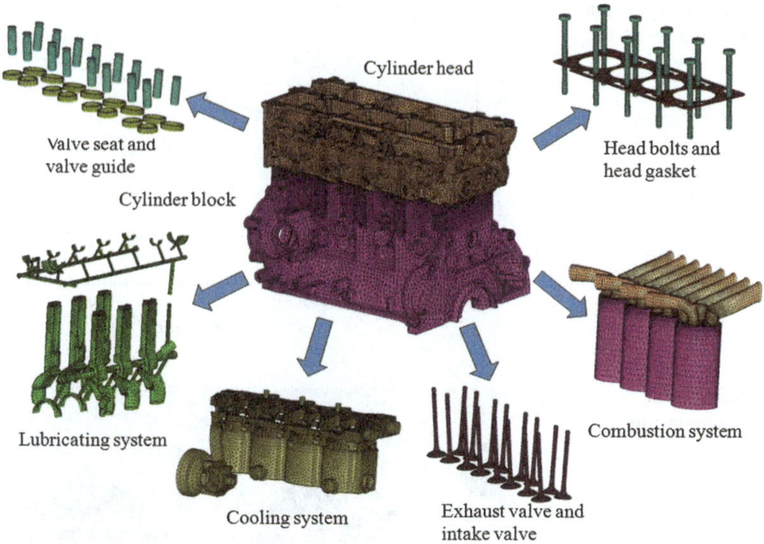

Fig. 7 FE-Model of GW4D20 diesel engine assembly

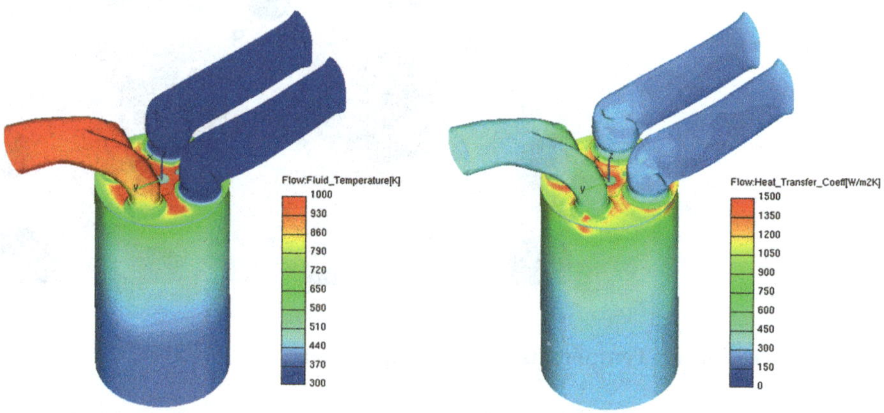

Fig. 8 Temperature and HTC distribution of gas side

structure which were subjected to the oil/air mix were given boundary condition in form of heat transfer with empirical values on the HTC and temperature. The interaction between different parts of the model was modelled with contact surfaces which were given gap conductance values [8].

Simulation results of heat exchange process at maximum load point were transformed into steady-state HTC and gas temperature of the FE mesh. And then, the wall temperature of coolant jacket was fed back to CFD-model as real wall boundary. The analysis was carried out with non-linear FEA-solver

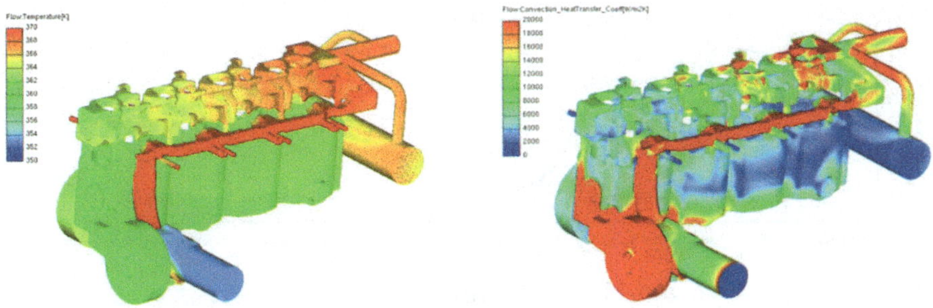

Fig. 9 Temperature and HTC distribution of coolant side

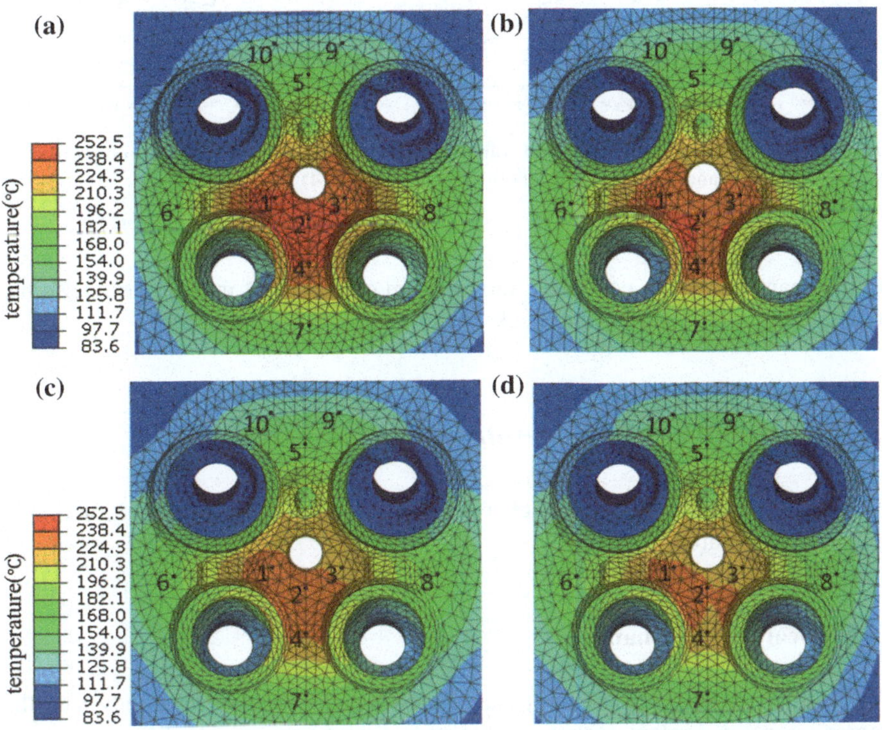

Fig. 10 Local temperature distribution on fire deck face of cylinder head (**a**, **b**, **c**, **d** shows temperature result of cylinder 1, 2, 3, 4)

ABAQUS (Fig. 10). Comparisons between the first experiment results and simulation results were shown in Fig. 11.

It can be observed that the simulated temperature field based on hardness plug method indicates a good agreement with experiment. Due to the fact that cylinder head operating at 250 °C with aluminium silicon alloy will gradually lose its

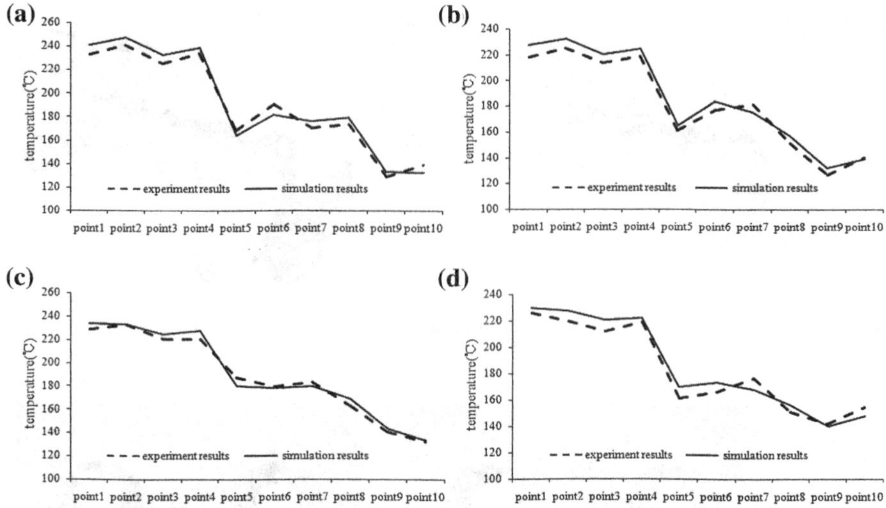

Fig. 11 Temperature field results and simulation results (**a**, **b**, **c**, **d** shows comparison of experiment results and simulation results of cylinder 1, 2, 3, 4)

attained hardness in T6 condition, the problem of superheat may contribute to the cracks in cylinder head bottom deck or in the exhaust valve bridge.

2.4 Improvement of Temperature by FE Analysis

The original scheme of coolant jacket needed to improve cooling condition due to the peak temperature (Fig. 10).

2.4.1 Local Design Change

Two cylinder heads with different characteristics were used in this study, and the improvement of coolant jacket as shown in Fig. 12 was applied to reduce the temperature of bottom deck face. The effects of increasing baffle for each cylinder were investigated by FE analysis.

The second temperature field experiment results and simulation results.

The same Fluid–Solid Coupled simulation analysis was also made for the improved one.

Compared with the first experiment results, temperature gradient of the second simulation results indicated significant reduction on fire deck face (Fig. 13).

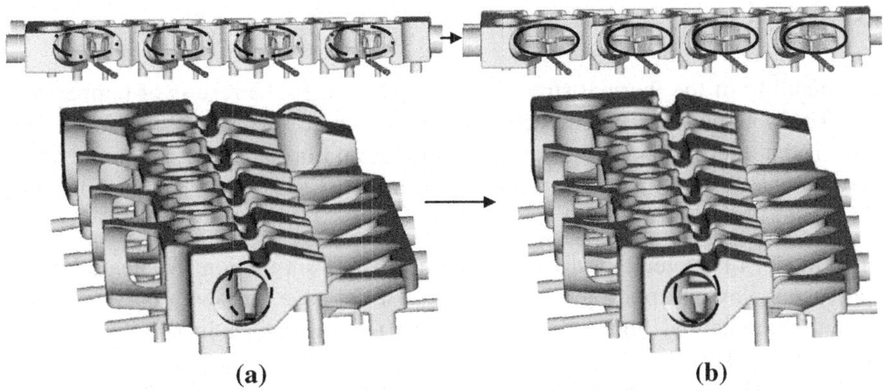

Fig. 12 Scheme of the geometry of the two coolant-jackets (**a**, **b** shows the initial and improved scheme respectively)

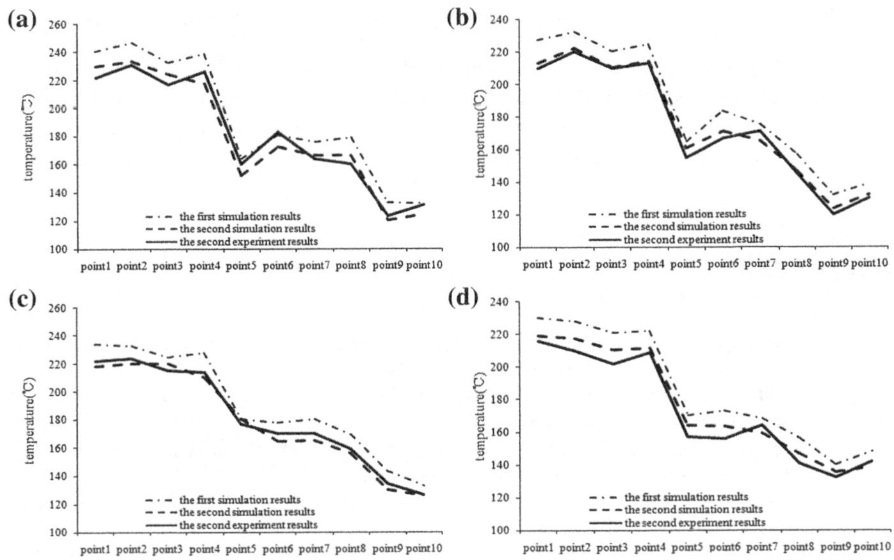

Fig. 13 The second experiment results and simulation results (**a**, **b**, **c**, **d** shows the two simulation result and the second experiment result of cylinder 1, 2, 3, 4)

2.5 *Discussion*

In addition to the study on head temperature field, it'll also need to consider the effect of other assembly parts and thermal mechanical fatigue in the next step. Simultaneously, the lack of experimental data limited the role of cylinder head design by FE analysis.

3 Conclusion

The reliability of the thermal stress is directly linked to the results of temperature field. The obtained agreement between simulation results and measured values confirms that the Fluid–Solid Coupled method is an improvement compared to the conventional method using constant temperature (120 °C) on the coolant side. This method has the following advantages:

- The model of nucleate boiling is applied to CFD simulation. Nucleate boiling processes play an important role in bridge of the nose area with only moderate flow velocities and high heat flux and can't be neglected in solving the heat transfer problem.
- Using the first temperature results, more actual heat boundary of coolant side will be mapped onto the FEA-model.
- In this paper, combustion is assumed to be at steady-state over the entire cycle with averaged HTC and local gas temperature on the time axes, and therefore the temperature and heat flow on chamber walls can be calculated easily.

A complete analysis process of temperature field has been presented with an improvement scheme for aluminium silicon alloy material. It has been indicated that cooling condition of the improvement scheme with increasing baffle for each cylinder is more feasible than the original one by reducing the temperature.

References

1. Zieher F, Langmayr F, Jelatancev A, Wieser K (2005) Thermal mechanical fatigue simulation of cast iron cylinder heads. SAE paper no. 2005-01-0796
2. Prince O, Morin G, Jouzeau C (2005) Validation test optimization based on a statistical approach for diesel engine. SAE paper no. 2005-01-1780
3. Steiner H, Kobor A, Gebhard L (2004) A wall heat transfer model for subcooled boiling flow. ASMEZSIS international thermal science seminar (ITSSII)
4. Steiner H, Kobor A, Gebnard L (2005) A wall heat transfer model for subcooled boiling flow. Int J heat mass transf 48:4161–4173
5. Forster HK, Zuber N (1955) Dynamics of vapor bubbles and boiling heat transfer. AIChE J 1(4):531–535
6. Jayamathy M, Vasanth R (2006) Aluminium piston alloy to retard age softening characteristics in motorcycle engines. SAE paper no. 2006-32-0030
7. Wimmer A, Pivec R, Sams Th (2000) Heat transfer to the combustion chamber and port walls of IC engines—measurement and prediction. SAE paper no. 2000-01-0568
8. Etemad S, Stain C, Eriksson S (2005) Heat transfer analysis and cycle averaged heat flux prediction by means of CFD and its validation for an IC-engine. SAE Paper NO. 2005-01-2029

Development and Validation of a Quasi-Dimensional Model for (M)Ethanol-Fuelled SI Engines

Jeroen Vancoillie, Louis Sileghem, Joachim Demuynck and Sebastian Verhelst

Abstract Methanol and ethanol are interesting spark ignition engine fuels, both from a production and an end-use point of view. Despite promising experimental results, the full potential of these fuels remain to be explored. In this respect, quasi-dimensional engine simulation codes are especially useful as they allow cheap and fast optimization of engines. The aim of the current work was to develop and validate such a model for methanol-fuelled engines. Several laminar burning velocity correlations and turbulent burning velocity models were implemented in a QD code and their predictive performance was assessed for a wide range of engine operating conditions. The effects of compression ratio and ignition timing on the in-cylinder combustion were well reproduced irrespective of the employed u_l correlation or u_{te} model. However, to predict the effect of changes in mixture composition, the correlation and model selection proved crucial. Compared to existing correlations, a new correlation developed by the current authors led to better reproduction of the effects of equivalence ratio and residual gas content and the combustion duration. For the turbulent burning velocity, the models of Damköhler and Peters consistently underestimated the influence of equivalence ratio and residual gas content on the combustion duration, while the Gülder, Leeds, Zimont and Fractals model corresponded well with the experiments. The combination of one of these models with the new u_l correlation can be used with confidence to simulate the performance and efficiency of methanol-fuelled engines.

Keywords Alcohols · Spark ignition engine · Thermodynamic · Modelling

F2012-A06-019

J. Vancoillie (✉) · L. Sileghem · J. Demuynck · S. Verhelst
Ghent University, Ghent, Belgium
e-mail: Jeroen.Vancoillie@UGent.be

SAE-China and FISITA (eds.), *Proceedings of the FISITA 2012 World Automotive Congress*, Lecture Notes in Electrical Engineering 190,
DOI: 10.1007/978-3-642-33750-5_12,

1 Introduction

The use of sustainable liquid alcohols in spark ignition engines offers the potential of decarbonizing transport and securing domestic energy supply while increasing engine performance and efficiency compared to fossil fuels thanks to a number of interesting properties [1, 2]. The most significant interesting properties of light alcohols include:

- High heat of vaporization, which causes considerable charge cooling as the injected fuel evaporates.
- Elevated knock resistance, which allows to apply higher compression ratios (CR), optimal spark timing and aggressive downsizing.
- High flame speeds, enabling qualitative load control using mixture richness or varying amounts of exhaust gas recirculation (EGR).

The potential of neat light alcohol fuels (methanol and ethanol) has been demonstrated experimentally in both dedicated and flex-fuel alcohol engines [1]. In dedicated alcohol engines high compression ratios enable peak brake thermal efficiencies up to 42 %, while throttleless load control using EGR allows to spread the high efficiencies to the part load regions [1, 3]. In flex-fuel engines the CR is limited due to knock constraints associated with gasoline operation, but still relative power and efficiency benefits of about 10 % and NO_x reductions of 5–10 g/kWh can be obtained over the entire load range thanks to more isochoric combustion, optimal ignition timing and reduced flow and dissociation losses [1].

Despite these promising experimental results, the full potential of alcohol blended fuels and their impact on engine control strategies remain to be explored. Today these issues can be addressed at low cost using system simulations of the whole engine, provided that the employed models account for the effect of the fuel on the combustion process.

In this respect quasi-dimensional (QD) engine simulation codes are especially useful as they are well suited to evaluate existing engines, performing parameter studies and predicting optimum engine settings without resorting to complex multidimensional models [4]. At Ghent University, a QD code for the power cycle of hydrogen fuelled engines has been developed and validated during earlier work (GUEST: Ghent University Engine Simulation Tool) [5, 6]. The current work aims to extend this code to alcohol blended fuels and to add models predicting the gas dynamics, knock onset and pollutant formation in engines running on these fuels [7]. In this chapter elements of this work pertaining to the power cycle simulation of engines running on neat methanol are presented.

2 Simulation Program

Framework and assumptions—The focus in this chapter is the validation of the power cycle model and turbulent combustion models for engine operation on neat methanol. Also, the in-house GUEST code was coupled to a commercial gas dynamics simulation tool during the current work, to enable simulation of the entire engine cycle (GT-Power [8]).

The current two-zone QD power cycle model was derived using several standard assumptions, as mentioned in [4, 5]. The equations for the rate of change of the cylinder pressure $dp/d\theta$, burned and unburned temperatures, $dT_b/d\theta$ and $dT_u/d\theta$, are derived from conservation of energy. Additionally, a number of models and assumptions are necessary to close these equations:

- Heat exchange is calculated separately for the cylinder liner, cylinder head and piston based on an extension of the Woschni model discussed in [9].
- The CFR (Cooperative Fuel Research) engine used for validation of the simulation (see later) has a simple disc-shaped combustion chamber and ran at a fixed speed of 600 rpm. Therefore, turbulence quantities are calculated using a very simple turbulence model based on measurements done in a similar engine [10]. The integral length scale L is kept constant at 1/5 of the minimum clearance height, and the rms turbulent velocity u' linearly decreases according to:

$$u' = u'_{TDC}\,[1 - 05(\theta - 360)/45] \tag{1}$$

Where u'_{TDC} is the rms turbulent velocity at top dead centre (TDC), taken to be 0.75 times the mean piston speed, θ is the crank angle (360 at TDC of compression).

- The mass burning rate is derived from a turbulent combustion model. The one used in this work is based on the entertainment framework, where the rate of entertainment of unburned gas into the flame front is given by

$$\frac{dm_e}{d\theta} = \rho_u A_f u_{te} \tag{2}$$

Where m_e is the entrained mass, A_f is the mean flame front surface, and u_{te} is the turbulent entertainment velocity (see later). The mass entertainment into the flame front is then supposed to burn with a rate proportional to the amount of entrained unburned gas, with a time constant τ_b:

$$\frac{dm_b}{d\theta} = (m_e - m_b)/\tau_b \tag{3}$$

$$\tau_b = C_3\lambda_T/u_l \tag{4}$$

Where C_3 is a calibration constant, u_l is the laminar burning velocity and λ_t is the Taylor length scale, given by:

$$\lambda_t = 1.5L/\sqrt{Re_t} \tag{5}$$

$$Re_t = u'L/\upsilon_u \tag{6}$$

Where L is integral turbulent length scale and $\upsilon_u$is the kinematic viscosity of the unburned gases. Eqs. (2) and (3) are used as a mathematical representation of the effects of a finite flame thickness [5].

- The quantities p, T_u, T_b, m_u, m_b, and m_e are initialized as mentioned in [5].
- Gas properties are taken from the standard GT-Power libraries [8], supplemented with data for methanol from the chemical oxidation mechanism of Li et al. [11].
- A flame propagating after spark ignition is first only wrinkled by the smallest scales of turbulence. For the simulations done in this work, a flame development multiplying factor for the turbulent entertainment velocity was used, based on work by Dai et al. [12]:

$$\left(1 - e^{-r_f/r_c}\right)\left(\frac{r_f}{r_c}\right)^{1/3} \tag{7}$$

Where r_f is the flame radius and r_c a critical flame radius given by $r_f = C_1 L$

- For simplicity, blowby rates and the influence of crevice volumes have been neglected.

Turbulent burning velocity model—A turbulent entertainment velocity u_{te} is needed for closure of Eq. (2). A number of turbulent burning models were selected and implemented in the form summarized below (for a full description of the models see the references or [6])

Damköhler [13]: $$u_{te} = C_2 u' + u_n \tag{8}$$

Gülder [14]: $$u_{te} = 0.62 C_2 u'^{0.5} Re_t^{0.25} + u_n \tag{9}$$

Leeds "K-Le" [15]: $$u_{te} = 0.88 C_2 u' (KaLe)^{-0.3} + u_n \tag{10}$$

Fractals [16]: $$u_{te} = u_n Re_t^{0.75(D_3-2)} \tag{11}$$

$$D_3 = \frac{2.35 C_2 u'}{u'+u_n} + 2.0 \frac{u_n}{u'+u_n}$$

Zimont [17]: $$u_{te} = C_2 u'^{0.75} L^{0.75} D_T^{-0.25} + u_n \tag{12}$$

Peters [18]: $$u_{te} = 0.195 C_2 u' Da\left[\left(1 + \frac{20.52}{Da}\right)^{0.5} - 1\right] + u_n \tag{13}$$

Where C_2 is a calibration constant, u_n is the stretched laminar burning velocity, Ka is the Karlovitz stretch factor [15], Da is the Damköhler number which is calculated using a laminar flame thickness based on the kinematic viscosity.

An alternative is to use a flame thickness correlation based on chemical kinetics calculations [19].

Laminar burning velocity correlation—Turbulent burning velocity models need (stretched) laminar burning velocity data of the air/fuel/residuals mixture at the instantaneous pressure and temperature. This implies the need for either a library of stretched flamelets or a model for the effect of stretch. However, calculating the local flame speed from stretch-free data and a stretch model requires stretch-free data, naturally. As of today, there are insufficient data on stretch-free burning velocities at engine conditions, for any fuel. Stretch and instabilities hamper the experimental determination of stretch-free data at higher (engine-like) pressures [20].

The authors have worked on the laminar burning velocity of methanol and ethanol mixtures, compiling data from the literature [21] and looking at numerical [21] as well as experimental [22] means to determine a suitable laminar burning velocity correlation. A laminar burning velocity correlation has been determined based on chemical kinetics calculations [21]. This correlation was used for the current work. No stretch model has been implemented in the code as of yet, partly because of a lack of reliable data regarding the effect of stretch on methanol flames at engine-like conditions [23].

The form of the developed correlation is given by:

$$u_l(\Phi, p, T_u, f) = u_{l0}(\Phi, p)\left(\frac{T_u}{T_0}\right)^{\alpha(\Phi,p)} F(\Phi, p, T_u, f) \tag{14}$$

Where ϕ, p, T_u and f are the fuel/air equivalence ratio, pressure, unburned mixture temperature and molar diluents ratio respectively. u_{l0} is the laminar burning velocity at T_0= 300 K. u_{l0} and α are third order polynomial functions of ϕ and p.

Published simulation work on methanol and ethanol fuelled engines often resorts to outdated laminar burning velocity correlations by Gibbs and Calcote [24] (used in [25]), Metghalchi and Keck [26] (used in [25, 27, 8]), Gülder [28] (used in [29, 30]). Some authors [27] prefer the more recent correlation of Liao et al. [31], which basically has the same form:

$$u_l = u_{l0}\left(\frac{T_u}{T_0}\right)^{\alpha}\left(\frac{p}{p_0}\right)^{\beta}(1 - \gamma f) \tag{15}$$

Where u_{l0} is a second order polynomial of ϕ, α and β are first order function of ϕ and γ is a constant.

Compared these older correlations, the form of Eq. (14) captures the strong interaction effects between equivalence ratio, pressure and temperature. Example the temperature and pressure effects are not assumed to be independent. The influence of residuals is incorporated into a separate correction term $F(\Phi, p, T_u, f)$ which is also a third order polynomial function of Φ, p, T_u and f. The term reflects the fact that the burning velocity is less sensitive to residual gases at higher temperatures. For a deeper discussion of the correlation and the polynomial functions, the reader is referred to [21].

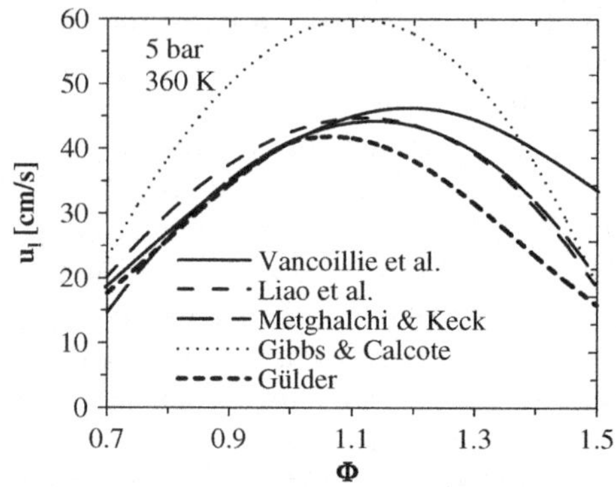

Fig. 1 u_l as a function of ϕ as predicted by several correlations

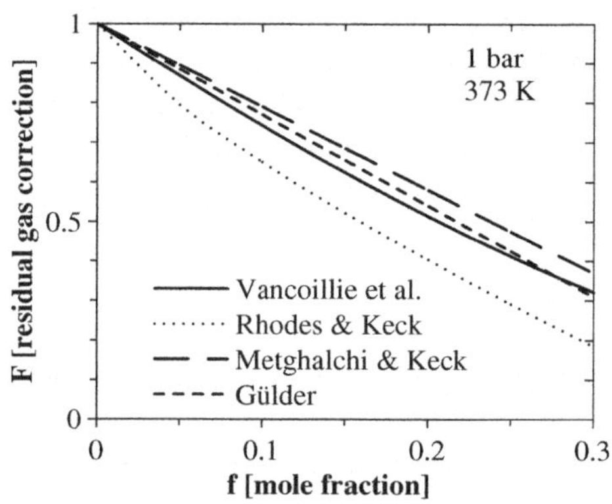

Fig. 2 Comparison of different residual gas correction terms F

In the light of the discussion, Figs. 1 and 2 highlight the main differences between the different methanol laminar burning velocity correlations in terms of equivalence ratio and residual gas content.

Compared to older correlations of Gibbs and Calcote [24], Metghalchi and Keck [26], Gülder [28], Liao et al. [31], the new correlation—Eq. (14)—places the peak laminar burning velocity at a richer equivalence ratio and predicts a less steep decrease in u_l for rich mixtures. As mentioned in [21] the correlation of Gibbs and Calcote [24] considerably overestimates u_l due to uncertainties in the experimental method employed by these authors.

The residual gas correction term F in the current correlation produces values comparable to other correlation developed for methanol, but its third order form avoids negative values for F at very high diluents ratios ($f > 0.5$). The correlation of Rhodes and Keck [32] predicts a steeper drop in burning velocity in terms of diluents ratio, but was developed for indolene/air/diluents mixtures.

Table 1 Characteristics of the single cylinder CFR engine

Bore	82.55 mm
Stroke	114.2 mm
Swept volume	611.7 cm^3
Geometry	Disc-shaped
Speed	600 rpm
IVO/IVC	17° CA ATDC / 26° CA ABDC
EVO/EVC	32° CA BBDC / 6° CA ATDC

3 Validation

Engine—To validate the combustion models' predictive capabilities, a series of measurements were done on a port fuel injected single cylinder CFR engine, described in [33]. The main characteristics of this engine are summarized in Table 1.

The measurements comprise variable fuel/air equivalence ratio ϕ, ignition timing (IT), compression ratio (CR) and throttle position (TP) (see Table 2). In order to allow distinction of the individual effects of these parameters, without resorting to a lot of one factor at a time sweeps, the experimental conditions have been chosen in such a way the Response Surface Methods can be applied to analyze the results [34]. This way, the resulting quantities of interest (e.g. IMEP, ignition delay) can be fit as a function of ϕ, CR, IT and TP.

Model setup—As the main focus of the current work was to evaluate combustion models, the employed engine model is limited to the closed part of the engine cycle (IVC to EVO). The initial conditions for mass fractions of air and fuel, the mean temperature and pressure at IVC are taken from the measurements. The residual gases (from the previous engine cycle) are estimated using a gas dynamics model of the entire intake and exhaust geometry.

Calibration—The calibration sets the coefficients for the heat transfer model, the flame development model (C_1) and the turbulent burning velocity model (C_2, C_3). For each model, the code has been calibrated at the first measurement condition in Table 2, as this condition is in the middle of the explored parameter space. The calibration constants are left constant for the other conditions.

The heat transfer calibration constant was set to 1.3 for all simulations, based on correspondence between the measured and predicted cylinder pressure during compression. The combustion model was calibrated by minimizing either the error in IMEP or the sum of squared differences between the measured and predicted normalized burn rate. These two methods lead to slightly different simulation results, as will be explained later. The measured burn rate was derived from the measured cylinder pressure by doing a reverse heat release analysis using the same cylinder model as used in the forward power cycle simulation [8]. The flame development constant C_1 is usually calibrated first in order to get a reasonable correspondence for the ignition delay. As mentioned in [5], increasing C_2 increases the mass entertainment rate, while increasing coefficient C_3 decreases the mass

Table 2 Measurement conditions on the CFR engine

No.	λ	$\phi = 1/\lambda$	CR	IT [°ca BTDC]	TP [°]	EGR (%)
1	1.00	1.00	9.0	15	75	6.1
2	1.00	1.00	9.0	15	87	12.5
3	1.00	1.00	9.0	5	75	6.3
4	1.30	0.77	9.0	15	75	6.6
5	1.15	0.87	8.5	20	81	10.0
6	0.85	1.18	8.5	10	81	8.7
7	1.15	0.87	8.5	10	69	5.8
8	0.85	1.18	8.5	20	69	5.7
9	1.15	0.87	9.5	10	81	8.5
10	0.85	1.18	9.5	20	81	8.4
11	1.15	0.87	9.5	20	69	5.7
12	0.85	1.18	9.5	10	69	5.2
13	0.70	1.43	9.0	15	75	6.2
14	1.00	1.00	9.0	15	75	6.5
15	1.00	1.00	9.0	25	75	6.6
16	1.00	1.00	9.0	15	63	5.6

burning rate. Consequently, both constants are simultaneously optimized using the Design of Experiments and Response Surface Methods embedded in GT-Power [8].

4 Results

Comparison of laminar burning velocity correlations—A first part of the work consisted of comparing the predictive capabilities of different laminar burning velocity correlations for methanol. Three correlations are considered here: the correlation of Gülder [35], the correlation for methanol implemented in GT-Power, which is based on the correlation of Metghalchi and Keck [26] and the residual gas

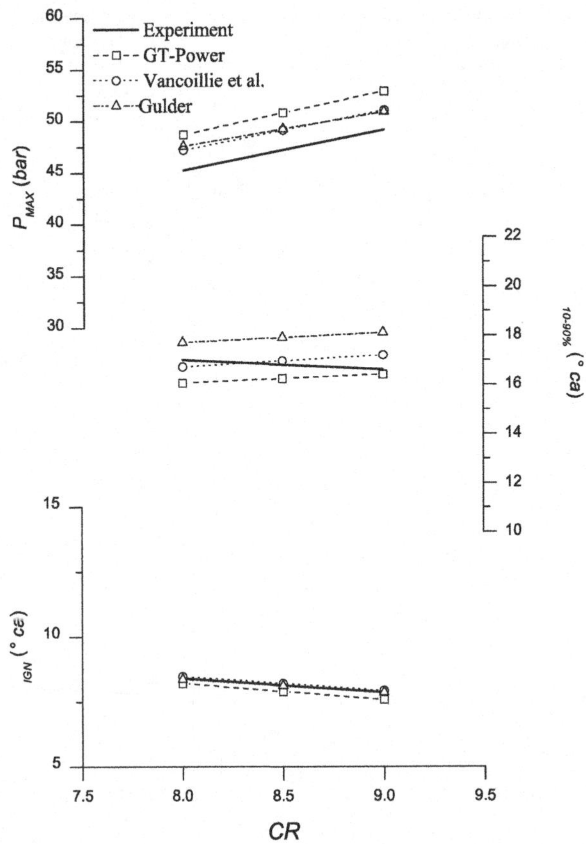

Fig. 3 Comparison of u_l correlations for varying compression ratio

term of Rhodes and Keck [32], and the new correlation—Eq. (14)—developed by the current authors [21]. To minimize the influence of the turbulent burning velocity model, the Damköhler model was used with the same calibration constants for all simulations discussed in this section ($c_1 = 0.5$, $C_2 = 0.46, C_3 = 1$

The simulation results are synthesized into graphs showing the ignition delay Δ_{IGN} (defined as ca from 0 to 2 % burned mass), the 10–90 % burn time ($\Delta_{10-90\,\%}$) and the maximum cylinder pressure (P_{MAX}). Figure 3 compares these quantities for variable compression ratio. Ignition delay slightly reduces at higher compression ratios. This is reproduced in by the model due to a reduced burn-up time constant τ_b at higher CR. This can be understood from looking at Eqs. (4) and (5). Increasing the CR decreases L (because of smaller top dead center clearance) and increases u_l (because temperature and pressure increase and residual gas content decreases). The experimental 10–90 % burn times is almost independent of CR, whereas all models predict a slight increase at higher CR. Additional measurements at more extreme compression ratios could help to better understand the effect of CR on the main combustion.

Fig. 4 Comparison of u_l correlations for varying ignition timing

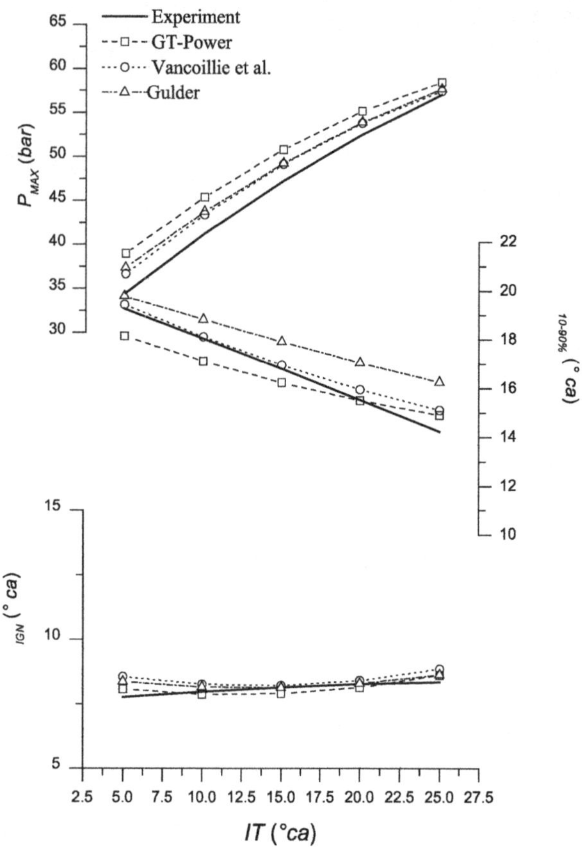

Figure 4 compares Δ_{IGN}, $\Delta_{10-90\%}$ and P_{MAX} for variable ignition timing. The ignition delay is barely influenced by spark timing. One might expect a slight reduction at more retarded spark timing due to the higher u_l (higher p, T). However, in the models this effect is counteracted by the reduction in u' and corresponding increase in burn-up time constant τ_b. The reduction in u' at more retarded ignition timing [see Eq. (1)] is also responsible for the increase in 10–90 % burn time, reflected in the measurements.

Note that in all of the simulation results discussed in this chapter, the maximum pressure is generally overpredicted. Possible reasons can be understood from looking at burn rates and corresponding pressure traces for condition 12 (Table 2) in Fig. 5. When comparing the simulated burn rate (dashed line) with those derived from the experimental pressure traces (full line), it can be seen that the latter ones are truncated at the peak and exhibit a tail. This is probably caused by an accumulation of inaccuracies. Blow-by and fuel in the crevice volume are neglected. Also, an identical heat transfer coefficient is used for the compression, combustion and expansion, leading to an underprediction of the heat transfer during combustion. These inaccuracies could be cured by further model

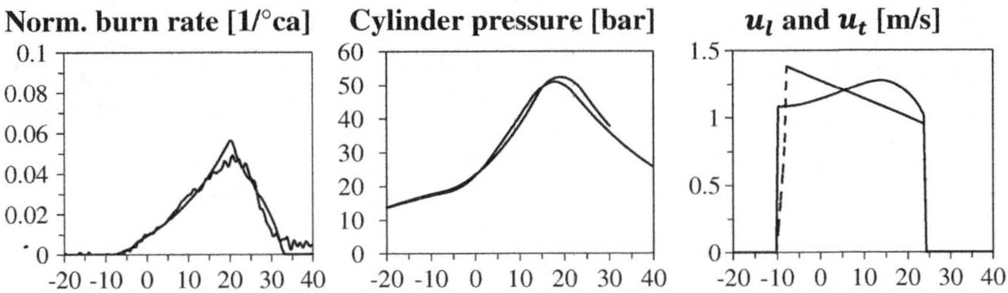

Fig. 5 Experimental (*full line*) and simulated (*dashed line*) normalized burn rate and corresponding cylinder pressure. The *rightmost* figure shows the simulated laminar and turbulent contribution to u_{te}

Fig. 6 Comparison of u_l correlations for varying equivalence ratio

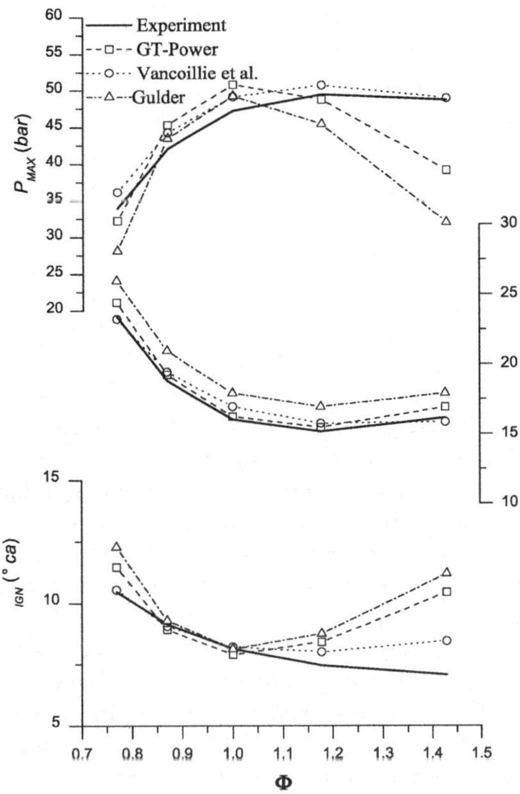

development, but do not impede the comparison of u_l correlations and turbulent burning velocity models aimed at here. Also shown in Fig. 5 are the laminar (u_l—full line) and turbulent (u_t—dashed line) contributions to turbulent entertainment velocity u_{te}. This demonstrates the large share of the laminar burning velocity in the entertainment velocity.

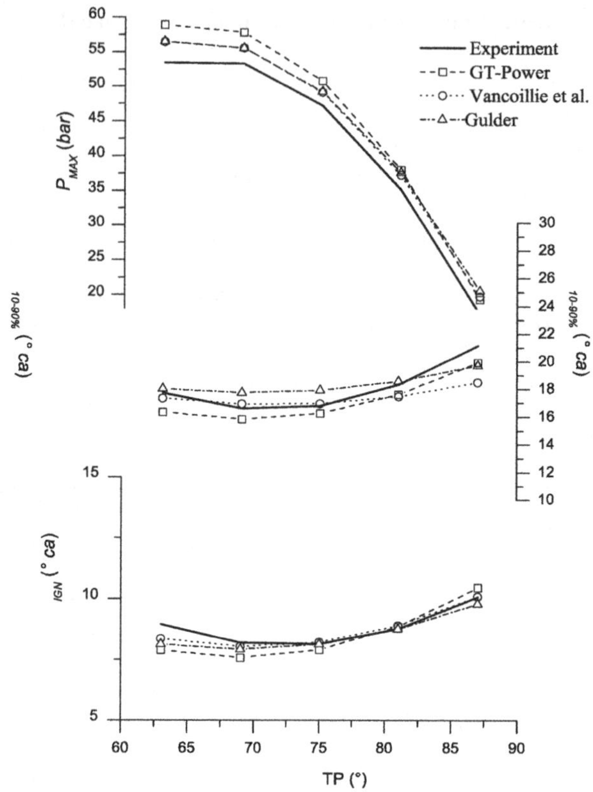

Fig. 7 Comparison of u_l correlations for varying throttle position

The predictive capabilities of the model for changes in CR and IT show little dependency on the employed laminar burning velocity correlation. As the laminar burning velocity groups the contribution of chemical kinetics to the combustion, its effects are most pronounced for changes in mixture composition.

Figure 6 displays the results for varying equivalence ratio ($\phi = 0.77 - 1.43$). For lean and rich mixtures, the ignition delay can be seen to be vastly overpredicted by traditional u_l correlations. As the predicted ignition delay is mainly determined by the laminar burning velocity, this is a direct effect of the erroneous evolution of these correlations as a function of ϕ (see Fig. 1). As the 10–90 % burn time is dominated by the turbulent combustion velocity, the difference between the correlations is less pronounced for $\Delta_{10-90\,\%}$. Still, the results indicate a better performance for the new correlation.

Finally, Fig. 7 shows the effect of throttle position. A more closed throttle (90° = closed) not only reduces the temperature and pressure of the cylinder charge, but also implies a higher residual gas percentage (see Table 1). The effect of TP on Δ_{IGN} is overpredicted by the GT-Power correlation, indicating that the residual gas correction term of Rhodes and Keck produces a too steep decrease of u_l as a function of diluents ratio (see Fig. 2). For the main combustion duration

Table 3 Calibration constants of turbulent burning velocity models

Model	C_1	C_2	C_3
Damköhler	0.8	0.75	0.15
Gülder	1	0.157	0.2
Fractals	1	0.943	0.156
Zimont	0.4	0.156	0.139
Leeds	–	0.23	0.5
Peters	–	0.44	0.78

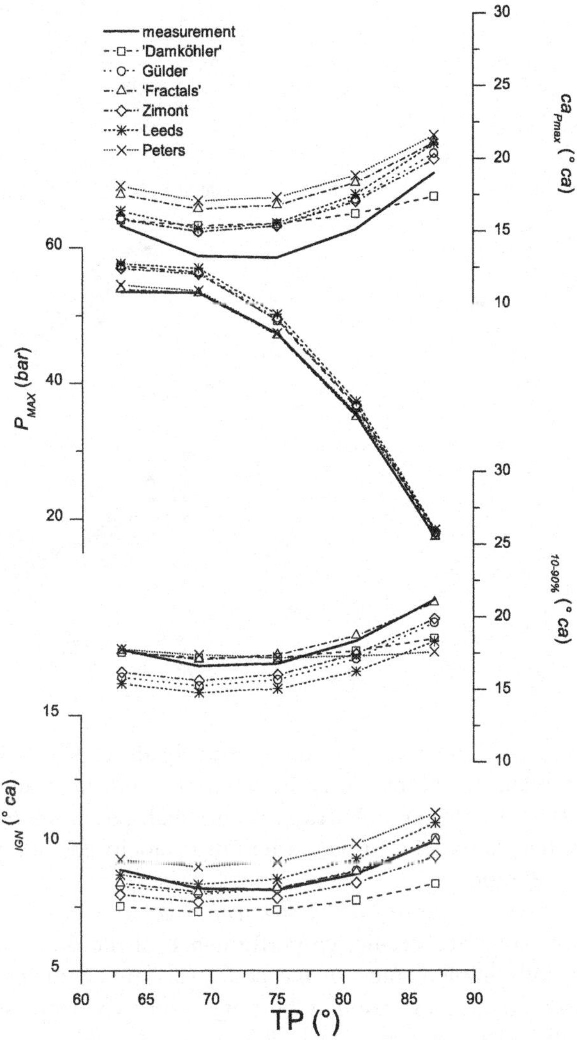

Fig. 8 Comparison of turbulent burning velocity models for varying throttle position

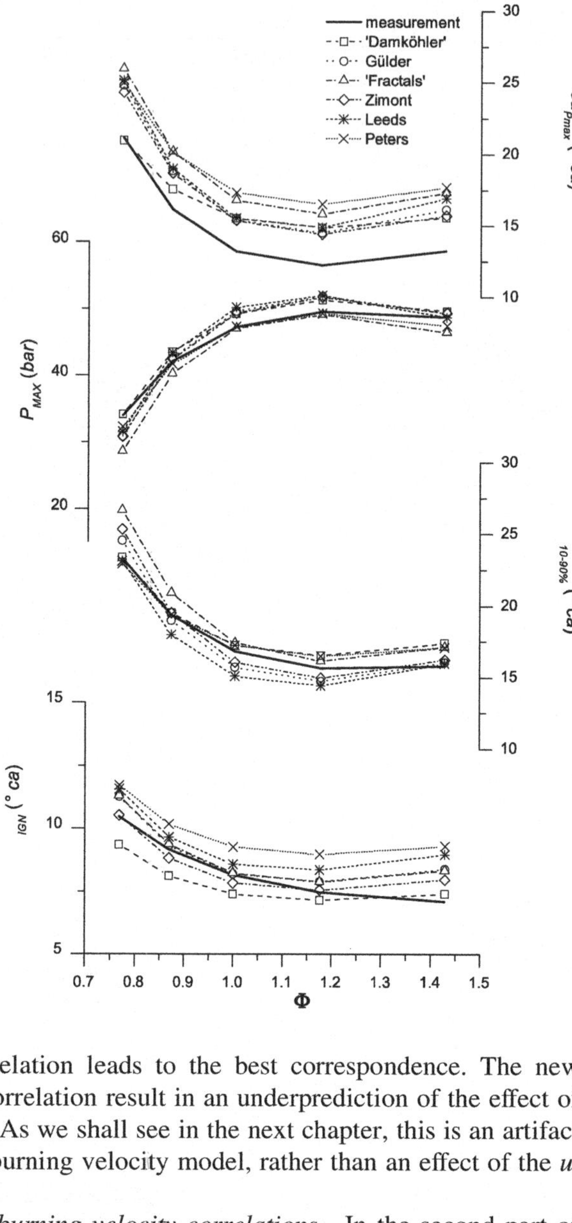

Fig. 9 Comparison of turbulent burning velocity models for varying equivalence ratio

$\Delta_{10-90\%}$, however, this correlation leads to the best correspondence. The new correlation and the Gülder correlation result in an underprediction of the effect of TP on the main combustion. As we shall see in the next chapter, this is an artifact of the Damköhler turbulent burning velocity model, rather than an effect of the u_l correlation.

Comparison of turbulent burning velocity correlations—In the second part of this work, the predictive performance of the different turbulent burning velocity models implemented in the code was compared. For all the simulations in this section, the new laminar burning velocity correlation was used [Eq. (14)]. The calibration constants for the six turbulent burning velocity models considered here are summarized in Table 3. Note that the Peters and Fractals models have been calibrated for minimal IMEP error, whereas the other models have been calibrated for best correspondence with the measured burn rate. As can be seen from Figs. 8 and 9, the former method leads to better peak pressure correspondence, while the latter leads to a better prediction of the crank angle of maximum pressure ca_{Pmax}.

The results for varying CR and IGN are very similar to those of the Damköhler model (see Figs. 5 and 3), irrespective of the applied turbulent burning velocity model and are not repeated here.

Differences between the turbulent burning velocity models begin to appear when simulating changes in mixture composition. Figure 8 displays the Δ_{IGN}, $\Delta_{10-90\,\%}$, P_{MAX} and ca_{Pmax} for varying throttle position. As mentioned before, a more closed throttle reduces the laminar burning velocity, leading to an increase in ignition delay, which is well predicted by all the models. However, the increased residual gas content also lengthens the main combustion period $\Delta_{10-90\,\%}$ and the ability to predict this effect clearly is the criterion to distinguish between models. The models of Gülder, Fractals, Zimont and Leeds follow the trend observed in the measurements, while the Damköhler and Peters model predict almost no dependency on the throttle position. The results for maximum pressure and crank angle of maximum pressure show that the models that have been calibrated for minimal IMEP error indeed lead to a better peak pressure correspondence but more retarded location of peak pressure.

As explained in [5] the appearance of different model groups can by understood from rewriting the models' equations using the definitions of Re_t, Da, Ka and Le.

Damköhler: $$u_{te} \sim Cu' + u_n \tag{16}$$

Gülder: $$u_{te} \sim Cu'^{0.75} u_n^{0.5} L^{0.25} \upsilon_u^{-0.25} + u_n \tag{17}$$

Leeds "K-Le": $$u_{te} \sim Cu'^{0.55} u_n^{0.6} L^{0.15} \upsilon_u^{-0.15} D_{T,u}^{-0.3} D_M^{0.3} + u_n \tag{18}$$

Fractals: $$u_{te} \sim Cu'^{0.25} u_n L^{0.25} \upsilon_u^{-0.26} + u_n \tag{19}$$

Zimont: $$u_{te} \sim Cu'^{0.75} u_n^{0.5} L^{0.25} D_{T,u}^{-0.25} + u_n \tag{20}$$

Peters: $$u_{te} \sim Cu' + u_n \tag{21}$$

As opposed to the other models, the Damköhler and Peters model do not contain the laminar burning velocity u_n in the turbulent contribution to u_{te}. This explain why these models are less sensitive to changes in mixture composition, which are reflected in u_n.

The results for varying equivalence ratio in Fig. 9 confirm the distinction between the Damköhler and Peters models, and the rest of the models. Compared to the other models, the ignition delay and 10–90 % burn times predicted by these models are less dependent on the changes in laminar burning velocity associated with the equivalence ratio. All models overpredict the ignition delay for rich mixture. During initial stages of combustion the flame resembles a stretched laminar flame. Rich methanol flames have been reported to exhibit negative Markstein numbers and cellular instabilities [36], which could result in a higher burning velocity during flame initiation. This effect cannot be capture by the simple burning velocity models considered here.

The influence of equivalence ratio on the main combustion duration is slightly overestimated by the Gülder, Fractals, Leeds and Zimont models. Especially very lean and rich mixtures seem to burn faster than the models predict. The experimental work on methanol-air turbulent burning velocity of Lawes et al. [36] might give an explanation for this behavior. During contained explosions in a spherical bomb, these authors found that the ratio of u_{te}/u_t is slightly dependent on the equivalence ratio. The ratio was 50 % higher at very lean ($\phi = 0.7$) and rich ($\phi = 1.6$) mixtures compared to the values at $\phi = 1.0 - 1.3$. This is partly due to the preferential diffusion effects and diffusive thermal effects discussed by Lipatnikov and Chomiak [17] and cannot be captured by the models used here, which have a ratio u_{te}/u_t which is independent of the equivalence ratio.

5 Conclusions

The focus of the current chapter was the development and validation of a quasi-dimensional model for the combustion of neat methanol in spark ignition engines. The predictive performance of several laminar burning velocity correlations and turbulent burning velocity models was assessed by considering a wide range of experimental conditions on a methanol-fuelled single cylinder engine.

The effects of compression ratio and ignition timing on the in-cylinder combustion were well reproduced by the simulation code, irrespective of the employed u_l correlation or u_{te} model. To predict the effect of changes in mixture composition, the choice of laminar burning velocity correlation proved crucial. Compared to existing correlations, a new correlation developed by the current authors [21] led to considerably less overprediction of the combustion duration and ignition delay for rich and lean mixtures. Also the effect of residual gas content was better reproduced by the new correlation.

The ability to predict the effects of mixture composition was also clearly the criterion to distinguish between turbulent burning velocity models. The models of Damköhler and Peters consistently underpredicted the influence of equivalence ratio and residual gas content on the combustion duration, while the Gülder, Leeds, Zimont and Fractals model corresponded well with the experiments. The combination of one of these models with the new u_l correlation can be trustfully used to simulate the performance and efficiency of methanol-fuelled engines and forms a solid base for further work regarding pollutant formation and knock modeling in alcohol engines.

Acknowledgments J. Vancoillie and L. Sileghem gratefully acknowledge a Ph. D. Fellowship of the Research Foundation—Flanders (FWO09/ASP/030 and FWO11/ASP/056). The authors would like to thank BioMCN for providing the bio-methanol used in this study.

References

1. Vancoillie J, Demuynck J, Sileghem L, Van De Ginste M, Verhelst S, Brabant L, Van Hoorebeke L (2012) The potential of methanol as a fuel for flex-fuel and dedicated spark ignition engines. Appl Energy
2. Pearson RJ, Turner JWG, Eisaman MD et al (2009) Extending the supply of alcohol fuels for energy security and carbon reduction, SAE paper no. 2009-01-2764
3. Brusstar MJ, Stuhldreher M, Swain D et al (2002) High efficiency and low emissions from a port-injected engine with neat alcohol fuels, EPA VW 1.9 TDI measurements, SAE paper no. 2002-01-2743
4. Verhelst S, Sheppard CGW (2009) Multi-zone thermodynamic modelling of spark ignition engine combustion—an overview. Energy Convers Manage 50(5):1326–1335
5. Verhelst S, Sierens R (2007) A quasi-dimensional model for the power cycle of a hydrogen-fuelled ICE. Int J Hydrogen Energy 32(15):3545–3554
6. Verhelst S (2005) A study of the combustion in hydrogen-fuelled internal combustion engines. PhD thesis, Department of Flow, heat and combustion mechanics
7. Vancoillie J, Verhelst S (2010) Modeling the combustion of light alcohols in SI engines: a preliminary study, presented at FISITA 2010 World Automotive Congress. Budapest, Hungary
8. GammaTechnologies (2009) GT-Suite version 7.0 user's manual
9. Morel T, Rackmil CI, Keribar R et al (1988) Model for heat transfer and combustion in spark ignited engines and its comparison with experiments, SAE paper no. 880198
10. Hall MJ, Bracco FV (1987) A study of velocities and turbulence intensities measured in firing and motored engines, SAE paper no. 870453
11. Li J, Zhao ZW, Kazakov A et al (2007) A comprehensive kinetic mechanism for CO, CH_2O, and CH_3OH combustion. Int J Chem Kinet 39(3):109–136
12. Dai W, Davis GC, Hall MJ et al (1995) Diluents and lean mixture combustion modeling for si engines with a quasi-dimensional model, SAE paper no. 952382
13. Blizard NC, Keck JC (1974) Experimental and theoretical investigation of turbulent burning model for internal combustion engines, SAE paper no. 740191
14. Gülder ÖL (1991) Turbulent premixed flame propagation models for different combustion regimes. Symp (Int) Combust 23(1):743–750
15. Abdel-Gayed RG, Bradley D, Lawes M (1987) Turbulent burning velocities: a general correlation in terms of straining rates, proceedings of the royal society of London. A Math Phys Sci 414(1847):389–413
16. Wu C-M, Roberts CE, Matthews RD et al (1993) Effects of engine speed on combustion in si engines: comparisons of predictions of a fractal burning model with experimental data, SAE paper no. 932714
17. Lipatnikov AN, Chomiak J (2005) Molecular transport effects on turbulent flame propagation and structure. Prog Energy Combust Sci 31(1):1–73
18. Peters N (2000) Turbulent combustion. Cambridge University Press, Cambridge
19. Vancoillie J, Demuynck J, Galle J et al (2011) A laminar burning velocity and flame thickness correlation for ethanol-air mixtures valid at spark ignition engine conditions. Fuel pp 160–169
20. Bradley D, Lawes M, Liu K et al (2007) Laminar burning velocities of lean hydrogen-air mixtures at pressures up to 1.0 MPa. Combust Flame 149(1–2):162–172
21. Vancoillie J, Verhelst S, Demuynck J (2011) Laminar burning velocity correlations for methanol-air and ethanol-air mixtures valid at SI engine conditions, SAE paper no. 2011-01-0846
22. Vancoillie J, Christensen M, Nilsson EJK et al (2012) Temperature dependence of the laminar burning velocity of methanol flames. Energy Fuels 26(3):1557–1564

23. Zhang Z, Huang Z, Wang X et al (2008) Measurements of Laminar burning velocity and Markstein lengths for methanol-air-nitrogen mixtures at elevated pressures and temperatures. Combust Flame 155(3):358–368
24. Gibbs GJ, Calcote HF (1959) Effect of molecular structure on burning velocity. J Chem Eng Data 4(3):226–237
25. Brown AG, Stone CR, Beckwith P (1996) Cycle-by-cycle variations in spark ignition engine combustion—Part I: flame speed and combustion measurements and a simplified turbulent combustion model, SAE paper no. 960612
26. Metghalchi M, Keck JC (1982) Burning velocities of mixtures of air with methanol, isooctane, and indolene at high pressure and temperature. Combust Flame 48:191–210
27. Pourkhesalian AM, Shamekhi AH, Salimi F (2010) Alternative fuel and gasoline in an SI engine: A comparative study of performance and emissions characteristics. Fuel 89(5):1056–1063
28. Gülder ÖL (1982) Laminar burning velocities of methanol, ethanol and isooctane-air mixtures. Symp (Int) Combust 19(1):275–281
29. Bayraktar H (2005) Experimental and theoretical investigation of using gasoline-ethanol blends in spark ignition engines. Renew Energy 30(11):1733–1747
30. Richard SBS (2009) Modelling and simulation of the combustion of ethanol blended fuels in a SI engine using a 0D coherent flame model, SAE paper no. 2009-24-0016
31. Liao SY, Jiang DM, Huang ZH et al (2007) Laminar burning velocities for mixtures of methanol and air at elevated temperatures. Energy Convers Manage 48(3):857–863
32. Rhodes DB, Keck JC (1985) Laminar burning speed measurements of indolene-air-diluent mixtures at high pressures and temperatures. SAE paper no. 850047
33. Demuynck J, De Paepe M, Huisseune H et al (2011) Investigation of the influence of engine settings on the heat flux in a hydrogen- and methane-fueled spark ignition engine. Appl Therm Eng 31(7):1220–1228
34. Box GEP, Draper NR (2007) Response surfaces, mixtures, and ridge analyses. Wiley, Hoboken
35. Gülder OL (1983) Laminar burning velocities of methanol, isooctane and isooctane/methanol blends. Combust Sci Technol 33(1–4):179–192
36. Lawes M, Ormsby MP, Sheppard CGW et al (2005) Variation of turbulent burning rate of methane, methanol, and iso-octane air mixtures with equivalence ratio at elevated pressure. Combust Sci Technol 177(7):1273–1289

A Development of Simplified Turbocharger Transient Heat Transfer Simulation Method (First Report)

Kyung Sub Sung, Kil Min Moon, Dong Ho Chu and Sang Joon Park

Abstract Repeated temperature inversion (compression/expansion) causes durability problem of twin-scroll type turbocharger during Motoring Mode thermal cycle durability test. The conventional simulation method takes 4 months to solve transient heat transfer problem of turbocharger including reverse engineering of turbine blade to have a blade CAD. Therefore, only experimental method was affordable to resolve the problem. From conduction of this study, By conduction of this study, we were able to reduce the transient simulation time up to 2 weeks and support CAE solutions for the durability problem, compression of development duration and reduction of number of experiments.

Keywords Turbocharger · Twin-scroll · Motoring mode thermal cycle · Transient heat transfer simulation

F2012-A06-020
Presented at 2011 Hyundai-Kia Motors internal Annual Conference

K. S. Sung (✉) · K. M. Moon
Power Train CAE Team, Gyeonggi, People's Republic of Korea
e-mail: kajenosks@hyundai.com

D. H. Chu
Gasoline Engine Design Team, Seoul, China

S. J. Park
Gasoline Engine R&D Team, Yongin -si, China

SAE-China and FISITA (eds.), *Proceedings of the FISITA 2012 World Automotive Congress*, Lecture Notes in Electrical Engineering 190,
DOI: 10.1007/978-3-642-33750-5_13,

1 Introduction

At recent, More Power, Better Fuel Efficiency, Lower Emission are requested for all vehicle engines, thus Turbocharger and Direct Injection technologies are adopted in Gasoline Engines.

However, Using turbocharger in gasoline engines causes a durability problem due to much higher exhaust gas temperature than Diesel engine's and N/A gasoline engine's. In thermal shock durability test, turbocharger housing and exhaust manifold were cracked by repeated temperature inversion and high exhaust gas temperature.

Turbocharger for damage caused by repeated thermal loading is repeated as the temperature inversion phenomenon attributed to a failure of these phenomena are caused by a temperature inversion to reproduce analytically predict the exact temperature distribution is the most important and only phenomenon Transient Analysis of Temperature Prediction can be implemented. It requires extra effort to solve a conventional analysis methods take about four months can not be interpreted to support the evaluation and problem solving was on endurance test with.

In this study, a new transient temperature prediction during repeated thermal loads acting simplify the analysis techniques developed/applied to the existing take about four months (including the blade reverse engineering) of the turbocharger transient analysis time dramatically reduced by about two weeks a new interpretation sikyeoteumyeo by applying the results of actual phenomena and failure to reproduce the same phenomenon contributes to the improvement haeotda endurance. However, when there is no blade turbine blade CAD reverse engineering, so to take three months through reverse engineering-based turbine blades that have similar displacement/output turbocharged engine, using the Transient temperature prediction method was developed to simplify that.

2 Main

2.1 Description of Turbocharger

Turbocharger (Exhaust gas) that is driven by the torque of the turbine compressor (Compressor) delivered to the engine intake stroke supercharged forced into the combustion chamber when the receiver to increase the engine output is the device (Fig. 1 reference).

The engine speed according to the turbocharger and exhaust flow rate of fluid velocity, mechanical friction losses, the efficiency is determined. The larger the turbine wheel size, increasing surface area in contact with the exhaust gas in the area of the high-speed shaft rotational speed increases, but increase the maximum output by increasing the weight of rotating turbo off the rack reactivity occurs.

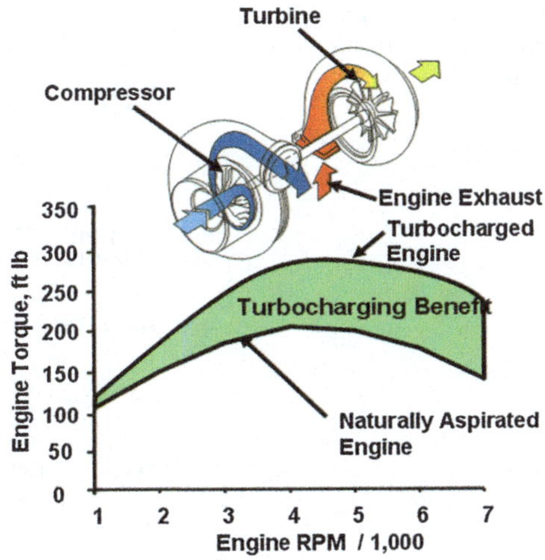

Fig. 1 Turbocharger working principle

Recently twin-scroll turbocharger has been applied to gasoline engines in order to minimize interference with the effective use of exhaust turbines to improve operational performance.

As shown in Fig. 2 chajyeoneun cylinder twin-scroll turbo type 1, 4 and 2, 3 are two entwined interference by the exhaust turbo charger with reflux symptoms prevent a reduction in output and increases the reaction rate in the low speed range turbo-lack should improve. However, bulkheads, blowing exhaust manifold and other parts of the scroll heat capacity relative to the repeated thermal loads acting less frequent phenomenon that occurs when damaged.

2.2 *Turbocharger Durability Test and Crack*

Turbocharger failure frequently occurred in developing the test conditions for developing the HMC exhaust system used to evaluate the duration is the mode of the motoring ten cycles.

Motoring ten cycle mode, as shown in Fig. 3 only the gasoline engine exhaust system as heat-shock survival mode, the engine driving condition sikidaga Rated rpm Fuel Cut state drive 4,000 rpm and makes the engine idling rpm.

Fuel Cut the receiver using the state rapidly cooling the exhaust system because the existing target of the engine, Thermal Shock durability mode (Fig. 4 reference) compared to the more intense the thermal shock given the short time available in the exhaust system is thermal shock endurance test.

However, twin-scroll turbo charger has split to prevent interference of exhaust gas. The split is thinner than other sites is even more sensitive to temperature. Gasoline engine's exhaust is over 100 °C higher than diesel engine's. If you apply for turbocharger to gasoline engine then the exhaust gas up to 950 °C. Therefore,

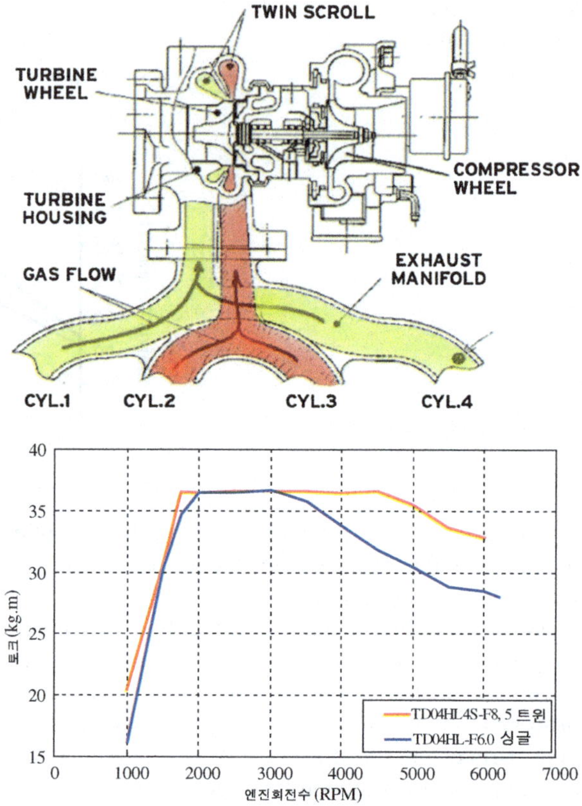

Fig. 2 Twin-scroll type turbocharger structure and performance

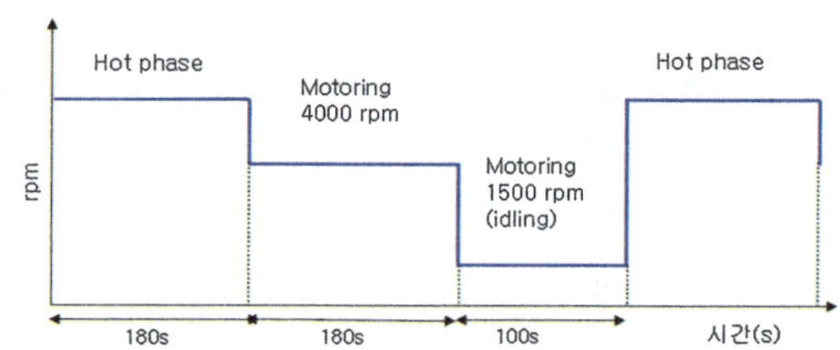

Fig. 3 Motoring thermal shock durability mode

the best exhaust system in each site have large low temperature differences, especially heating (Rated rpm) and upon cooling (Fuel Cut) with different temperature gradient occurs because the temperature inversion phenomenon causes the damage (see Figs. 5, 6).

In Fig. 5 the split 2, 5, Ex Out flange look at heating slit 2, 5, Ex Out Flange order has the highest temperature level and the temperature gradient is reversed when anti-cooling can be seen. In other words, Ex Out Flange partition walls 2, 5, compressive strength when heated to the point that the tension during the cooling effect as a condition repeatedly, so as in Fig. 6 Breakage occurs.

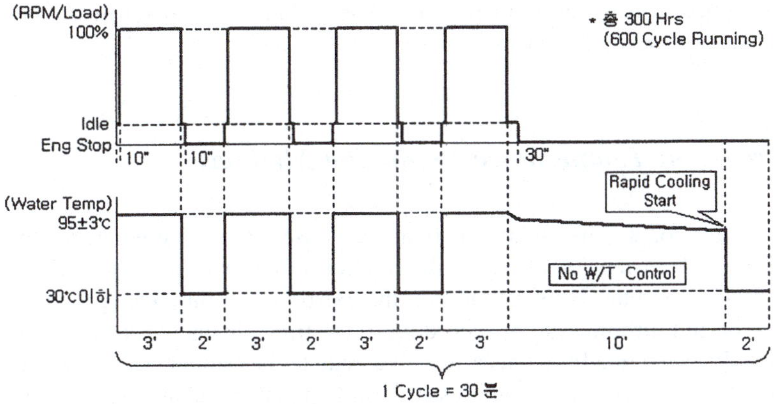

Fig. 4 Regular thermal shock durability mode

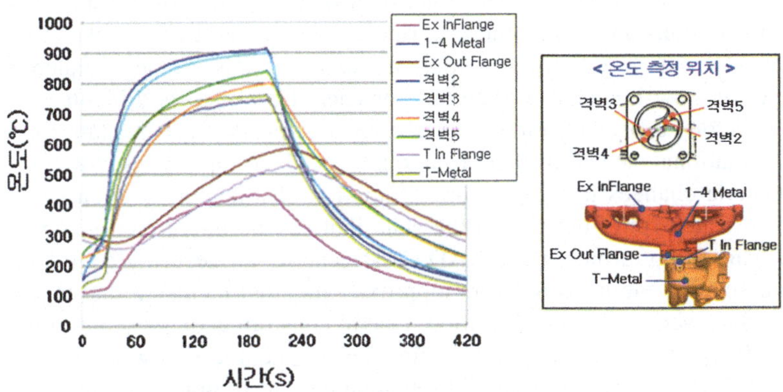

Fig. 5 Exhaust temperature theta TGDI motoring mode

Fig. 6 Theta motoring mode when testing TGDI breakage

If the outer wall of the housing turbocharger housing and scroll turbocharger with bulkheads are damaged by temperature inversions.

2.3 Transient Temperature Prediction Methods

Figures 5 and 6, as shown in the damage caused by a temperature inversion phenomenon Transient temperature distribution in order to reproduce the prediction is most important. How to interpret the existing Full turbocharger including a compressor and turbine system according to the time change after modeling Solid Flow and Transient analysis was a way to calculate the temperature. If you use this method, the turbine and compressor rotation effect chain of fast rotation (approximately 150,000 rpm) to simulate very small for the Time-Step and requires approximately 1.5 million of the mesh is the interpretation that requires a lot of time to Computing. Turbocharger Transient Analysis of Temperature Distribution with conventional methods at runtime near about three weeks of time. In addition, the nature motoring mode, the exhaust pressure as frequently and therefore the flow boundary conditions for each Time-Step is difficult to obtain. Heating sectors, especially emissions and exhaust pressure, increasing the turbocharger rotational speed turbine rotational speed is rapidly growing segment unpredictable changes in temperature simulation of a heating section can not be conducted Rated rpm Fuel Cut in the state that can be interpreted guganman. In addition, the turbo turbine blade CAD CAD vendors for securing confidential, so the blade reverse engineering hours (approx. 3 months) inclusion that takes about four months a year is the interpretation. Due to constraints on these interpretations twin-scroll type turbocharger petrol engine development testing with interpretation without the support was on durability evaluation and troubleshooting. This analysis evaluated the test of times and duration time is increased turbocharger can contribute substantially to the development of analysis methods became necessary.

In this study, analysis methods could not be resolved in the following analysis support by addressing two aspects were possible. A. Transient Analysis of Temperature Distribution simplifies the turbocharger Two. Transient temperature turbine can be interpreted in the absence of CAD.

2.3.1 Transient Temperature Prediction Simplify

Figure 7 Transient turbocharger for the sake of simplicity, such as temperature forecasts for each rpm motoring mode, the blade geometry and flow conditions with rotation Steady state temperature distribution analysis is performed. Steady state flow analysis performed after each rpm and the exhaust from the exhaust system Solid heat transfer coefficient (HTC) at the contact surface and T (temperature) is extracted. New Analysis Transient temperature distribution in contrast to conventional methods for analysis and time-consuming flow will not be

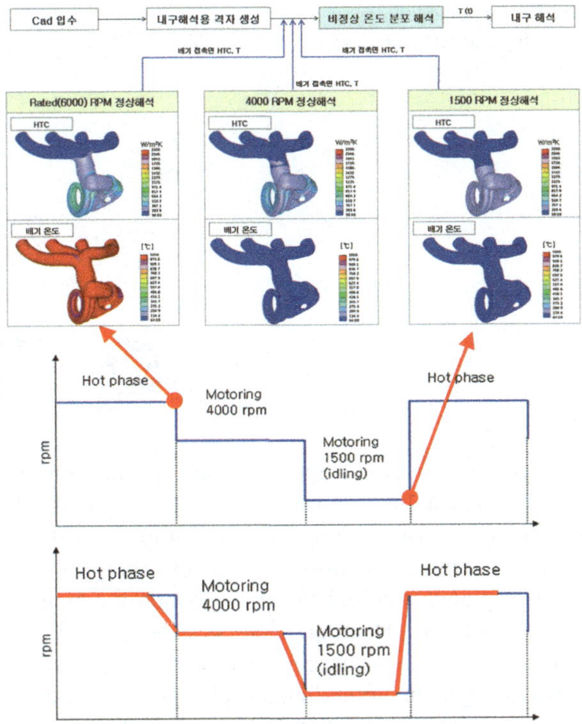

Fig. 7 Transient analysis of temperature distribution simplifies the overview

counted. Steady state, instead of extracted from HTC and T at each rpm exhaust boundary conditions given on the contact surface and changes in rpm, depending on the time and these values by interpolation to simulate the transient effects. Due to using only Solid for heat transfer calculation, Exhaust System Transient Analysis of Temperature Distribution can be performed in a few hours.

2.3.2 Transient Temperature Prediction Simplification Apply/Review

Transient temperature predictions that were described earlier simplification exhaust manifolds and turbocharger housings TGDI theta types and integral types, each separated by applying a temperature distribution and temperature distribution in the accuracy of the results when applying survival analysis to reproduce the actual degree of damage phenomena were investigated.

Figure 8 Transient Figs. 5 and 6, as shown in the temperature inversion caused by the breakage phenomena Transient temperature distribution in order to reproduce the prediction is most important. How to interpret the existing Full turbocharger including a compressor and turbine system according to the time change after modelhan Solid Flow and Transient analysis was a way to calculate the temperature. If you use this method, the turbine and compressor rotation effect chain of fast rotation (approximately 150,000 rpm) to simulate very small for the

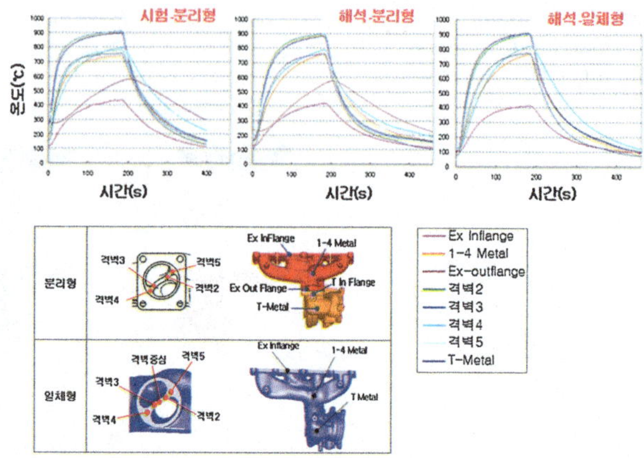

Fig. 8 Detachable/integral geometry and test/analysis

Time-Step and requires approximately 1.5 million of the mesh is the interpretation that requires a lot of time to Computing. Turbocharger Transient Analysis of Temperature Distribution with conventional methods at runtime near about three weeks of time. In addition, the nature motoring mode, the exhaust pressure as frequently and therefore the flow boundary conditions for each Time-Step is difficult to obtain. Heating sectors, especially emissions and exhaust pressure, increasing the turbocharger rotational speed turbine rotational speed is rapidly growing segment unpredictable changes in temperature simulation of a heating section can not be conducted Rated rpm Fuel Cut in the state that can be interpreted guganman. In addition, the turbo turbine blade CAD CAD vendors for securing confidential, so the blade reverse engineering hours (approx. 3 months) inclusion that takes about 4 months a year is the interpretation.

Due to constraints on these interpretations twin-scroll type turbocharger petrol engine development testing with interpretation without the support was on durability evaluation and troubleshooting. This analysis evaluated the test of times and duration time is increased turbocharger can contribute substantially to the development of analysis methods became necessary. Prediction of Temperature simplification compared with the measured value is Analysis of all test results are similar to values and damage to the turbocharger housing, temperature inversions are a major offender is represented well.

Figure 9 Transient temperature predictions of the discrete theta TGDI Durability Analysis conducted by applying the actual phenomena and reproduce the degree of damage were investigated. Stress and strain analysis in the red color is the biggest part of its distribution is in good agreement with the experimental results. As shown in Fig. 10 Theta TGDI damaged areas through integrated interpretation of the test also showed good agreement with.

Not included in this chapter Tau TGDI Removable/Integrated Exhaust System Test and Analysis of the other damaged areas also were made to reproduce well.

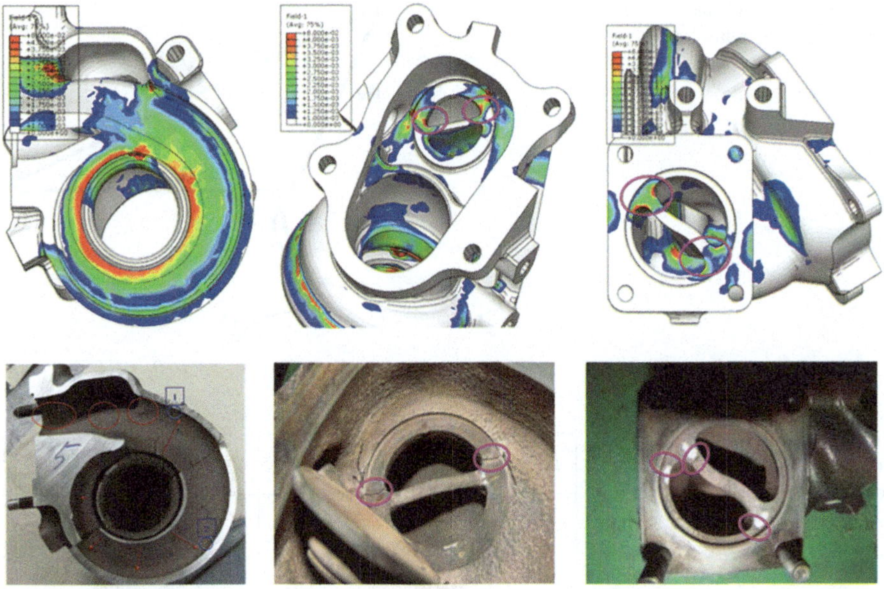

Fig. 9 Durability analysis and failure to reproduce discrete theta TGDI

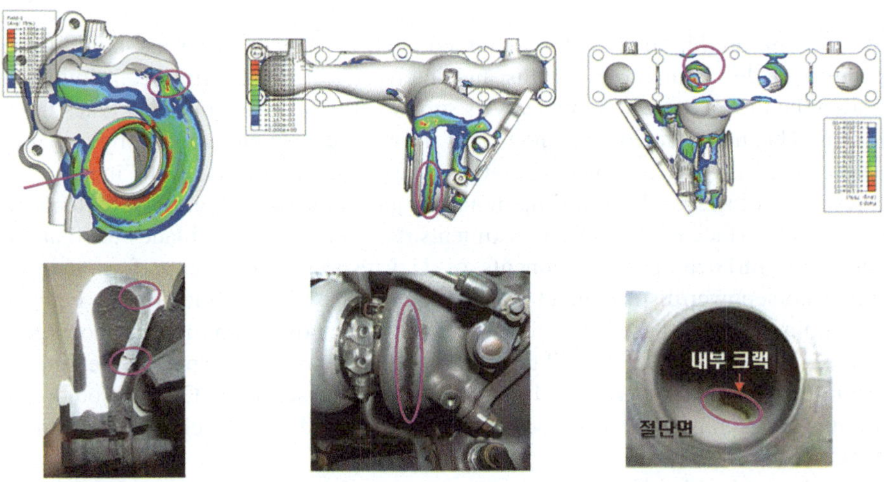

Fig. 10 Integrated durability analysis and failure to reproduce theta TGDI

2.3.3 Commonization Turbine Blade

An analysis of temperature distribution during turbocharger is the most difficult to secure the turbine blade is CAD. CAD turbine blade, so each turbo from the vendor company's confidential information to provide you with CAD is difficult.

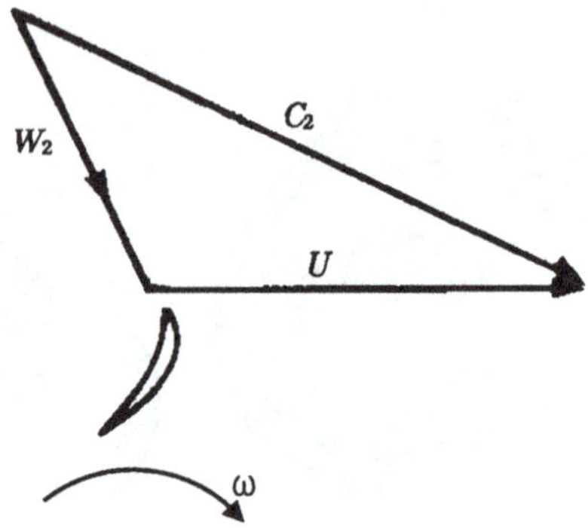

Fig. 11 Turbine blade velocity component

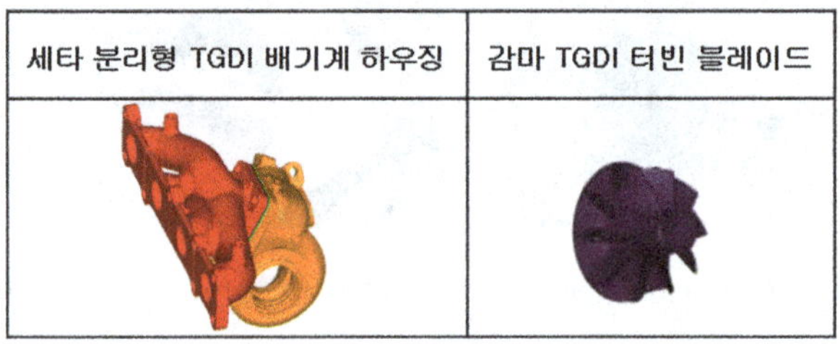

Fig. 12 Similar displacement engine blades commonization

Thus, in interpreting the new turbocharged CAD reverse engineering should be to secure a blade about three months and interpretation of the time-consuming to shorten the time even when the analysis support still is a restaurant that takes three months. This notation Co., turbine blades through reverse engineering by utilizing CAD to perform analysis of temperature distribution was to take this measure.

As shown in Fig. 11 the turbine housing, gas velocities, U and T to determine the contact surface of the HTC constituents. Rated rpm turbine blades to rotate at high-speed turbocharger components of U irrespective of the performance of similar horsepower/displacement engines are similar in the blade. This is held by the turbine blades based on other similar horsepower engine turbochargers Analysis of Temperature Distribution in Steady progress by applying the simplification Transient temperature distribution in the absence of CAD through the turbine blade temperature distribution analysis method was developed that can be carried out.

As shown in Fig. 12 with a similar engine displacement size of the blade because it has a similar but different type of blade is easier to Commonization. In this study, theta and gamma discrete TGDI TGDI Turbo Exhaust Housing Commonization possibility was verified by the blade.

Figure 13 motoring mode, the temperature distribution of the theta TGDI separate test data and turbo housing and Gamma Theta TGDI TGDI Transient temperature distribution with Blades motoring mode is the result. Other similar engine displacement as a result even with the blade shows the results of similar

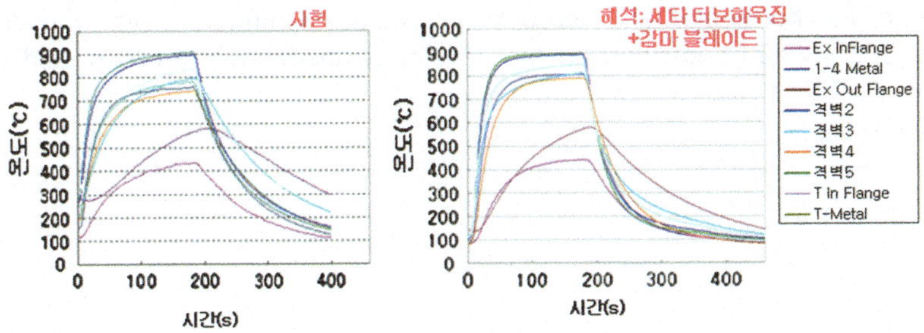

Fig. 13 Similar displacement engine blades commonization

tests. As mentioned earlier, turbocharged ten motoring endurance in cycle mode of liner failure due to repeated phenomenon, so the temperature distribution is similar to the actual damage similar phenomenon can be reproduced. Thus, simplification and Transient Analysis of Temperature Distribution Using the blade under the public more actively in the development stage turbocharger durability analysis support for problem solving is considered to be.

3 Conclusion

Computing needs a lot of time in the turbine blade motion, blade motion, instead of considering the Transient Analysis of Temperature Distribution Analysis Based on Steady motoring mode through the exhaust contacts at each rpm HTC, T by extracting the exhaust flow by using the contact boundary conditions alone, without Transient Solid calculating the temperature distribution was a good agreement with the test results. Transient Analysis of Temperature Distribution in an existing turbocharger that used to take about 3 weeks time was reduced to just a few hours interval rpm increase compared to traditional methods could not be interpreted whole mode temperature distribution can be interpreted. In addition, survival analysis using the results of the temperature distribution at small stress and strain distribution in good agreement with the actual damage phenomena contribute to the development of the turbocharger can substantially simplify the analysis technique has been developed. In addition to acquiring CAD turbine blade, so it takes a lot of time to overcome the flow through the turbine blades Commonization Transient temperature distribution analysis and test results of applying the simplified scheme similar to the temperature behavior over time was get the results. Parts Turbo split temperature behavior over time is directly related to the actual damage phenomena. Transient temperature distribution in the turbine blades Commonization and simplified technique when applied to the test values show a similar temperature behavior by damage similar to the value prediction exam will also be judged. Turbocharger for future development Transient temperature prediction developed in this study simplifies the analysis method developed for horizontal and turbocharger blades Commonization reduce the number of tests and contribute to the development schedule are expected to be reduced.

A Detailed Analysis of the Initiation of Abnormal Combustion with Reaction Kinetics and Multi-cycle Simulation

Michael Heiss, Nikola Bobicic, Thomas Lauer, Bernhard Geringer and Simon Schmuck-Soldan

Abstract For highly boosted gasoline engines with direct injection (DI) the operating conditions with lowest fuel consumption are restricted by irregular combustion like knocking. Therefore, the initiation mechanism for knocking was the subject of this research work. A 4-cylinder DI test engine that was provided by GM Europe was set up at the institute's test bench. Experimental data at low speed at the knock limit (2–3 knock events per 100 cycles) were the basis for numerical investigations. A 1D multi-cycle simulation of gas-exchange and combustion was applied to calculate the in-cylinder properties and charge composition for the knocking cycles. Additionally, a CFD-simulation was carried out in order to obtain the spatial in cylinder distribution of temperature, mixture and residual gas. These results served to set up a stochastic reactor model including the full chemistry of the low and high temperature combustion. The model was initialised with the boundary conditions of the knocking cycle and the temperature and concentration distributions from the CFD-simulation. This method enabled the analysis of knocking combustion on the basis of chemical principles. The results of the multi-cycle analysis showed that important charge properties like the charge temperature at inlet valve closing or the internal residual gas fraction were within a narrow range over all cycles. Furthermore it could be shown that the burn duration for converting 2 % of the fuel mass correlated with cycles showing autoignition. All cycles with an accelerated early flame development showed an irregular heat

F2012-A06-021

M. Heiss (✉) · N. Bobicic · T. Lauer · B. Geringer
Institute for Powertrains and Automotive Technology, Vienna University of Technology, Vienna, Austria
e-mail: michael.heiss@ifa.tuwien.ac.at

S. Schmuck-Soldan
GM Powertrain Europe, Rüsselsheim, Germany

SAE-China and FISITA (eds.), *Proceedings of the FISITA 2012 World Automotive Congress*, Lecture Notes in Electrical Engineering 190,
DOI: 10.1007/978-3-642-33750-5_14, © Springer-Verlag Berlin Heidelberg 2013

release later on during the combustion phase. A detailed modelling of this behaviour was carried out for the stochastic reactor model. It could be shown that only with an initial reactor temperature distribution according to the CFD simulation, which included hot regions, the autoignition occurred as early as in the measurement. The formation of typical intermediate species for the low temperature oxidation leading to autoignition could be described. Finally, a possibility was shown to introduce quasi-spatial information of the flame propagation into the zero dimensional reactor model. In this way critical regions for knocking were identified in accordance with optical measurements of knock sources on the test bench. Most theoretical investigations on abnormal combustion are based on a simplified approach considering the mean cycle for a given operating point. However, abnormal combustion never occurred for the mean cycle but rather for a very early cycle. For that reason, this work focused especially on the detailed reconstruction of a measured knocking cycle with reaction kinetics. Considering the effect of temperature inhomogeneities and flame propagation on knock initiation in a stochastic reactor model is thereby a new and innovative approach.

Keywords Reaction Kinetics · Knocking · Autoignition · Cyclic variations · Multi-cycle

1 Introduction

The need for gasoline engines with lowest fuel consumption and CO_2 emission leads to concepts with high boost pressure and direct injection. For such engines, the combustion with highest efficiency is restricted by irregular combustion like knocking with critical pressure peaks and oscillations [1]. Common remedies to avoid knocking combustion like a delayed spark timing or fuel enrichment are inevitably associated with an efficiency loss due to lowered combustion temperatures [2]. Therefore, the responsible initiation mechanism is still subject of extensive research work.

For the presented work GM Europe provided a turbocharged 4-cylinder DI test engine that was set up at the institute's test bench in order to investigate the phenomena of irregular combustion. Experimental data was carried out at low speed and high load with spark timing at the knock limit. These measurements were the basis for the following numerical investigations.

2 Experimental Investigations and Multi-cycle Analysis

The knock limit for the test bench measurements was defined as 2–3 detected knock events per 100 cycles. To find out the particular conditions that were different only for the knocking cycles a multi-cycle process simulation for 150

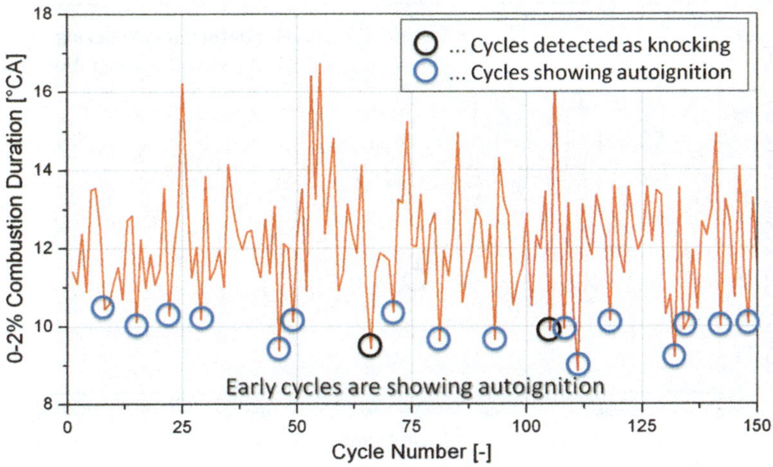

Fig. 1 0–2 % combustion duration of 150 measured consecutive cycles

consecutive measured cycles was carried out with the 1D simulation method three-pressure-analysis (TPA) in GT-Power. For the TPA, the measured pressure profiles of intake, exhaust and cylinder were applied as model boundaries to reproduce the cyclic fluctuations of the cylinder charge properties.

The simulation results of the intake and compression stroke showed that the fluctuations of obvious knock critical properties like the cylinder pressure and temperature at intake valve closing and at spark timing as well as the residual gas fraction lay within a narrow range without any noticeable outliers that would indicate the knock cycles. In contrast, the combustion phase showed considerable differences already at the very beginning of the flame propagation. Figure 1 shows the 0–2 % burn duration. Although the applied knock sensor on the test engine only detected knocking for the two black marked cycles the multi-cycle simulation showed an irregular heat release on the basis of the measured cylinder pressure signal for all early cycles in the range of 10 °CA or below (blue circles). This was interpreted as a weak autoignition that was not strong enough to be detected via knock sensor but already showed exothermic pre-reactions with an irregular pressure increase and consequently can be seen as a pre-stage to knocking.

In contrast, the 10–90 % combustion duration in Fig. 2 makes clear that there was no correlation between the entire combustion duration and the autoigniting cycles. This means that fast flame propagation in the early combustion phase is critical for knocking but not a certain burn duration of the entire combustion.

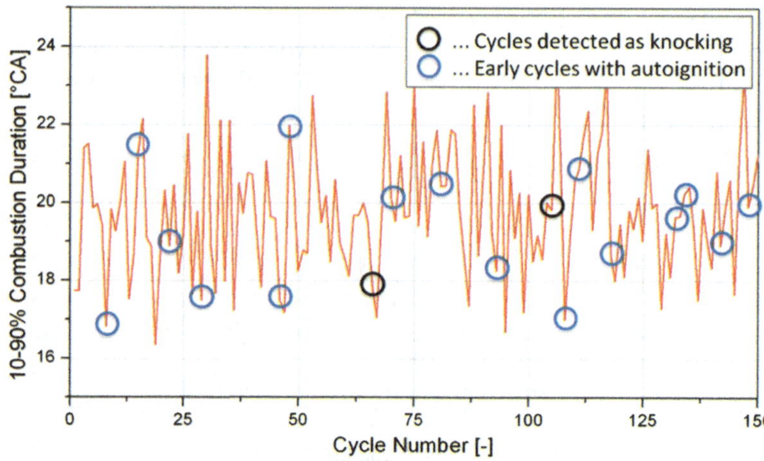

Fig. 2 10–90 % combustion duration of 150 measured consecutive cycles

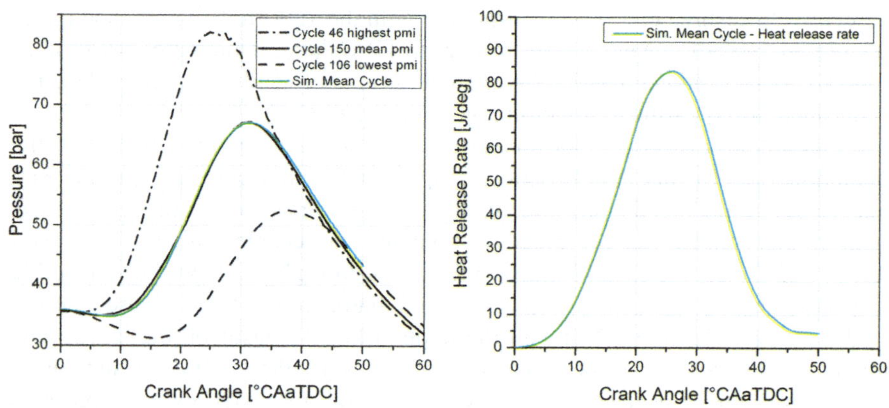

Fig. 3 Measured cylinder pressure for the cycles with minimum, maximum and mean *pmi* (*left*) and the simulated pressure and heat release rate for the mean cycle with the reactor model (*left and right, green*)

2.1 Mean Cycle and Cyclic Variations

A common method and simplification in simulations is to evaluate knock probability by means of the mean cycle at a certain operating point. For that reason, cycle no. 150 which is a representative mean cycle of the 150 consecutive measured cycles was recalculated with reaction kinetics. Looking at Fig. 3 the pressure trace and the heat release of the calculated mean cycle (green lines) showed no autoignition or other combustion irregularities. This makes clear that at the knock

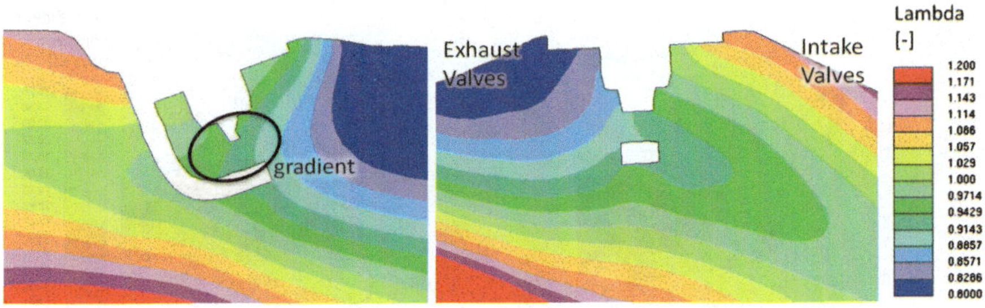

Fig. 4 Lambda distribution at ignition timing in orthogonal sectional views through the spark plug

limit the mean cycle is very unlikely to show a critical behaviour with respect to autoignition and is therefore inapplicable for knock investigations.

Further analysing the measurements, the cycles with minimum and maximum indicated mean effective pressure were added to the left chart in Fig. 3. Although the associated coefficient of variation COV_{pmi} did not exceed 2.2 % the differences between these cycles in pressure rise and maximum pressure were considerable. Keeping this fact in mind an investigation of knock initiation be means of the mean cycle will not yield reasonable results. Therefore, a cycle which was at the autoignition limit was used for the reaction kinetic calculations.

In order to describe the spatial distribution of cylinder charge properties a CFD-simulation with implemented fuel direct injection was carried out in STAR-CD. A model fuel consisting of seven hydrocarbons with different boiling temperatures was used for best representation of the real fuel evaporation behavior and mixture cooling [3, 4]. The CFD wall temperatures were pre-calculated with a wall temperature solver in the 1D cycle simulation that used a simplified FE model to calculate the surface temperatures.

The multi-cycle simulation revealed that the global in-cylinder properties could be excluded as reason for the cyclic variations because they remained nearly constant. In contrast to this, the CFD results showed that a possible explanation for the cyclic fluctuations of the early flame development in Fig. 1 was the local distribution of fuel vapour, velocity magnitude and turbulence around the spark plug gap and hence the condition for the early flame development [5]. The detailed modelling of the spark plug geometry in the CFD model showed that the spark plug electrodes acted as an obstacle to the flow field leading to steep gradients, especially in turbulence and lambda, see Figs. 4 and 5.

This means that the fluctuations of the early combustion duration of Fig. 1 can be explained by means of variations in the turbulent flow field as well as in the spark development. Already minor variations of the spark's shape and position have a considerable effect on the ignition conditions and hence the early combustion velocity, the direction of flame propagation and the knock probability.

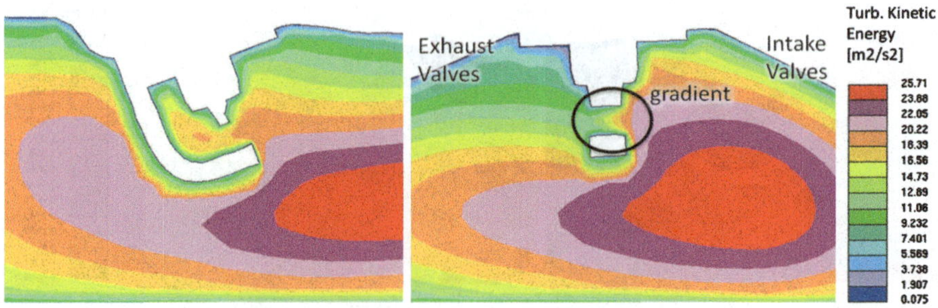

Fig. 5 Turbulence distribution at ignition timing in orthogonal sectional views through the spark plug

Table 1 Octane numbers of different toluene reference fuel mixtures

Toluene Reference Fuel (TRF)				
iso-Octane [Vol %]	n-Heptane [Vol %]	Toluene [Vol %]	RON	MON
100.0	0.0	0.0	100.0	100.0
0.0	100.0	0.0	0.0	0.0
0.0	0.0	100.0	120.0	109.0
95.0	5.0	0.0	95.0	95.0
20.64	18.82	60.54	95.0	85.0

3 Analysis of Knock Initiation with Reaction Kinetics

3.1 Principle of the Stochastic Reactor Model

The software SRM suite is a specialised 0D simulation tool for in-cylinder combustion which is solving thermo, chemical and fluid dynamic processes simultaneously including detailed chemical kinetics [6]. An ensemble of so called "stochastic particles" is used to approximate the distribution of in-cylinder properties such as the composition of chemical species and the temperature. Each particle represents a statistical portion of the in-cylinder volume. At each time-step, the chemistry, heat transfer and turbulent mixing processes are solved for each stochastic particle [6].

A so called Toluene-Reference-Fuel (TRF) consisting of toluene, iso-octane and n-heptane features a different RON to MON and is therefore sensitive regarding self ignition to changes in pressure and temperature. The associated mechanism of Andrae et al. [7] contains the reaction mechanisms of all three components as well as their cross reactions for low and high temperature combustion (137 species, 633 reactions). The appropriate mixture ratio of the three TRF components to satisfy the RON of 95 and the MON of 85 was determined according to the RON/MON mapping described by Morgan et al. [8] for an optimal representation of the test bench premium fuel and is listed in Table 1 next to the pure components.

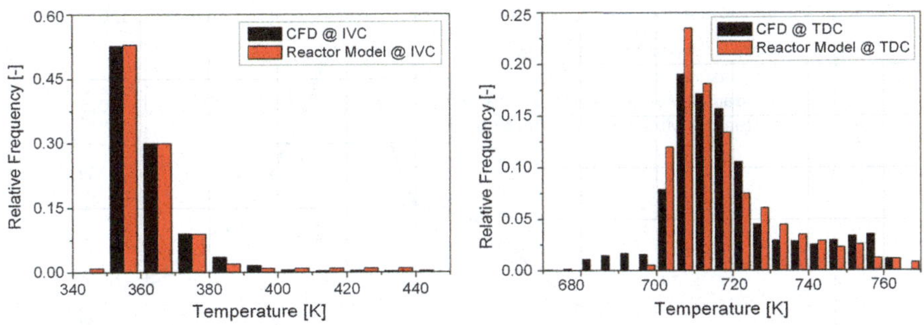

Fig. 6 Temperature distributions at inlet valve closing (*left*) and top dead centre (*right*)

3.2 Setup of the Reactor Model with Temperature Inhomogeneities

Gradients in temperature were shown in experiments to be strongly influential on autoignition [9]. For that reason a 3D-CFD simulation was carried out to deliver the spatial distribution of temperature in the cylinder. Temperature inhomogeneities in the combustion chamber can have several sources like hot wall temperatures, especially close to the exhaust valves or spark plug electrodes, but can also be caused by an insufficient mixture preparation due to a low charge motion or incomplete fuel spray evaporation. The consideration of all these effects was possible with the performed CFD simulation. The results at intake valve closing were extracted to initialise the particles with the cylinder temperature distribution. The temperature spectrum for about 6,20,000 cells of the CFD model at intake valve closing (IVC) is indicated by black bars in the left diagram of Fig. 6. The distribution for the initialisation of the 100 stochastic particles (red bars) derived from this data.

After calculation start, the reactor temperature distribution during compression stroke is dependent on stochastic heat exchange, turbulent mixing and cylinder wall heat transfer. Therefore, the turbulent mixing parameters of the reactor model had to be adjusted to meet the temperature distribution from CFD. The good correlation with the CFD result at top dead centre (TDC) which is depicted in the right diagram of Fig. 6 revealed that a proper setup was found for the stochastic heat exchange and the turbulent mixing between the particles. Especially the formation of knock sensitive areas in the mid and high temperature range was represented properly. The lack of cold particles in the range of 680–700 K played only a minor role for autoignition reasons and was therefore accepted [10]. However, the final mean compression temperatures of CFD simulation and reactor model were equal.

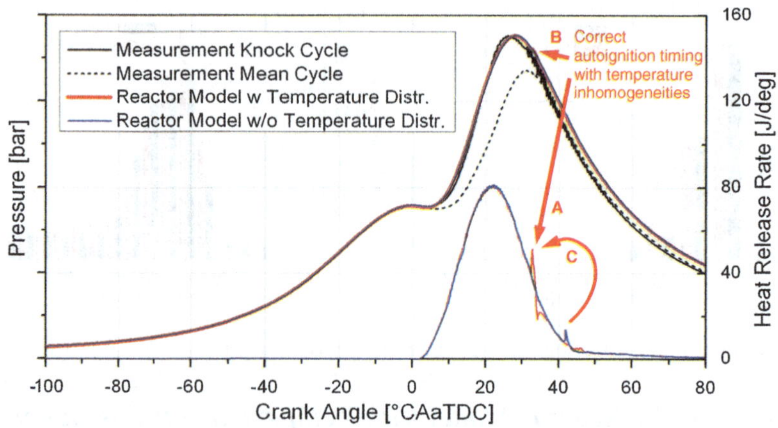

Fig. 7 Cylinder pressure and heat release with an initial temperature distribution (*red*) and with a uniform initial reactor temperature (*blue*)

3.3 Verification of the Autoignition Sensitivity

Figure 1 showed a correlation between the early combustion duration and autoignition. Cycle 102 showed a fast 2 % burn duration and additionally was identified as a knocking cycle by the knock sensor. Therefore, it was denoted as a representative cycle with an autoignition close to the knock limit. In Fig. 7 the measured cylinder pressure over crank angle of knock cycle 102 is depicted as a full black line. As one can see, the pressure rise and the associated oscillations due to knocking were relatively moderate what was typical for an operation at the knock limit. Using a Wiebe approach the heat release of this cycle was rebuilt with the reactor model in order to find the best fit to the measured cylinder pressure. In this way the conditions in the unburned zone and therefore the low temperature chemistry for the autoignition process were comparable to that of the experiment.

Looking at the results of cylinder pressure and heat release with an initial temperature distribution (Fig. 7, red lines), the time of autoignition was in a good agreement with the measurements (full black line). This can be seen by means of the peak in the heat release (arrow A) occurring at the same time as the measured oscillations in the cylinder pressure (arrow B). In contrast, a uniformly initialised reactor temperature led to an unrealistic evenly distributed temperature during compression and combustion what made the reactor too resistant against autoignition. This behaviour can be seen by means of the rather late autoignition timing in Fig. 7 (blue lines) and the arrow C which indicates the shift of the spontaneous heat release dependent on the type of initialisation. This revealed that introducing the temperature inhomogeneities from CFD in the SRM was an important measure for the autoignition sensitivity of the reactor model.

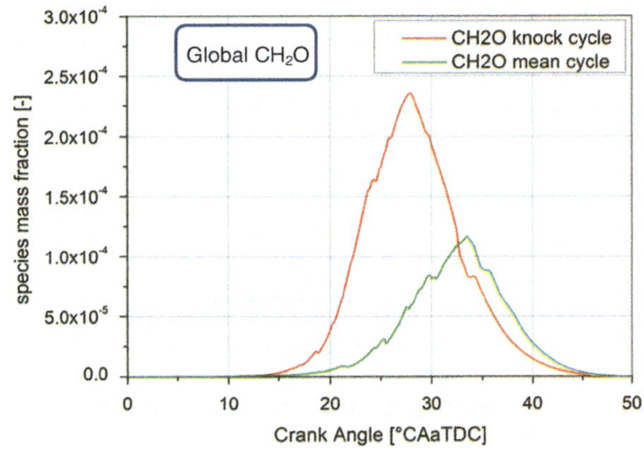

Fig. 8 Calculated global mass fraction of formaldehyde CH_2O for the knocking and the mean cycle

3.4 Formation of Chemical Species and Intermediates

The formation of certain chemical species and intermediates is typical for the self-ignition in knocking cycles [10]. Formaldehyde CH_2O is a representative indicator occurring in the low temperature oxidation and is known because of its spectroscopic detectability in LIF measurements [11]. Figure 8 shows the calculated crank angle resolved CH_2O mass fraction. The CH_2O formed in the knocking cycle (red) reaches its maximum already several degrees before the moment of autoignition what confirmed that it is a precursor to knocking. However, CH_2O could also be observed on a lower level for the mean cycle (green) without autoignition. To clarify this behaviour, a closer look to the key branching species in each particle was carried out. The hydroperoxy radical HO_2 is known as an important intermediate that can lead to chain branching under the desired temperature and pressure conditions in the unburned zone [11].

Figure 9 shows the HO_2 mass fractions of each particle to give a better insight into the radical formation process. The left diagram shows the HO_2 formation for the 100 particles of the mean cycle. That pre-reactions with HO_2 were already taking place for the mean cycle revealed that the desired operating point was at the knock limit. In comparison to the knocking cycle (right diagram) the mass fractions of all particles remained under a certain threshold. This can be explained by a later onset of HO_2 formation due to a slower combustion of the mean cycle leading to lower pressure and temperature in the unburned zone. Additionally, the propagating flame entrained the particles early enough and avoided a further increase of local HO_2. This becomes obvious through the curves which showed an increase of HO_2 and then dropped rapidly. A similar behaviour was noticed for the knocking cycle. Several particles showed an increase to a rather high level but were entrained, with the exception of one particle (red line), before the self-ignition set in. Since each particle represented a portion of the charge volume the flame propagation showed to be a crucial influencing factor on autoignition. Therefore it was analysed in the following.

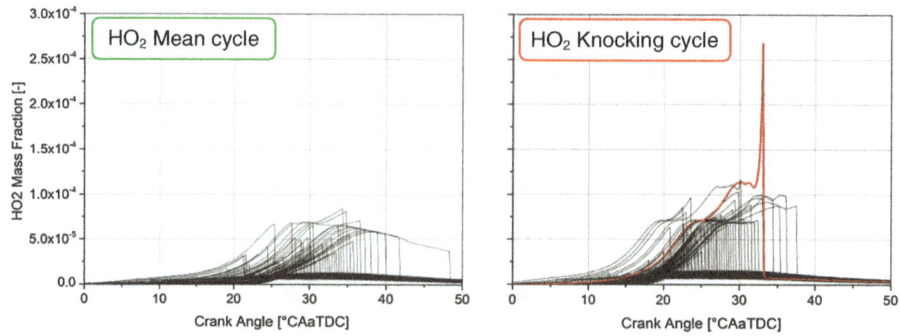

Fig. 9 Calculated HO_2 mass fractions of all 100 particles for the mean cycle (*left*) the knocking cycle (*right*)

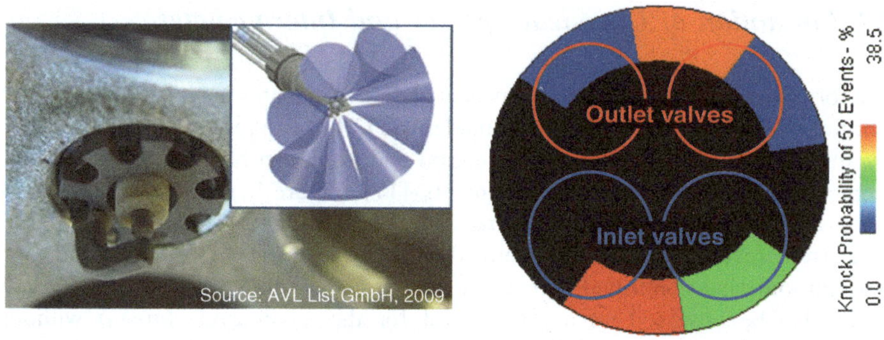

Fig. 10 AVL VisioKnock spark plug (*left*) and its application in optical measurement of the spatial knock probability on the test bench (*right*)

3.5 Knock Probability and Flame Propagation

The sources of knock initiation are of high interest and, in combination with reaction kinetics, can be used to identify knock critical conditions. For this purpose, an optical spark plug, as depicted in the left image in Fig. 10, was installed in the test engine. With eight optical channels the knock critical regions were covered for recording the knock events. The right image in Fig. 10 shows the result of the relative knock occurrence for 52 recorded knock events. Because the CFD simulation calculated a single hot area in the proximity to the exhaust valves, knocking was expected to occur mainly there. However, the optical measurement identified the irregular combustion to occur with a similar probability at the inlet and the outlet region. Consequently, the SRM was applied to find out the reason for that.

The SRM is a zero dimensional approach without solving the spatial propagation of the flame. Anyhow, the reproduction of the measured knocking scenarios was possible by specifying the particle combustion order dependent on the particle temperature. In this way, knocking on the hot exhaust side was simulated via

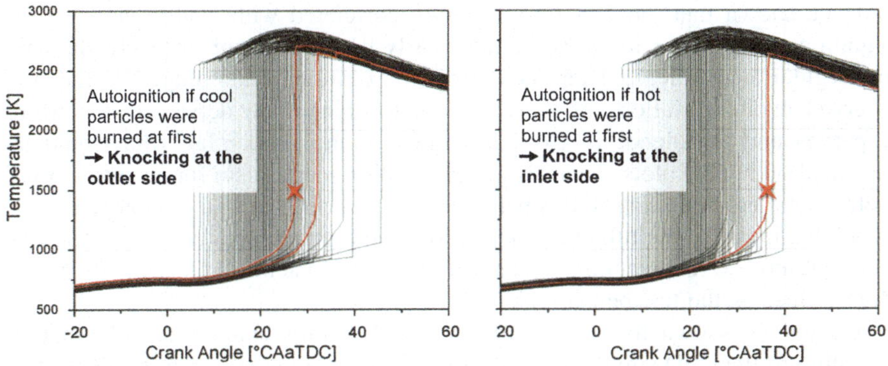

Fig. 11 Particle temperatures for two different particle combustion order scenarios: cool particles burned at first (*left*); hot particles burn at first (*right*)

burning the cool particles of the intake first, whereby the hot side remained as unburned zone and vice versa. Figure 11 shows the resulting particle temperature histories of both scenarios. All regular burning particles are depicted as black lines. They are characterised by an almost vertical increase to a high temperature level above 2,500 K which means that the particle has been entrained into the hot flame. In contrast to this, self-igniting particles showed a considerably different behaviour. Marked with red colour, these particles were passing a kind of heating process with a progressive rise in temperature until their self-ignition set in followed by a spontaneous combustion.

The autoignition in both cases made it clear that knocking occurred independent on the flame propagation. The unburned zone at this operating point is also critical for knocking if the hot regions burn at first. This is in accordance with the CFD simulation as well as the optical knock detection.

The presented method gives the possibility to analyse the influence of inhomogeneities and flame propagation on autoignition probability with reaction kinetics. Via modifying the particle combustion order a quasi-spatial information of the flame propagation can be introduced into the zero dimensional reactor model. If a critical knock behaviour is ascertained for a particular combustion order the demand for an intensified charge motion or tumble level to avoid the corresponding flame propagation with knocking can be stated.

4 Conclusions

Experiments at low engine speed at the knock limit on a 4-cylinder DI test engine, which was provided by GM Europe, were the basis for numerical investigations. A multi-cycle analysis of gas-exchange and combustion showed that important charge properties like the charge temperature at inlet valve closing or the internal residual gas fraction were within a narrow range over all cycles. Furthermore it could be shown that the 2 % burn duration correlated with cycles showing autoignition. All cycles with an accelerated early flame development were showing an irregular heat release later on during combustion. This suggested that the observed cyclic variations of the early combustion phase depended on charge properties and turbulence in close proximity to the spark plug what could be explained with CFD plots. To analyse the initiation mechanism for an autoigniting cycle a detailed modelling of this behaviour was carried out in a stochastic reactor model that included the full chemistry of low and high temperature combustion. A toluene reference fuel reaction mechanism was used to achieve the same RON and MON values as the test bench fuel.

It could be shown that only with an initial reactor temperature distribution according to the CFD simulation that included hot regions, the autoignition set in as early as in the measurement. The crank angle resolved formation of typical intermediate species for the low temperature oxidation leading to autoignition could be described.

Finally, a possibility was shown to introduce quasi-spatial information of the flame propagation into the zero dimensional reactor model. In this way critical regions for knocking were identified in accordance with optical measurements of knock sources on the test bench.

For the analysis of knock initiation, the combination of engine cycle simulation, CFD and chemical reaction kinetics proved to be an appropriate combination to find out the complex mechanism behind the knocking combustion.

References

1. Heywood JB (1988) Internal combustion engine fundamentals. McGraw-Hill, New York. ISBN 0-07-100499-8
2. Eichlseder H (2008) Grundlagen und Technologien des Ottomotors - Der Fahrzeugantrieb. In: List H (ed) Springer Verlag, Wien, ISBN 978-3-211-25774-6
3. Batteh JJ, Curtis EW (2005) Modeling transient fuel effects with alternative fuels. In: SAE paper 2005-01-1127
4. Heiss M, LauerT (2012) Simulation of the mixture preparation for an SI engine using multi-component fuels. In: STAR global conference 2012, Amsterdam
5. Holmström K, Denbratt I (1996) Cyclic variation in an SI engine due to the random motion of the flame kernel, In: SAE paper 961152
6. CMCL Innovations (2012) SRM suite user manual, version 7.2.1. Cambridge
7. Andrae JCG et al (2008) HCCI experiments with toluene reference fuels modeled by a semidetailed chemical kinetic model. Combust Flame 155(4):696–712
8. Morgan N et al (2010) Mapping surrogate gasoline compositions into RON/MON space. Combust Flame 157:1122–1131
9. Pöschl M, Sattelmayer T (2008) Influence of temperature inhomogeneities on knocking combustion. Combust Flame 153:562–573
10. Compton RG, Hancok G (1997) Low-temperature combustion and autoignition comprehensive. Chemical kinetics, vol 35. Elsevier, Amsterdam
11. Azimov U et al (2012) UV–visible light absorption by hydroxyl and formaldehyde and knocking combustion in a DME-HCCI engine. Fuel 98:164–175

Modeling of Six Cylinder Diesel Engine Crankshafts to Verify Belt Load Limits

Kumar B. Dinesh, M. Nagarajan, Patil Shankar, P. Mahesh and N. Muralitharan

Abstract This chapter presents the advantage of using system based model analysis to assess the dynamic behaviour of an internal combustion engine crankshaft. The crankshaft model couples the structural dynamics and the main bearing hydrodynamic lubrication using a system approach. The scope of this chapter is the study of the crankshaft strength and main bearing structural integrity which limits the front end belt loads. Basically there are two limiting factors for the belt loads: 1. The crankshaft strength limits the belt loads in vertical upward direction due to the superposition with the maximum gas load. 2. The front end main bearing limits the belt loads in horizontal direction due to the restricted load carrying capacity in the area of the split line. Crank Train simulation for the following was performed: 1. Torsional analysis. 2. Bearing analysis. 3. Strength analysis. In this work Torsional Vibration (TV) trials were conducted on 5.7l 6-cylinder Euro3 160HP Turbocharged Intercooled Diesel Engine in test bed and Crankshaft fatigue testing was done. The data of these experimental investigations were used for model validation and correlation of Crank Train Simulation results. According to design criteria, regarding structural integrity, the maximum recommended vibration amplitudes per order at the crankshaft front end are in the range of 0.20–0.25° for in-line six cylinder engines. The analysis considers a Torsional Vibration Damper (TVD) assembled to the crankshaft front-end along with additional pulley for power take-off, this is used to drive AC compressor and other auxiliary units (FEAD-Front End Auxiliary Drive) as shown in Figs. 1 and 2.

F2012-A06-022

K. B. Dinesh (✉) · M. Nagarajan · P. Shankar · P. Mahesh · N. Muralitharan
Product Development-Engines, Ashok Leyland, Chennai, India
e-mail: DINESHKUMAR.B@ASHOKLEYLAND.COM

SAE-China and FISITA (eds.), *Proceedings of the FISITA 2012 World Automotive Congress*, Lecture Notes in Electrical Engineering 190,
DOI: 10.1007/978-3-642-33750-5_15, © Springer-Verlag Berlin Heidelberg 2013

Keywords Torsional vibration · Torsional vibration damper · Crankshaft strength · Crankshaft design · FEAD

1 Introduction

Market pressures on internal combustion engine design calls for various FEAD without major modification in engine. Also, efforts to reduce vibration while improving reliability have become increasingly important to the automotive industry due to more stringent requirements for higher performance, low cost and fast to market engine designs. To realise competitive designs optimization of engine components is must. This calls for sophisticated analytical tools which greatly enhance the understanding of various physical phenomena associated with operation critical engine components. This is particularly true of crankshafts, the most analysed engine components [1].

The bearing pressures are very important in the design of crankshafts. The maximum permissible bearing pressure depends upon the maximum gas pressure, journal velocity, amount and method of lubrication and Change of direction of bearing pressure.

The following two types of stresses are induced in the crankshaft.

1. Bending stress; and 2. Shear stress due to torsional moment on shaft.

A crankshaft-block subsystem consists of the crankshaft and the engine block coupled by the hydrodynamically lubricated main bearings. The loading on the system comes from the cylinder pressure and the piston-connecting rod inertia. The cylinder pressure applied on the piston crown is transmitted to the crankshaft rod assembly. The inertia of the piston-connecting rod provides a load on the crankpin as well. The crankpin loads deform the crankshaft and are transmitted to the engine block at the main bearing locations through main bearing hydrodynamics. Both the deformation of crankshaft and engine block affect the main bearing oil film thickness and therefore, the bearing hydrodynamics. For this reason, the mathematical model of the engine system requires three individual models which are coupled together; a structural model of the crankshaft, a structural model of the engine block and a tribological model of the main bearings [2].

The large number of studies reported in the literature on the crankshaft-block interaction problem shows its importance to the industry. A representative sample is given in Ref. [1]. Due to complexity of the mathematical calculations involved, many simplifying assumptions have been used.

The first attempt to solve the problem is reported in Ref. [3] where a calculation of bearing performance in statically indeterminate crankshaft system is described. A static (not dynamic) crankshaft analysis is used and the block elasticity is represented by linear springs. The significance of the crankshaft-block interaction problem in internal combustion engine design is due to a variety of reasons. The

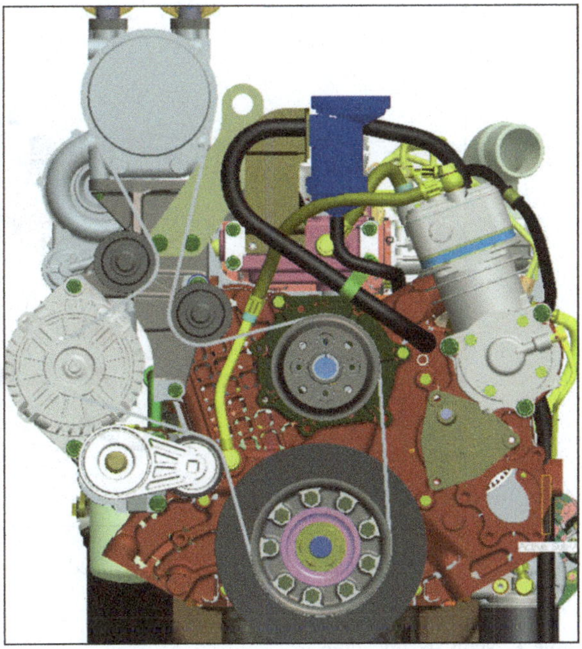

Fig. 1 FEAD with engine mounted AC compressor for a front engine application

accurate prediction of dynamic stress levels is important for durability and high fatigue life [4]. The crankshaft dynamic response is also needed for optimizing crankshaft accessories as pulleys [5] and flywheels [6]. Also the reproduction of the oil film hydrodynamics in the crankshaft-block interaction problem provides an enhanced bearing load prediction [6].

This chapter describes a crankshaft-block interaction model. Subsequent sections describe the experiments done for crankshaft strength analysis and AVL Excite coupled crank-shaft-engine block model. The accuracy and efficiency of crankshaft structural dynamic analysis is demonstrated. Comparison with experimental results shows the good accuracy of the proposed method. Finally, the capabilities and significance of the proposed crankshaft-engine block system model in engine design are illustrated.

2 Specifications of Experimental Engine

The test engine used for the measurement of torsional vibration angular displacement is the 6 cylinder in-line, automotive high speed diesel engine. TV amplitudes were measured with and without additional pulley. The main specifications are shown in Table 1.

Fig. 2 Rear engine application with remotely mounted AC compressor and radiator fan drive

Table 1 Main specifications of experimental engine

Parameter	Content
Max power (HP)	160 @ 2400 rpm
Max torque (Nm)	550 @1400–1900 rpm
No. of cylinders	6
Bore × Stroke(mm)	104 × 113
Capacity (cc)	5700
Compression ratio	17.5
Aspiration	TCI
FIE	Mechanical in-line FIP
Emission norms	Euro III

3 Methodology

3.1 Crankshaft Fatigue Tests

In terms of fatigue failures of a crankshaft, bending is considered to be the most critical loading. Therefore only bending tests were performed. The most critical areas are the fillets of the crankpins. At the test rig, the crankshaft was clamped at a crankpin and a main journal and the bending load was applied at another main journal as shown in Fig. 3.

The following parameters were used:

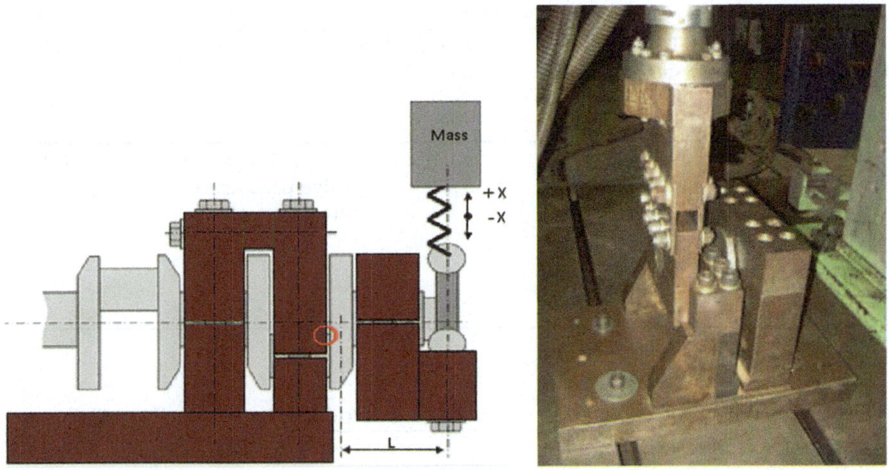

Fig. 3 Crankshaft fatigue test setup

Table 2 Vertical compressive and tensile force

Loading percentage	Compression, kN (peak gas force)	Tensile, kN (inertial force)
250 %	116.725	1.05
300 %	140.1	1.05
Reference (1G = 100 %)	46.69	1.05

- Ultimate number of load cycles: 5 million
- Wave form: sinusoidal
- Test frequency: 35 Hz
- Const. load amplitude and load ratio (R = −0.14)

With the given mass forces and an assumed PFP of 120 bar, Vertical compressive force corresponding to 250 and 300 % of the peak gas pressure and tensile force corresponding to inertial force was applied through one pin with adjacent journal and the next pin restrained (Tables 2, 3).

Based on the listed results a statistical evaluation was done. A fatigue limit model was used which allows for derivation of a fail save stress level from the measured fatigue data. The result of the model is a fail save level at 970 Nm with a confidence level of 95 %. Since the maximum calculated load under engine operation is at 900 Nm the crankshafts are expected to sustain a PFP of 135 bar with a safety factor of 1.08.

Table 3 Crankshaft fatigue testing observation

Crankshaft ID	Test sample throw	Test sample ID	Test load ratio (%)	Target cycles	Results
CS02	2-Web	S4	250	5 million	All the samples passed the test. Checked by dye penetrant spray and no cracks were observed
CS03	2-Web	S7	300		
CS01	4-Web	S2	250		
CS04	4-Web	S11	300		
CS05	6-Web	S15	250		
CS03	6-Web	S9	300		

Table 4 Measured torsional vibration amplitude data at critical orders with rubber TVD

TV test data on 160HP engine		200–215 Hz					
		3 order		4.5 order		6 order	
		W/O pulley	With pulley	W/O pulley	With pulley	W/O pulley	With pulley
Hot	Ramp up	0.22	0.27	0.21	0.24	0.18	0.16
	Ramp down	0.23	0.27	0.22	0.25	0.16	0.16
Cold	Ramp up	0.22	0.33	0.21	0.33	0.17	0.22
	Ramp down	0.22	0.3	0.23	0.31	0.15	0.2

3.2 Torsional Vibration Analysis

TVA analysis were done for H6, 160HP, BS3 engine with In-line mechanical Fuel Injection pump to investigate the TV amplitudes while mounting the newly developed additional pulley provided for to drive remotely mounted AC compressor. Initial trials were done with rubber TVD with and without additional pulley, refer Table 4.

Based on initial test results it was clear that the 3rd and 4.5th order TVA with rubber TVD were around 0.30° which is not within acceptable limit.

Second set of trials were done with newly developed viscous TVD to suit this particular engine application (with and without additional pulley) and TVA were measured, results are shown in Table 5. Results clearly indicate TVA at critical order with Viscous TVD is well within design limit of 0.25°. According to design criteria, regarding noise level and structural integrity, the maximum recommended vibration amplitudes, per order, at the crankshaft front end are in the range of 0.20–0.25° for in-line six cylinder engines. By this way, the rubber TVD is not recommended for the considered engine and only viscous damper is suitable for the mentioned application vis-à-vis the mentioned criteria.

Table 5 Measured torsional vibration amplitude data at critical orders with viscous TVD

TV test data on 160HP engine		3rd order		4.5th order		6th order	
		W/O pulley	With pulley	W/O pulley	With pulley	W/O pulley	With pulley
Hot	Ramp up	0.1	0.136	0.17	0.21	0.12	0.14
	Ramp down	0.1	0.11	0.16	0.2	0.12	0.146
Cold	Ramp up	0.1	0.138	0.18	0.22	0.12	0.13
	Ramp down	0.1	0.13	0.18	0.217	0.129	0.139

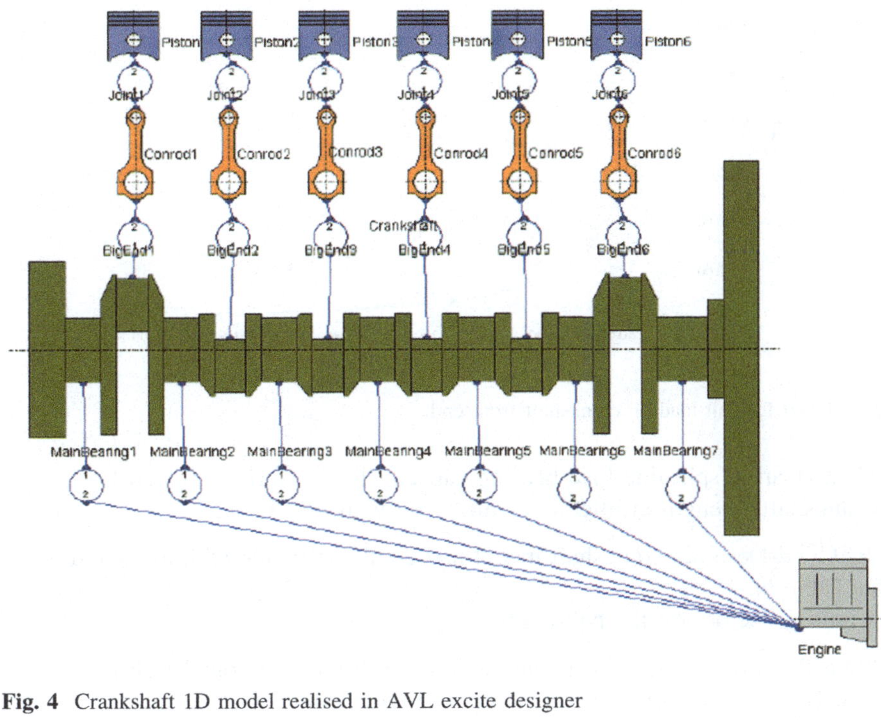

Fig. 4 Crankshaft 1D model realised in AVL excite designer

3.3 1D Simulation

Crankshaft model is similar to a simple spring-damper combination. The flexibility of engine block is represented by it stiffness at each main bearing location. The main output from simulation is the crankshaft dynamic response in terms of TVA at critical orders, damper dissipated power, Damper hub deflection, cyclic speed irregularity and Max Total torque. The experimental test results were used to validate the crankshaft model realised using AVL Excite designer (Fig. 4).

Limiting factors for the belt loads:

Table 6 Excite analysis of TVA at critical orders with viscous TVD

EXCITE analysis data	160HP					
	3 order		4.5 order		6 order	
	W/O pulley	With pulley	W/O pulley	With pulley	W/O pulley	With pulley
TVA	0.13	0.195	0.2	0.22	0.18	0.22

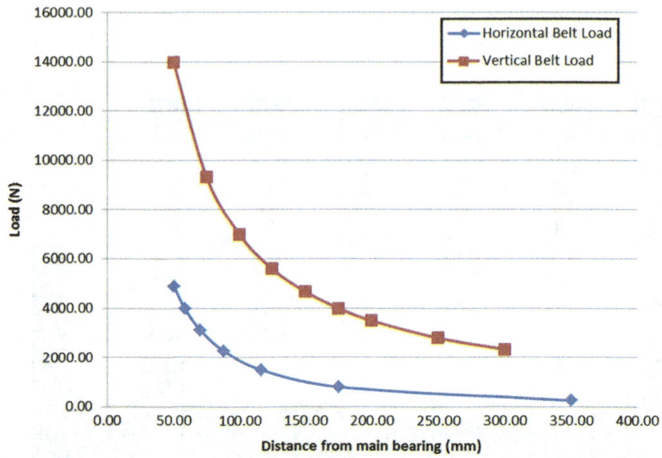

Fig. 5 Limit for belt load on crankshaft front end

- Main Bearing split line load bearing capacity limits Horizontal Belt Load.
- Crankshaft Strength in fillet areas limits Vertical Belt Load.

EXCITE-Designer crankshaft model had good correlation with measured data, refer Table 6.

Limiting factors for the belt loads:

- Main Bearing split line load bearing capacity limits Horizontal Belt Load.
- Crankshaft Strength in fillet areas limits Vertical Belt Load.

To incorporate the front end loading into the web loads user can define the loads definition in the load data which would combine the gas or inertia load with front end loads and apply it with web load selection. So, every web in the shaft modeler has, when the strength is checked in for analysis, the web load definition. Belt loads are applied based upon bending moment limit obtained through simulation. Similarly calculations are done for vertical loading. Permissible load for both horizontal and vertical direction were calculated using Excite simulation results and plotted as shown in Fig. 5.

Simulation result for various belt loads in horizontal direction and corresponding minimum oil film thickness are tabulated in Table 7.

Table 7 Excite analysis for MOFT and horizontal load

Max Belt load (N)	MB load (N)	Distance (mm)	Moment (Nmm)	MOFT (mm)
4891.57	7000.00	50.00	350000.00	0.0015
3992.35	6000.00	58.33	350000.00	0.0015
3118.28	5000.00	70.00	350000.00	0.0015
2280.10	4000.00	87.50	350000.00	0.0015
1495.70	3000.00	116.67	350000.00	0.0015
797.25	2000.00	175.00	350000.00	0.0015
248.93	1000.00	350.00	350000.00	0.0015

4 Conclusions

In this work, the crankshaft torsional properties of a 6 cylinder in-line diesel engine were identified with the help of a single point measurement of angular speed. 1D model of the crankshaft was modelled with assumption of a flexible crankshaft. Initial model parameters were determined by analysing experimental test data done on the engine. The model was then optimized with an actual measurement in steady state condition. Analysing the obtained results and comparing them to the measured ones it is possible to conclude that a good match was obtained between simulation and experimentation. The validated model was then used to study permissible belt loads based upon the maximum bending moment permissible at first main bearing and MOFT corresponding to that bending moment.

Acknowledgments The Authors wish to express their gratitude to Engine R&D Testing and Design team colleagues for their technical support and guidance.

References

1. Mourelatos ZP (2001) A crankshaft system model for structural dynamic analysis of internal combustion engines. Comput Struct 79(2001):2009–2027
2. Kolchin A, Demidov D (1984) Design of automotive engines. MIR Publishers, Moscow
3. Welsh WA (1982) Dynamic analysis of engine bearing systems. MS Thesis, Cornell University
4. Heath AR, McNamara PM (1989) Crankshaft stress analysis-the combination of finite element and classical techniques. ASME ICE 9
5. Okamura H, Morit T (1993) Influence of crankshaft-pulley dimensions on crankshaft vibrations and engine-structure noise and vibrations. In: SAE international proceedings of the 1993 noise and vibration conference, SAE paper no. 931303, pp 329–338
6. Mourelatos ZP (1995) An analytical investigation of crank-shaft-flywheel bending vibrations for a V6 engine. In: SAE international proceedings of the 1995 noise and vibration conference. vol 1, SAE paper no. 951276

Model-Based Control and Calibration for Air-Intake Systems in Turbocharged Spark-Ignition Engines

Kunihiko Suzuki and Seiji Asano

Abstract The purpose of this study is to develop model-based methodologies, which employ thermo-fluid dynamic engine simulation and multiple-objective optimization schemes, for engine control and calibration, and to validate the reliability of the method using a dynamometer test. In our technique, creating a total engine system model begins by first entirely capturing the characteristics of components affecting the engine system's behaviour, then using experimental data to strictly adjust the tuning parameters in physical models. Engine outputs over the engine operation conditions determined by design of experiment (DOE) are simulated, followed by fitting the provided dataset using a nonlinear response surface model (RSM) to express the causal relationship among engine operational parameters, environmental factors and engine output. The RSM is applied to an L-jetronic® air-intake system control logic for a turbocharged engine. Coupling the engine simulator with a multi-objective genetic algorithm, the optimal valve timings are investigated from the viewpoints of fuel consumption rate, emissions and torque. The calibrations are made over all the operation points; the control map is implemented in the turbocharged air-intake system control logic. The validation of the control logic was demonstrated using a model-in-the-loop simulation (MILS). The logic output of the charging efficiency transition due to the varying throttle valve opening angle and variable valve timing was compared with the simulator output. According to the results of the MILS, in-cylinder air mass estimations are in good agreement with the engine simulator under several transient operations.

F2012-A06-026

K. Suzuki (✉)
Hitachi, Ltd, Chiyoda, Japan
e-mail: kunihiko.suzuki.fk@hitachi.com

S. Asano
Hitachi Automotive Systems, Ltd, Chiyoda, Japan

SAE-China and FISITA (eds.), *Proceedings of the FISITA 2012 World Automotive Congress*, Lecture Notes in Electrical Engineering 190,
DOI: 10.1007/978-3-642-33750-5_16, © Springer-Verlag Berlin Heidelberg 2013

Keywords Engine · Model-based control · Calibration · Turbocharger · Simulation

1 Introduction

To meet worldwide emission and fuel consumption regulations, the latest engine systems in passenger cars have been equipped with many electronic technologies, such as actuators and sensors in the air intake system, which allow engine multitasking. For more sophisticated control of these actuators in response to dynamic, time-varying factors, the number of controllable parameters and their variable ranges have also increased. With so many matters to be decided during engine system design, the structure of the engine control software has become increasingly complex, and the time and cost of development have dramatically increased.

From this viewpoint, many model-based calibration tools, which mainly consist of mathematical algorithms and test bench automation techniques, have been produced. Using one available calibration toolbox, Gurerrier et al. [1] established methodologies for a variable-valve SI engine calibration. Their model-based calibration concept was developed using advanced DOE and a radial basis function network (RBF), and the concept was demonstrated through its application to steady-state dynamometer engine mapping. They show that their technique covers the EGR optimization problem with regard to combustion stability, fuel consumption rate, and NOx and HC emissions. The use of a multi-objective optimization scheme allows an informed decision to be made under the multiple restrictions. These methodologies seem to have potential for saving time and allowing the calibrators to understand the complex relation between the control parameters and engine performance [1–4].

However, they rely on a precondition of acquiring large amounts of experimental data from a prototype engine before beginning optimization procedures. Hence they do not appear to be sufficiently useful in avoiding time lost when changes are made in the engine hardware specifications. If the system-level simulation can help provide an understanding of an engine system's behaviour early in the control design process, and can also be used in the engine calibration process, more significant timesaving can be realized. We have developed model-based methodologies for turbocharger engine control and calibration and to validate the reliability of the proposed method.

2 Turbocharged Engine Simulation

2.1 Engine Modeling

A one-dimensional compressible-gas dynamics flow model was used to calculate the intake and exhaust pipe flows. Mass, momentum and energy conservation equations were solved using the finite difference method [5, 6].

$$\begin{cases} \frac{\partial \rho}{\partial t} + \frac{\partial \rho u}{\partial x} = -\frac{\rho u}{A}\frac{dA}{dx}, p = \rho RT \\ \frac{\partial \rho u}{\partial t} + \frac{\partial \rho u^2}{\partial x} = -\frac{\partial p}{\partial x} - \left(\frac{C_f}{D} + C_p\right)\frac{\rho u|u|}{2} - \frac{\rho u^2}{A}\frac{dA}{dx} \\ \frac{\partial \rho e_t}{\partial t} + \frac{\partial}{\partial x}\left\{\rho u\left(e_t + \frac{p}{\rho}\right)\right\} = -\frac{4h_c(T-T_w)}{D} - \frac{\rho u}{A}\left(e_t + \frac{p}{\rho}\right)\frac{dA}{dx} \end{cases} \quad (1)$$

where ρ is the density; u is the velocity of the pipe flow; A is the cross-sectional area of the pipe; D is the diameter of the pipe; p is the pressure; C_f is a friction-loss coefficient; C_p is a pressure-loss coefficient; $e_t(= c_v T + u^2/2)$ is the total energy; c_v is the specific heat at a constant volume; and h_c is the heat transfer coefficient.

In-cylinder gas mass and energy were obtained from the following two differential equations:

$$\begin{cases} \frac{dm_z}{dt} = \frac{dm_{in}}{dt} - \frac{dm_{ex}}{dt}, p_z V_z = m_z R_z T_z \\ \frac{dm_z e_z}{dt} = -p_z \frac{dV_z}{dt} + \frac{dQ_b}{dt} + \frac{dQ_w}{dt} + h_{in}\frac{dm_{in}}{dt} - h_{ex}\frac{dm_{ex}}{dt} \end{cases} \quad (2)$$

where subscript z represents the value of the in-cylinder gas; m is the mass; e is the internal energy; V is the volume; dQ_b/dt is the heat release rate due to combustion; dQ_w/dt is the heat transfer to the cylinder wall; dm_{in}/dt and dm_{ex}/dt are the mass flow rates into the cylinder via the intake valve and out of the cylinder via the exhaust valve, respectively; and h_{in} and h_{ex} are upstream enthalpies at the intake valve and exhaust valve, respectively. The details of models concerned with the physical values mentioned above are summarized in the literature [7].

2.2 Turbocharger Modeling

According to the turbocharger dynamics equation, a change of rotor speed, N_T, can be calculated by the excess power (or power deficiency) required to drive the compressor and delivered by the turbine [5, 6].

$$\begin{cases} \frac{dN_T^2}{dt} = \frac{2}{J_T}\left(\frac{60}{2\pi}\right)^2\left(\dot{W}_T - \dot{W}_C - \dot{W}_M\right) \\ \dot{W}_C = \frac{c_p T_{Cup}}{\eta_C}\frac{dm_C}{dt}\left\{\left(\frac{p_{Cdn}}{p_{Cup}}\right)^{(\kappa-1)/\kappa} - 1\right\}, \dot{W}_T = c_p T_{Tup}\eta_T\frac{dm_T}{dt}\left\{1 - \left(\frac{p_{Tdn}}{p_{Tup}}\right)^{(\kappa-1)/\kappa}\right\} \\ \dot{W}_M = c_0 + c_1 N_T + c_2 N_T^2 \end{cases} \tag{3}$$

where subscript C and T represent the values of the compressor and turbine, respectively; J_T is the rotational inertia of the turbocharger; $\dot{W}_C$ is the work-transfer rate to drive the compressor; $\dot{W}_T$ is the power delivered by the turbine; η_C and η_T are the efficiencies of the compressor and the turbine, respectively; and dm_C/dt and dm_T/dt are the mass flow rates of the compressor and the turbine, respectively. These values for the turbocharger characteristics were calculated from the turbocharger specifications and its performance map. $\dot{W}_M$ is the mechanical friction loss of the rotor as a function of N_T.

3 Response Surface Methodology

3.1 Regression Model

First, a model-based calibration is carried out to acquire the causal relationship between engine input and output, then the database is fitted using a nonlinear RSM, and finally the model is implemented into the control logic. Our model-based design procedure was developed using regression analysis as described in this section [8, 9].

The regression model, which consists of a regress and y, referring charging efficiency, and x_1, x_2, x_3, x_4, and x_5 as regressors, was set up. Here, five regressors represent the engine speed, intake manifold gas pressure, intake valve open timing, exhaust valve close timing, and intake manifold gas temperature, respectively. The values of these regressors were normalized using their maximum and minimum values. Two types of regression models, nonlinear polynomial equation and RBF, were analyzed.

The 4th-order polynomial equation was defined in the following form.

$$\begin{aligned} y = \beta_0 + \sum_{i=1}^{5}\beta_i x_i + \sum_{i=1}^{5}\sum_{j=1}^{5}\beta_{ij}x_i x_j \\ + \sum_{i=1}^{5}\sum_{j=1}^{5}\sum_{k=1}^{5}\beta_{ijk}x_i x_j x_k + \sum_{i=1}^{5}\sum_{j=1}^{5}\sum_{k=1}^{5}\sum_{l=1}^{5}\beta_{ijkl}x_i x_j x_k x_l \end{aligned} \tag{4}$$

Each of the terms was converted as, $\xi_1 = 1$, $\xi_2 = x_1$, $\xi_3 = x_1^2$, $\xi_4 = \cdots$, and thus, the multiple linear regression model is given. In the matrix notation, the regression model is given in the following form:

$$\begin{cases} Y = X\beta + E \\ Y = \begin{bmatrix} y_1 \\ y_2 \\ \vdots \\ y_m \end{bmatrix}, \beta = \begin{bmatrix} b_1 \\ b_2 \\ \vdots \\ b_n \end{bmatrix}, X = \begin{bmatrix} \xi_{11} & \xi_{12} & \cdots & \xi_{1n} \\ \xi_{21} & \xi_{22} & \cdots & \xi_{2n} \\ \vdots & \vdots & \ddots & \vdots \\ \xi_{m1} & \xi_{m2} & \cdots & \xi_{mn} \end{bmatrix}, E = \begin{bmatrix} \varepsilon_1 \\ \varepsilon_2 \\ \vdots \\ \varepsilon_m \end{bmatrix} \\ \beta = (X^T W X)^{-1} X^T W Y, W = diag[w_1, w_2, \cdots w_m] \end{cases} \quad (5)$$

where m is the number of datasets; n is the number of model parameters to be identified; Y is an $m \times 1$ vector of the simulation results; X is an $m \times n$ matrix of the levels of the regressor variables; β is an $n \times 1$ vector of the partial regression coefficients; and E is an $l \times 1$ vector of random errors. According to the robust linear least squares method, partial regression coefficients can be calculated using the above relations. Here W is an $m \times m$ diagonal matrix of weights with diagonal elements $w_1, w_2, \cdots w_m$. The values of the weights were taken from the literature [7, 10].

On the other hand, the normalized Gaussian radial basis function kernel was defined as follows:

$$\sum_{i=1}^{m} \{\beta_i \cdot \varphi(x, x_i)\}, \varphi(x, x_i) = \exp\left(-\frac{\|x - x_i\|^2}{r^2}\right) \quad (6)$$

where r is the radius of the Gaussian kernel. In the matrix notation, the regression model is given in the following form:

$$\begin{cases} Y = \beta + E \\ Y = \begin{bmatrix} y_1 \\ y_2 \\ \vdots \\ y_m \end{bmatrix}, \beta = \begin{bmatrix} b_1 \\ b_2 \\ \vdots \\ b_m \end{bmatrix}, = \begin{bmatrix} \varphi_{11} & \varphi_{12} & \cdots & \varphi_{1m} \\ \varphi_{21} & \varphi_{22} & \cdots & \varphi_{2m} \\ \vdots & \vdots & \ddots & \vdots \\ \varphi_{m1} & \varphi_{m2} & \cdots & \varphi_{mm} \end{bmatrix}, E = \begin{bmatrix} \varepsilon_1 \\ \varepsilon_2 \\ \vdots \\ \varepsilon_m \end{bmatrix} \\ \beta = ({}^T + \lambda I)^{-1} {}^T Y \end{cases} \quad (7)$$

where m is the number of datasets, which is equal to the number of model parameters to be identified; Y is an $m \times 1$ vector of the simulation results; is an $m \times m$ matrix of the levels of the regressor variables; β is an $m \times 1$ vector of the kernel weights; and E is an $m \times 1$ vector of random errors. To minimize the influence of outliers, the regularized least squares algorithm was adopted for the present analysis. λ represents the regularization parameter and was calculated by the cross-validation method.

3.2 Design of Experiments

From a practical viewpoint, it is inefficient to carry out the model-based calibration in all combinations of a full factorial design. In the present analysis, a D-optimal design was adopted to collect data as parsimoniously as possible while providing sufficient information to accurately estimate the model parameters [8, 9].

According to the linear least squares method, partial regression coefficients can be obtained by the expression transformation of normal equations. If $|X^T X|$ is a maximum, the variance of the regression coefficients, β, is minimized. Following this statistical relation, the index D_{eff} was taken as the optimality criterion:

$$D_{eff} \equiv \frac{\ln\{\det(X^T X)\}}{n} \rightarrow Maximum \tag{8}$$

If the database is generated by an iterative search of the parameter settings, then as the value of D_{eff} becomes larger, the data-producing process is actively manipulated to improve the quality of information and to eliminate redundant data for the specified model.

In the present analysis, sixteen sampling points were taken for each of the five regressor variables in the full factorial design; 10^5 candidate combinations were prepared; then 1024 data points were chosen from the candidates using a genetic algorithm (GA), with the value of D_{eff} as a maximum. The final iteration was taken as the specified design.

4 Results and Discussion

4.1 Regression Analysis

Engine calibrations were carried out under 160 different operating conditions, which consisted of engine speeds of 800–6000 rpm, and intake manifold pressures of 0.03–0.20 MPa in a variable intake and exhaust valve engine. Using the measured data to strictly adjust the tuning parameters was followed by providing the dataset. The valve timings were also varied for each condition. The simulation results are shown in Fig. 1 plotted against the experimental results. As shown in the figure, the simulation results are in good agreement with the experimental results, with a correlation of 0.9992. It can be said that the simulation accuracy is good enough to be adapted to a model-based calibration preceding regression modeling.

Using this simulator, the charging efficiency was calculated for the different valve timings. The case where the intake valve timing is retarded and the exhaust valve timing is advanced is shown in Fig. 2a; the case where the intake valve timing is advanced and exhaust valve timing is retarded is shown in Fig. 2b.

Fig. 1 Relationship between experimental results and simulation results

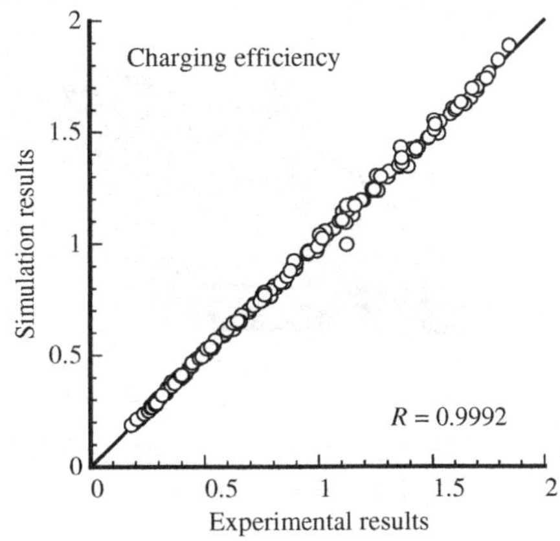

According to the comparison of the results, the charging efficiency increases as the valve overlap duration increases. This tendency can be seen notably at lower engine speeds. This is due to the fact that the relation between intake and exhaust gas pressures reverses due to the balance characteristics of the pressure ratio to mass flow rate in the compressor and the turbine. It can be said that the valve timing calibration in a turbocharged engine must consider the effect of these turbocharger characteristics.

A database is provided to acquire the static causal relationships among the charging efficiency and engine inputs, followed by fitting the dataset using the polynomial equation and the RBF with a least squares method. The regression model results are shown in Fig. 3 plotted against the experimental results. As shown in these figures, the regression model results are in good agreement with the experimental results; the correlations of the polynomial equation and the RBF are 0.9984 and 0.9992, respectively. Thus, the RBF was taken as the regression model for charging efficiency.

4.2 Valve timing Calibration

Using the engine simulator in conjunction with available optimization tool, modeFRONTIER®, a model based calibration was demonstrated through its application to variable intake and exhaust valve engine mapping. In the practical wide open throttle (WOT) calibration procedure, the valve timing, spark timing and equivalence ratio are simultaneously satisfied as torque improves under WOT with regard to acceptable limits of CO emissions and exhaust gas temperature. In the present analysis, the intake/exhaust valve timings and the equivalence ratio were set as independent variables; and brake mean effective pressure (BMEP), NOx and CO emissions, and exhaust gas temperature were set as the dependent variables. Setting these parameters in this way on the workflow of modeFRONTIER, a multi-objective genetic algorithm (MOGA) was run.

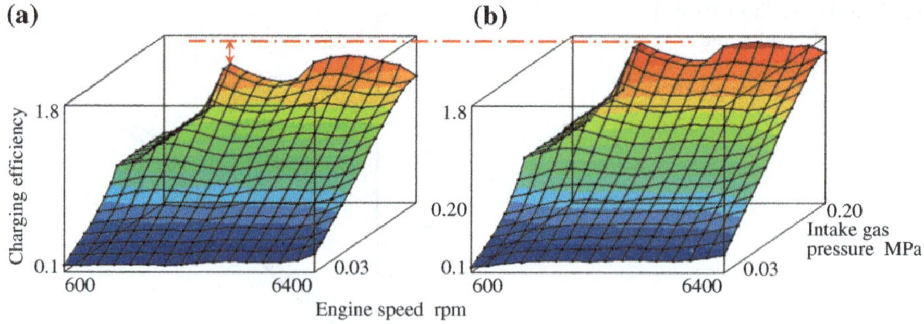

Fig. 2 Charging efficiency for different valve timings. (**a**) Intake: retarded, exhaust: advanced (**b**) Intake: advanced, exhaust: retarded

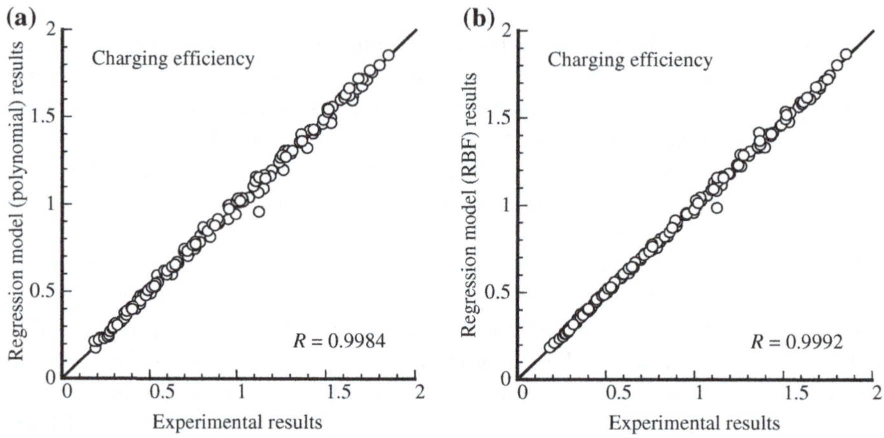

Fig. 3 **a** Relationship between simulation results and polynomial equation results; **b** Relationship between simulation results and RBF results

Figure 4a shows the heuristic processes of intake valve timing, equivalence ratio, CO emissions, and exhaust gas temperature at 3200 rpm WOT, and (b) shows contour maps of the mean effective pressure, CO and NOx emissions, and exhaust gas temperatures for different equivalence ratios and intake and exhaust valve timings. According to Fig. 4a, each set of variables converges to unity, which is satisfied with objectives and restrictions. As shown in Fig. 4b, these are remarkably affected by the equivalence ratio. This is due to the fact that the fuel enrichment changes the equilibrium composition and reduces the flame temperature and knock occurrence. It can be said that this visualization of the solution space allows an informed robust decision to be made in engine mapping.

Figures 5a, b, and c show the contour maps of the BMEP and EGR for different valve timings at 2000 and 3200 rpm WOT; and Fig. 5d shows the contour maps of the brake specific fuel consumption rate (BSFC) for different valve timings under a 1600 rpm partial load. The manual calibration results are also plotted in the

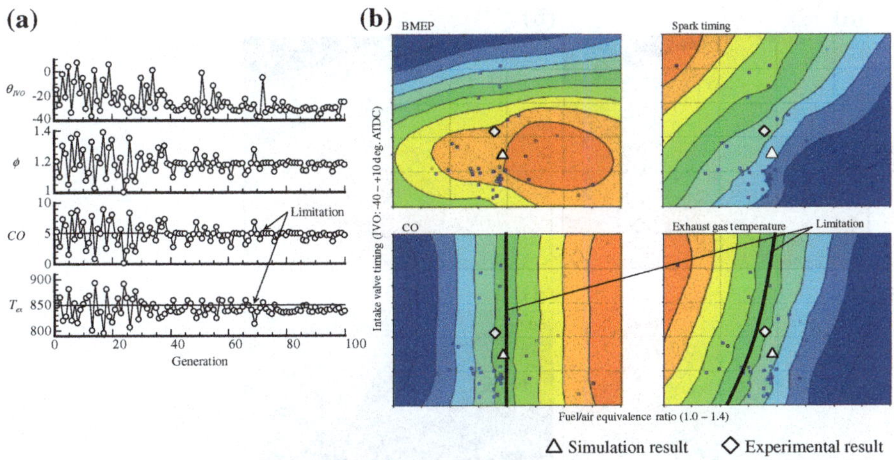

Fig. 4 **a** Heuristic pressures of intake valve timing, equivalence ratio, CO emissions, and exhaust gas temperature at 3200 rpm WOT; **b** Contour maps of BMEP, CO and NOx emissions, and exhaust gas temperature for different equivalence ratios and intake valve timings

figures. According to the figure, the optimal intake valve timing is advanced at lower engine speeds, and retarded at higher engine speeds. On the other hand, the optimal exhaust valve timing is retarded monotonically as engine speed increases. EGR decreases considerably at 2000 rpm WOT when valve overlap is large because EGR is scavenged. The results of the model-based calibration is in good agreement with that of the manual calibration results. Thus, the valve timing calibration was also carried out under other operating points in the same way, and the results were stored in the lookup tables.

4.3 Intake Air Estimation Logic

When the throttle valve opening angle changes rapidly, the mass flow detected by an air-flow sensor differs considerably from the actual mass flow rate into the cylinder via the intake valve due to intake manifold gas dynamics. To control the air–fuel ratio during these transient situations, the in-cylinder air mass must be accurately estimated [11–15]. In variable valve engines, valve timing also needs to be considered.

The intake manifold gas pressure and temperature can be solved by four differential equations, which are derived simultaneously from the mass and energy conservation equations and the state equation, as follows:

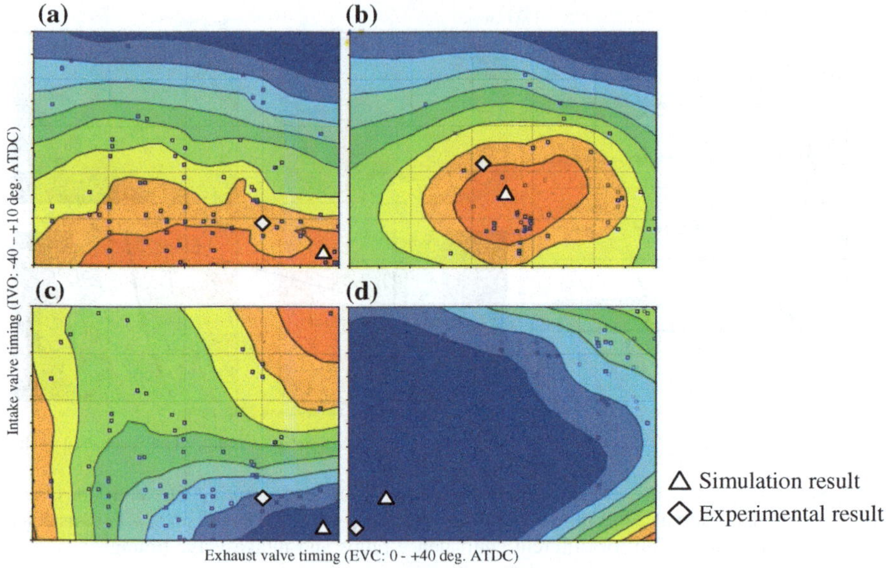

Fig. 5 **a**, **b**, **c** Contour maps of BMEP and EGR for different intake and exhaust valve timings at full load; **c** Contour maps of BSFC for different intake and exhaust valve timings at partial load. (**a**) BMEP: 2000 rpm WOT, (**b**) BMEP: 3200 rpm WOT, (**c**) EGR: 2000 rpm WOT, (**d**) BSFC: 1600 rpm Partial

$$\begin{cases} \frac{dp_{up}}{dt} = \frac{\kappa R}{V_{up}} \left\{ T_{atm} \frac{dm_{afs}}{dt} \left(\frac{p_{up}}{p_{atm}} \right)^{(\kappa-1)/\kappa} - T_{up} \frac{dm_{th}}{dt} \right\} + \frac{\kappa-1}{V_{up}} \frac{dQ_{up}}{dt} \\ \frac{dT_{up}}{dt} = \frac{T_{up}}{p_{up}} \left\{ \frac{dp_{up}}{dt} - \frac{RT_{up}}{V_{up}} \left(\frac{dm_{afs}}{dt} - \frac{dm_{th}}{dt} \right) \right\} \\ \frac{dp_{dn}}{dt} = \frac{\kappa R}{V_{dn}} \left(T_{up} \frac{dm_{th}}{dt} - T_{dn} \frac{dm_{cyl}}{dt} \right) \\ \frac{dT_{dn}}{dt} = \frac{T_{dn}}{p_{dn}} \left\{ \frac{dp_{dn}}{dt} - \frac{RT_{dn}}{V_{dn}} \left(\frac{dm_{th}}{dt} - \frac{dm_{cyl}}{dt} \right) \right\} \end{cases} \quad (9)$$

where the subscripts *up* and *dn* represent the values upstream and downstream of the throttle valve, respectively; κ is the specific heat ratio; T_{atm} is the atmospheric temperature; dm_{cyl}/dt is the mass flow rate into all the cylinders, which can be calculated using the RBF; dm_{th}/dt is the mass flow rate via the throttle valve; dm_{afs}/dt is the mass flow rate detected by the air flow sensor; and dQ_{up}/dt is the heat transfer to the intercooler, calculated from the lookup table as a function of the mass flow rate. The intake and exhaust valve timings are calculated from the lookup table, which is provided by the model based calibration procedure. Figure 6 shows a block diagram of the control logic for air mass flow rate estimation in turbocharged variable valve engines. The charging efficiency obtained from the implementation logic is applied to controlling the spark timing and fuel injection.

Model verification and logic validation was demonstrated using a MILS in which the engine simulator was used as a plant model. The logic output of the charging efficiency transition due to the varying throttle valve opening angle and

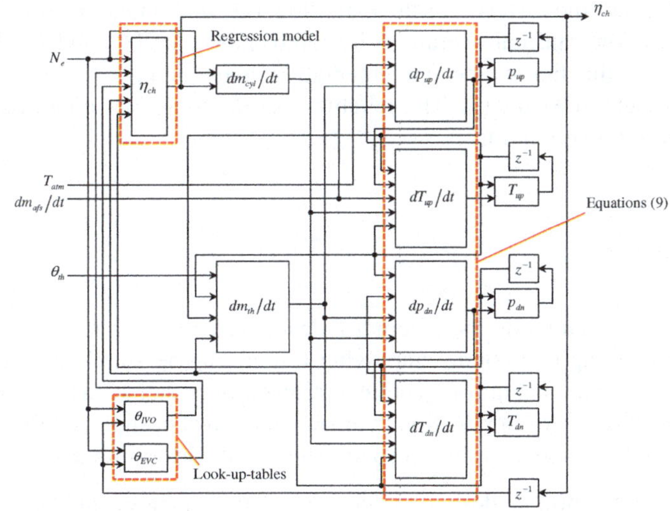

Fig. 6 Block diagram of the control logic for air mass flow rate estimation in turbocharged variable valve engines

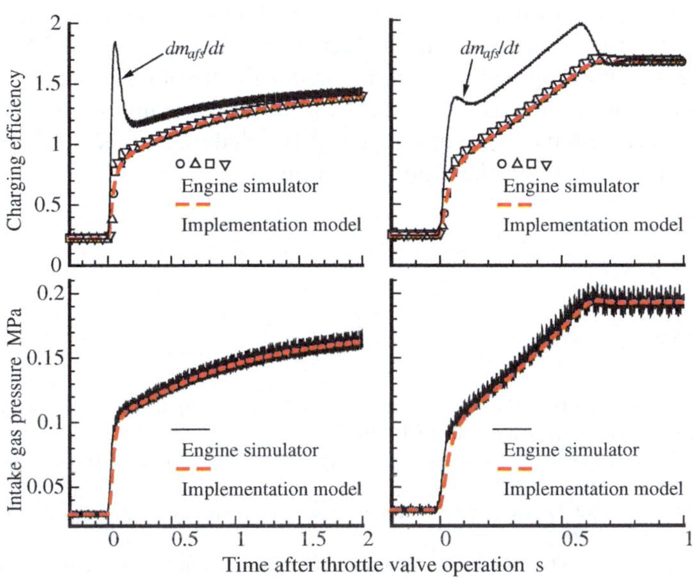

Fig. 7 Control logic validation using MILS

variable valve timing was compared with the simulator output. In order to take the mechanical responses of the actuators and sensors into account, a 2nd order transfer function was used to curve-fit the actual behaviour.

Figure 7 shows the histories of the mass flow rate and intake manifold pressure under acceleration transient operations in step response at 2000 and 3200 rpm. As can be seen in the figure, the implementation logic outputs follow the engine simulator outputs in each case. The reliability of the logic was confirmed using a model-in-the-loop simulation.

5 Conclusion

We have developed a model-based control design environment for practical use in a variable valve engine system, employing a total engine simulation, a response surface method, and a multi-objective optimization scheme. The concept was demonstrated through its application to an actual control design, and the reliability of the proposed method was investigated. The results are summarized as follows:

1. Applying a D-optimal design to the providing database, and a radial basis function and regularized least squares method to fit the database, the methodology for a model-based control design in an air intake system equipped with a turbocharger was investigated.
2. Coupling the turbocharged engine simulator with MOGA, a model-based calibration for optimal valve timing was carried out under multiple objectives and restrictions. The tendency of the optimal valve timing calculated by the simulator was in good agreement with the manual calibration results.
3. L-jetronic control logic using the RSM for the intake manifold in turbocharged variable valve engines was developed. The reliability of the logic was confirmed using a model-in-the-loop simulation.

References

1. Guerrier M et al (2004) The development of model based methodologies for gasoline IC engine calibration. SAE paper 2004-01-1466
2. Holliday T et al (1998) Engine—mapping experiments: a two stage regression approach. Techno metrics 40:120–126
3. Rose DW et al (2002) An engine mapping case study—a two stage regression approach. ImechE C606/025/2002
4. Morton TM et al (2002) Radial basis functions for engine modeling. ImechE C606/022/2002
5. Heywood JB (1988) Internal combustion engine fundamentals. McGraw-Hill Inc,
6. Merker GP et al (2005) Simulating combustion. Springer
7. Suzuki K et al (2011) Model-based technique for air-intake-system control using thermo-fluid dynamic simulation of SI engines and multiple-objective optimization. SAE paper 2011-28-0119
8. Montgomery DC et al (2006) Introduction to linear regression analysis. Wiley
9. Montgomery DC (2008) Design and analysis of experiments. Wiley

10. DuMouchel W et al (1989) Integrating a robust option into a multiple regression computing environment. Proceedings of the 21st symposium on the interface, American Statistical Association, pp 297–301
11. Turin R et al (2009) Low-cost air estimation. SAE paper 2009-01-0590
12. Tanabe H et al (2007) Fuel behavior model-based injection control for motorcycle port-injection gasoline engines. SAE paper 2007-32-0045
13. Iwadare M et al (2009) Multi-variable air path management for a clean diesel engine using model predictive control. SAE paper 2009-01-0733
14. Okazaki S et al (2009) Development of a new model based air-fuel ratio control system. SAE paper 2009-01-0585
15. Sekozawa T et al (1992) Development of a highly accurate air-fuel ratio control method based on internal state estimation. SAE paper 920290

Study on Dynamics Modeling and Analysis of Valvetrains

Caiyun Guan, Wenjie Qin and Xiaobo Wang

Abstract In this paper, the one-mass model of the valve train of one engine is developed and parameters of the model are determined. Finite element method is used to calculate the nonlinear contact stiffness of between the cam and the follower. It is observed that the contact stiffness changes little with the rotation angle of the cam and have a nonlinear relationship with the contact load. Dynamic performance testing equipment is used to measure the motion characteristics of the valve train and the simulation results of the dynamics model are compared with measured data at the same speeds. The simulated results show that both cases of applications of the linear and nonlinear contact stiffness are in good agreement with the experimental results at low speeds. But at high speeds, model of nonlinear contact stiffness is more closely meet the experimental results.

Keywords Valve train · Dynamics model · Nonlinear · Contact stiffness · Simulation

1 Introduction

The valve train has important influence on the performance of an engine as it controls the intake and exhaust process. In the early stages, gearing chain of the valve train was considered to be rigid. However, as the speed increase of the internal combustion engine, it was found that the movement of the valve was not

---

F2012-A06-028

C. Guan (✉) · W. Qin · X. Wang
School of Mechanical Engineering, Beijing Institute of Technology, Beijing, China
e-mail: gcyff@163.com

SAE-China and FISITA (eds.), *Proceedings of the FISITA 2012 World Automotive Congress*, Lecture Notes in Electrical Engineering 190,
DOI: 10.1007/978-3-642-33750-5_17,

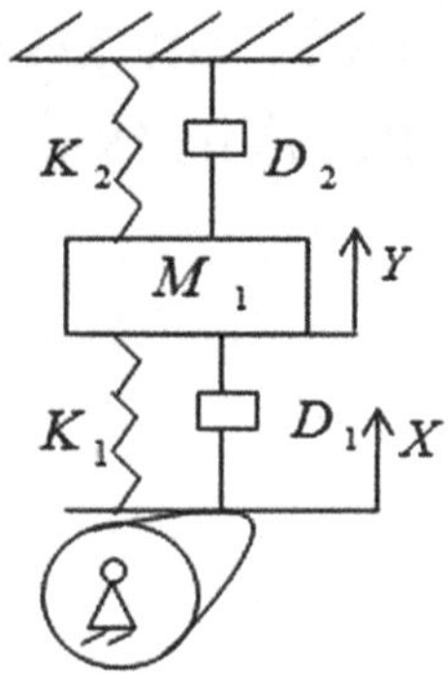

Fig. 1 One-mass dynamic model

conform to the cam profile. It is due to elastic deformation existed in the chain [1]. Therefore, elastic dynamics analysis must be carried out. The dynamics system is simplified to lumped massed connected by springs and dampers. Systems can be divided as one-mass model and multi-mass model. Among them, the one-mass model consists of one-mass and the corresponding connection springs and dampers. In this paper, dynamic performance testing equipment is established based on the valve train of a diesel engine. Further, the one-mass dynamics model is developed. Contact stiffness of the cam and its follower and other parameters of the model are determined, and the relationship between the contact stiffness and the cam rotate angle and the contact load is studied in this paper. Motion characteristics of the mechanism under variable speeds are simulated and compared with the test data. By the comparison of simulation and experimental results in both cases of applications of the linear and nonlinear contact stiffness, the effect of the nonlinear contact stiffness on the dynamic characteristics of the valve train can be verified. This may have certain significance on the prediction of the valve train under high speeds.

2 Dynamic Model

The one-mass dynamic model developed to represent the valve train systems used in the testing equipment is shown in Fig. 1. This mode uses the motion of the lumped mass M_1 to describe the motion of the follower. One of its end is connected to the ground through the valve spring with the stiffness indicated by K_2 and the other end is connected to the spring with stiffness of K_1 which is controlled by the 'equivalent' cam whose motion characteristics is known. K_1 In the model represents the stiffness of the whole gearing chain and D_1 is the damping. D_2 is the damping of the spring.

The lift function of the cam follower need to be determined in the simulation, that is the displacement of M_1 denoted by Y, $Y = Y(\alpha)$.

The force focused on M_1 is defined as:

$$F = M_1 a = M \cdot \omega^2 \cdot \frac{d^2Y}{d\alpha^2} \tag{1}$$

Where, α is the rotation angle of the camshaft and ω is the angular velocity of the camshaft. The force includes:

(a) The elastic restoring force $K_1 \cdot J$, where

$$J = \begin{cases} X(\alpha) - Y(\alpha), & X(\alpha) - Y(\alpha) > 0, \\ 0, & X(\alpha) - Y(\alpha) \le 0 \end{cases} \tag{2}$$

(b) The internal damping force $D_1 \cdot \omega \cdot J_v$, where

$$J_v = \begin{cases} \frac{dX}{d\alpha} - \frac{dY}{d\alpha}, & X - Y > 0, \\ 0, & X - Y \le 0 \end{cases} \tag{3}$$

(c) The spring force of the valve $-K_2 \cdot Y(\alpha)$
(d) The spring preload of the valve $-F_0$
(e) The external damping force $-D_2 \cdot \omega \cdot \frac{dY}{d\alpha}$

The motion differential equation is as follow:

$$M \cdot \omega^2 \cdot \frac{d^2Y}{d\alpha^2} = K_1 \cdot J + D_1 \cdot \omega \cdot J_v - K_2 \cdot Y(\alpha) - F_0 - D_2 \cdot \omega \cdot \frac{dY}{d\alpha} \tag{4}$$

The initial condition for the equation is:

$$Y|_{\alpha=\alpha_0} = \frac{dY}{d\alpha}|_{\alpha=\alpha_0} = 0 \tag{5}$$

Where α_0 is the angle of camshaft when the follower reaches the peak.

3 Determination of Parameters

3.1 Contact Stiffness

Stiffness is important to the dynamic characteristics and the nonlinear contact stiffness is difficult to calculate [2]. Generally, the contact deformation and contact stiffness are calculated according to Hertz theory [3–5]. But to the flat cam follower, the radius of curvature is infinite. So it is not fully meet the application conditions of the Hertz formula.

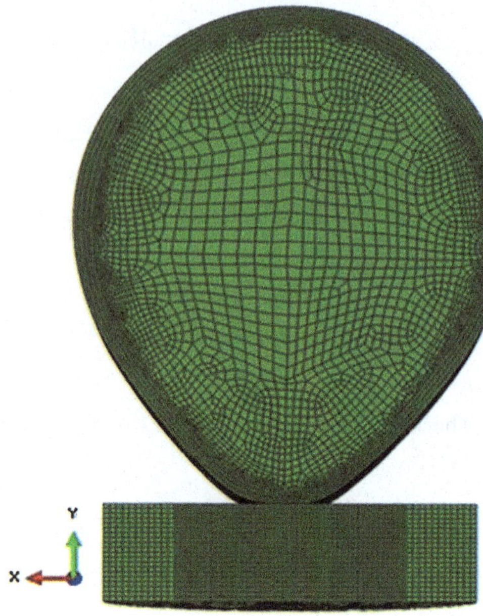

Fig. 2 The mesh model of the cam and its follower

In this paper, the contact between the cam and the tappet is simplified to the contact between the cam and a flat follower with the diameter of 45.5 mm and thickness of 12 mm. Finite element model is established as a 1/2 model due to the symmetry of the structure. The model is meshed by hexahedral element and edges of the cam and its follower are refined. Fixed constraint is imposed on the central point of the base circle of the cam and translation constraint is imposed on the follower except Y direction. Finally, uniform load is imposed on the plane of the flat follower. Figure 2 shows the mesh model of the cam and its follower. The elastic deformation can be obtained by the displacement value of the follower in Y direction. The contact stiffness value is calculated in the following two circumstances. Firstly, the load value imposed on the follower is kept invariant, the contact stiffness value is calculated under various rotation angle of the cam. Figures 3, 4 and 5 show the displacements in Y direction with the load of 1,500 N and the cam rotate angle of 0, 45 and 90°. Secondly, keep the position of the cam relative its follower unchanged, calculate the contact stiffness values under different load values. Figure 6 shows the curve of the contact stiffness versus the cam rotate angle under the load of 1,500 N. Figure 7 shows the fitting curve of the contact stiffness changing with the load at the cam rotate angle of 0°. From Figs. 6 and 7, it can be seen that contact stiffness fluctuates little versus the cam rotate angle and has a nonlinear relationship with the contact load.

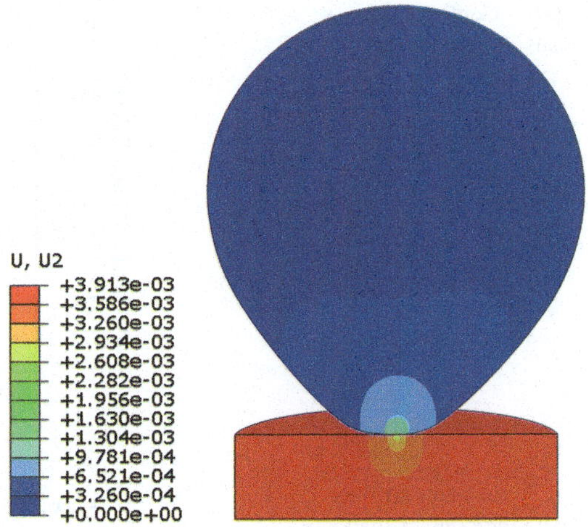

Fig. 3 The displacement in Y direction at the cam angle of 0°

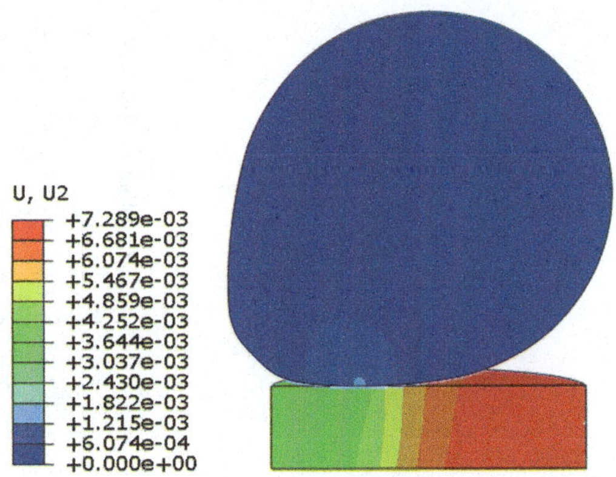

Fig. 4 The displacement in Y direction at the cam angle of 45°

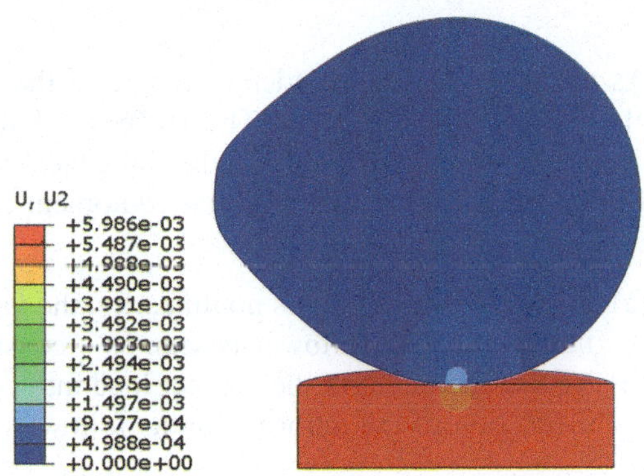

Fig. 5 The displacement in Y direction at the cam angle of 90°

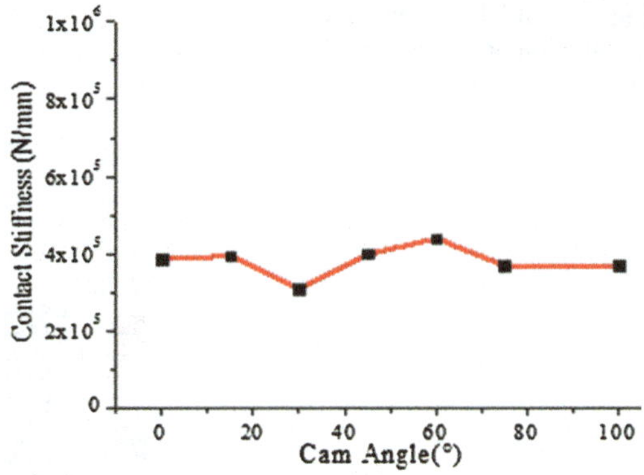

Fig. 6 Contact stiffness versus the cam angle

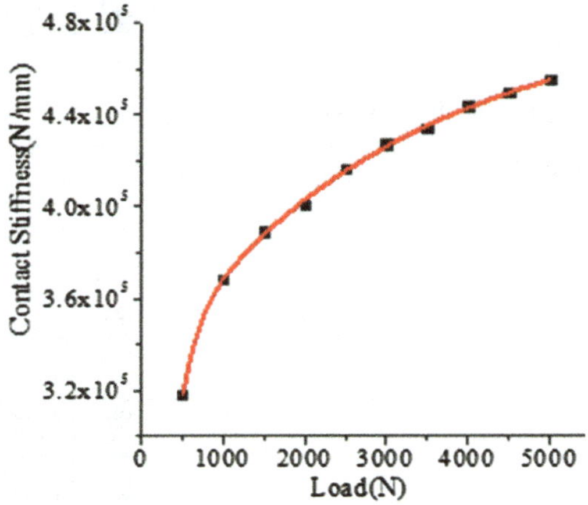

Fig. 7 Contact stiffness versus the contact load

3.2 Gearing Chain Stiffness

As the follower is a slender cylinder, so the stiffness can be obtained from the elasticity formula directly. The stiffness of the latch piece in the experimental device can be calculated using the finite element method. The stiffness of the total gearing chain of the experimental equipment can be obtained under the following two cases:

(a) Without considering its nonlinearity, the value of the contact stiffness between the cam and its follower is seen as a constant value.
(b) Taking the effect of the contact load into consideration, the contact stiffness is a function of the deformation as follow:

$$K_c = F(\delta) \tag{6}$$

Where, K_C is the contact stiffness, δ is the deformation of the cam and its follower.

Numerical interpolation method is used to calculate the contact stiffness value under different loads. The stiffness of the gearing chain is calculated as follow:

$$K_1 = \frac{1}{\frac{1}{K_t} + \frac{1}{K_c} + \frac{1}{K_S}} \tag{7}$$

Where, K_1 is the stiffness of the total gearing chain, K_t is the stiffness of the follower, K_C is the contact stiffness between the cam and its follower, K_S is the stiffness of the latch piece in the experimental device.

3.3 Contact Damping

Damping in the system is calculated, using the formula introduced in literature [6].

$$D_1 = 1.07\sqrt{(K_1 + K_2)M_1} \tag{8}$$

Where, K_1 is the overall stiffness of the gearing chain,K_2 is the stiffness of the valve spring and M_1 is the lumped mass of the model.

4 The Dynamic Performance Testing Equipment

The dynamic performance testing equipment is established as shown in Fig. 8 Acceleration signals of the cam follower under different speeds can be measured by the testing device [7]. The output of the motor is controlled by the frequency changer. The cam is connected to the motor by the coupler and the camshaft is driven by the motor. The follower is pushed by the cam and is connected to the restoring spring. Measured data of the acceleration can be obtained through the sensor fixed at the bottom of the follower passed to the amplifier and dealt with the computer software.

5 Comparision Between the Simulation and Experiment Results

The one-mass dynamic model established in this article is analyzed in both cases of the constant contact stiffness and nonlinear contact stiffness. At the same time, the acceleration of the cam follower is measured by the dynamic performance testing equipment at different speeds. Finally, the simulation and experimental

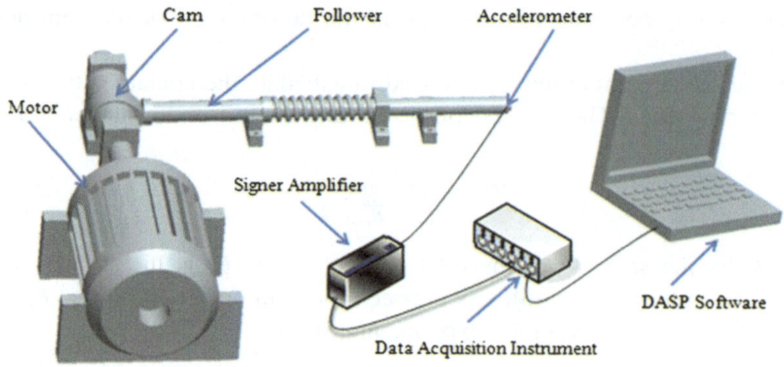

Fig. 8 Schematic of the testing equipment

Fig. 9 Results with the nonlinear contact stiffness at 300 rpm

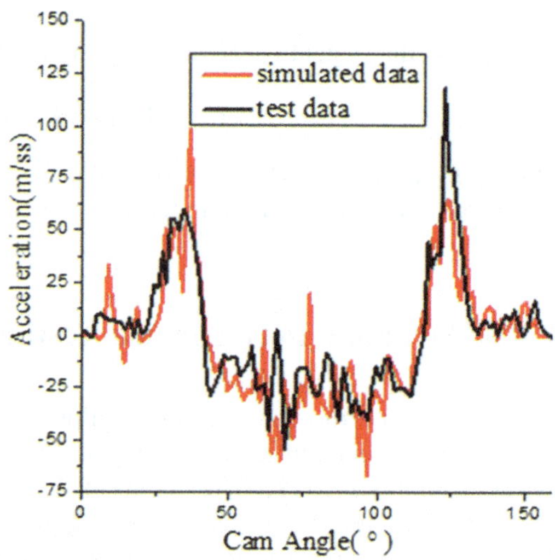

results are compared at the same speed. The Figs. 9, 10, 11, 12, 13, 14 are the comparative results at 300, 600, and 900 rpm.

From figures above, it can be seen that with the speed increasing of the cam, accelerations of the cam follower are also proportionally increased. Amplitudes fluctuation of the acceleration is also significantly increased, because the elastic vibration of the system becomes more severe. At low speeds, the simulation results and experimental results are in good agreement in both cases of applications of the linear and nonlinear contact stiffness. That means, the nonlinearity of the contact stiffness has small effect on the dynamic characteristics of the follower at low speeds. But at high speed, simulation results with nonlinear contact stiffness are more consistent with the experimental results.

The accuracy of the one-mass dynamic model can be verified from the above comparison. Comparing with the experimental data, the dynamic model with the

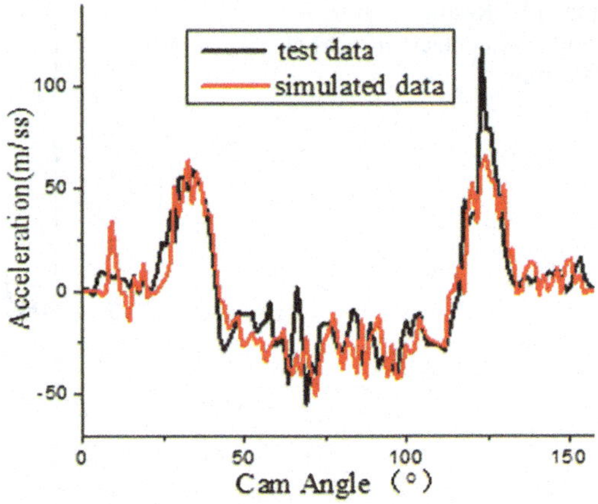

Fig. 10 Results with the constant contact stiffness at 300 rpm

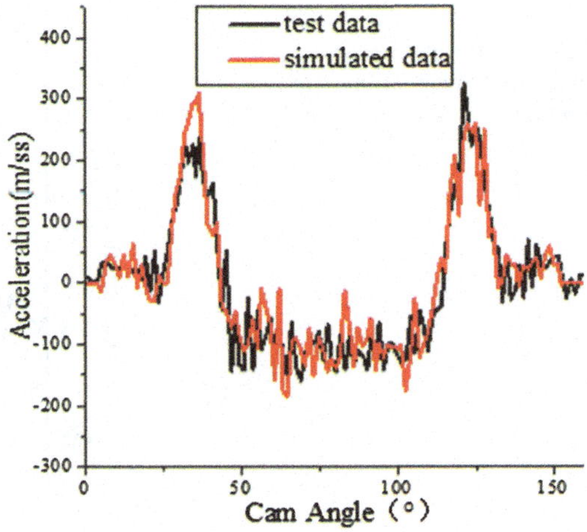

Fig. 11 Results with the nonlinear contact stiffness at 600 rpm

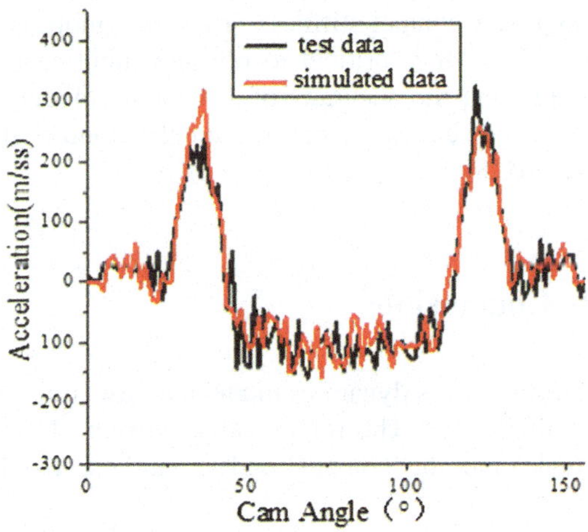

Fig. 12 Results with the constant contact stiffness at 600 rpm

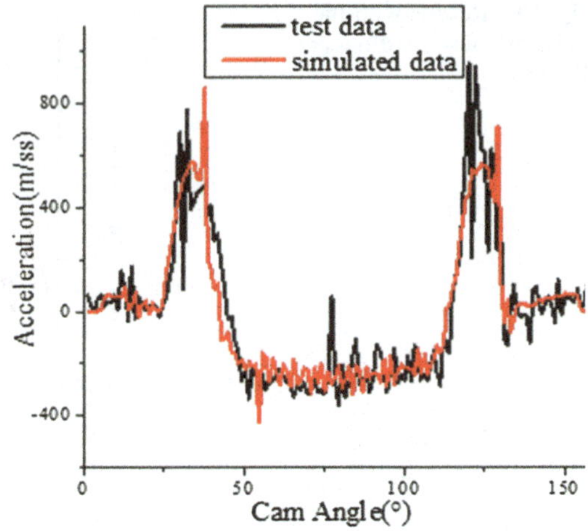

Fig. 13 Results with the nonlinear contact stiffness at 900 rpm

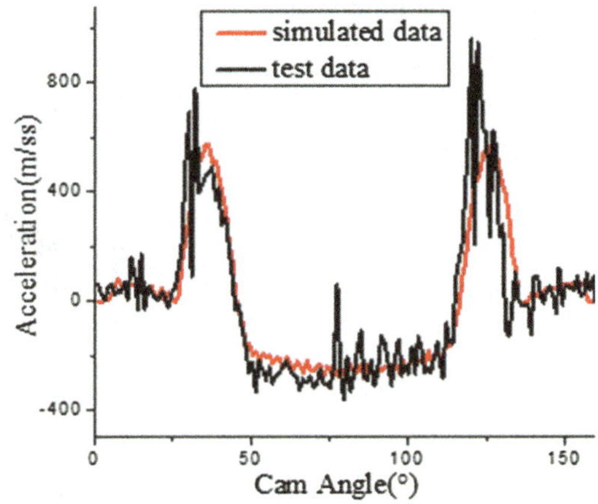

Fig. 14 Results with the constant contact stiffness at 900 rpm

nonlinear contact stiffness provides a more accurate simulation result at the high speed. So it is critical to consider nonlinear contact stiffness of valve train for improving the accuracy of a dynamic model, especially, for a high-speed engine since the elastic vibration and the contact impact of the valve train are more severe [8].

6 Conclusions

The one-mass dynamics model and test equipment for the valve train is established in this paper. The relationship between the contact stiffness and the cam rotate angle and the contact load is studied. Its dynamic performance is measured.

Finally, the simulation results are compared with the measured results. The following conclusions can be drawn from above:

a) The contact stiffness between the cam and flat follower fluctuates little versus the rotation angle of the cam. But it has a nonlinear relationship with the contact load.
b) The comparison results between the simulation data and test data show that the nonlinearity of the contact stiffness has small effect on the dynamics characteristics of the valve train at low speeds, but the dynamic model of nonlinear contact stiffness provides a more accurate simulation result at the high speeds.

References

1. Liu Z, Yu X, Shen Y-M (2005) Comparison of valve train dynamic models. J Z U (Engineering Science), (In chinese)
2. Teodorescu M, Votsios V, Rahnejat H (2006) Jounce and Impact in cam-tappet conjunction induced by the Elastodynamics of valve train system. Meccanica 41:157–171
3. Moreno D, Mucchi E (2007) Multibody analysis of the desmodromic valve train of the ducati motogp engine. Multibody Dyn, pp 1–13
4. Lee J, Patterson DJ (1997) Nonlinear valve train dynamics simulation with a distributed parameter model of valve springs. J Eng Gas Turbines Power 119:692–698
5. Shang H-J (1988) Valve train of the internal combustion engine—design and calculation shanghai. Fudan University Press (In chinese)
6. Yan HS, Tsai MC (1996) An experimental study of the effects of cam speeds on cam-follower systems. Mech Mach Theory
7. Kisu L (2004) Dynamic contact analysis for the valvetrain dynamics of an internal combustion engine by finite element techniques. Automobile Eng
8. Robert L, Norton (1998) Analyzing vibrations in an IC engine valve train. SAE 980570

Mechanistic Modeling in System Engineering: Real-Time Capable Simulation of a TGDI Engine Powered Vehicle

Johann C. Wurzenberger, Titina Banjac, Roman Heinzle and Tomaz Katrasnik

Abstract A mechanistic overall vehicle model is presented. Besides the mechanical driveline domain it comprises a mean value based air path, a crank-angle resolved cylinder description, exhaust gas aftertreatment, cooling and lubrication models and a software ECU to run transient operating conditions. An innovative coupling of different domains and sub-domains that allows for a very good ratio between modeling depth and computational expenses is presented. The capability of the overall model to adequately capture the interaction of different domains is demonstrated by a certification cycle simulation for cold start engine conditions. Enhancements on the engine performance are investigated by electrifying the propulsion of the water pump. The study highlights that even small changes in the powertrain topology and their assessment already call for a comprehensive multi domain system engineering model capable to mechanistically describe all essential effects.

Keywords Modeling · Thermo-management · System engineering · Real-time simulation · TGDI

F2012-A06-032

J. C. Wurzenberger (✉) · T. Banjac
AVL List GmbH, Graz, Austria
e-mail: johann.wurzenberger@avl.com

R. Heinzle
MathConsult GmbH, Linz, Austria

T. Katrasnik
University of Ljubljana, Ljubljana, Slovenia

SAE-China and FISITA (eds.), *Proceedings of the FISITA 2012 World Automotive Congress*, Lecture Notes in Electrical Engineering 190,
DOI: 10.1007/978-3-642-33750-5_18,

1 Introduction

The development of future powertrains is guided by steadily decreasing emission limits, CO2/fuel consumption targets and additional requirements on a maximum level of performance, drivability and safety. One promising approach to meet these requirements is to downsize the internal combustion engine (ICE) and to apply super/turbochargers, variable valve timing, EGR, advanced combustion systems and comprehensive exhaust aftertreatment. The challenges in the development of future powertrains do not only lie in the design of individual components but in the assessment of the powertrain as a whole. On a system engineering level it is required to optimize individual components globally and to balance the interaction of different sub-systems by applying tailored control algorithms.

Computer simulation as complementary part to test-bed is a well-accepted methodology to support the powertrain development process. The simulation of whole powertrains configured in various topologies incorporating multiple energy sources/sinks requires multi-physical models in a flexible simulation environment. The system engineering models still need to feature maximum physical depth to ensure predictability in early development phases where experimental data are rarely available. Besides predictability, especially computational speed and ease of parameterization strongly improve the potential of system engineering simulations. Computational efficient approaches are not only beneficial during the design phase. They are the inevitable prerequisite for supporting the reuse of models also during the late control calibration and testing phase. By using the same models for SiL and HiL simulations the often time-consuming re-development of models can be avoided.

This study presents a system engineering model following the requirements on predictability and computational constrains. The approach is discussed based on the example of a conventional compact class vehicle equipped with a turbocharged, direct injecting gasoline engine. The system engineering model comprises several different modeling domains. (1) These are the vehicle chassis and driveline, (2) the engine air path including combustion and exhaust aftertreatment, (3) cooling and lubrication, and (4) control. Special emphasis is put on the selection of appropriate modeling depths in order to have reliable and numerically efficient models available in early stages of the engine/vehicle development process. This is reached by additionally considering characteristic time scales of particular domains as discussed by Katrašnik et al. [1]. The capability of the overall model to adequately capture the interaction of different domains is demonstrated by an NEDC drive cycle simulation for cold start engine conditions. In this study special emphasis is put on the enhancements on the engine performance by electrifying the propulsion of the water pump. The study does not only show the impact on reduced energy demand and fuel consumption by modifying the powering of the water pump. It also highlights the mutual interaction-between the different domains. The interaction of the oil and water circuit e.g. influences the temperatures of the coolant media, heat transfer and heat-up in the engine and further the catalytic converter light-off and tail pipe emissions. Similarly, the given oil temperature directly interacts with the driveline domain by engine friction losses.

2 Model

The basic characteristics of the different modeling domains (vehicle, engine, cooling and control) are outlined in the following sections, where a special focus is put on a more in-depth discussion of the applied cooling and lubrication models.

2.1 Vehicle

The vehicle model describes a compact class car with a gross weight of 1.2 tons equipped with a manual 6 speed gear box. As shown in Fig. 1, the model is assembled out of base components for wheels, brakes, differential, single ratio transmission, gear box, engine, chassis, control components etc. The required input data (e.g. tire size, transmission ratios, etc.) were taken from manufacturer data sheets. The block called ICE is a placeholder representing a full thermodynamic engine air path and cooling model depicted in Fig. 2.

The wheel/tire model covers longitudinal loss forces that depend on vehicle mass and load distribution, the location of the wheel, rolling resistance, wheel slip, dynamic variation of the rolling radius and moment of inertia effects. The model of the brakes calculates a braking torque as a function of given brake signals considering the influence of piston surface, friction and efficiencies. When the vehicle drives through the corner, the differential unit compensates for the different rotational wheel speeds. The single ratio transmission translates the speed/torque input from the gear box passed further to the differential.

The vehicle and the driveline can be seen as a mechanical multi-body system represented mainly by the rotational degree of freedom. According to Pfau et al. [2] such systems can be represented by

$$\begin{pmatrix} \mathbf{M} & \mathrm{B} \\ \mathbf{C} & \mathbf{D} \end{pmatrix} \cdot \begin{pmatrix} a \\ m \end{pmatrix} = \begin{pmatrix} 0 \\ \tau \end{pmatrix} \tag{1}$$

where **M** is the diagonal mass matrix and **B** represents the acting of momentums between the components and on the components. **C** describes constraints between the mechanical parts and **D** covers the relation between the momentums. The accelerations of the individual components are represented by a and m describes the momentums in the system. τ comprises all momentums calculated in the individual components which act on the mechanic system as well as losses. τ is evaluated within the individual components and may depend on all of the states (speed, acceleration, controls, voltage, current, etc.).

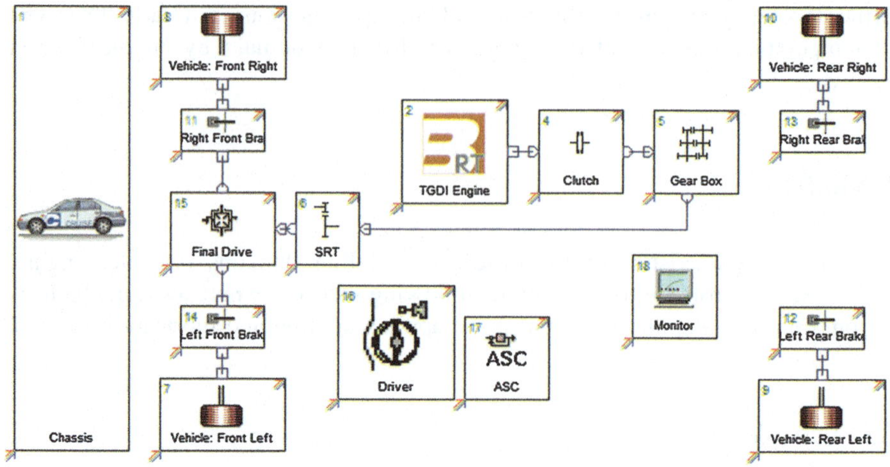

Fig. 1 Conventional vehicle configuration

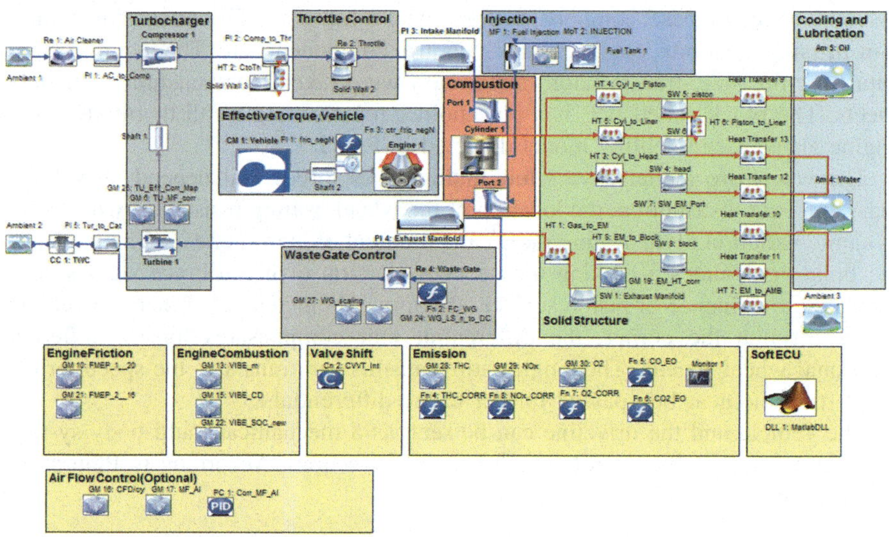

Fig. 2 Air path, combustion, control and cooling (partly) of TGDI engine

2.2 Engine

The engine model describes a 4 cylinder turbocharged direct injection gasoline (TGDI) engine. As shown in Fig. 2 the air path, combustion, control and cooling model is assembled out of base components like air cleaner, compressor, intercooler, port, cylinder, walls, injection, turbine, catalyst, vehicles etc. The latter represents a placeholder comprising the model given in Fig. 1. Similarly, the

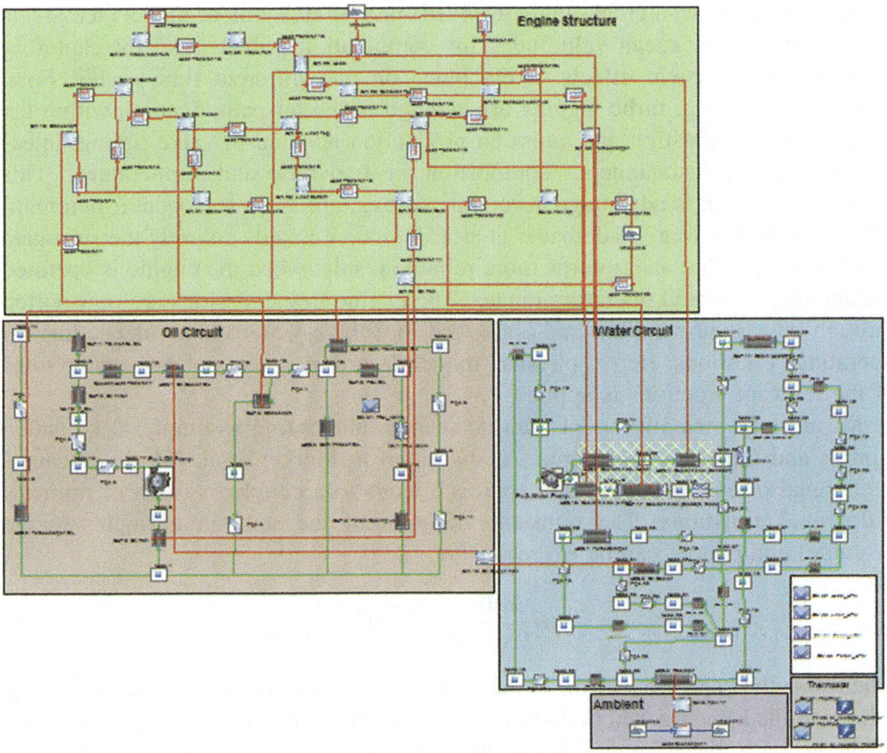

Fig. 3 Engine block structure, oil, water circuit and radiator air flow

blocks "Oil and Water" represent detailed models as given in Fig. 3. A software ECU model is applied to control engine air aspiration and injection by steering throttle and waste gate position, intake valve phasing and injection shape.

The internal combustion engine model given in Fig. 2 typically comprises several different characteristic time scales. The shortest time scale is associated with the in-cylinder phenomena, wave dynamics in the engine manifold and torque oscillations at the engine shaft. A larger time scale is associated with filling and emptying the engine manifolds during transient operation of turbocharged engines or during changed throttle position. Third, the largest time scale is given by the thermal response of the engine. This is associated with the thermal inertia of the solid structure and coolant media in the oil and water circuits.

Engine models resolving the effects of the shortest time scale certainly lead to the most accurate results. However, with the goal to apply a real-time capable system engineering model with a maximum level of physical predictability several dedicated model simplifications are applied in the present work. Assuming that wave dynamic effects influencing the volumetric efficiency can be neglected to a certain extent (see [3]) a mean value based description of the air path can be applied. In order to ensure a high level of predictability particularly of the combustion process, a crank-angle resolved state-of-art cylinder model (see [4]) is embedded into the mean value air. This combination is beneficial for capturing appropriately transient effects taking place on two different time scales. First, phenomena like e.g. turbo lagging are covered in the air path. Second, within the cylinder the combustion and emission production respond to valve timing, injection timing and instantaneous combustion air–fuel ratio and temperatures. This physical approach is advantageous compared to simplifying surrogate (often map) based approaches (e.g. Stadelbauer et al. [5]) since it avoids not only the surrogate training process but also returns more reliable results when the engine is operated outside the calibrated operating ranges. The engine raw emissions are transported through the exhaust system and converted in a three-way-catalyst, depending on operating conditions. Here a physical model describing chemical conversion rates of the relevant reactions is applied.

Mean value air bath models can be seen as alternating assembly of so called storage and transfer components. As discussed in Merker et al. [6], this a-causal (not signal oriented) bond-graph approach leads to a coupled system of ordinary differential equations. The transient variation of the state in a single storage (volume) component can be expressed by

$$B \cdot \frac{d\boldsymbol{\Phi}}{dt} = +\sum \mathbf{F}_{\mathrm{k}} \tag{2}$$

where $\mathbf{B}$ is the capacity matrix, is the state vector and $\mathbf{F}_{\mathrm{k}}$ represents the flux vector related to the k'th attached bond (transfer component). Focusing on the description of the air path, the state and flux vectors comprise the conservation variables for mass, energy and species with the corresponding intensive variables of pressure, temperature and species composition.

2.3 Cooling

The cooling and lubrication models describe the transport of heat generated mainly in the engine block and several other components (e.g. turbine, gear box, etc.) to dedicated heat sinks. This approach is widely used in literature, using different non-commercial and commercial tools (see [7–9]). As shown in Fig. 3, the overall model is assembled out of base components for masses of the structure, conductive and convective heat transfer, pipes, valves, heat exchanger, junction, pumps and nodes. The overall cooling domain comprises the area of heat conduction within the solid structure of engine block and the gear box to the oil and water circuit and further to the surrounding ambient. The oil circuit transfers heat from piston, liner and gear box to the oil–water-cooler, whereas some heat is also dissipated through the engine structure. The water circuit transfers the heat from the head, liner, ports, turbocharger and the oil–water-cooler to the radiator (water–air heat exchanger), whereas some heat is also dissipated through the engine structure and on demand supplied for the cabin heating. Depending on the operating conditions and the thermostat opening the water either runs in a short circuit without entering the radiator or it passes through the radiator to be cooled by the surrounding air. The oil and water pumps in the baseline engine are mechanically driven, whereas for the purpose of this study an additional model with an electrically driven water pump is also investigated to highlight the influences of a more advanced cooling strategy.

The heat distribution within the solid structure of the block can be seen as a 3D heat conduction problem. This can be described by any level of spatial resolution with the help of generic mass nodes and heat conduction components. This approach is also based on the bond-graph concept given in Eq. (2) where the state represents the local temperature and the fluxes are evaluated following Fourier's law.

The oil and water circuits can be modeled as incompressible, non-isothermal flow networks. The assumption of incompressibility is based on typical pressure ranges of cooling circuits and the required simulation accuracy. High frequency wave propagation effects, as they are essential especially in high pressure injection systems, are not considered in the present cooling model. Thus, a simplified equation system can be applied with the corresponding faster numerical solution.

The mass balance in all node components is given by a steady-state correlation

$$\sum\nolimits_{i} mf_i = 0 \tag{3}$$

where the sum of all incoming and outgoing (determined by the sign) mass flows mf_i sums up to zero. Similarly, the enthalpy balance in the node point is given by

$$\sum\nolimits_{\mathrm{i}} h_{i,\mathrm{in}} \cdot mf_{i,\mathrm{in}} + \sum\nolimits_{j} h_{j,\mathrm{out}} \cdot mf_{j,\mathrm{out}} = 0 \tag{4}$$

where an explicit distinction between the summation of incoming $mf_{i,\mathrm{in}}$ and outgoing $mf_{j,\mathrm{out}}$ flows is applied. This is needed in order to associate the correct specific enthalpies. For incoming flows the enthalpy $mf_{i,\mathrm{out}}$ from the upstream component is taken. For outgoing flows the enthalpy $mf_{j,\mathrm{out}}$ is evaluated by a mass flow averaging.

With the node balances given, pressure drop and heat transfer correlations can be applied for the different components. Taking into consideration inertia effects when accelerating/decelerating the fluid, the following momentum balance can be applied

$$\frac{\mathrm{d}mf}{\mathrm{d}t} = \frac{A_{\mathrm{cross}}}{l_{\mathrm{comp}}} \cdot \left(p_{\mathrm{up}} - p_{\mathrm{dn}} - \Delta p_{\mathrm{loss}} - \rho \cdot \mathrm{g} \cdot \Delta z\right) \tag{5}$$

where the time derivate of the mass flow mf equals the difference of the upstream p_{up} and downstream p_{dn} pressure reduced by the pressure drop Δp_{loss} and forces due to geodetic height differences Δz. A_{cross} represents the fluid cross section, l_{comp} is the flow length in the component, ρ is the fluid density and g represents the gravity force. The pressure loss Δp_{loss} individually depends on flow characteristics of the different components. The present work applies e.g. the Hagen-Poiseuille law for pipe flow in laminar regimes and the Colebrook correlation in turbulent regimes. The pressure drop in heat-exchangers is either approached by empirical correlations from literature or by input tables (pressure loss versus mass flow) filled with manufacturer data of the considered component.

Assuming that thermal inertia effects on the fluid and wall side and also the 1D heat transfer characteristic are of importance, balance equations for a transient 1D two-phase heat transfer problem can be applied. This is given by

$$\begin{aligned}\rho \cdot c_{\mathrm{p}} \cdot \frac{\mathrm{d}T}{\mathrm{d}t} &= -\frac{mf}{A_{\mathrm{cross}}} \cdot \frac{\mathrm{d}h}{\mathrm{d}z} + \frac{P}{\mathrm{A_{cross}}} \cdot \alpha \cdot (T - T_{\mathrm{w}}) \\ \rho_{\mathrm{w}} \cdot c_{\mathrm{p,w}} \cdot \frac{\mathrm{d}T_{\mathrm{w}}}{\mathrm{d}t} &= \frac{\mathrm{d}}{\mathrm{d}z}\left(\lambda_{\mathrm{w}} \cdot \frac{\mathrm{d}T_{\mathrm{w}}}{\mathrm{d}z}\right) - \frac{P}{\mathrm{A_{cross,w}}} \cdot \alpha \cdot (T - T_{\mathrm{w}})\end{aligned} \tag{6}$$

where ρ, ρ_{w}, c_{p}, $cp_{,\mathrm{w}}$, T, T_{w}, A_{cross}, $A_{\mathrm{cross,w}}$, are the density, heat capacity, temperature and cross section of the fluid and wall, respectively. P represents the perimeter of the fluid-to-wall surface, λ_{w} is the wall heat conduction and z is the coordinate in space domain. The heat transfer coefficient α is evaluated individually for each element. Here e.g. Gnielinski and Colburn correlations are used for pipes, whereas, for highly 3D flow situations also tables (Nusselt vs. Reynolds) populated with manufacturer data are applied.

2.4 Control

In order to run the engine/vehicle in transient operating conditions several simple closed and open loop controls were applied. For example, the target boost pressure is closed/open loop controlled by throttle and waste gate opening, the fuel injection angle and duration is open loop controlled and depending on operating conditions overlaid by an idle speed controller. As shown in Fig. 1 a soft ECU model takes care of this task where ECU parameters are based on experimental values measured for a wide range of parameter variations.

2.5 Overall System

The model equations discussed above can be summarized to a global set of algebraic, ordinary and partial differential equations. The latter are discretized in space domain, following methods of line reducing all partial to ordinary differential equations. The integration in time domain of the resulting DAE system is performed by tailored solver techniques suited to account for characteristic time scales of individual domains. These are the crank-angle cylinder, air path and mechanical system, cooling flow and cooling enthalpy network. Within these domains the equations are simultaneously integrated as this supports the robustness of the solution. All equations are time integrated with explicit time (crank-angle) increments. The cylinder typically applies crank-angle steps in the range of 1degCRA, the air path and mechanical system integrated with steps of 1 ms and the cooling system applies integration steps ranging up to 0.1 s considering pre-set solver tolerances. Special emphasis is taken to preserve flux conservation while coupling different domains.

3 Results

The results in this study focus on the comparison of different water pump configurations and their impact on fuel consumption. The underlying system engineering model of the whole vehicle is extensively investigated and validated in previous studies. The capability of the vehicle and engine model to adequately describe transient operating conditions in terms of engine and vehicle performance as well as in terms of fuel consumption was demonstrated in Wurzenberger et al. [10]. Extensive validation of the investigated engine and of the corresponding three-way catalytic converter is presented in Wurzenberger et al. [11]. Moreover, calibration of the investigated vehicle and engine model was performed in Vihar [12] by comparing fuel consumption on the NEDC for cold start engine conditions, maximum vehicle velocity and time for full load acceleration from 0 to 100 km/h. The discrepancy between measured and simulated values of all three parameters is below 1 %, which also indicates adequacy of the cooling and lubrication models as they significantly influence the fuel consumption during engine heat-up period.

The system engineering model of the complete vehicle including the propulsion unit and the cooling and lubrication circuits is used to investigate potential enhancements of the engine performance by electrifying the propulsion of the water pump. Therefore, a baseline vehicle model using a mechanically driven pump in the water circuit (MWP—Mechanical Water Pump) was alternatively equipped with an electrically driven pump (EWP—Electrical Water Pump). The pump size was chosen to provide sufficient cooling capability at maximum engine power additionally considering a safety margin. The model considers an adequate powering of the electric pump via the electrical network. This is also reflected by drawing more power from the mechanical network to produce on-board electricity. In addition, the thermostat characteristics of the EWP configuration were slightly adapted to account for different coolant mass flows. All other components of the cooling and lubrication circuit remained unchanged to obtain an insight into potentials of the retrofitted EWP system.

Figure 4a confirms the capability of the vehicle and the driver model to adequately follow the NEDC velocity profile. Moreover, Fig. 4b shows power outputs of both engines, which are, as expected, almost identical. This is imposed by the application of identical vehicle parameters and identical desired velocity profiles. Minor discrepancies arise from the forward facing basis of the model as different

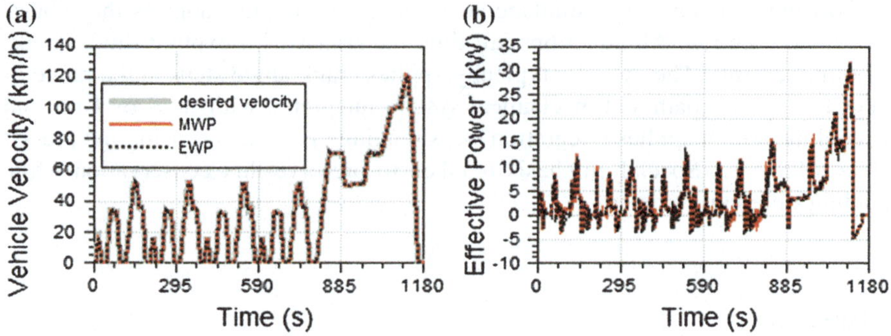

Fig. 4 Vehicle velocity (**a**), and effective engine power (**b**), during NEDC for an engine equipped with a mechanical and an electrical water pump

heat-up characteristics of the engines and different auxiliary power consumptions slightly influence the engine power output and thus the response of the driver.

Multiple strategies of steering electric water pumps are discussed in literature. One of the obvious solutions is not to operate the pump at all until a certain water temperature that is close to the target water temperature is reached. This approach definitively minimizes the energy needed for powering the pump. However, it also has some drawbacks. One of them arises from increased thermo-mechanical stresses on the components that are exposed to heterogeneous temperature fields when the engine operates at high loads with no flow of cooling water, e.g. through the exhaust ports. Moreover, it should also not be overseen that during the engine heat-up period the water temperature generally rises faster than the oil temperature. Thus, oil is heated by the water in the oil–water-cooler during this phase. Additionally, no water flow in the engine during the heat-up period might negatively influence the overall energy balance. The positive contribution of lower energy consumption for pump propulsion might be compensated by increased frictional losses arising from lower oil temperatures. Due to these facts it was decided to always maintain a small mass flow of the cooling medium also in the case of the electrically driven pump as given in Fig. 5a.

Figure 5a clearly shows that the mass flow through the mechanically driven pump follows the engine speed variation. Such a system is typically dimensioned in a way that it provides sufficient cooling capability in the worst case scenario. Thus, it is frequently operated in an inefficient manner. This is related to the fact that the mass flow directly depends on the engine speed and thus the pump supplies excessive water mass flows most of the time. Figure 5a also highlights that the electric pump is operated independently of engine speed. During the initial phase of the cold start drive cycle a small water mass flow maintains temperature gradients at a reasonable level and also ensures sufficient mass flow through the oil–water-cooler. As a result it can be observed in Fig. 5b that water is heated-up faster in the case of the electrically driven water pump. However, the difference is not as significant as it would be in the case of no flow at low coolant temperatures.

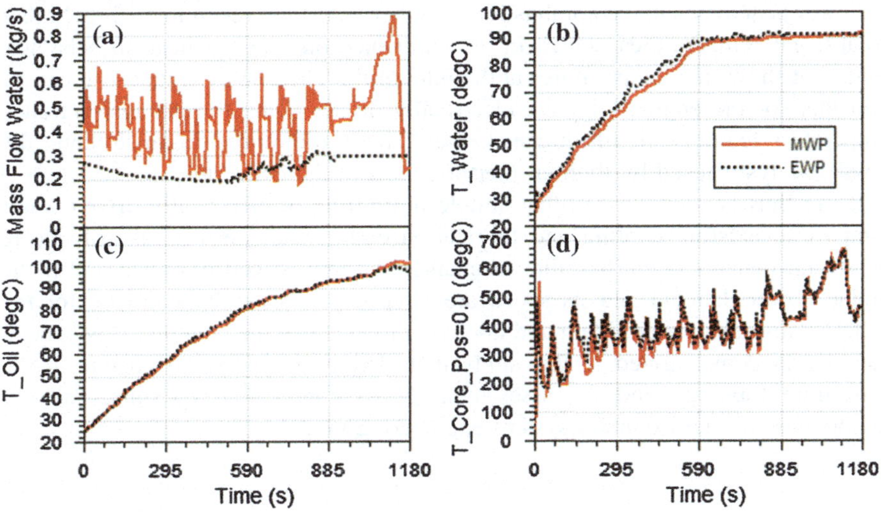

Fig. 5 Mass flow of the cooling water (**a**), temperature of the cooling water (**b**), temperature of the oil (**c**) and temperature at catalyst inlet (**d**), during NEDC for an engine equipped with a mechanical and an electrical water pump

Consequently, a small water mass flow in the heat-up phase also allows for adequate oil heating. This leads to similar oil heat-up trends in the EWP system compared to the engine equipped with the mechanically driven water pump (Fig. 5c). Additionally, also the cylinder head and the upper part of the liner that are predominantly in contact with the cooling water also feature higher temperatures during the heat-up phase of the engine with EWP. This is mainly due to the smaller heat flow from these components to the cooling medium as lower mass flows result in lower heat transfer coefficients. Higher water temperatures additionally reduce the heat transfer from the structure to the coolant medium. It can be seen in Fig. 5a that the mass flow of cooling water slightly increases when the thermostat opens and water reaches an approximately constant temperature.

All above described effects certainly influence the fuel consumption. However, the comparison of the analyzed case reveals that just a simple retrofitting of the electric water pump only leads to 1.2 % benefit in fuel economy. Although the energy needed for powering the water pump was reduced by 86 %, this reduces the fuel consumption only by 0.4 %. This is mainly due to the fact that the baseline vehicle is already equipped with a well dimensioned water pump. Especially the low average velocity given by the NEDC, combined with the 6-speed transmission, additionally ensures low engine speeds and correspondingly low energy consumption of the mechanically driven water pump. The remaining fuel economy enhancement of 0.8 % is mainly due to higher effective engine efficiency. The engine efficiency is enhanced by lower heat losses to the combustion chamber walls and lower friction losses due to slightly higher oil temperature.

Lower heat losses to the combustion chamber walls also result in slightly higher exhaust gas temperatures, which in turn influence the catalyst heat-up behavior. Figure 5d shows the temperatures at the inlet of the Three-Way-Catalyst. It can be seen that for the engine with the electrically driven pump the catalyst entrance temperature features slightly higher values. These differences diminish along the length catalyst caused by thermal inertia effects of the converter.

It can thus be summarized that simple retrofitting of electrically driven water pumps into existing cooling and lubrication circuits does not necessarily bring significant benefits if the baseline mechanical system is well designed and if the engine power demand and its speed are relatively low as characteristic for the NEDC. This calls for an integrated approach to enhance the benefits. However, it was clearly demonstrated in this study that already a very small change in the powertrain topology and the assessment of its consequences requires a very comprehensive multi domain system engineering model.

4 Conclusions

The conclusions and innovative aspects of the present study are manifold. A mechanistically based model of the complete powertrain including a crank-angle resolved cylinder description and dedicated cooling and lubrications models is presented in the paper. Innovative coupling of different domains and sub-domains allows for a very good ratio between the modeling depth and the computational times. Moreover, a novel, numerically efficient solver technology allows performing vehicle performance analyses of the entire system with CPU times below real-time being an essential requirement for further investigations on HiL systems. Compared to many system simulations published in literature being either often purely map based or a sophisticated co-simulation of various tools, the present study applies a fully integrated approach. The mechanistic models are—unlike map models—also able to cover highly transient operating regimes that significantly deviate from the steady-state regimes making them very suitable for performing a large variety of powertrain performance studies. Additionally, the physical nature of the models ensures adequacy of results when optimizing power trains, since a mechanistic modeling basis enables adequate response to changed parameters of individual components and adequate interaction between the components in different domains.

Acknowledgments This work was co-funded by the Austrian Ministry of Economics and the Upper Austrian Government.

References

1. Katrašnik T , Wurzenberger JC (2011) Optimization of hybrid power train by mechanistic system simulations. International Scientific Conference on Hybrid and Electric Vehicles RHEVE 2011
2. Pfau RU, Schaden T (2011) Real-time simulation of extended vehicle drivetrain dynamics. Multibody Dyn Comput Methods Appl Sci 23:195–214
3. Hrauda G, Strasser R, Aschaber M (2010) Gas exchange simulation from concept to start of production: avl's tool chain in the engine development process. THIESEL, Conference on Thermo- and Fluid Dynamic Processes in Diesel Engines
4. Wurzenberger JC, Bardubitzki S, Bartsch P, Katrašnik T (2011) Real time capable pollutant formation and exhaust aftertreatment modeling–HSDI diesel engine simulation. SAE Technical Paper 2011-01-1438
5. Stadlbauer S, Alberer D, Hirsch M, Formentin S, Benatzky C, del Re L (2012) Evaluation of virtual NOx sensor models for off road heavy duty diesel engines. SAE Technical Paper 2012-01-0538
6. MerkerGP, Schwarz C, Teichmann R (2012) Grundlagen Verbrennungsmotoren, Vieweg + Teubner, 6th edn
7. Cho H, Jung D, Assanis DN (2005) Control strategy of electric coolant pumps for fuel economy improvements. Int J Aut Tech 6:269–275
8. Kim KB, Choi KW, Lee KH, Lee KS (2010) Active coolant control strategies in automotive engines. Int J Aut Tech 11:767–772
9. Lodi FS (2008) Reducing cold start fuel consumption through improved thermal management. University of Melbourne, Master Thesis
10. Wurzenberger JC, Bartsch P, Katrašnik T (2010) Crank-angle resolved real-time capable engine and vehicle simulation – fuel consumption and performance. SAE Technical Paper 2010-01-0748
11. Wurzenberger JC, Prah I, Tominc P, Katrašnik T (2012) Optimization of Hybrid Power Trains—Physical Based Modeling For Concept Design. SAE Technical Paper 2012-01-0359
12. Vihar R (2012) Optimization of the energy efficiency and exhaust emissions of the personal vehicle. Diploma Thesis, University of Ljubljana, Slovenia

A Multi Zone Spray and Combustion Model for Formation of Polycyclic Aromatic Hydrocarbons and Soot in Diesel Engines

Ali Salavati-Zadeh, Vahid Esfahanian, Asghar Afshari and Mahdi Ramezani

Abstract Considering the certain impact of exposing to soot particles in increasing the risk of lung cancer and mortality in human being and the inevitable role of Polycyclic Aromatic Hydrocarbons (PAHs) on these carcinogenic effects, formation of selected PAHs during diesel cycle combustion is investigated. To accomplish this, a detailed chemical mechanism is embedded with a multi zone diesel cycle model. In development of the multi zone diesel cycle model, emphasis is given to including most of the processes taking place in the cylinder during the closed cycle and to describing as many as possible of these processes by correct modelling of physical laws. Diesel fuel combustion chemistry is modelled under the light of a detailed mechanism consisting of 1696 reactions, forward and backward counted separately, and 155 species. The mechanism is capable of modelling combustion of hydrocarbons up to C_{10} and considers most recent developments in gas-phase reactions and aromatic chemistry. It includes sub-mechanisms for normal (n-decane and n-heptane), ramified (iso-octane) and aromatic (benzene and toluene) species to correctly model the combustion feature of engine relevant fuels. Formation of selected PAH species (Phenanthrene and Pyrene) in a diesel engine and how they are affected with change in engine speed and load are studied. Both these PAHs are cancerogenic themselves. Meanwhile many studies have shown that soot particle nuclei, smallest visible solid particles, are dimers which form through collision of Pyrene molecules. Hence, correct prediction of Pyrene concentration will be a large step toward correctly solving the

F2012-A06-037

A. Salavati-Zadeh (✉) · V. Esfahanian · A. Afshari
School of Mechanical Engineering, University of Tehran, Tehran, Iran
e-mail: alisalavati@ut.ac.ir

M. Ramezani
Department of Mechanical Engineering, Iowa States University, Ames, USA

SAE-China and FISITA (eds.), *Proceedings of the FISITA 2012 World Automotive Congress*, Lecture Notes in Electrical Engineering 190,
DOI: 10.1007/978-3-642-33750-5_19, © Springer-Verlag Berlin Heidelberg 2013

first mode of bimodal soot Number Density Function (NDF), which can be regarded as one of the engines sooting characteristics. Despite lower computational costs of the new model comparing to multidimensional simulations, the predicted data are in good agreement with the results obtained experimentally. This proves that soot nucleation is a chemically controlled process, therefore bearing in the mind that coupling detailed reaction schemes with multidimensional simulations is not feasible due to very high computational cost, multi zone methodology could serve as a powerful tool.

Keywords Polycyclic aromatic hydrocarbons · Soot · Diesel · Detailed chemical kinetics · Multi zone model

1 Introduction

Diesel engines, despite suffering from drawbacks of high nitrous oxides and particulate emissions, are the most practical and efficient power sources available for ground transportation and power plants. However, recent experiments have revealed the possibility to reduce or eliminate particle emissions in these engines. There are still many uncertainties in realizing phenomenology of forming soot precursors and particles. Meanwhile, growing stringent regulatory limits on nitrous oxide and soot emissions from diesel engines can be regarded as one of the most difficult combustion based problems with which diesel engine designers are confronted. Obviously, development of numerous prototypes and other conventional solutions are not economically feasible anymore. Hence, Numerical models can now be powerful tools to meet the rapid responses required by regulatory agenda on the part of combustion designers.

Soot particles and PAHs such as Phenanthrene and Pyrene, which are proven to be also the main precursors of soot particles, are serious dangers to human health [1, 2]. On the other hand, soot formation will enhance the heat loss in combustion systems, due to its similar transmission behaviour to black bodies [3] causing waste of energy.

Although multi-dimensional models are very useful for understanding detailed spatial information and complex interactions of many simultaneous phenomena, coupling CFD computations with detailed chemical models for in-cylinder phenomena simulations are limited or sometimes impossible, since they are very time consuming and computationally expensive. There are though advances in reduction of chemical kinetic models to skeletal schemes, but reducing a chemical scheme will inherently exclude its capability to simulate some features of combustion. On the other hand, soot formation is controlled by chemical reactions governing the formation and conversion of small hydrocarbon radicals and formation of soot odorous precursors from these species. Hence, one should keep in the mind that using a valid chemical mechanism for fuel oxidation provides the necessary platform for correctly representing the complex processes of soot formation and oxidation.

Multi zone models [4–7], capable of fairly well approximating the geometric details and the effect of engine operation on the combustion process seem to be reasonable choices. Detailed chemical schemes can be embedded with these models requiring much lower computational cost. This is the theme of the reported work here.

This work is aimed to couple detailed chemistry with a diesel cycle multi zone model to study the formation of PAHs and other key soot precursors. Large PAHs, such as Phenanthrene and Pyrene are of the main importance here, since they can give a good approximation of soot nuclei formed during diesel combustion.

2 Multi Zone Model Description

Hiroyasu Correlations are employed for modeling spray phenomena. Fuel spray is considered as a cone divided to multiple axial and radial zones. As injection continues, the spray penetrates into cylinder. To predict the spray penetration, a correlation has been proposed by Jung [8] based on Hiroyasu correlation. In order to characterize the cone, the cone angle should be calculated along with the penetration length, which constructs the cone length. Different equations are proposed for computing of the cone angle. In this research, volume of each zone is calculated instead of total volume of the cone. Air entrainment to the spray depends on the diffusion velocity also. The volumes of the zones can be computed considering cone angle [9] and momentum conservation equation [6].

To solve for combustion related parameters and ignition delay, a reaction scheme previously proposed by Salavati-Zadeh et al. [10] has been employed. The base mechanism has been developed for fossil fuel surrogates and accounts for both low- and high-temperature oxidation of aromatic compounds and large normal and ramified alkanes up to C_{10}. Special concern was paid to correctly predict the key soot precursors such as Acetylene, Ethylene, the C_3H_4 isomers, Propyne and aromatic species from Benzene to Pyrene. This mechanism has been proven to accurately simulate the ignition delay times under a wide range of shock tube conditions and laminar burning velocities of different hydrocarbon species typically found in conventional diesel fuels. The base scheme can be divided into two parts. First is the main reaction chain describing the combustion environment features and the formation of the first aromatic ring, benzene. This part of the mechanism consists of 15 reaction classes of 25 classes previously used by Curran et al. [11, 12] and Westbrook et al. [13] for large alkanes:

1. High temperature mechanism: Reaction classes 1–3, 5 and 6.
2. Low temperature (high pressure) mechanism: Reaction classes 10, 12–17, 22–24.

In addition, the mechanism is embedded with sub-mechanisms for considered aromatic species (benzene and toluene) and for C_1–C_4 species. The latter which accounts for small HC, controls many characteristics of the combustion process like heat release, ignition delay and burning velocity. On the other hand, it plays role in predicting the intermediate species, which link the small HC to formation of benzene ring.

In this research, air swirl is taken into account based on recommendations of Hiroyasu and Kadota [14]. Also, it is assumed that the spray continues its penetration after impingement on the cylinder walls, with a slower speed. This reduction in speed will cause air entrainment to improve in axial momentum equations.

The instant pressure of the cylinder is calculated using thermodynamics first law. The cylinder temperature is computed using this pressure based on ideal gas assumption. Before combustion and after that, the temperature and pressure inside cylinder are calculated using isentropic process relations.

2.1 Equations Description

The penetration of spray into the cylinder before break-up assuming constant penetration speed is calculated using Eq. (1). After break-up to fuel droplets, Eq. (2) is employed [14]. The spray break-up time is yielded by equating the two correlations, Eq. (3).

$$S = C_D(2\Delta p/\rho_{\mathrm{f}})^{0.5}t \quad (1)$$

$$S = 2.95(2\Delta p/\rho_{\mathrm{a}})^{0.25}D_{\mathrm{N}}^{0.5}t^{0.5} \quad (2)$$

$$t_{\mathrm{br}} = 4.351\rho_f D_{\mathrm{N}}/\left[\mathrm{C}_{\mathrm{D}}^2(\rho_a \Delta p)^{0.5}\right] \quad (3)$$

where D_{N} stays for injector nozzle hole diameter, indices f and a stays for fuel and air, respectively. The calculations are based on the assumption that when injection period comes to end, the spray continues its penetration. Also, one should be aware that the difference between air and fuel pressures and densities are calculated at the beginning of the injection period, and the same value will be considered for the rest of injection time. In case of impingement of the spray with the wall, the term $t^{0.5}$ in Eq. (2) is changed to $t^{0.48}$.

In Hiroyasu and Kadota Model [14] it is possible to simulate the effects of air swirl by considering a coefficient in spray penetration correlation and using Eq. (4).

$$S = \left[1 + (\pi R_s NS)/(30u_{\mathrm{inj}})\right]^{-1}S \quad (4)$$

In Eq. (4) swirl ratio is calculated by dividing air swirl speed to engine speed. Fuel injection rate can be computed by nozzle hole discharge equations. When the injection rate curve is not known, Eq. (5) is employed.

$$dm_{\mathrm{f}}/dt = C_{\mathrm{D}}A_{\mathrm{N}}\sqrt{}(2\rho_{\mathrm{f}}\Delta p) \tag{5}$$

Fuel spray momentum remains constant as it penetrates, hence employing momentum conservation equation, mass of entrained air to each zone is calculated using (6) [14].

$$m_{\mathrm{a}} = m_{\mathrm{f}}\left(u_{\mathrm{inj}}dt/ds - 1\right) \tag{6}$$

Annand Eq. [15] is employed to account for heat losses during diesel cycle, e. g.

$$dq/dt = a\,C_p/B\,Re^b\left(T_{\mathrm{w}} - T_{\mathrm{g}}\right) + c\left(T_{\mathrm{w}}^4 - T_{\mathrm{g}}^4\right) \tag{7}$$

Piston mean speed is used to compute Reynolds number. The empirical factors in this correlation are considered as $a = 0.35$, $b = 0.7$ and $c = 0.6$.

The liquid droplets begin to evaporate after the breakup period. The model proposed by Jung and Assanis [6] is employed. This model combines the equations governing the mass diffusion of vapor [16] and convective heat transfer from hot gas to liquid fuel droplets [17] to obtain the rate of evaporation rate. It should be noted that the total number of droplets are known within the cylinder knowing the total fuel mass and the droplet size. To know the droplet diameter after breakup the model proposed in Ref. [6] is used.

3 Model Setup

The model is solved by using a marching technique. The time step size of 1×10^5 s is found suitable for reaction scheme and is converted to degrees crank angle considering the engine speed. The chemical mechanism begins to be taken into account from the end of spray break up and is solved both in the air zone to evaluate ignition delay time and spray area to predict the emission. The design characteristics as well as the operating data at the beginning of the cycle, i. e. at the IVC event are given at the start of the calculation. The calculation stops when EVO occurs.

4 Results

The spray penetration sub-model is validated against several experimental data sets available in the literature in constant volume chambers. Figure 1 shows the experimental and calculated data for spray with diameter of 0.2 mm, discharge

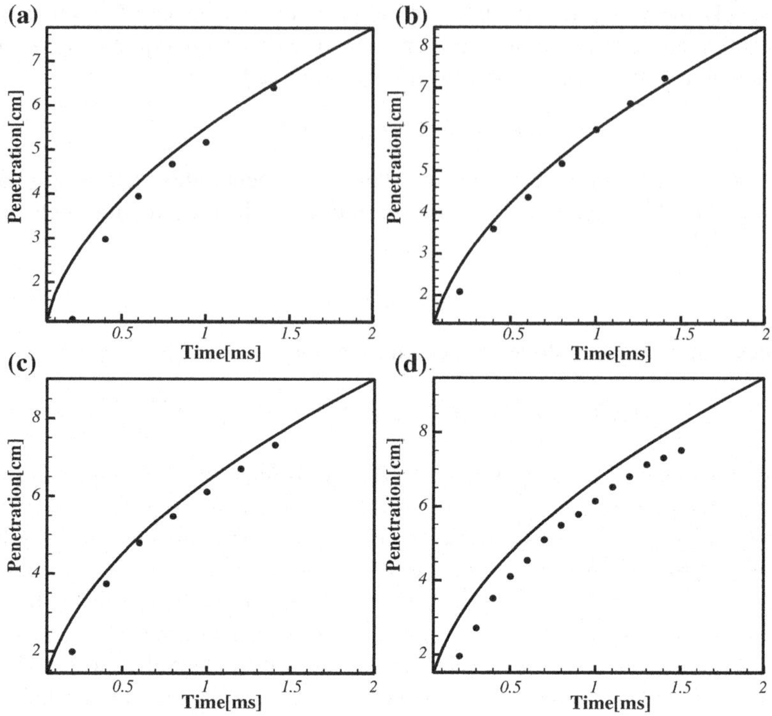

Fig. 1 Spray Penetration in a constant volume chamber for injection pressure of **a** 55 MPa, **b** 77 MPa, **c** 99 MPa, **d** 120 MPa. In all the cases symbols stand for experimental data while solid lines are the calculated results with the current model

coefficient of 0.66 and injection pressure of 55, 77, 99 and 120 MPa. In all the cases good agreement with experimental data was observed.

Same analyses were performed for air entrainment in the spray. To this end, a spray with diameter of 71 μm in a 1.2 L chamber with spray discharge coefficient of 0.86 is considered [6]. Initial temperature and pressure inside the chamber is equal to 1000 K and 4.274 MPa, which are very close to those of diesel engines before injection. Apart from a little deviation in the peak pressure between experimental and calculated results, which can be due to the heat transfer model, the results show acceptable agreement. These results although excluding the engine speed as an important factor, indicate that the model shows good performance.

To consider the engine speed, the effects of load and injection timing on a diesel engine performance are being investigated. The engine was previously studied experimentally by Rakopoulos et al. [18]. All the engine specifications are provided in the same reference. The composition of the fuel is assumed to be 38.5 % toluene, 6.5 % n-Heptane and 55 % n-Decane. The cetane number of this composition, which is considered to be equal to the mole fraction-weighted sum of the

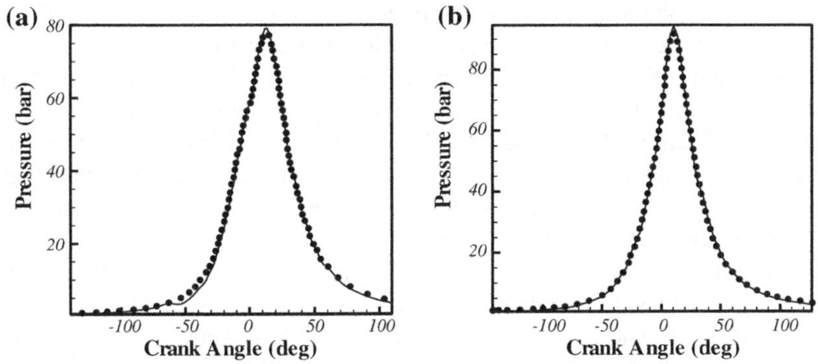

Fig. 2 Calculated and experimental results for the specified single cylinder diesel engine at the engine speed of 2500 rpm and 80 % load. In both figures symbols stand for experimental data [18] while solid lines are the calculated results with the current model

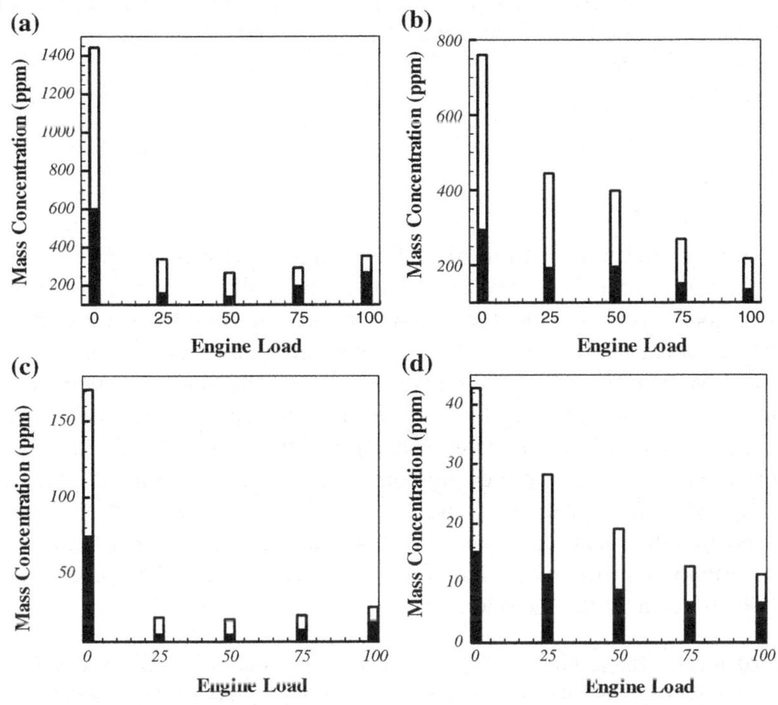

Fig. 3 Calculated and experimental mass concentration of selected PAHs versus engine load **a** phenanthrene at 1000 rpm, **b** phenanthrene at 3000 rpm, **c** pyrene at 1000 rpm and **d** pyrene at 3000 rpm. In all the cases full bars are the calculated results and empty bars represent experimental findings

cetane numbers of its constituent compounds is found to be 52.98. The lower heating value of the considered mixture is also 43150.2 kJ/kg. Both cetane number and heating value of the mixture are very close to those of the fuel employed in the experiment. The calculated pressure diagrams from the model at 80 % of full load for injection timing of 15 and 20 CA BTDC are illustrated in Figs. 2a, b respectively. The results show very good agreement with experimental observations and it seems that the heat transfer model performance is much better in the case of diesel engines.

The validated model is then employed to study the formation of Phenanthrene and Pyrene in a diesel engine. This engine has been experimentally studied by Collier et al. [19]. The engine is a 4-cylinder Perikins Prima DI diesel engine. Specifications of this engine are brought in Ref. [19].

The composition of the initial fuel here is assumed to be 11 % Propane, 7 % Benzene, 22 % Toluene, 20 % n-Heptane and 40 % n-Decane. Propane here account for small alkanes, and all the aromatic content are assumed to be Toluene and Benzene, because smaller size of reaction set is desired. Figure 3 illustrates the amount of the selected PAHs in the exhaust of the engine in different loads in engine speeds of 1,000 and 3,000 rpm. The calculated result shows good agreement with the experiment.

5 Conclusion

A multi zone model of direct injection diesel engine cycle is developed in which most of the in-cylinder processes are included. The multi zone model is embedded with a comprehensive kinetic scheme which accounts for both low- and high-temperature oxidation of a wide range of saturated alkanes and aromatic species. The model is employed in predicting the amount of selected cyclic odorous compounds formed during the diesel cycle. Despite good agreement of the calculated results and the experimental observations, it is found that the formation of soot precursors, including large cyclic compounds such as Pyrene, is inherently a chemically controlled process. These large compounds will further form soot nuclei and therefore one can conclude that nucleation of soot as the onset of particle formation in diesel engines is completely dominated by chemical processes which occur in the cylinder.

Acknowledgments The authors would like to thank University of Tehran for supporting this research. This work has benefited from the support of Vehicle, Fuel and Environment Research Institute, University of Tehran.

References

1. Berube KA, Jones TP, Williamson BJ, Winters C, Morgan AJ, Richards RJ (1999) Physicochemical characterization of diesel exhaust particles: factors for assessing biological activity. Atmos Environ 33:1599–1614
2. Richter H, Howard JB (2000) Formation of polycyclic aromatic hydrocarbons and their growth to soot–a review of chemical reaction pathways. Prog Energy Combust Sci 26:565–608
3. Tree DR, Svensson KI (2007) Soot processes in compression ignition engines. Prog Energy Combust Sci 33:279–309
4. Xiaoping B, Shu H, Dai Z, Yin S, Duan C (1994) A Multi-Zone Model for Prediction of DI Diesel Engine Combustion and Soot Emission. SAE Paper 941900
5. Liang Huang Y, Wang Z, Hu Y (1996) The Application of Phenomenological Combustion Model for Experimental Study of D. I. Diesel Engines. SAE Paper 961051
6. Jung D, Assanis DN (2001) Multi-Zone DI Diesel Spray Combustion Model for Cycle Simulation Studies of Engine Performance and Emissions. SAE Paper 2001-01-1246
7. Saeed A, Saeed K, Ahmad A, Malik KA (2001) Multizones Modeling of the Combustion Characteristics of Oxygenated Fuels in CI Engines. SAE Paper 2001-01-0051
8. Jung D (2001) A Multi-Zone Direct-Injection Diesel Spray Combustion Model for Cycle Simulation Studies of Large-Bore Engine Performance and Emissions. PhD Dissertation, University of Michigan
9. Rakopoulos CD, Rakopoulos DC, Giakoumis EG, Kyritsis DC (2004) Validation and sensitivity analysis of a two zone diesel engine model for combustion and emissions prediction. Energy Convers Manage 45:1471–1495
10. Salavati-Zadeh A, Esfahanian V, Afshari A. Detailed kinetic modeling of soot particles and key precursors formation in laminar premixed and counterflow diffusion flames of fossil fuel surrogates. ASME J Energy Res Technol (Under Review)
11. Curran HJ, Gaffuri P, Pitz WJ, Westbrook CK (1998) A comprehensive modeling study of n-heptane combustion. Combust Flame 114:149–177
12. Curran HJ, Gaffuri P, Pitz WJ, Westbrook CK (2002) A comprehensive modeling study of iso-octane oxidation. Combust Flame 129:253–280
13. Westbrook CK, Pitz WJ, Herbinet O, Curran HJ, Silke EJ (2009) A comprehensive detailed chemical kinetic reaction mechanism for combustion of n-alkane hydrocarbons from n-octane to n-hexadecane. Combust Flame 156:181–199
14. Hiroyasu H, Kadota T (1983) Development and use of a spray combustion modeling to predict diesel engine efficiency and pollutant emissions. Bulletin JSME 26:569–575
15. Annand WJD (1963) Heat transfer in the cylinders of reciprocating internal combustion engines. Proc Inst Mech Eng 177:973–990
16. Borman GL, Johnson JH (1962) Unsteady vaporization histories and trajectories of fuel drops injected into swirling air. SAE Paper 598C
17. Gosman AD, Johns RJR (1980) Computer analysis of fuel-air mixing in direct injection engines. SAE Paper 800091
18. Rakopoulos CD, Rakopoulos DC, Kyritsis DC (2003) Development and validation of a comprehensive two-zone model for combustion and emissions formation in a DI Diesel engine. Int J Energy Res 26:1221–1249
19. Collier AR, Rhead MM, Trier CJ, Bell MA (1995) Polycyclic aromatic compound profiles from a light-duty direct injection diesel engine. Fuel 74:362–367

Evaluation of Crank Mounted Fan for TV and Crankshaft First Journal Bearing MOFT Analysis for 3.8 L 4 cyl. Diesel Water Cooled Engine

V. Sandeep, Singh Harsumel, D. Patil Shankar, P. Mahesh and N. Muralitharan

Abstract Engine crankshaft used to validate for Torsional Vibrations by limiting the amplitude within the safe working design limit. The same is radially measured with non-contact optical laser sensor equipment mounted on Crankshaft front end. With the test bed measured Peak firing Pressure Vs Crank angle data, crank train has been modelled in AVL EXCITE. The model is then calibrated for TV and MOFT measured on actual engine. To this, a fan model has been added and results were derived for TV and MOFT with crankshaft mounted rigid fan and Viscous drive Fan. The bearing life is predicted considering extra bending load due to fan's weight. The effect of this extra load on the First journal bearing oil film thickness is derived by adding equivalent gas pressure on the crankshaft and thereby predicting the bearing life. This paper addresses the effect on Torsional Vibration Characteristics and Minimum Oil Film Thickness by applying viscous fan and mounting of Fan on Crankshaft as opposed to that on Water Pump for 3.8 L 4 cyl. 140 HP Engine.

Keywords Diesel engine · Crank shaft · Viscous fan · Torsional vibrations (TV) · Minimum oil film thickness (MOFT)

F2012-A06-038

V. Sandeep (✉) · S. Harsumel · D. Patil Shankar · P. Mahesh · N. Muralitharan
Ashok Leyland Limited, Chennai, India
e-mail: sandeep.v@ashokleyland.com

SAE-China and FISITA (eds.), *Proceedings of the FISITA 2012 World Automotive Congress*, Lecture Notes in Electrical Engineering 190,
DOI: 10.1007/978-3-642-33750-5_20,

1 Introduction

The objective of this work is to simulate and analyse the engine torsional vibration level in different orders with and without fan in hot and cold conditions and simulate the minimum oil film thickness using AVL EXCITE software. This work is part of application of 4 cyl. 3.8 L Engine for ICV 7.5 T Truck.

2 Viscous Fan Drive

Air sensing viscous fan drives normally used to test on water to air and air to air cooled engines and the results indicate that these air sensing viscous fan drives can be successfully implemented to track the radiator coolant temperature on Vehicle and with the proper selection of drive calibration, viscous fan drive provides a simple acceptable method of controlling engine coolant temperature [1].

Viscous clutch drive is fluid coupling using friction to control radiator fan speed. Basically clutch assembly is divided into two parts: Driving part and Driven part. Driving part is mounted on engine drive (crankshaft pulley) which is input speed for fan clutch. Hence this part always rotates at same speed as engine drive. Driven part is connected to Fan. Driven part and driving part are placed with gap as small as 0.5 mm to all sides. During normal drive when temperature is below set temperature, clutch always rotates at idle speed i.e. disengaged condition [2].

When air temperature starts increasing, bimetal senses & starts bending to form concave shape which allows pin to move and valve lever starts opening. Thus silicon oil stored in reservoir chamber starts flowing to working chamber (Ref Fig. 1). Oil fills gap and speed from driving part get transfer to driven part i.e. fan speed starts increasing. Oil return to reservoir chamber through provided path. This continuous cycle gives uniform speed of fan. Further increase in temperature allows full opening of valve lever, all oil come into working chamber and fan starts rotating at input speed with some specified slip i.e. full engagement condition. When temperature drops bimetal become straight which pushes pin to close valve lever. Thus oil returns to reservoirs and fan speed drops to idle speed.

Viscous fan finalized for this application with performance characteristics as mentioned in Fig. 2 to achieve intake manifold temperature differential (IMTD) $\leq$ 25 °C and limiting ambient temperature (LAT) $\leq$ 45 °C. Also vehicle has passed De-aeration test.

Engine details along with cooling prediction (CAC) for this application are provided in Table 1. The IMTD is within limits of less than 25 °C.

Table 1 Cooling prediction, CAC

Engine inputs	140 hp	
Parameters	Max torque	Rated power
Engine speed (RPM)	1,600	2,800
CAC heat rejection	9.0	14.16
Condenser heat rejection (KW)	11.9	11.9
Charge air flow (Kg/s)	0.108	0.174
Charge air pressure (Kpa) abs	2.15	2.15
Ambient air temp (°C)	45	45
Specifications		
CAC inlet temp (°C)	140.0	147.0
IMTD (°C)	<25	<25
Max pressure drop across CAC (kpa)	12	12
Prediction results		
Predicted IMTD (°C)	25	24.9
CAC heat rejection (KW)	7.8	14
Predicted CAC pressure drop (kPa)	8	11
Fan diameter (mm)	560	
Fan ratio	1:1	
Fan slip (%)	1%	1%

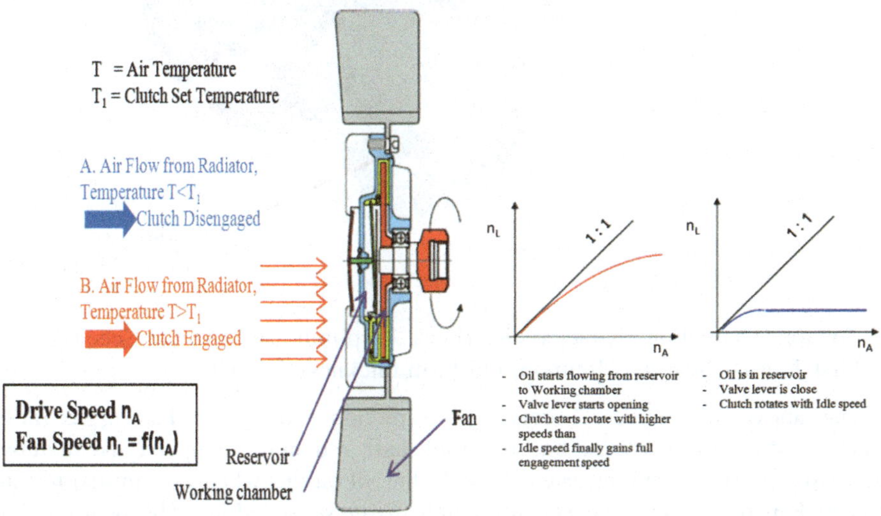

Fig. 1 Viscous drive cut section and function

3 Effects of Crank Shaft Mounted Fan on Engine

Once the fan has been finalised (based on IMTD, LAT and de-aeration criteria), the effects on engine have to be studied. The effect of crankshaft mounted fan on engine will be felt on:

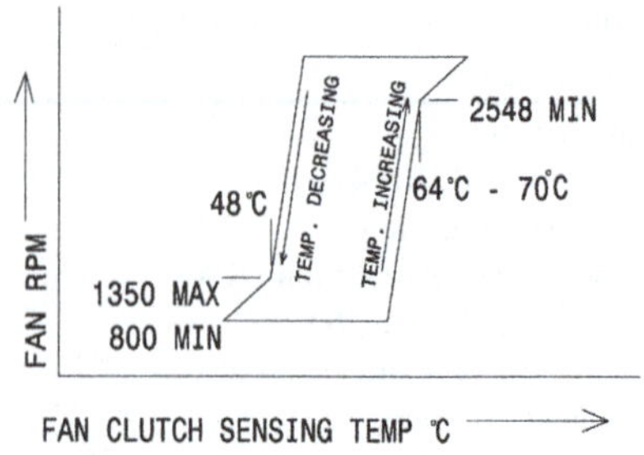

Fig. 2 Performance characteristics of on–off fan clutch

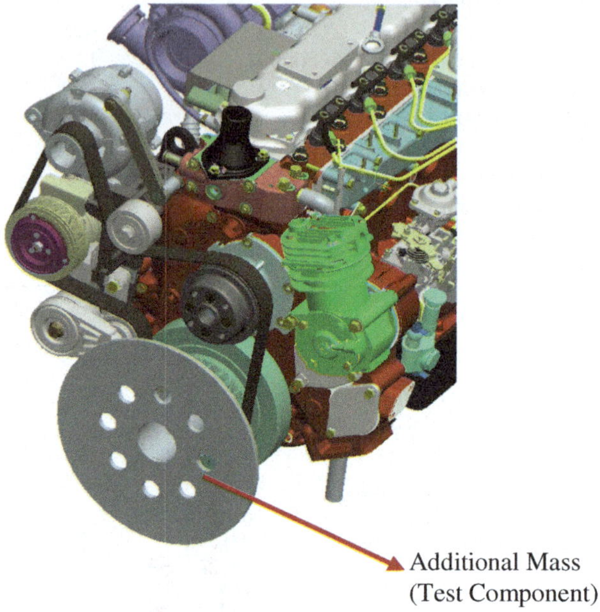

Fig. 3 TV test with additional mass (*Axial TV measurement*)

1. Torsional Vibration characteristics (Peak Amplitude and Critical Speeds).
2. First Journal Bearing Minimum Oil Film thickness (MOFT).

The above said effects become more pronounced as the Fan drive ratio decreases due to direct mounting on crankshaft, reducing fan speed and thereby increasing its size. If not arrested properly the vibrations when transmitted to fan lead to Fan blade breakage and ultimately Crankshaft failure. The bearing life drastically comes down if the first journal bearing MOFT falls below design limits.

Another important factor to be understood here is the effect of viscous fluid of fan on the TV of Engine. One can easily suggest that the viscous fluid in the Fan will act as an additional damper and reduce the TV of the engine further. To gain experimental backup to the assumption, TV test has been conducted on Engine with the following:

Fig. 4 TV test with non-contact type optical laser sensor (*Radial TV measurement*)

Table 2 Engine related inputs for analysis

Power	140 hp
Rated speed	2,800
PFP limit	120 bar
Bore	104 mm
Stroke	113 mm
No. of cylinder	4
Swept volume	3.83 L
Max torque (Nm)	580
Max torque speed (rpm)	1,200–1,600
Fuel	diesel

1. Additional Mass (Fig. 3) (Test component with Mass and Moment of Inertia same as that of Viscous Fan corresponding to complete engaged Viscous Fan).
2. Viscous Fan (Fig. 4).

Test results with the damper with additional mass (Axial TV measurement) show that the TV has exceeded the limits 0.25° whereas with the viscous fan it is 0.19° (well within limits). This shows that the viscous fluid acts as a damping agent.

Fig. 5 Fan drive assembly

- Mass: 5.3 kg
- CG: 59.4 mm
- MOI: 58951.69 kg-mm^2
- Polar Moment of Inertia: 7.78 x 10^8 mm^4

4 Engine Crankshaft Modelling in AVL Excite

The Engine crankshaft, a continuous system is discretized into finite set of rigid bodies that are interconnected with springs and dampers, known as lumped-mass model. The Torsional Vibration and MOFT characteristics of the engine can be predicted with cylinder pressure, crankshaft geometry and Block geometry as inputs to the model. Engine and Fan related inputs are shown in Table 2 and Fig. 4.

5 Torsional Vibration Analysis

COLD and HOT conditions are relevant and for rubber dampers as the viscosity of rubber i.e. the dissipating capacity of rubber between inertia ring and hub changes drastically with temperature. This is verified at test rig when a variation of 60 Hz between rubber damper at room temp and at 70 deg is observed i.e., the damper becomes largely less effective when it is not operating at its desired temperature. Thus, we cannot ignore the TV value under COLD conditions for cold starting and vehicle applications in cold countries Fig. 5.

For Viscous dampers the viscosity does not changes as drastically as it does in the case of Rubber; though it gets affected by temperature variation, but the variation is of smaller centistokes Figs 6, 7, 8, 9.

Comparison between the Peak values and trend of TV for Rigid fan Model analysis and Proto test results is shown in Table 3.

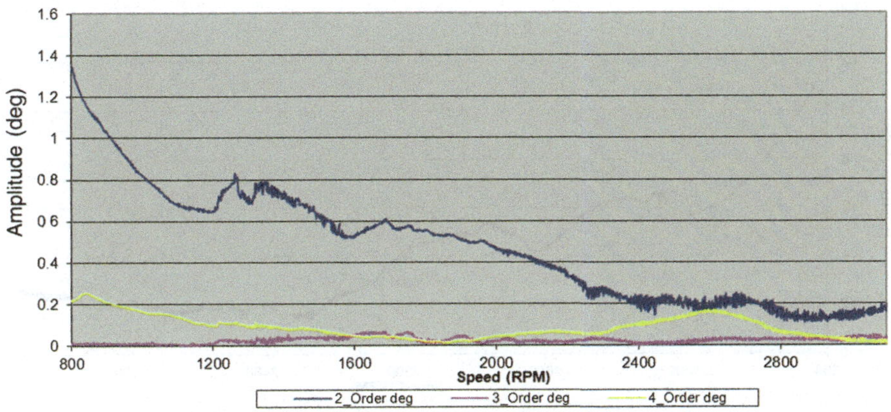

Fig. 6 Engine test bed results for engine torsional vibrations without fan

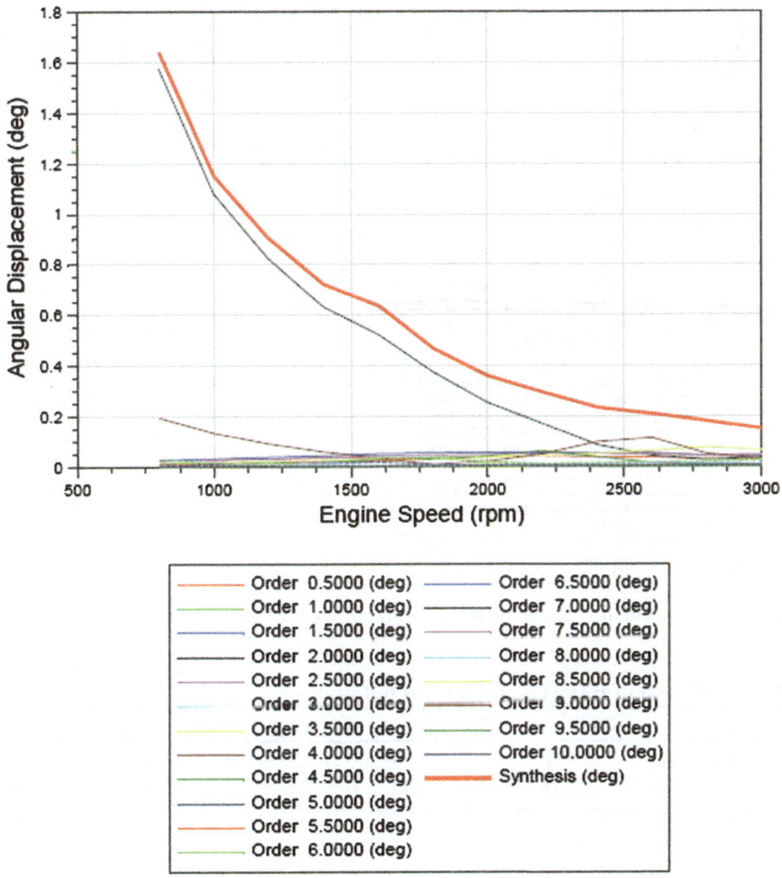

Fig. 7 EXCITE results for engine torsional vibrations without fan

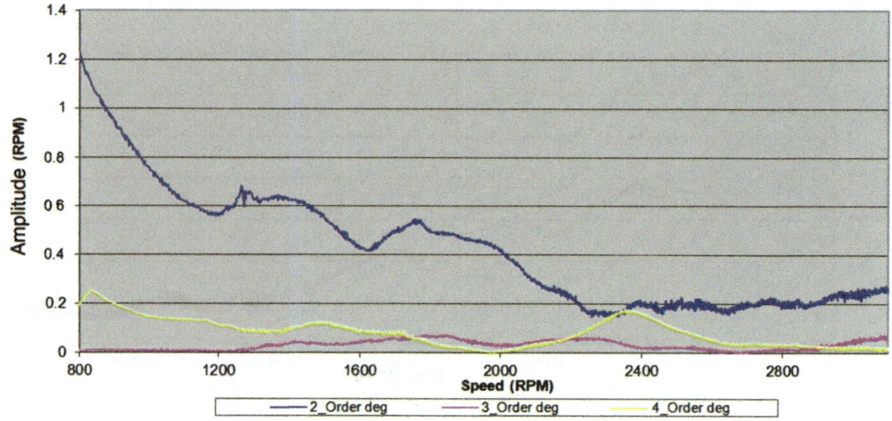

Fig. 8 Engine test bed results for engine torsional vibrations with fan

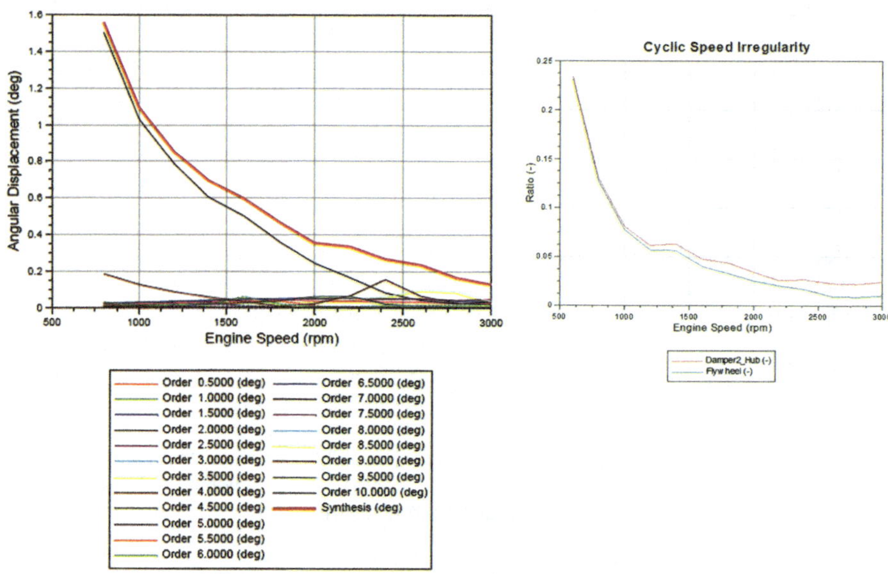

Fig. 9 EXCITE results for engine torsional vibrations with fan

Table 3 Comparison of EXCITE analysis and test results

	2nd order	3rd order	3.5 order	4th order	4.5 order
EXCITE analysis	1.2/800	0.03/1,700	0.2/2,200–2,600	0.25/800	0.19/2,000
Actual test on engine	1.23/800	0.07/1,800	0.23/2,640	0.25/840	0.18/2,100

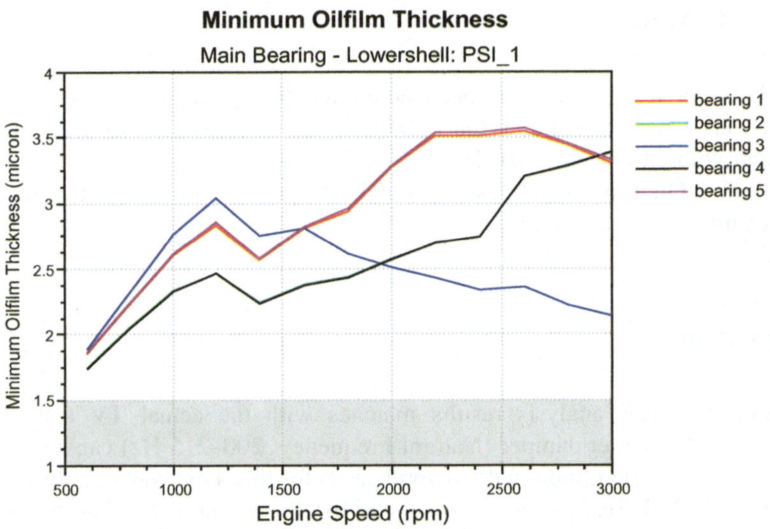

Fig. 10 EXCITE results for MOFT without fan

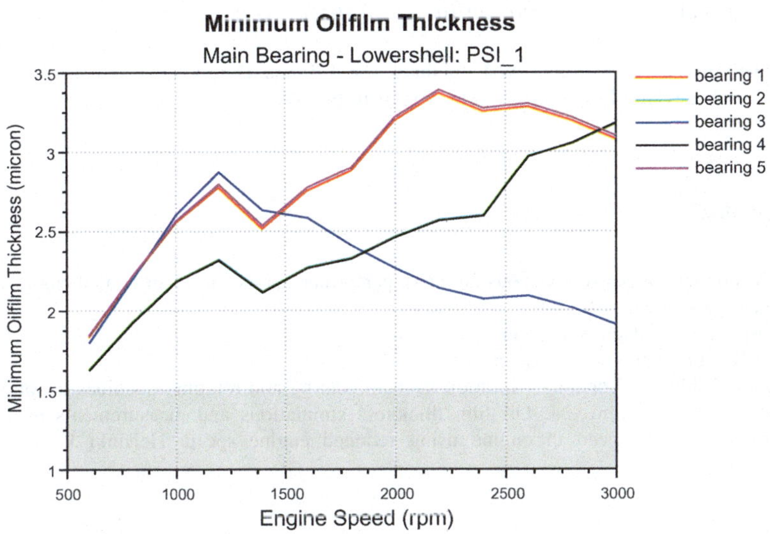

Fig. 11 EXCITE results for MOFT with fan

Comparison of Analysis and TV test data:

The results mentioned in Table 3 shows that Excite analysis results matches with Torsional Vibration test results on engine.

6 MOFT Analysis

The additional load of fan has been converted to equivalent gas pressure and analysis has been carried out for MOFT. Little decrease in MOFT with Fan compared without Fan Figs 10, 11.

Minimum Oil Film Thickness is found acceptable and within the specified limits of minimum 1.5 micron.

7 Conclusion

The AVL EXCITE analysis results matches with the actual TV test results, proving that the rubber damper (Natural Frequency: 200–215 Hz) can be used for engine with crank shaft mounted fan application for this 4 cyl water cooled engine.

Maximum Peak Firing Pressure measured for this engine is 119 bar. Simulation results correlates well with test results.

This study is limited by the AVL EXCITE's capability of simulating only 1D torsional vibrations. Hence the effect on the bearing life has been predicted by adding equivalent gas pressure on the crankshaft.

Acknowledgments The authors wish to thank the ER and D test lab Engineers, Technical Centre, Ashok Leyland Limited, for their support in this work.

References

1. John M Pouder Air Sensing viscous fan drive performance on charger air cooled engines. SAE technical paper series 861944
2. Behr M/S Product technical sheet
3. AVL EXCITE analysis software manual
4. Ronald L Eshleman Torsional vibration in reciprocating and rotating machines handbook
5. Tamminen J, Sandstom C-E Oil film thickness simulations and measurements of a main bearing in a high speed di engine using reduced engine speed. Helsinki University of Technology

1-D Simulation of a Four Cylinder Direct Injection Supercharged Diesel Engine Equipped with VVT Mechanism

Cristian Soimaru, Anghel Chiru and Daniel Buzea

Abstract Present paper exhibits the influences that variable valve timing concept has on a four cylinder turbocharged diesel engine. In this scope, using Amesim 10 1-D simulation software the engine was created and after two main strategies of valve timing were studied in order to observe the behaviour of intake mass flow rate and pumping losses diagram during 100 cycles.

Keywords Engine · Diesel · Valve timing · Mass flow · Pumping losses

1 Introduction

Automotive engineers, especially the ones who deal with internal combustion engines, in the last 20 years have been more and more pressured to develop new solutions that would improve energetically and environmental performance parameters like: reduction of CO, HC and NO_x emissions; lowering fuel consumption; higher power and torque. Among the solutions developed there may be listed: common-rail high pressure fuel injection systems, variable geometry turbochargers, after treatment systems, variable compression ratio engine, variable valve timing and lift etc.

F2012-A06-039

C. Soimaru (✉) · A. Chiru · D. Buzea
Department D02–High Tech Products for Automotive,
Transilvania University of Braşov, Braşov, Romania
e-mail: cristian.soimaru@unitbv.ro

SAE-China and FISITA (eds.), *Proceedings of the FISITA 2012 World Automotive Congress*, Lecture Notes in Electrical Engineering 190,
DOI: 10.1007/978-3-642-33750-5_21,

From solutions presented above, variable valve timing represents the focus of this paper. This strategy has been studied for a long time and its theoretical approach is presented in Atkinson and Miller cycles. Paper [1] presents the principle of obtaining variable valve actuation in two ways: by modifying the exhaust valve closing, according to Atkinson cycles, and modifying the intake valve closing, Miller cycles. Both of the techniques have the same main improvement, expansion stoke bigger than compression stroke so that higher torque could be obtained and a decreasing value of pumping looses also.

Within last 20 years many studies have been conducted on spark ignition engines equipped with variable valve timing system so that the influence that it manifests over the energetically and environmental parameters could be determined. Scientific papers like [2–4] present the influences that the variable valve timing involves over engine's energetically performances, emissions and new fuel burning concepts like CAI—Controlled Auto Ignition.

Regarding diesel engines the variable event valve timing has not been appropriate to be studied due to reasons like: wide area spreading of turbochargers due to their improvements to engine's efficiency, internal and external gas recycling efficiency and improvements for internal combustion engines, and the third one huge solution like high pressure fuel injection systems and injection characteristics [5].

2 Paper's Objective

The objective of this paper is to present the influence that dimension of valve overlapping has on the internal combustion engine's intake mass flow rate and on cylinder pressure, by modifying the moment of intake valve opening. This objective was possible to achieve by using 1-D Amesim 10 software simulation tool.

3 Methodology

3.1 Theoretical Approach of the Simulation

The theoretical approach of the simulation is supplied by [1, 6–8].

3.1.1 Volumetric Efficiency

$$\eta_v = \frac{2 \cdot \dot{m}_a}{\rho_a \cdot V_h \cdot N} \tag{1}$$

$\dot{m}_a$ = *air mass flow inducted into the engine*, ρ_a = *air density*, N = *engine speed*

V_h = *engine displacement.*

3.1.2 Pseudo: Flow Velocity

$$v_{ps} = \frac{1}{A_m} \cdot \frac{dV}{d\theta} = \frac{\pi \cdot B^2}{4 \cdot A_m} \cdot \frac{ds}{d\theta} \tag{2}$$

V = *cylinder volume*, B = *cylinder bore*, s = *distance between crank axis and wrist pin*, A_m = *valve area.*

3.1.3 Air Flow Rate

$$\dot{m} = \frac{C_D \cdot A_R \cdot p_0}{(R \cdot T_0)^{1/2}} \cdot \left(\frac{p_T}{p_0}\right) \left\{ \frac{2 \cdot \gamma}{\gamma - 1} \cdot \left[1 - \left(\frac{p_T}{p_0}\right)^{\left(\frac{\gamma-1}{\gamma}\right)} \right] \right\}^{1/2} \tag{3}$$

C_D = *discharge coefficient*, p_0 = *upstream stagnation pressure*, T_0 = *upstream stagnation temperature*, ρ_T = *pressure at the restriction*, A_R = *reference area of the valve*, γ = *adiabatic coefficient.*

3.1.4 Equivalence Ratio

$$\phi = \frac{\left(A/F\right)_{acutal}}{\left(A/F\right)_{teoretic}} \tag{4}$$

$$A/F = Air - fuelratio$$

3.1.5 Intake Mass Speed

$$p_{atm} - p_c = \sum \Delta p_j = \rho_a \cdot \bar{S}_p^2 \cdot \sum \xi_j \cdot \left(\frac{A_p}{A_j}\right)^2 \tag{5}$$

p_{atm} = *atmospheric pressure*, p_c = *cylinder pressure*, $\bar{S}_p$ = *mean piston speed*, A_p = *piston area*, A_j = *component minimum flow area*, Δ_{pj} = *total quasy steady*

ξ_j = *resistance coefficient for that component which depends on its geometric details*

A_p = *piston area*, A_j = *component minimum flow area*, Δ_{pj} = *total quasy steady pressure loss.*

3.1.6 Mass Flow Rate

$$\frac{dm}{dt} = A_{eff} \cdot p_{0I} \cdot \sqrt{\frac{2 \cdot \psi}{R_0 \cdot T_{0I}}} \tag{6}$$

dm/dt = *mass flow rate*, A_{eff} = *effective flow area*, p_{0I} = *port upstream static pressure*, T_{0I} = *port upstream static temperatrue*, R_0 = *gas constant*

3.1.7 Combustion Model

$$\frac{dx}{d\alpha} = \frac{a}{\Delta\alpha_c} \cdot (m+1) \cdot y^m \cdot e^{-a \cdot y \cdot (m+1)} \tag{7}$$

$$dx = \frac{dQ}{Q}$$

$$y = \frac{\alpha - \alpha_c}{\Delta\alpha_c}$$

Q = *total heat amount received*, α = *Crank Angle Degree*, α_0 = *corresponding angle for begining of combustion*, $\Delta\alpha_c$ = *combustion duration*, m = *form coefficient*, $a = 6.9$, *Wiebe coefficient*

3.1.8 Burnt Fuel Quantity

$$x = 1 - e^{-a \cdot y \cdot (m+1)} \tag{8}$$

3.2 Realizing Virtual Model of the Engine

Main parameters of the engine, used in simulation, are presented in Table 1.

In Fig. 1 is presented the virtual model of the engine which consists in pictograms that have a specific function beginning with fluid properties, engine definition and finishing with in cylinder parameters definition. In this simulation the variability characteristic of the turbine was not used, also the EGR was kept at zero value.

Table 1 Simulation's parameters

Parameter	Value
Stroke	86.5 mm
Bore	86.5 mm
Compression ratio	18
Connecting rod length	150 mm
Engine speed	3,000 rpm
Engine load	100 %
Injection duration	0.0016 s
Injection timing	15° CAD
Injection pressure	1,000 bar
Combustion model	Wiebe
Intake valve opening BTDC	46.5° CAD
Intake valve opening duration	277° CAD
Exhaust valve closing ATDC	39.5° CAD
Exhaust valve opening duration	257° CAD
Turbocharging pressure	1.35 bar
Turbocharger speed	148,500 rpm
Atmospheric temperature	21 °C
Air temperature after the intercooler	85 °C
Atmospheric pressure	1 bar

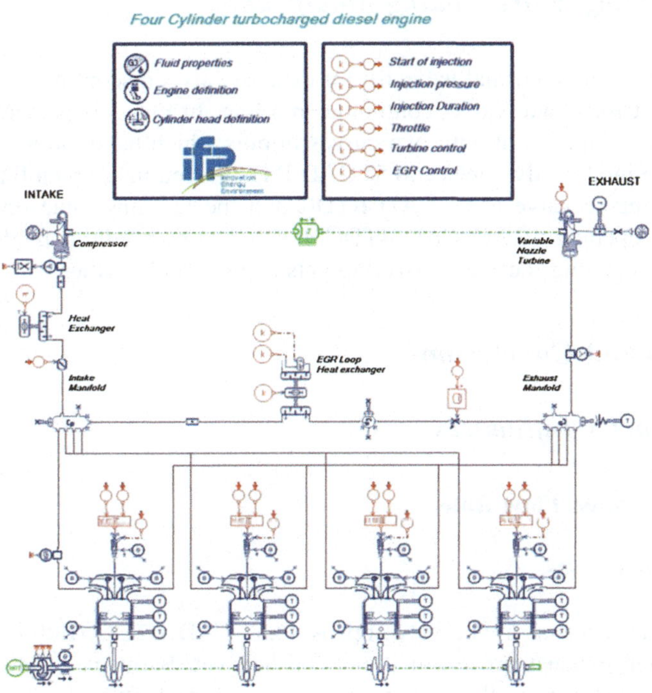

Fig. 1 Virtual model of the four cylinder turbocharged diesel engine (Adapted from [7])

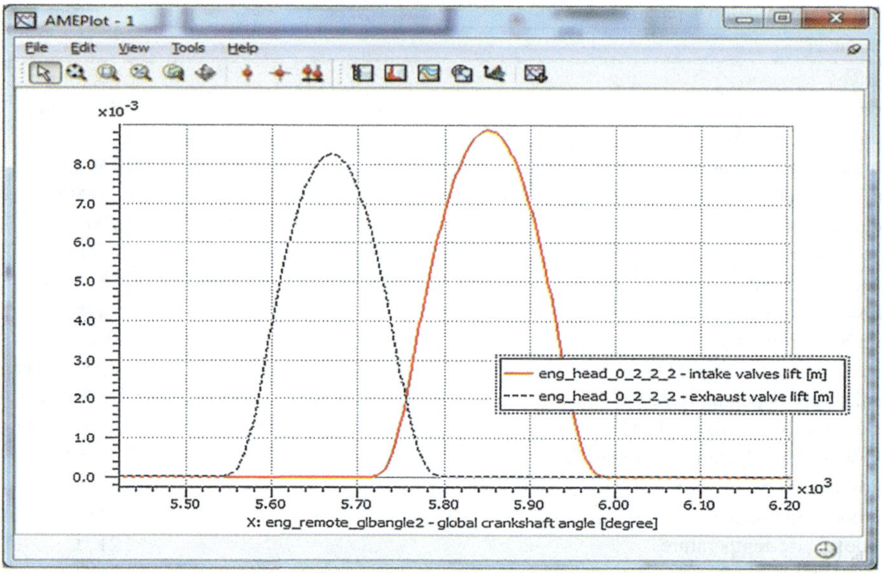

Fig. 2 Normal valve lift

3.3 Presenting Valve Timing Modification

In Fig. 2 is presented normal valve lift for exhaust valve, discontinuous black line, and valve lift for intake valve, continuous red line. In Fig. 3 is presented normal situation, as in Fig. 2, but with two curves in plus which have the next meaning: normally the intake valve opens 46.5° CAD BTDC—red line, green line indicates opening of intake valve at 0° CAD BTDC and the last line, blue one indicates intake valve opening −37.5° CAD BTDC. It may be observed that by delaying the intake valve opening the valves overlap gets smaller and smaller.

4 Results and Conclusions

4.1 Parameter Influences

4.1.1 Intake Mass Flow Rate

There are three cases:

First case:

Due to the fact that intake valve opens 46.5° CAD BTDC (Before Top Dead Centre), when exhaust process it is not finished yet, burnt gases pressure p_2 is superior to p_1 pressure of the air, this is the reason for which it may be observed a backflow of gases from cylinder to the intake pipe. This trend of gases is stopped

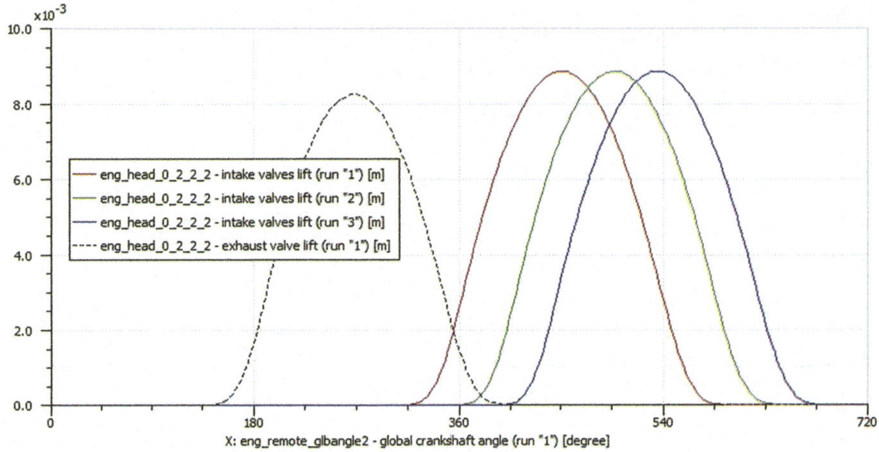

Fig. 3 Modification of opening timing for intake valve

when the piston reaches Top Dead Centre, after when it goes towards Bottom Dead Centre there appear positive values of the intake flow within the cylinder which reach maxim value of approximately 80 [g/s].

Positive values are encountered even in compression stroke but only until pressure in cylinder has the same values as pressure p_1 of fresh air. When p_2 gets bigger than p_1 then it may be observed a second tendency of fresh air to backflow, red curve, which will stop when the intake valve will be completely closed.

Second case:

In this case the green curve shows the evolution of intake mass flow when the intake valve opens with 0° CAD BTDC. Due to the fact that there is no advance for the valve to open there will be no backflow. Maxim value of the intake mass flow close to 90 [g/s] and has the next reason: by opening the intake valve right in TDC, air pressure is equal with pressure in the cylinder, but only for a moment because now the piston goes to BDC which decreases p_2 in such a manner that mass flow increases rapidly. Another reason would be air speed.

In Fig. 4 it may be observed that by the end of intake stroke, which is extend long in compression stroke than first case, mass flow decreases dramatically because of the backflow. Beside this, the area under the curve is smaller than in the first case, this shows that inducted air is smaller in quantity.

Third case:

In this case intake valve opens late, with 31.5° CAD ATDC. Due to this large delay there is no backflow, but the mass flow reaches its maximum flow approximately 135 [g/s]. This is motivated by the depression caused from piston motion towards BDC. When the piston turns towards TDC the backflow appears even sooner due to late extension of intake process in the compression stroke. The area under the blue curve is the smallest, hence a proportional inducted fresh air.

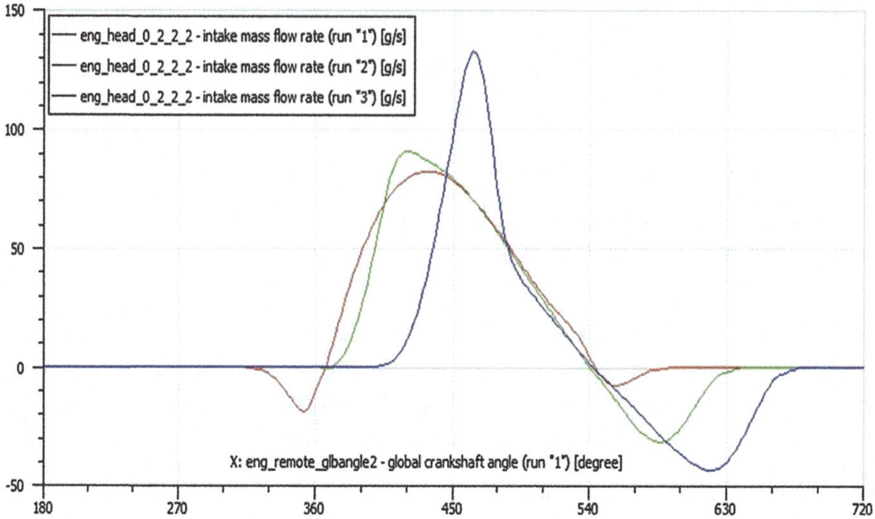

Fig. 4 Intake mass flow rate

4.2 Pressure in Cylinder

Pressure in cylinder is proportional with volumetric efficiency and it is demonstrated by the next formula:

$$P = N \cdot V_h \cdot \rho_a \cdot \left({}^{F}/_{A}\right) \cdot Q_{HV} \cdot \eta_f \cdot \eta_v \tag{9}$$

P = engine power, N = engine speed, V_h = cylinder displacement
ρ_a = air density, (F/A) = fuel–air ratio, Q_{HV} = fuel lower heating value,
η_f = combustion efficiency, η_v = volumetric efficiency

Volumetric efficiency depends on intake mass flow. Thus, in order to keep air–fuel ratio equal to 1, for a less quantity of air inducted in the cylinder there is a proportional quantity of fuel which leads to less pressure in the cylinder and brake power. This is how the descending trend of pressure from approximately 130 bar, to 110 bar and 70 bar may be motivated (Fig. 5).

This statement is supported also by pumping losses presented in Fig. 6. The main argument according to which pumping losses area increases with the late opening of the intake valve is that the dimension of the backflow occurring in compression stroke needs a proportional work effectuated by the piston.

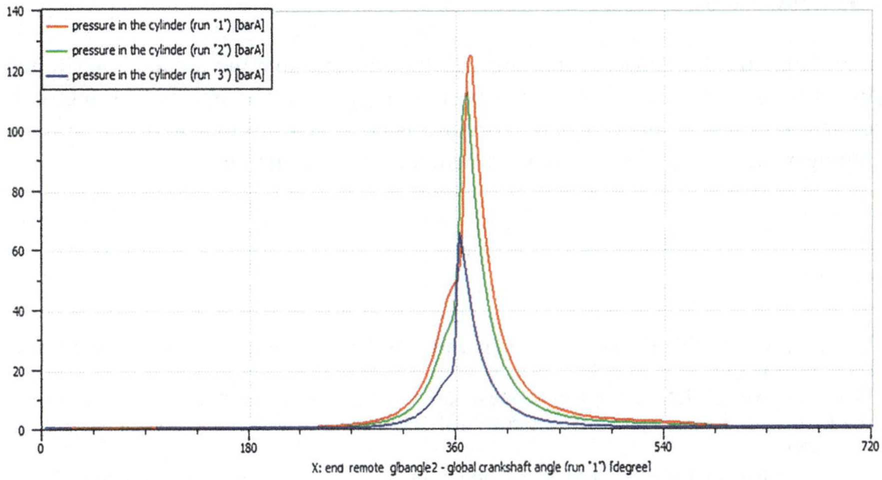

Fig. 5 Pressure in cylinder

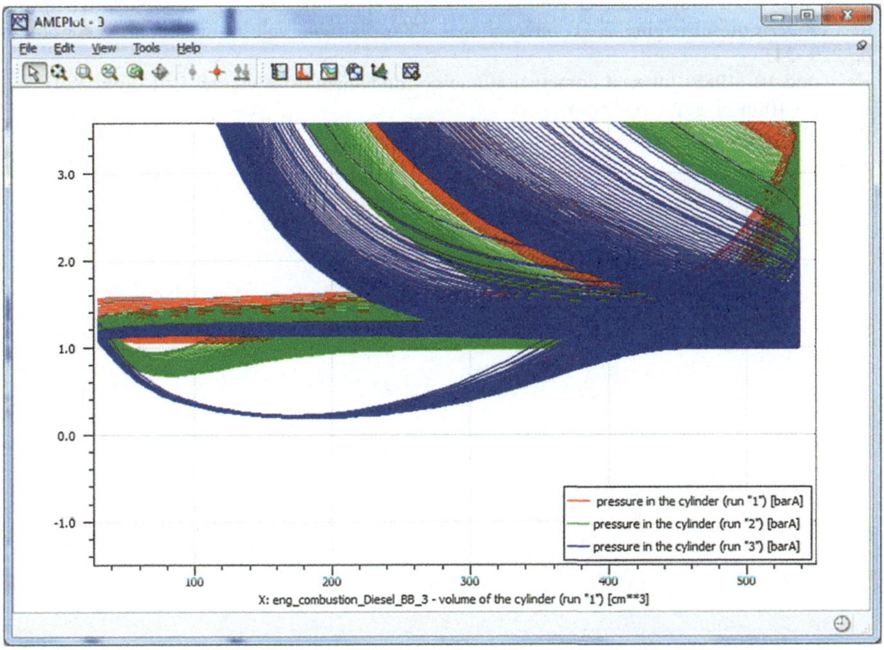

Fig. 6 Pumping losses

4.2.1 Conclusion

It is obvious the fact that by opening late the intake valve the air inducted in the cylinder is also influenced. Thus, the best strategy to obtain an efficient engine is open late the intake valve but also to close it near BDC so that will not be extra pumping work and hence less power available for the engine.

References

1. Pulkrabek WW (2003) Engineering fundamentals of the internal combustion engine. Prentice Hall, New Jersey
2. Bohac SV, Assanis DN (2006) Effect of exhaust valve timing on gasoline engine performance and hydrocarbon emissions. SAE international 2006, variable vale train systems technology, pp 19–29
3. Leroy T, Chauvin J, Petit N (2009) Motion planning for experimental air path control of a variable valve timing spark ignition engine. Control engineering practice, vol 17. Elsevier, pp 1432–1439
4. Milovanovic N, Chen R, Turner J (2004) Influence of variable valve timings on the gas exchange process in a controlled auto–ignition engine. IMechE Part D, 2004, pp 567–583
5. Lancefield T, Methley I, Räse U, Kuhn T (2006) The application of variable event valve timing to a modern diesel engine. SAE international 2006, variable vale train systems technology, pp 229–241
6. Heywood JB (1988) Internal combustion engine fundamentals. McGraw Hill, New York
7. Amesim 10, user guide pdf book
8. Boost 2010, user guide pdf book

In-Cylinder Flow Oriented Intake Port Development of Diesel Engine

Kang Li, Haie Chen, Wei Li, Huayu Jin and Jiaquan Duan

Abstract With the increasingly stringent emission standards and economical efficiency, it is an inevasible challenge to improve higher efficiency combustion of the diesel engine. Due to the development of engine design and accessorial technology, the research direction of the modern diesel engine turns to accurate control and optimization of the engine system. The design of the intake ports has direct and significant impact on in-cylinder air motion intensity and distribution. The diesel engine's combustion is relative to the conditions of the fuel spray, the shapes of the combustion chamber and the air flow intensity in cylinder. The traditional development of the intake ports of the diesel engine was just focused on the steady-state characteristics of the intake ports, the main parameters to evaluate the intake ports were flow coefficient, swirl ratio and tumble ratio etc. But those macroscopic parameters could not truly estimate the influence of the intake ports performance during the diesel engine dynamic process. Especially, we can not get the accurate information about the movement and distribution state of the engine dynamic process from the stationary flow test bench, and so it is not effective to accurately control the combustion and emission. During the development of intake ports of the diesel engine, we found that the influence on the dynamic intensity of the swirl motion in cylinder was obvious when the same intake port matching with different combustion chambers. Hence, it is not imagine the reality mixture flow conditions in cylinder if we just insist on the intensity of the swirl motion measured from the stationary flow test bench. Based on those reasons, we bring for-

F2012-A06-044

K. Li (✉) · H. Chen · W. Li · H. Jin · J. Duan
CHINE FAW CO., LTD R&D Center, Mainland, China
e-mail: lk_rdc@faw.com.cn

SAE-China and FISITA (eds.), *Proceedings of the FISITA 2012 World Automotive Congress*, Lecture Notes in Electrical Engineering 190,
DOI: 10.1007/978-3-642-33750-5_22, © Springer-Verlag Berlin Heidelberg 2013

ward the conception of in-cylinder flow oriented intake port development. With the help of the stationary flow test, steady and dynamic CFD, the method of accurately controlling the air flowing in-cylinder can instruct the intake port design and development.

Keywords Intake port · In-cylinder flow · CFD

1 Define the Combination and Orientation of Two Intake Ports

It has been known that swirl motion in-cylinder had a crucial effect on the mixture formation and combustion for DI diesel engines. But the air flow in-cylinder of four valves DI diesel engines is much more complicated in comparison with two valves engine, because the flows discharging from two adjacent intake valves will interfere to each other. The combination and orientation of two intake ports have important influences on the flow interference. Swirl ratio and flow coefficient of intake system. Before the process of dynamic CFD analysis, we should ensure the kind of combination and orientation of two intake ports, which can fulfil the requirement of the swirl level for combustion system.

The experiment was carried out on a steady flow test rig. To easily change the combination and orientation of the intake ports, a steel plate with two valve seat holes on it was designed to install intake ports through two valve seats (in Fig. 1). Two identical helical ports and two identical directed ports were made by a kind of rapid casting resin based on the same design of directed intake port and the same design of helical port [1]. These four ports were used to combine each other into four different combined ports: helical and helical, helical and directed, directed and directed, directed and helical. Each port can rotate freely around its valve axis. The results of the mean swirl ratio and mean flow coefficient of the difference combination and rotated angle were in Fig. 2. There were rapid cast apart two resin models for helical port and tangent port, so could interchange the ports each other. There would have 4 kinds of combination of the intake ports: the left is helical and the right is tangent (LH–RT), the left is tangent and the right is helical (LT–RH), two all helical (LH–RH) and two all tangent (LT–RT). To be convenient for describing the combination and orientation of two intake ports, the following symbol is used: H and T represents that the helical port and tangent port, the left and right direction is defined from the top view.

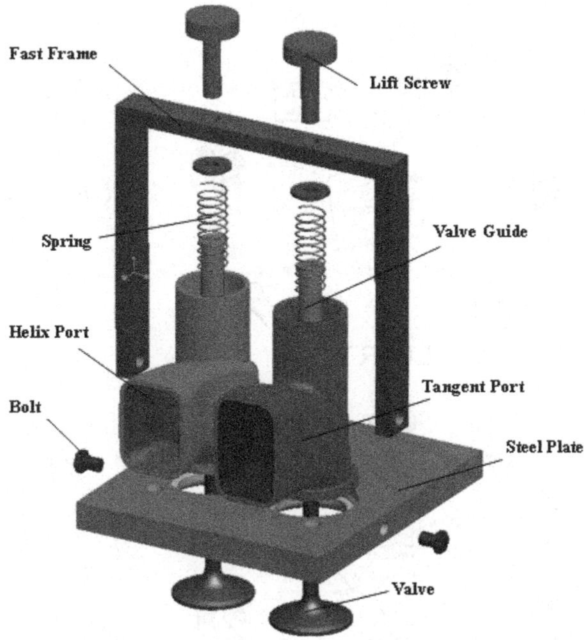

Fig. 1 Assembled equipment for intake ports combination and orientation

2 Experiment and Evaluation

The experiment was carried out on a steady flow test rig. A paddle wheel borne by two low friction bearings was pivoted on the cylinder centerline to measure the swirl. All measured parameters such as the paddle speed, the flow rate, the valve lift, the ambient temperature, the pressure drop across the intake valve would be collected and processed automatically by PC computer. Through respectively rotating each separate inlet and adjusting different Inlet-Port combination can be obtained different in cylinder swirl intensity changes. The requirements of test results and the combustion system determine the combined angle and port type.

There are several types of parameter to evaluate the flow resistance of the intake port and its swirl capability based on the steady flow test. The non-dimensional flow coefficient C_f and the non-dimensional swirl ratio R_s at certain valve lift are two parameters commonly used [2].

$$\mathrm{Cf} = \frac{\mathrm{m}_{real}}{m_{th}} \tag{1}$$

$$\mathrm{Rs} = \frac{n_D}{n} = \frac{n_D \rho V_h}{30 m_{real}} \tag{2}$$

Where m_{real} is real mass flow rate and m_{th} is theoretic mass flow rate of air through intake valves (kg/s); n_D is the speed of the paddle wheel (1/s), n is engine speed (1/s); V_h is cylinder displacement, ρ is air density.

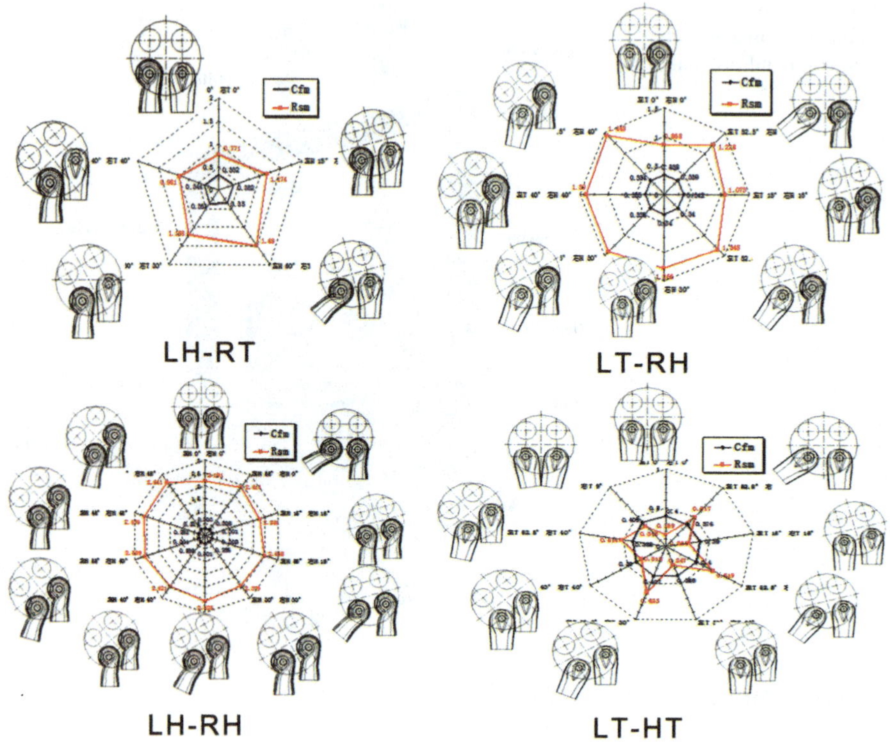

Fig. 2 The results of the different combination and orientation of the intake ports

Swirl ratio at the end of intake process $(R_s)_m$ and mean flow coefficient during intake process $(C_f)_m$ can be calculated using the AVL method.

$$(\mathrm{Cf})_m = \frac{1}{\sqrt{\frac{1}{\pi}\int_0^{\pi} \left(\frac{c(\alpha)}{c_m}\right)^3 \frac{1}{C_f^2}\, d\alpha}} \tag{3}$$

$$(\mathrm{Rs})_m = \frac{1}{\pi}\int_0^{\pi} \frac{n_D}{n}\left(\frac{c(\alpha)}{c_m}\right)^2 d\alpha \tag{4}$$

Where α is the crank angles as the intake valves open and close respectively; C_m is mean piston speed and $c(\alpha)$ denotes the actual piston speed. Equations (3) and (4) give evaluations of swirl intensity and flow resistance generated by intake ports.

Figure 2 show double helical ports combination can get more higher swirl intensity but lower flow capability, the range of mean swirl ratio is 2.191 ~ 2.641; Double tangent ports combination can get more higher flow capability but lower swirl intensity, the range of mean swirl ratio is 0.159 ~ 0.623. Just one helical port and one tangent port combination is appropriate combination, the range of

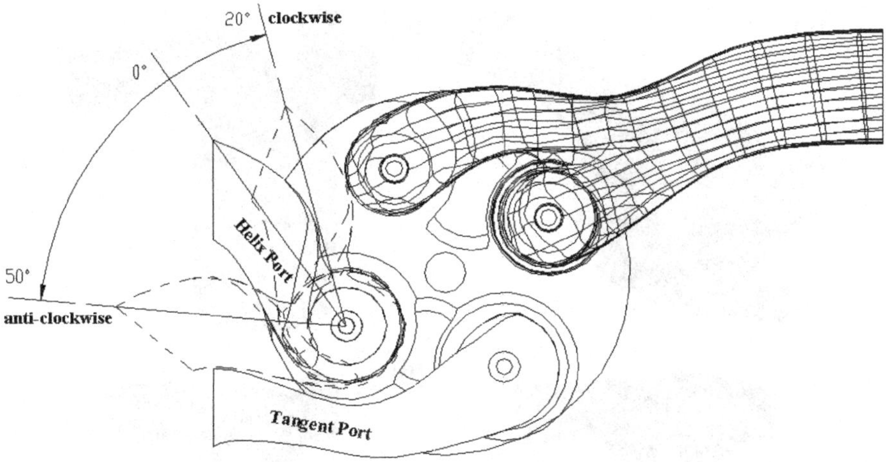

Fig. 3 The ports layout and the rotatable angle range of the helical port

mean swirl ratio is 0.771 ~ 1.448. According to the requirement for the swirl intensity in cylinder of the combustion system, and refer to the above experimental results can determine the reasonable combination of the intake port type and the relative angle relative to the cylinder center line, which more further to provide the base layout for the CFD dynamic analysis.

3 Dynamic Flow Analysis in Cylinder

With the further development of simulation, CFD analysis can more clearly show the precise changes of flow field and turbulent kinetic energy distribution in cylinder. Especially, the flow field in cylinder at the injection timing has a great influence on the combustion. So by organizing it, the combustion state can be controlled well. The analysis of flow field forming process and forming reason in cylinder can instruct the design of the intake ports. Through the changes of some key geometric parameters including the intake ports combination, relative angle to each other and swirl chamfer, suitable state of the swirl intensity and the distribution of the turbulent kinetic energy in cylinder can be achieved. On the basis of meet the above requirements, the uniform distribution of the fuel injection from each nozzle hole which was influenced by air movement in cylinder also should be ensured. It is good for the emissions and fuel consumption Fig. 3.

The computational grid of the 6DL2-35E3 diesel engine was build. Its compression ratio was 17.1 and the total grid number was 9,78,000. Intake port combination of a single helical port and a single tangent port was adopted. Due to its great influence on the swirl intensity in cylinder when the rotation angle of the helical port relative to the helical valve center line was changed, whereas the

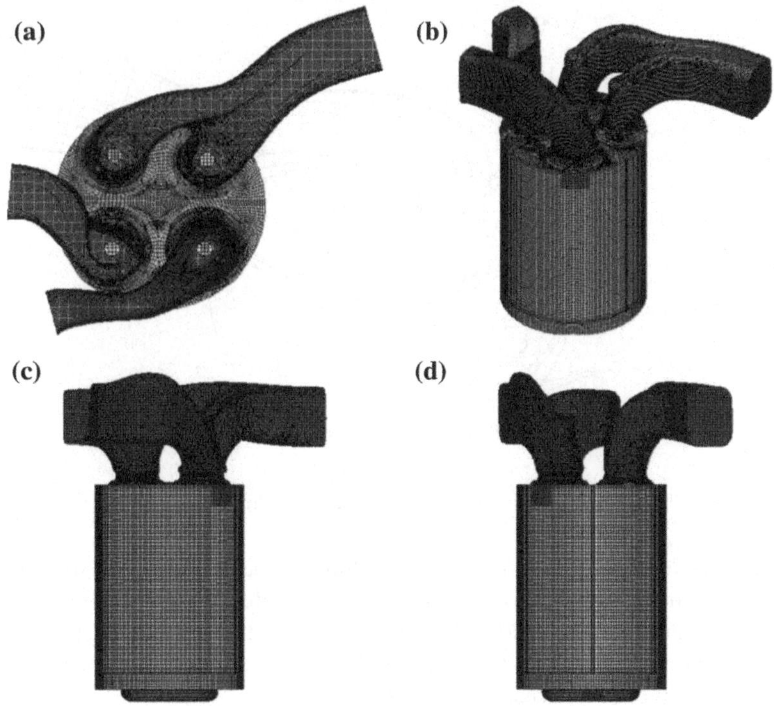

Fig. 4 Show of the computational grids

tangent intake port was not very sensitive, it was just decided to only adjust the helical port rotation angle in CFD analysis. Limited by actual structure of the cylinder head, the rotatable maximum angle range of helical intake port was 50° counter clockwise and 20° clockwise, so based on the original layout of the computing grid, the rotation angles of the cases were respectively clockwise rotations of 10 and 20° and counter clockwise rotations of 10, 30, 50° relative to the valve center line for the helical intake port. The dynamic pressure was set as the inlet and outlet pressure boundary conditions Fig. 4.

In the process of dynamic flow simulation, the parameters of transient swirl ratio were used to evaluate the air motion intensity in the cylinder. The transient swirl ratio (R_{sT}) was defined as the ratio of the mean angular speed relative to the cylinder center line divided by angular speed of engine crankshaft.

$$R_{ST} = \frac{\bar{\omega}}{\omega_n} \tag{5}$$

$$\bar{\omega} = \frac{\sum m_i v_i r_i}{\sum m_i r_i^2} \tag{6}$$

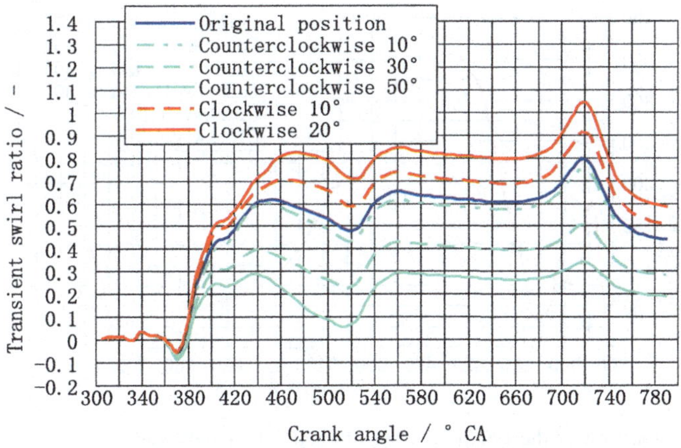

Fig. 5 Contrast of the transient swirl ratio of each case

In which $\bar{\omega}$ is calculated for each cell of the grids in the cylinder and m_i means cell mass, v_i means tangential velocity around the cylinder center line and r_i means the distance from the cell centroid to the cylinder center line.

Transient swirl ratio is the integrated embodiment of the magnitude of the rotation speed and the extent deviation of swirl center to cylinder center line. The bigger its value is, the stronger the flow is. Figure 5 is the contrast of the transient swirl ratio of each case. It can be seen the more the helical port clockwise rotation is, the greater the transient swirl ratio is, and the more the helical port counter clockwise rotation is, the smaller the transient swirl ratio is. Therefore the helical port clockwise rotation is benefit for vortex forming and counter clockwise rotation is opposite.

$$k = \frac{\sum m_i k_i}{\sum m_i} \tag{7}$$

The mass-average turbulent kinetic energy is defined as Eq. (7) and calculated for each cell of the grid in cylinder, where m_i is the mass and k_i is the turbulent kinetic energy of each cell respectively. Turbulent kinetic energy is another important criterion parameter to evaluate the flow intensity in cylinder. The bigger turbulent kinetic energy is, the more advantage for mixing of the fuel and air is. Figure 6 is the comparison of the mass-average turbulent kinetic energy of each case, and it can be seen the turbulent kinetic energy of counter clockwise rotation case is larger than that of clockwise ones during intake process. The distinction of the turbulent kinetic energy in each case will be weakly after the intake valves closed, especially near the compression TDC, 720 °CA. The fuel is injected at 714.5 °CA, and it makes the turbulent kinetic energy increase rapidly. So the distinction of the kinetic energy of each case becomes smaller.

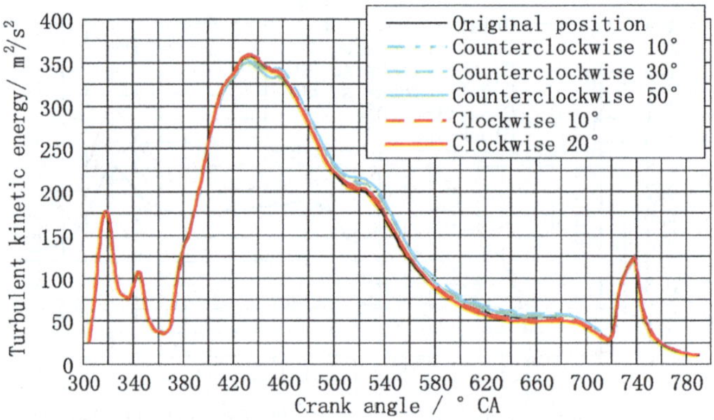

Fig. 6 The comparison of the kinetic energy of each calculated cases

As shown in Fig. 7 ("−22 mm section" is 22 mm distant from the cylinder head bottom downward), the direction of helical intake port rotating counter clockwise 50° determines a significantly strong clockwise speed around the cylinder center, which reduces the transient swirl ratio in cylinder. So the transient swirl ratio of this case is the lowest. But the helical port rotating clockwise 20° increases the counter clockwise swirl in cylinder. From vertical section, the section 1 of the counter clockwise 50° case shows the direction of combined speed built by two different intake velocities is lean to the left, this strong effects the holdback of the counter clockwise rotating speed, but the clockwise 20° case shows the direction of combine speed is lean to the right, and this strengthens the speed of the counter clockwise rotation. Section 6 shows that the air forward velocity is relatively strong in the counter clockwise case, which make the combined flow of the exhaust side against convolute to the helical port side, but the clockwise 20° case is not. All of the above explain the reason why the transient swirl ratio of the clockwise 20° is much stronger than the counter clockwise 50°.

For transient calculation results, optimal cases are determined through the analysis and evaluation of several key macroscopic parameters. Figure 8 shows that helical port has a little influence on cylinder pressure and heat release rate but more influence on emission. In the case of the helical port clockwise 20°, whose swirl level is higher, the NO emission is obviously increased, but the Soot is reduced, the case of counter clockwise 50° is just opposite. All above means that combustion state is improved for the case of clockwise 20°. After helical port clockwise rotating 20°, the transient swirl ratio is increased and the eccentricity ratio δ becomes smaller. When the eccentricity ratio becomes smaller, the mixture of the fuel and air is more homogenized. When the swirl ratio increase, the fuel spray beam bends and the contact area with air becomes larger, especially the combustion of the windward side is improved, and the combustion temperature becomes higher. So the NO emission is obviously increased and the Soot is reduced Fig. 9

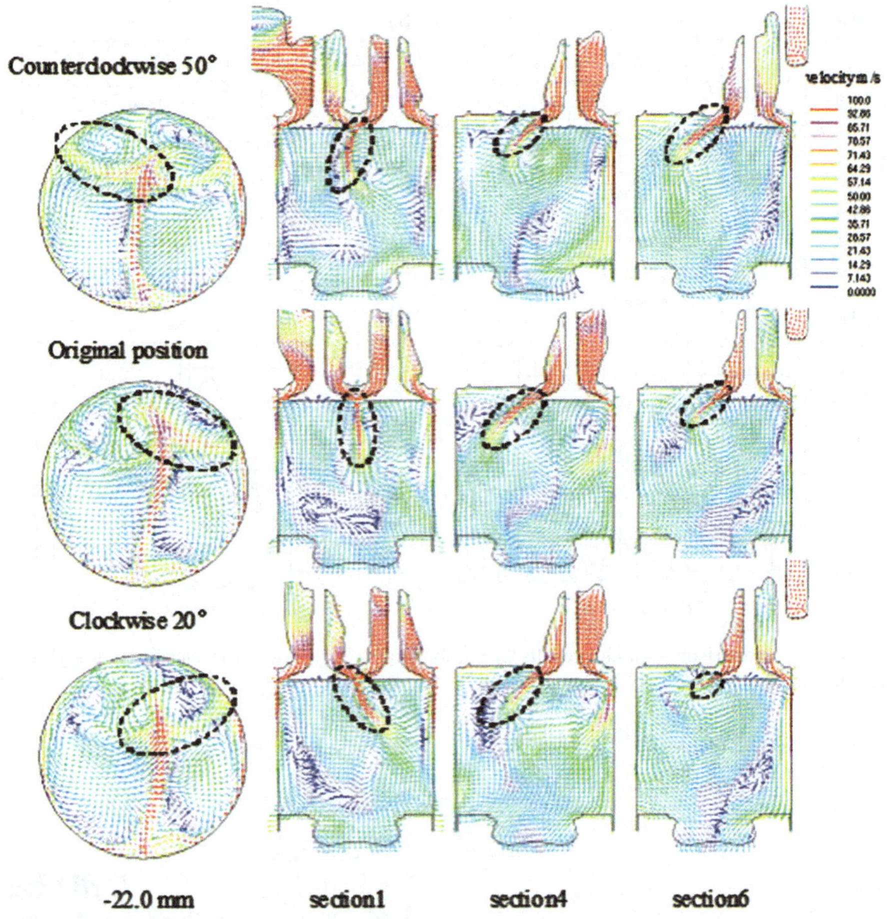

Fig. 7 The comparison and analysis of the flow field at crank angle 460 CA°

$$\delta = \frac{\mathrm{R}_w}{R} \tag{8}$$

δ is eccentricity ratio, Rw is eccentricity radius of the swirl center, R is cylinder radius. General, we just compare the eccentricities at vortex center near TDC just before injection. At that moment the vortex becomes rigid, and the vortex center is basically stable. When the eccentricity ratio becomes smaller, the distribution of spray is more uniform.

Figure 10 shows the comparing and analysing of the flow field at different sections below the bottom cylinder head, the distances between the bottom cylinder head and the five sections are −3, −5.9, −8.9, −11.9 and −14.9 mm, the injector has 8 holes. When the helical port rotates counter clockwise 50°, the

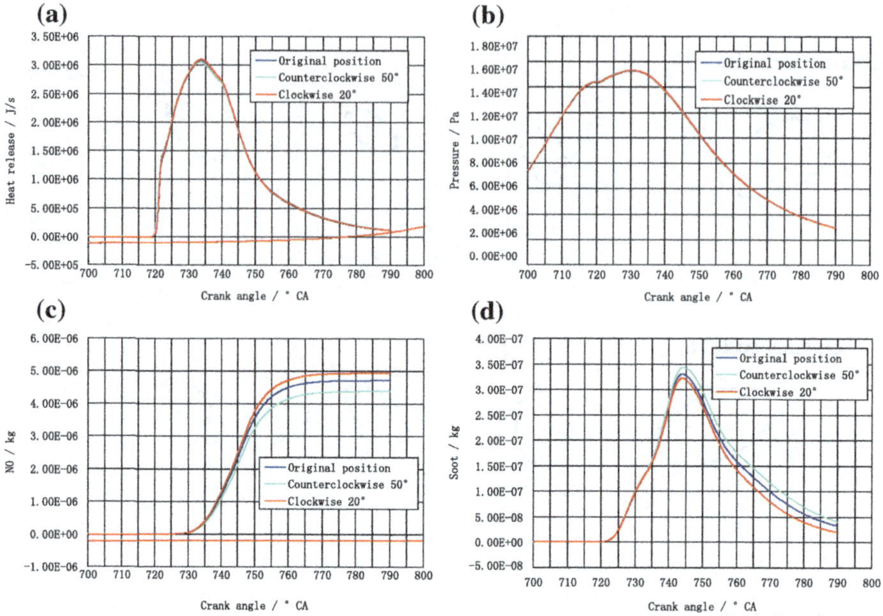

Fig. 8 The comparison simulate results of combustion and emission for each case. **a** Cylinder pressure. **b** Heat release rate. **c** No. **d** Soot

Fig. 9 Definition of the eccentricity ratio

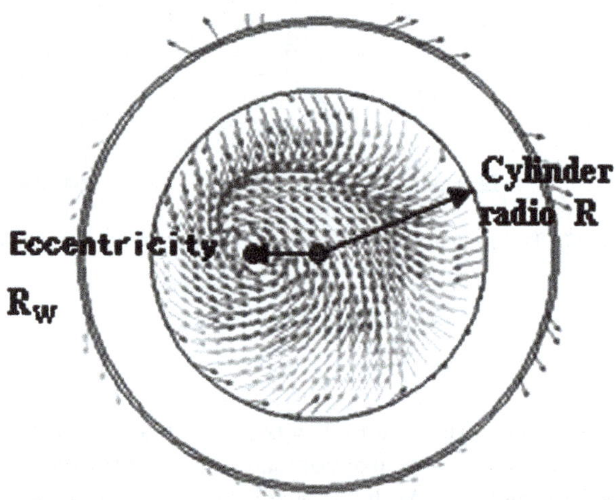

mixture formation of the fuel and air is not great uniform. The intake flow through the inlet valves hits the wall and hinders the formation of cylinder swirl, which makes the swirl ratio reduce. The distribution of clockwise 20° is more uniform because of the more uniform air flow in-cylinder.

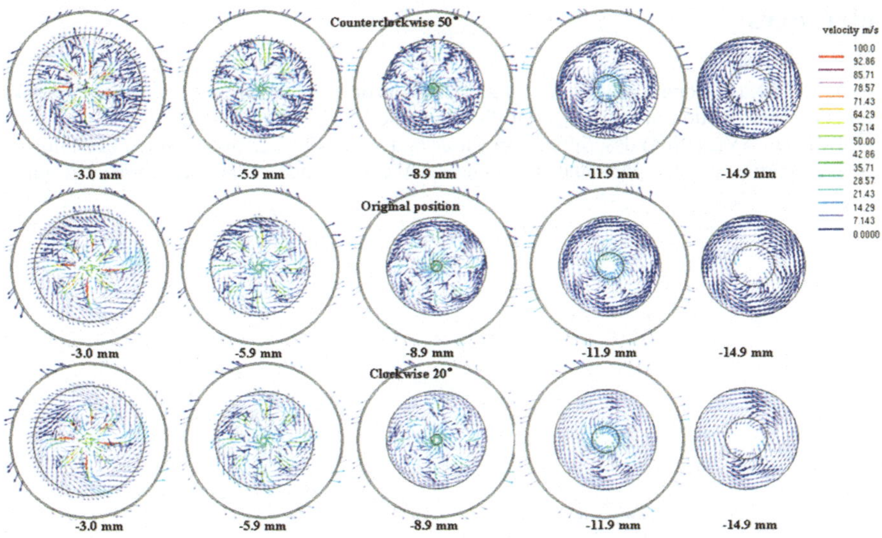

Fig. 10 The comparison and analysis flow field at difference projection at 730 °CA

4 Conclusions

1. It is very important to confirm the position of the four valves' center and the combination of the intake ports. Choosing a proper combination and relative angle of intake ports could ensure much better characteristics of mixture formation in-cylinder at a certain extent.
2. An intake swirl coinciding with the cylinder axis can be achieved by changing intake ports design. Transient CFD simulation shows that, with this intake flow, the mixing of fuel spray and air can be improved, which is not possible to be evaluated in steady state simulation and testing. The CFD simulation results are validated by flow measurement in steady state.
3. Intake flow has a significant impact on engine performance and emission. Besides steady state evaluation, it is necessary to consider the transient intake flow and its effect on mixture formation and combustion with the development of intake ports.
4. With the help of CFD tool, it is possible to consider the in-cylinder transient flow process during intake ports development and the steady flow evaluation on steady flow bench.

References

1. Li Y, Li L, Xu J, Gong X, Liu S, Xu S (2000) Effects of combination and orientation of intake ports on swirl motion in four-valve DI diesel engines. SAE 2000-01-1823
2. G. E. Thien AVL (1964) description of measuring methods for the investigation of stationary flow properties at valve ports and statements for the evaluation of the measurement results

The Development of FAW New 3L Diesel Engine

Fanchen Meng, Jun Li, Fujian Hou, Wenlie Pi and Qun Chen

Abstract For the next generation of FAW light duty truck, a LD diesel engine with high power, low emission and low fuel consumption is required. A new 3 L diesel engine has been developed to fulfill the requirement, which can meet EURO4 and EURO5 emission standards. Special attention has been paid to the reduction of the operating and service costs. Following FAW's 'W' engine product development process, engine performance, structure and subsystem design were optimized and confirmed by CAD and CAE during concept phase, premium design and detail design phase. DOE method was applied to optimize combustion system and engine performance. The reliability and durability were validated according to FAW diesel engine mechanical development processes. By using cooled EGR, DOC and POC technologies, EURO4 emission standards has been achieved with a maximum rated power of 125 kW, maximum torque of 380 Nm and satisfactory fuel consumption. The engine features Double Over Head Camshaft (DOHC) driven by rear chain. Balance shaft module benefits NVH performance. Double layer cylinder head water jacket and M-type block water jacket structure enables this engine to endure a peak fire pressure more than 18 Mpa, which has been validated by reliability and durability test. It is unique in the market for FAW 3 L LD diesel engine to have a configuration with air compressor. By thorough development, FAW 3 L LD diesel engine realized the power and torque targets with low emission, satisfactory fuel consumption, high reliability and durability, and excellent NVH performance. The engine is expected to be the main power for light duty commercial vehicles for FAW.

F2012-A06-045

F. Meng (✉) · J. Li · F. Hou · W. Pi · Q. Chen
CHINA FAW Co. Ltd., R&D CENTER, No.1063, Chuangye Street, Changchun, China
e-mail: mengfanchen@rdc.faw.com.cn

SAE-China and FISITA (eds.), *Proceedings of the FISITA 2012 World Automotive Congress*, Lecture Notes in Electrical Engineering 190,
DOI: 10.1007/978-3-642-33750-5_23, © Springer-Verlag Berlin Heidelberg 2013

Keywords LD diesel engine · Emission · Fuel consumption · Structure · NVH

1 Summary

For the next generation of FAW light duty truck, a LD diesel engine with high power, low emission and low fuel consumption is required. A new 3 L diesel engine has been developed to fulfill the requirement, which can meet EURO4 and EURO5 emission standards. Special attention has been paid to the reduction of the operating and service costs.

2 Conceptual Design

By using cooled EGR, DOC and POC technologies, EURO4 emission standards has been achieved with a maximum rated power of 125 kW, peak torque of 380 Nm and satisfactory fuel consumption. The engine features double overhead camshaft (DOHC) driven by rear chain. Balance shaft module benefits NVH performance. Double layer cylinder head water jacket and M-type block water jacket structure enables this engine to withstand a peak fire pressure more than 18 Mpa, which has been validated by reliability and durability test. The common-rail system delivers an injection pressure of 1,800 bar. It is unique in the market for FAW 3 L LD diesel engine to have a configuration with air compressor. Engine data is listed in Table 1.

3 Structure Design

3.1 Layout

The 3L engine features many advanced structural designs (Fig. 1), i.e. 4-valve, DOHC, cylinder head with integral double layers water jacket, cracked connecting rod, rear timing drives etc., the cylinder block is still made of cast iron.

The engine accessories are driven by two belt drives that are driven off the outer hub of crankshaft torsional vibration damper. The first is used to drive the cooling fan, alternator, water pump, and steering pump, and has six ribs. The second e is used to drive the air conditioning compressor only and has four ribs.

Oil filter and oil cooler combined module is built in cylinder block. Oil and water is directly connected to cylinder block oil gallery and water jacket respectively.

Table 1 Technical data

		Parameter
Cylinder arrangement/number	[–]	I4
Number of valves	[–]	4
Displacement	[cm3]	2,999
Bore	[mm]	95.4
Stroke	[mm]	104.9
Stroke-to-bore ratio	[–]	1.1
Compression ratio	[–]	17:1
Power at engine speed	[kW]	125
	[rpm]	3200–3600
Torque at engine speed	[Nm]	380
	[rpm]	1400–2800
Emission standard	[–]	Euro4

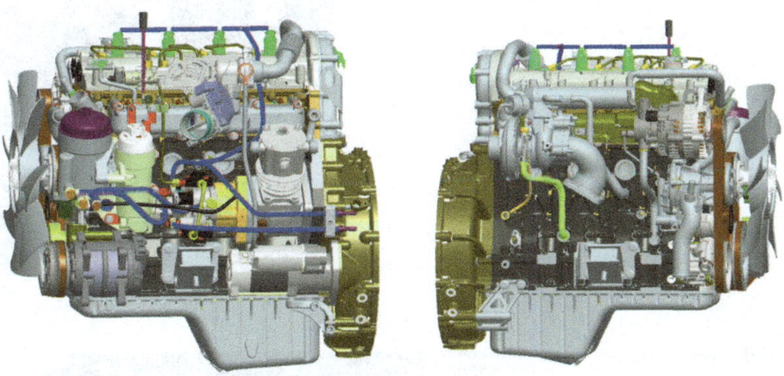

Fig. 1 FAW 3L engine

The timing drive is mounted inside the rear part of the cylinder block and cylinder head. It's beneficial to timing drive system reliability and engine noise and vibration.

One of the key requirement specifications for the FAW LD vehicle engine equipped with air compressor installation makes it different from its competitors.

3.2 Cylinder Block

The cylinder block is sand-cast in grey iron and does not utilize liners (Fig. 2). It appears to be an evolution of a previous engine design. A bedplate is used, instead of main bearing caps, to increase structure stiffness and optimize the vibration level.

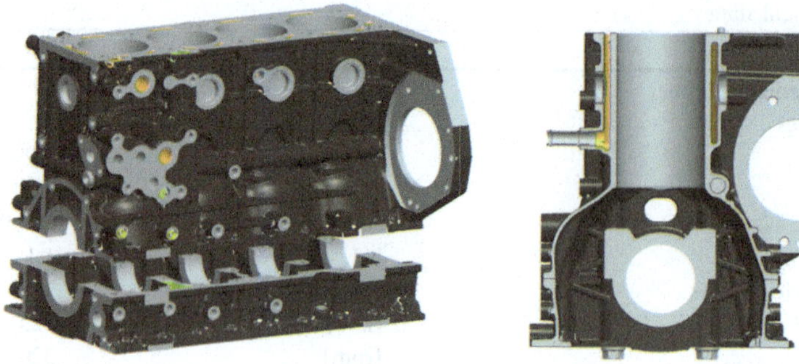

Fig. 2 Cylinder block

Fig. 3 “M” water jacket

Fig. 4 High cycle fatigue analysis of cylinder block

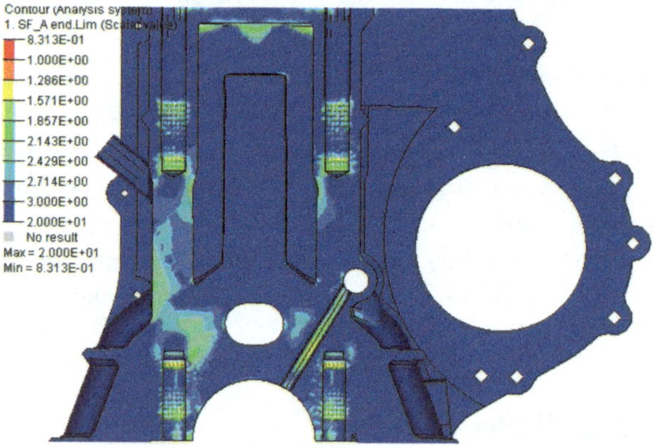

Special “M” water jacket design (Fig. 3) optimizes water flow around liners and also reduces cylinder bolt force effect to liner distortion. The selection of main structural design and main parameters of cylinder block is made based on large amount of database analysis and structural FEM analysis (Fig. 4).

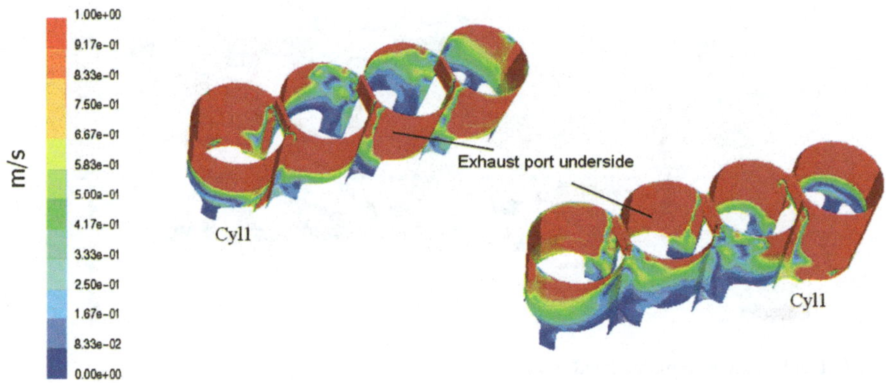

Fig. 5 Analysis of cooling water flow velocity distribution of cylinder liner water jacket

Fig. 6 Profile of cylinder head

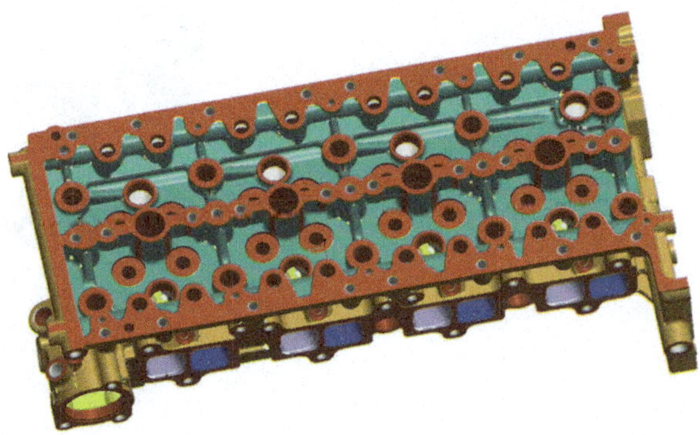

Cylinder block water jacket design fully considers cooling of the first piston ring at BDC. By CFD analysis, cooling cycle inside cylinder block has been examined and optimized (Fig. 5).

3.3 Cylinder Head

3L engine cylinder head design features a unitary structure (Fig. 6). Cylinder head water jacket is designed to a double layer structure, which divides cooling water inside cylinder head into double layer control (Fig. 7). The area in vicinity of valve bridge which demands rigorously on cooling is located in the lower cooling jacket, in which water flows similarly "W", to ensure high velocity of flow at such critical part. Whereas water flows longitudinally in the upper water jacket which ensures smooth water return inside cylinder head, to achieve the optimal water flow effect of the whole cylinder head. Meanwhile, the double layer water jacket helps to improve structural strength and stiffness of the whole cylinder head.

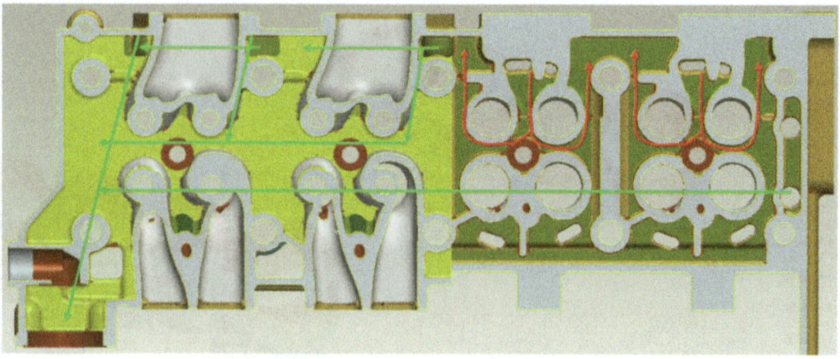

Fig. 7 Double-layer cylinder head water jacket

Fig. 8 Crank train

With 4-valve structural design, pollutant emission can be reduced effectively. One spiral port and one tangent port are adopted in the air intake design, these two ports combined with a swirl chamfer to obtain the required flow rate coefficient and swirl ratio, thus satisfying engine performance requirement.

3.4 Crank Train

The cracked connecting rod design improves locating accuracy of connecting rod, ensuring cylindricity of crank end during mounting of connecting rod, meanwhile eliminating some key procedures, i.e. locating and fine machining of joining surface of connecting rod, thus reducing cost.

Eight balance weights of forged steel crankshaft are adopted (Fig. 8). The dynamics analysis ensures sufficient strength safety factor. In addition, the sufficient stiffness of forged steel crankshaft contributes to good acoustic quality of engine. And the second order inertial forces are completely eliminated through the adapted Lanchester balancer.

Fig. 9 Valve train

The aluminium piston is designed jointly by MAHLE and FAW, it features internal cooling oil duct structure, and the applicable PFP (peak firing pressure) is up to 180 bar. The composite insert of piston ring land improves wear resistance of piston ring groove. Coated graphite technique is applied to piston skirt, which greatly improves engine running-in performance in initial use phase of engine.

3.5 Valve Train

DOHC structure is adopted in engine valve train, and combined with 4-vavle, roller rocker arm follower, hydraulic clearance adjust structure,the optimal design could be achieved in stiffness, reliability and friction performance of the valve train. Adopting rear-end timing gear and chain drive train structure, with the gear drive air compressor and fuel injection pump (Fig. 9).

TYCON program is used in the whole valve train design validation and optimization for engineering calculation and analysis.

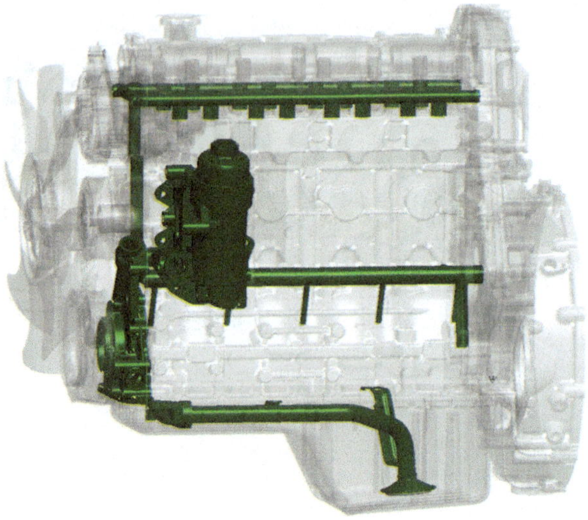

Fig. 10 Lubricating System

3.6 Lubricating System

Rotor oil pump is driven by crankshaft front end. Oil enters directly to the oil cooler and filter module through oil gallery on cylinder block and then enters into main gallery (Fig. 10).

Analysis with Flow master program is performed to check and optimize the oil distribution of the lubricating system and performance of various valves (Fig. 11).

3.7 Fuel System

To meet requirements of China IV and V emission regulations, 3L engine is equipped with Denso third generation electronically-controlled common-rail fuel injection system, with fuel injection pressure at pump end exceeds 1,800 bar. The electronically-controlled fuel injection pump is mounted with flange; high pressure rail is directly mounted on the cylinder head cover. Overall configuration is compact, proper and has high match adaptability (Fig. 12).

3.8 Crankcase Ventilation

An active oil and gas separation structure is applied in 3L engine crankcase ventilation system. The centrifugal impeller is driven by timing gear train. Analysis with CFD program is performed to optimize the oil and gas separation efficiency (Fig. 13).

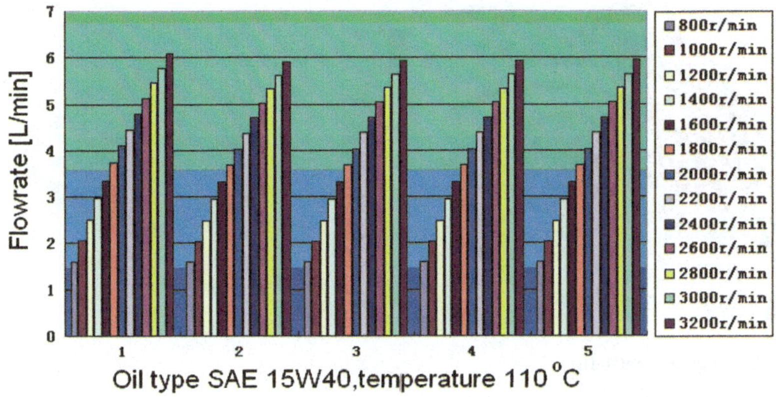

Fig. 11 Calculation results of oil flow rate of the main bearing

Fig. 12 Fuel system

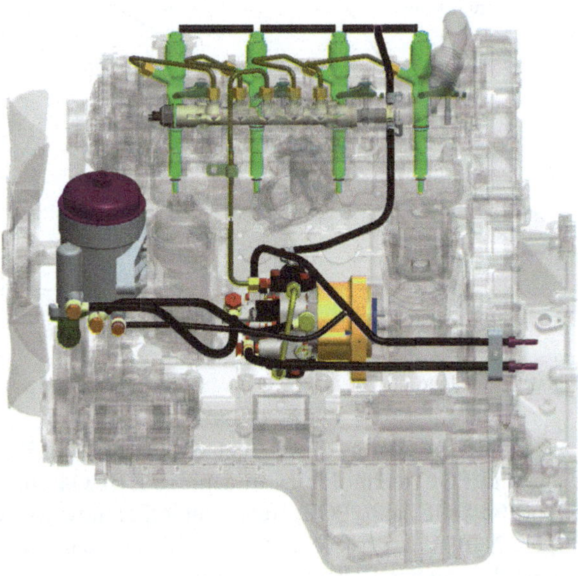

4 Power, Torque and Emissions

The targets of product development have been fully achieved. Refer to Fig. 14 for development results for power and torque. The 3L engine delivers a nominal output of 125 kW at 3,200 rpm. The maximum torque of 380 Nm is available between 1,400 and 2,800 rpm. The full load curve shows the improvements over the previous model. Particular importance was achieving a robust torque characteristic. Maximum torque is available across a broad rpm range, which guarantees an agile response.

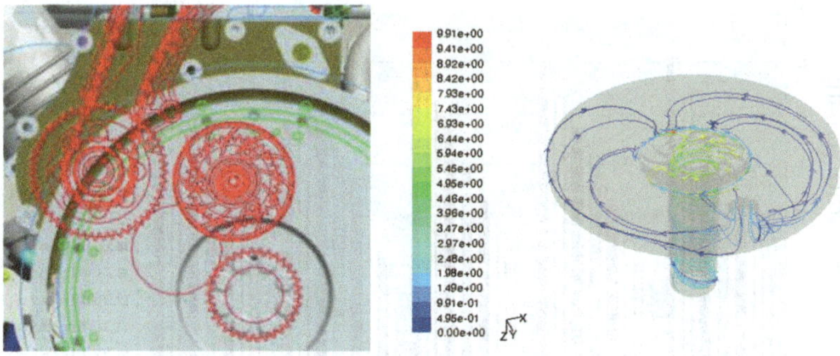

Fig. 13 Crankcase ventilation

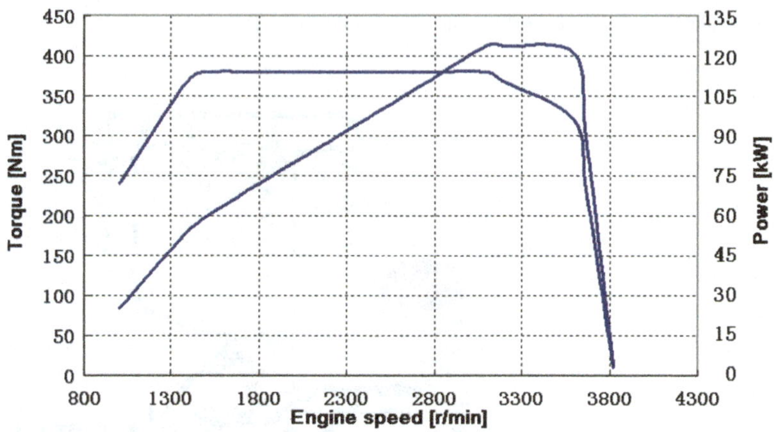

Fig. 14 Full load curve

Considering China diesel fuel quality status, 3L engine adopts cooled EGR, DOC and POC technologies to achieve EURO4 emission (Fig. 15). According to 50–60 % DOC and POC treatment efficiency to define engine inner emission target: NOx < 3.5 g/kWh, PM < 0.036 g/kWh. The engine test emission result (Table 2) satisfies EURO4 emission.

5 Noise, Vibration

The 3L engine acoustic optimizations were carried out by means of simulation and measurement on engine dynamometers. A main objective was to optimize the vibration level at the component bearings of the powertrain. In accordance with the engine's comfort requirements, the 3L engine was equipped with Lanchester shafts. To optimize the level in the all-frequency range, cylinder block ladder

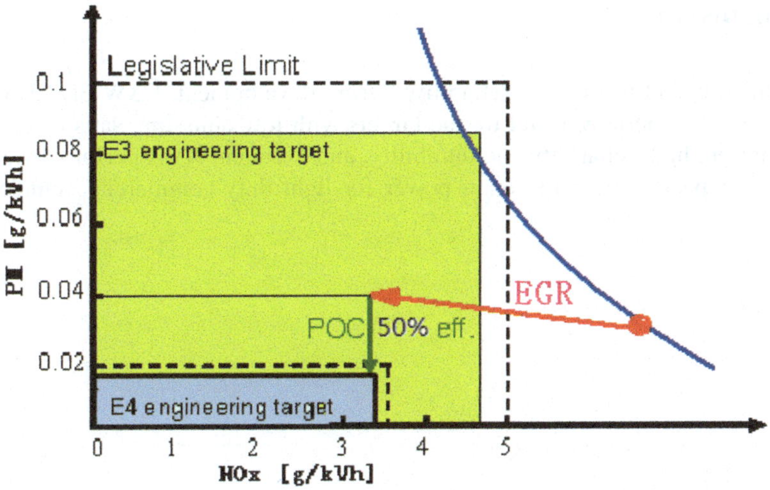

Fig. 15 Euro4 emission route

Table 2 Engine emission result

Emission	Unit	Value
NOX	g/kw.h	3.31
CO	g/kw.h	0.86
HC	g/kw.h	0.14
SOOT	g/kw.h	0.0076
BSFC	g/kw.h	237.4

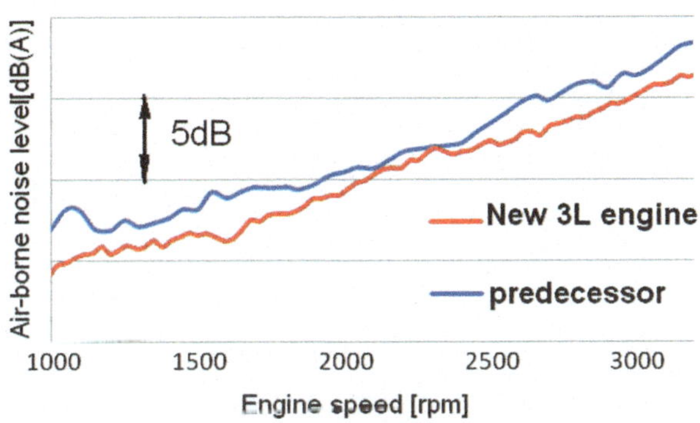

Fig. 16 Air-borne noise compared with predecessor engine

frame, hydraulic clearance adjust rocker arm and rear-end timing system are applied in the engine.

The noise level of the new 3L diesel engine is again significantly improved over the predecessor in the total level (Fig. 16). In terms of the air-borne noise level, the new 3L engine is within the very favorable range of NVH.

6 Conclusion

Through sufficient test on the reliability during development, FAW 3L LD diesel engine realized the power and torque targets with low emission, satisfactory fuel consumption, high reliability and durability, and excellent NVH performance. The engine is expected to be the main power for light duty commercial vehicles for FAW.

Transient Behavior Study of HD Diesel Engine and the Effects of Turbochargers

Yanbin Shi, Guangyong Zheng, Haie Chen and Lei Wang

Abstract The steady performance of diesel engine is always focused on during the development. But to some extent, the trade-off between steady and transient behavior has to be considered. In order to evaluate the transient performance during engine development phase, both simulation and dyno testing methods are needed, especially for the effect of turbocharger. Focusing on the transient behavior CA6DL2-35 turbocharged HD diesel engine, the authors developed the transient simulation method based on traditional thermodynamic software, in which the effects of turbocharger, the strategy of fuel injection and inertia of engine were included. Dyno measurements of torque increase at constant speed and speed increase at WOT were compared and commented. Simulation was validated by dyno testing. The transient performance of turbocharged engines depends significantly on the response of intake path. The transient simulation and testing was used to support the selection of turbocharger. In this chapter, different turbochargers were investigated by simulation and testing for transient performance. Comparison showed that the simulation of transient is able to predict the behavior of turbocharged diesel engine and can be used to support performance development.

Keywords Diesel engine · Turbocharger · Transient behavior · Simulation · GT-Power

---

F2012-A06-047

Y. Shi (✉) · G. Zheng · H. Chen · L. Wang
China FAW Co., Ltd. R&D Center, Changchun, China
e-mail: shiyanbin@rdc.faw.com.cn

SAE-China and FISITA (eds.), *Proceedings of the FISITA 2012 World Automotive Congress*, Lecture Notes in Electrical Engineering 190,
DOI: 10.1007/978-3-642-33750-5_24,

1 Introduction

The economic performance of the commercial vehicles is one of the key factors to determine their market competitiveness, and the engine performance is the most important for the vehicle economic performance. Up to now lots of researches have proved that good engine static performance can not reflect good vehicle fuel consumption. Because most vehicle condition is transient, good engine transient performance is required to meet marketing and customer requirement. At the transient operating conditions, the response of the fuel injection is much faster than air intake, especially for the turbocharged engine. Therefore, reasonably matching the turbocharger and enhancing engine air intake response is the key factor to improve the transient performance of the engine. As acceleration the engine combustion is not complete and low efficiency since lack of air resulting in the deterioration in vehicle power and economic performance. For the trucks operated in medium and low speed the effective method to decrease the vehicle fuel consumption is the improvement of the engine's low speed response and effectively shortening of the vehicle acceleration time. By the simulation and experimental test the turbocharger performance is evaluated and the experimental results prove the feasibility of the method of simulation analysis to improve the transient performance of the engine.

2 Setting Up the Engine Model and Calibration

2.1 Main Features of the Test Engine

The main parameters of the test engine are listed in Table 1. The engine is a 6 cylinders, in line, 8.6 L, turbocharged, heavy duty diesel engine. The turbocharger is waste gate with a maximum boost pressure of about 1.7 bar. Experimental investigations were carried out under steady state operating conditions. The selected modes are full loads from the engine speed 800 to 2100 rpm. For each mode the running parameters are recorded including cylinder pressure.

2.2 Setting up and Calibration of the Engine Model

Computational simulation is based on GT-Power and a one dimensional fluid-dynamic code developed by Gamma Technologies for engine performance prediction. The base model establishment requires accurate experimental data of the engine and detailed the intake and exhaust system geometry, which have to be carefully analyzed to properly set heat transfer and friction losses coefficients, combustion heat release profiles and engine friction estimation. And then a

Table 1 Main design features of the test engine

Type	Diesel\6 cylinders in line\4 stroke
Displacement(L)	8.6
Compression ratio(−)	17.2
Maximum torque(N·m)	1500(1200–1600 r/min)
Maximum power(kW)	258(2100 r/min)
Fuel injection system	Common rail direct injection
Valves	4 valves/cylinder

detailed validation process is required to assess the engine model accuracy and reliability [1]. The simulation results have therefore to be compared with the experimental results over the whole engine operating range including air intake, brake power, and cylinder pressure history. A comparison between experimental and simulated values of A/F and brake mean effective pressures and BSFC and boost pressure is shown in Figs. 1, 2, 3 and 4. The error of simulated and measured values can be observed and lower than 2 % on the whole speed range. The simulation also provides good prediction for both the maximum pressure value and the crank angle position. As an example, the results obtained at A100 are shown in Fig. 5. Almost same traces between measured and calculated values were observed over the whole speed range.

3 Setting Up the Simulation Model and Analysis

In the transient process, the response of the fuel injection is much faster than the air intake, especially for the turbocharged engine. Therefore, reasonably matching the turbocharger and enhancing engine air intake response is the key factor to improve the transient performance of the engine. At the early acceleration stage, due to the lagged response of the turbocharger, the building up of the boost pressure is far slower than the increased amount of fuel injection resulting in air-fuel ratio will drop suddenly. Therefore, the engine combustion is not complete and low efficiency since lack of air resulting in the deterioration in vehicle power and economic performance.

The engine simulation model established should have the ability to reflect this air-lag effect, so it is required that all components related to the gas flow use the physical or semi-physical model. The combustion process should be able to reflect the combustion deterioration caused by less air. The rotating parts have to take into account the moment of inertia and reflect the influence to acceleration.

Based on the static calibrated model, A map is established including combustion model, exhaust back pressure, friction, intercooler parameters and turbo correction coefficient, which is changed with speed and load and has close relationship with the engine transient performance. The model is coupled into Simulink and upgraded to the engine transient simulation platform (Fig. 6).

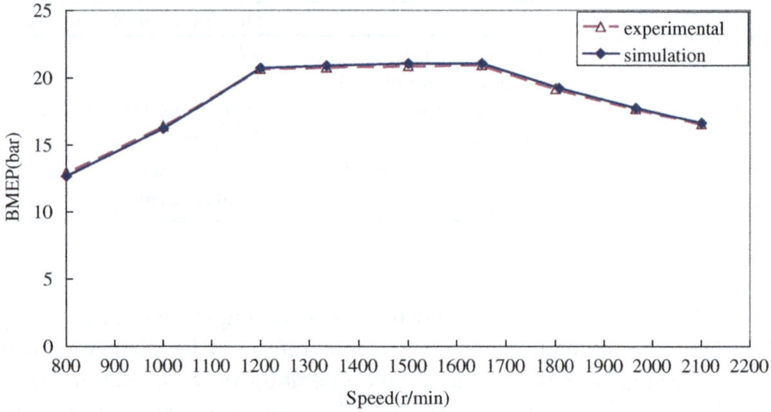

Fig. 1 Comparison between experimental and simulation results for BMEP at full load

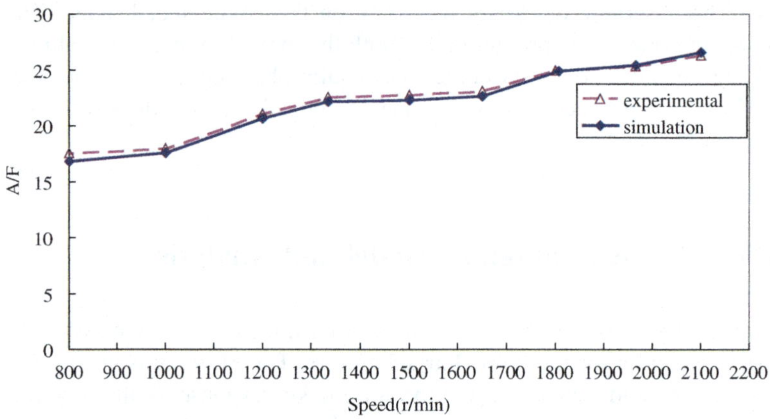

Fig. 2 Comparison between experimental and simulation results for A/F at full load

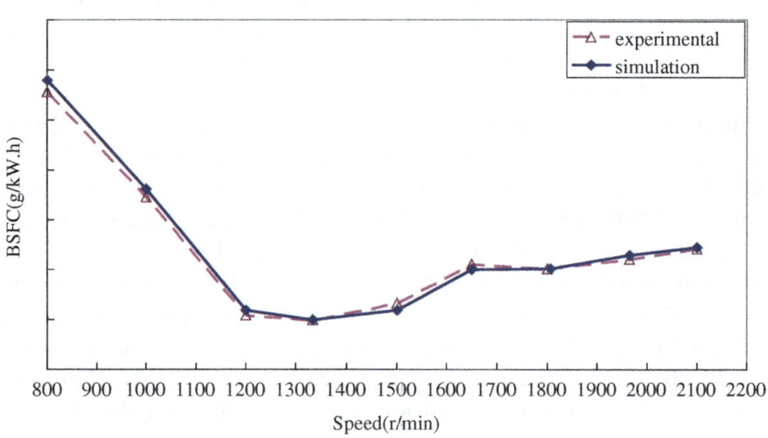

Fig. 3 Comparison between experimental and simulation results for BSFC at full load

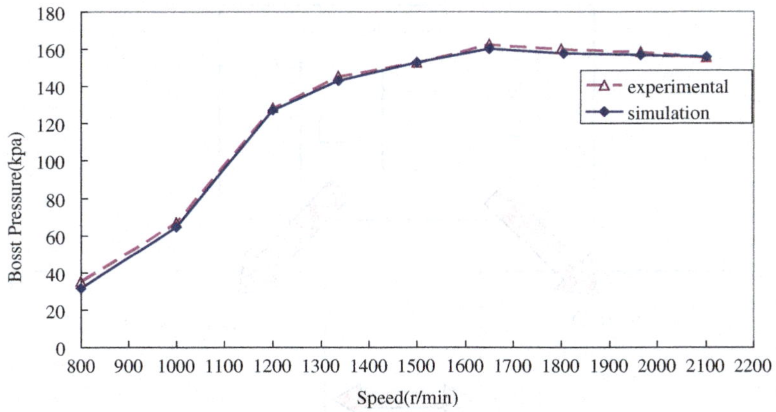

Fig. 4 Comparison between experimental and simulation results for boost pressure at full load

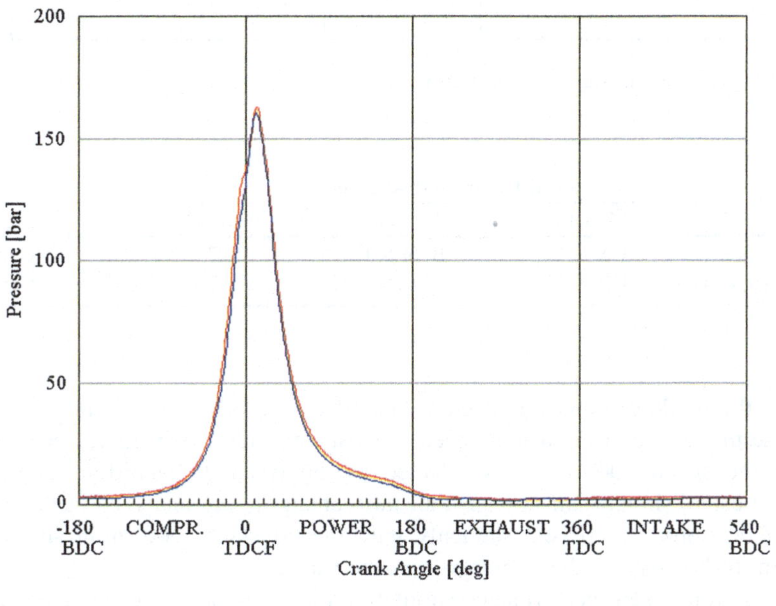

Fig. 5 Comparison between experimental and simulation results for in-cylinder pressure at full load, A100

3.1 Comparison of the Engine Transient Process

Corresponding to the vehicle transient process, the engine condition can be roughly divided into the two types of constant speed increased torque process and constant torque increased speed. With regard to the vehicle acceleration, the engine operation is close to the constant speed increased torque process [2]. In this

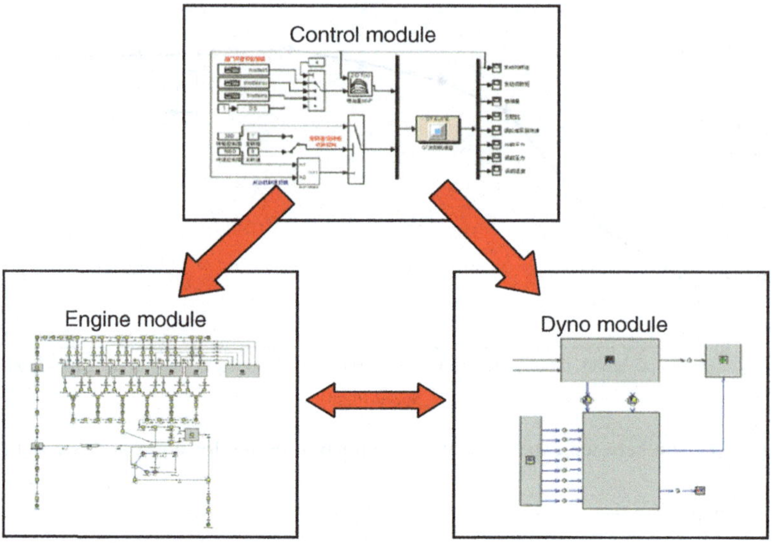

Fig. 6 Engine transient simulation platform

Table 2 Main design features of the test turbocharger

	Model name	Turbine	Compressor	Supplier
Turbocharger A	HX40 W	B76L84FH18	8597-B85J71GL	Cummins
Turbocharger B	HX40 W	A76L84FH14	8597-B85J71GL	Cummins

chapter the mode of constant speed increased torque is simulated and evaluated. Because the acceleration at low speed is most important for vehicle acceleration ability the speed 1000 r/min is selected as the research condition. The boost pressure establishment and the air–fuel ratio changing are analyzed at the condition of the torque increased suddenly with the constant 1000 r/min speed. For different turbochargers the compressor is the same.

Based on transient performance simulation platform the accelerating process is optimized. The parameters of the turbocharger is shown in Table 2, the turbine flow capacity comparison is shown in Fig. 7. The flow capacity of different turbochargers at the same speed is different up to 22 %. The application of small turbine can improve the engine transient performance since better utilization of the exhaust gas energy at the low speed and smaller moment of inertia [3].

In the engine transient simulation process, the fuel injected into the cylinder increased from 1 mg per cycle to 137 mg per cycle within 1 s keeping the constant speed of 1000 r/min. The history of air intake and torque is shown in Figs. 8 and 9.

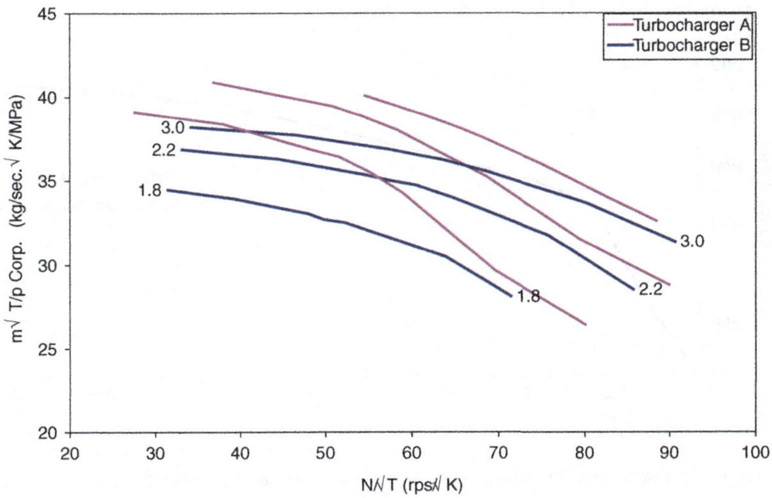

Fig. 7 Turbine flow capacity comparison

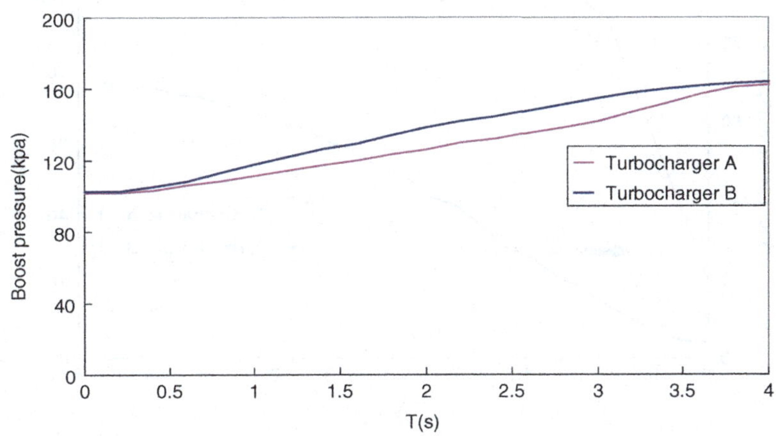

Fig. 8 Boost pressure response at 1000 r/min—Engine transient simulation results

From the calculation results, case B is much faster in building up the boost pressure resulting in faster air response and shorter fuel injection lag. The engine air–fuel ratio can be recovered to 15 quickly and it is helpful to shorten abnormal combustion duration caused by lack of air. The time reaching 1000 Nm is less than 1 s compared with case B and the engine torque response is improved by 24.5 %.

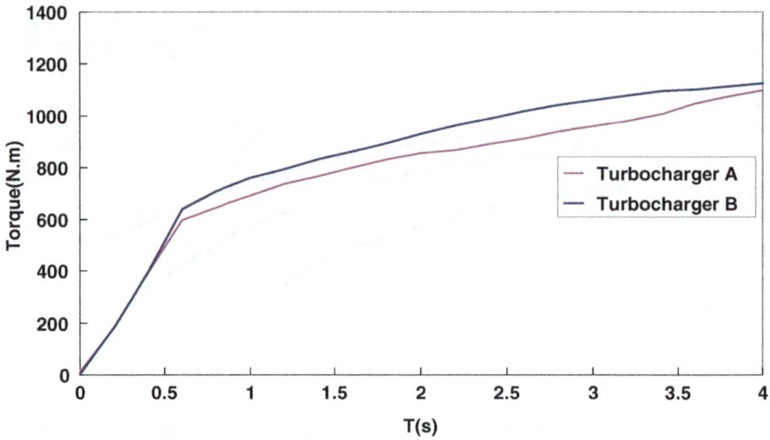

Fig. 9 Torque response at 1000 r/min—engine transient simulation results

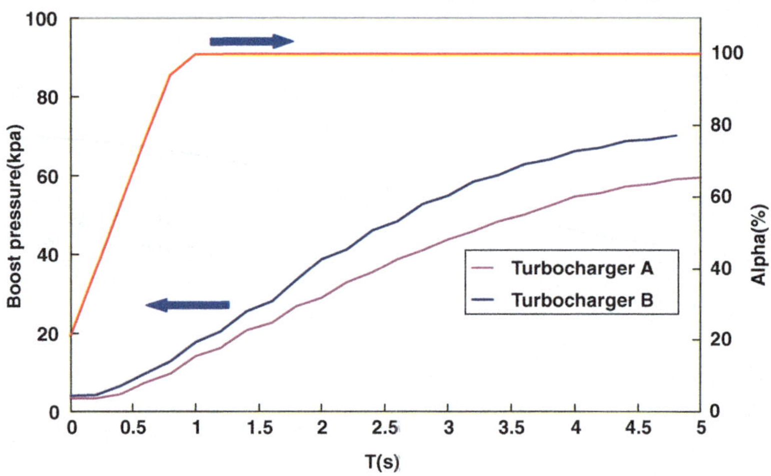

Fig. 10 Boost pressure response at 1000 r/min—engine transient experimental results

4 Engine Test Validation

The experimental test is conducted at test bench to verify the accuracy of the simulation and evaluate the effect of different turbocharger cases on the real engine performance.

The compared tests are conducted for the engine transient performance. The engine is firstly motored at 1000 rpm and increased fuel injection to full load within 1 s. Response results can be seen from Fig. 10, the boost pressure start to increase from 0.2 s and be stable from 4.5 s after the throttle response. Case B has

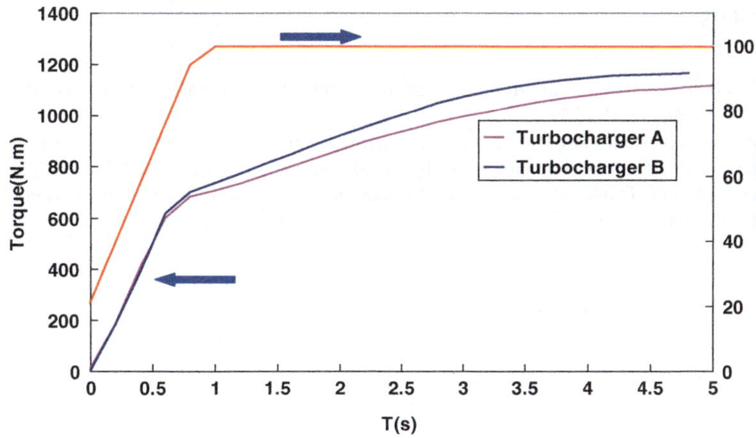

Fig. 11 Torque response at 1000 r/min—engine transient experimental results

faster response compared with case A and the time of reaching 60 kPa is decreased about 1.5 s regarding to the original turbocharger.

From Fig. 11, the torque response is faster than the boost pressure, but since air intake lag serious and combustion deterioration the engine torque increases slowly. Case B has the shorter time of 2.6 s to build up torque of 1000 Nm and improvement of 25 % than Case A. For both pressure and torque response, simulation and experimental results in the trends are in good agreement indicating the accuracy of the simulation method to predict the transient response of the engine.

5 Conclusion

The GT-Power and Simulink co-simulation platform is established and engine transient simulation is carried out. The transient features of different turbochargers are analyzed. The validation is conducted at the engine test bench. Case B has the shorter time of 2.6 s to build up torque of 1000 Nm and improvement of 25 % than Case A. Test result showed great improvement in engine low speed performance and engine transient response The application of simulation methods can analyze and evaluate the transient performance of the engine and vehicle at the early stage of design so as to avoid vehicle acceleration deterioration only based on the static development. The application of simulation methods for engine transient performance evaluation can shorten the development time, minimize test matrix and save cost.

References

1. Millo F (2005) The potential of dual stage turbocharging and miller cycle for HD diesel engines. SAE Paper, 01-0221
2. Ling-Ge S (2010) Realization of the simulation platform of turbocharged diesel engine with EGR loop under transient conditions. In: ICCASM 2010 Shanxi, Taiyuan, China. V8274–279
3. Watson N (1981) Transient performance simulation and analysis of turbocharged diesel engines. SAE paper, 810338

Combustion System Development of Direct-Injection Diesel Engine Based on Spatial and Temporal Distribution of Mixture and Temperature

Jun Li, Kang Li, Haie Chen, Huayu Jin and Fang Hu

Abstract This study developed a Distributed Combustion Design (DCD) method, based on spatial and temporal distribution of mixture and temperature in cylinder, to optimize the combustion of high-pressure common-rail direct-injection diesel engine. Due to the significant improvements of fuel atomization and evaporation by high-pressure common-rail system, mixing control of fuel and air becomes a critical factor to improve efficiency and emissions. Bowl shape, spray angle, multiple-injection and swirl ratio are effective approaches to control the fuel distribution and movement in cylinder both spatially and temporally, therefore can be used to improve the mixing. The whole space in cylinder was divided into three zones, Chamber, UPcham and Outerclearance. And the start of combustion, the timing of injection impacting on bowl lip, the end of injection and the end of combustion were selected as critical timings for temporal zoning controls. Studies proved that the history of fuel quantity distributed to Chamber was an important parameter affecting mixture and temperature distribution, then thermal efficiency and emissions. Combustion and emissions histories and their relationship with mixture distribution were studied under different approaches and schemes and were validated by single cylinder optical engine tests. Furthermore, the DCD method was put forward. As a result of its application, 2 % reduction of fuel consumption was achieved on a 9 L diesel engine.

Keywords Distributed combustion design · Spatial and temporal zoning control · Fuel distribution · Air–fuel mixing · CFD

F2012-A06-048

J. Li (✉) · K. Li (✉) · H. Chen (✉) · H. Jin (✉) · F. Hu
FAW Co., Ltd. R&D Center, Changchun, China
e-mail: lj_qy@faw.com.cn

SAE-China and FISITA (eds.), *Proceedings of the FISITA 2012 World Automotive Congress*, Lecture Notes in Electrical Engineering 190,
DOI: 10.1007/978-3-642-33750-5_25,

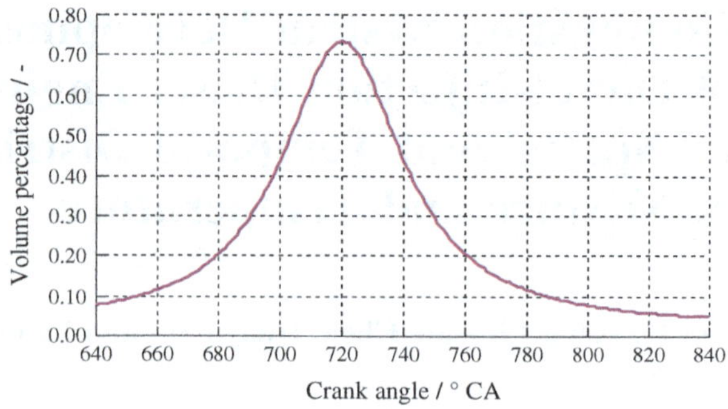

Fig. 1 Volume percentage of bowl to cylinder

1 Introduction

The main goal of engine combustion design is high efficiency and low emissions. Conventional diesel engine combustion is mainly the diffusive combustion. If the mixing is not sufficient, some local mixture will be over-rich and the other over-lean, which will not only affect combustion efficiency, but also produce a large amount of NOx and soot emissions. Studies have shown that the main harmful emissions from engine combustion are generated in the specific mixture concentration and the combustion temperature [1, 2]. Therefore, for high-pressure common-rail direct-injection diesel engine, a reasonable mixture distribution, which can be achieved by optimizing air flow and fuel distribution, is the critical factor to improve combustion and emissions. The DCD concept is using spatial and temporal zoning controls of fuel distribution in cylinder to control the history of mixture concentration distribution so as to achieve efficient combustion and low emissions.

2 Concept of Spatial and Temporal Zoning Controls

2.1 Characteristics of Different Regions in Cylinder

A lot of fresh air flows into cylinder during intake stroke and its pressure and temperature increase during compression process. Then near TDC, fuel is injected and combustion happens. The cylinder volume increases and the ratio of bowl to cylinder volume decreases gradually with the combustion process (Fig. 1). Velocity and turbulence distribution in cylinder change all the time and are very uneven. The turbulent kinetic energy distribution in cylinder induced by flow only (no injection) is shown in Fig. 2. The strongest turbulent kinetic energy region is in bowl at compression TDC, and then it moves outside.

In order to achieve higher thermal efficiency, combustion should close to TDC. According to the change of volume and turbulent kinetic energy distribution, fuel

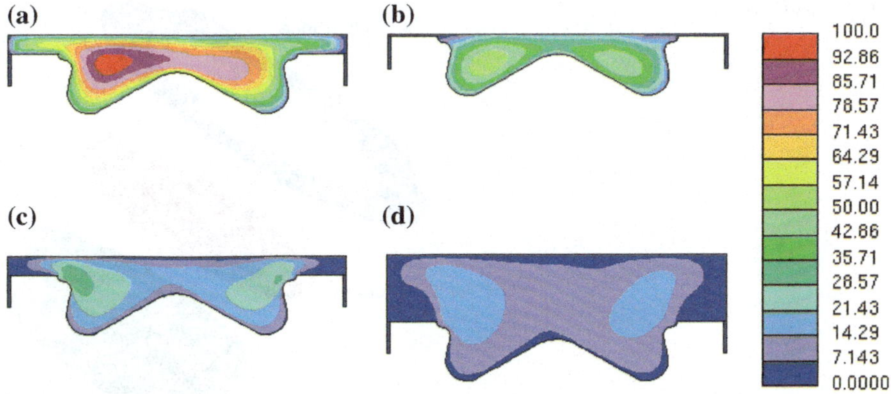

Fig. 2 Turbulence kinetic energy distribution (m^2/s^2). **a** 20 °CA BTDC, **b** TDC, **c** 20 °CA ATDC, **d** 40 °CA ATDC

should be mainly distributed to the bowl where there is more oxygen and higher turbulent kinetic energy at early period, so the mixing and combustion efficiency are better. However, due to the space of bowl, the fuel distributed to it has to be limited. Otherwise, it will lead to combustion deterioration. When the piston goes down, more and more oxygen moves to the region above the piston, so does the strongest turbulent kinetic energy region. Combustion will be more efficient there. Also because temperature is not so high there, NOx emissions decrease. So at that time some fuel should be distributed to this region. But the ratio of this part of fuel has to be controlled due to its late combustion phase. Turbulence kinetic energy and temperature is lower near cylinder wall, therefore, the fuel should not be distributed to this region. So, the proportion of fuel distributed to different regions at different time should be optimized by all kinds of control approaches. Reasonable fuel distribution in whole cylinder is very important to achieve high efficiency and low emissions.

2.2 Spatial Zoning

Due to the different combustion phase and contribution of each region, breakdown of the whole space in cylinder and subdivision of the entire process must be used to control mixture concentration history and to achieve precise control of combustion. The whole space in cylinder is divided into three zones: (1) Chamber: the space of bowl; (2) UPcham: the space just above bowl; (3) Outerclearance: the space of clearance above piston excluding UPcham, as illustrated in Fig. 3.

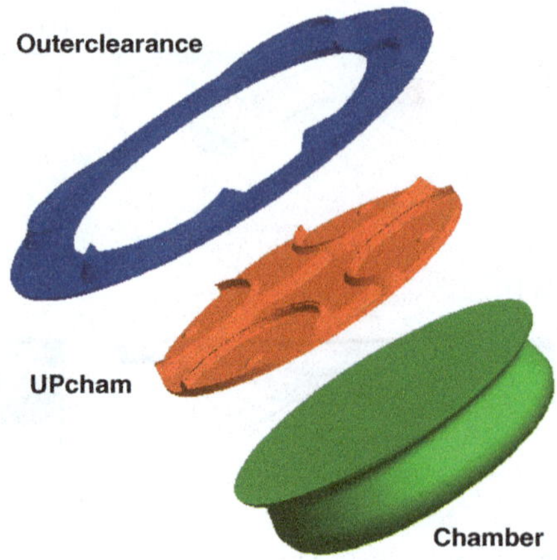

Fig. 3 Spatial zoning

2.3 Temporal Zoning

The fuel injection, atomization, combustion of diesel engine is a transient process. In addition to initial distribution, fuel and air will move from one zone to another during the process, so the proportion will change with time. Chamber accounts for the maximal proportion of the total cylinder volume near TDC, so most of the fuel should be injected to this zone at this time. Volume proportion of Chamber decreases, but those of UPcham and Outerclearance increase after TDC. UPcham accounts for about 40 % of the total volume at 385 °CA, more than that of Chamber, so it is rather reasonable that a considerable part of the fuel originally assigned to Chamber but still unburnt should move to UPcham rapidly.

Therefore, not only the spatial zoning controls of fuel distribution, but also the temporal zoning controls are needed. The time is divided into four stages from the beginning of fuel injection to the end of combustion as shown in Fig. 4: (1) Ignition delay period when fuel is injected into the cylinder but burning has not yet occurred and fuel mainly concentrated in bowl. It primarily affects the initial fuel preparation and the initial combustion heat release. (2) The period of main fuel consumption in bowl. It is from the start of combustion to the time when injection impacts on bowl lip, which is the main time segment to determine the combustion phase. Under the precondition of guarantee that fuel can move outside bowl successfully latter, most of fuel should be distributed to bowl. Soot and NOx generate quickly from this period. Considering the different requirements of EGR and SCR solution to engine-out NOx emissions, the appropriate amount of fuel distributed to bowl is different for the same combustion system configuration. (3) The main fuel distribution period. It is from the time when injection impacts on chamber lip to the end of injection. On the one hand, fuel injected to bowl is enough, on the other hand, the spatial geometric relationship of spray and chamber also decides that a considerable part of fuel will be injected into the space above piston. Most of this part of fuel should be distributed into UPcham, not into Outerclearance, especially not into the region near cylinder wall. (4) Fuel in

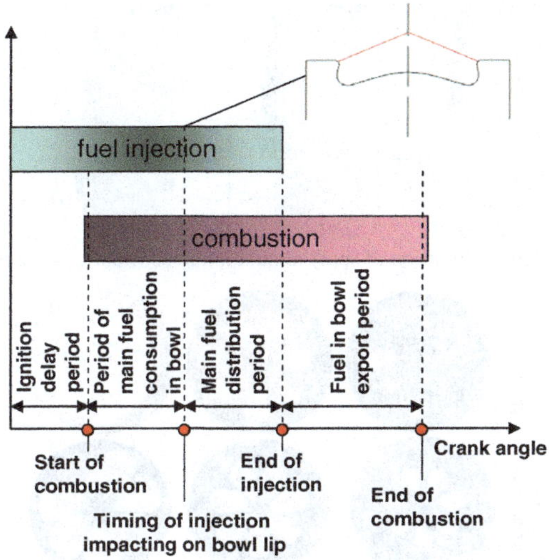

Fig. 4 Temporal zoning

bowl export period. It is from the end of injection to the end of combustion (the time of 95 % of fuel quantity burnt out). This period is very important to soot emissions. A part of fuel should move outside the bowl quickly in order to take full advantage of oxygen in cylinder to achieve more soot oxidation. Of course, a certain quantity of fuel should be kept in bowl to ensure combustion efficiency. So the start of combustion, the timing of injection impacting on bowl lip, the end of injection and the end of combustion are important timings and selected as critical timings for temporal zoning controls. By zoning controls of the fuel distribution at these critical timings, the combustion can be optimized throughout the process.

3 The Principle and Approaches of Spatial and Temporal Zoning Controls

As mentioned above, whether the fuel distribution proportion to each zone is reasonable strongly affects the distribution and combustion of the mixture. So it can be realized by controlling these proportions at critical timings to control the concentration distribution in cylinder, therefore to control the temperature distribution indirectly.

A large number of CFD simulations were done to study zoning controls of fuel distribution. Firstly, spray measurement in constant volume bomb were used to calibrate spray model. And then the optical engine tests were used to calibrate mixing and combustion simulations. The corresponding relationship of mixture concentration distribution in cylinder, thermal efficiency and emissions was concluded based on engine performance tests. The result comparison of CFD, constant volume bomb tests, optical engine tests and engine performance tests is shown in Figs. 5, 6 respectively.

(a) Calibration of spray simulation in constant volume bomb

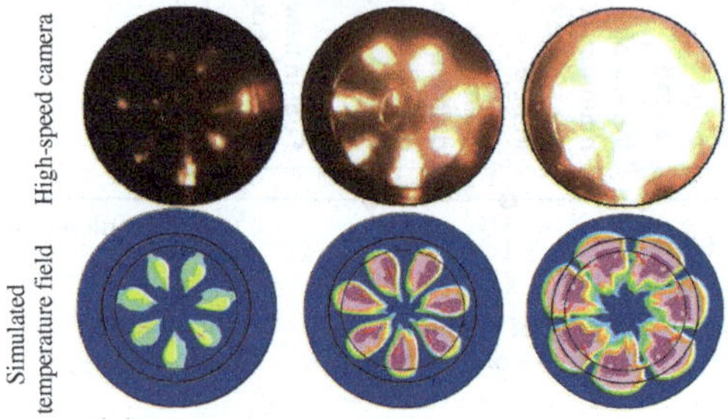

(a) The result comparison of CFD and optical engine tests

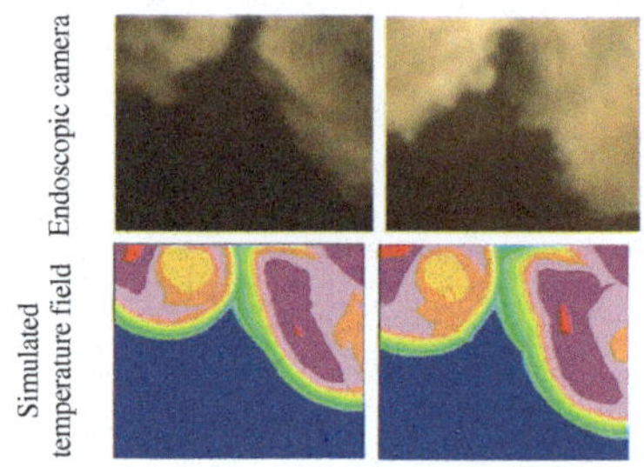

(b) The result comparison of CFD and endoscopic view

Fig. 5 The result comparison of CFD, constant volume bomb tests and optical engine tests

3.1 The Principle of Spatial and Temporal Zoning Controls

The history of fuel distribution proportion to each zone can be obtained by CFD simulation as shown in Fig. 7. MallFuel means the proportion of fuel vapor mass in cylinder to all quantity of fuel injection. MFuelchamber, MFuelUPcham, MFuelouterclearance means the proportion of fuel vapor mass in Chamber, UPcham, Outerclearance to all in cylinder respectively, which represents the distribution of fuel vapor in cylinder.

Engine bench tests and CFD simulations showed that the lip guidance chamber of the 9 L engine could reduce the fuel quantity injected to Chamber at the latter period of injection, which made the fuel distribution more reasonable. The effects of lip guidance chamber were more remarkable when the lip thickness increased at the 11 L diesel engine. Moreover, more fuel could be injected to Chamber having larger diameter and a higher thermal efficiency was achieved. In the optimizations of 6, 9, 11, 13 L high-pressure common-rail direct-inject diesel engines, the fuel quantity distributed to Chamber was the most important and sensitive factor to control the mixture concentration distribution, combustion and emissions. This quantity was different with various combustion systems, for example, the residence time in Chamber of oxygen was longer at low speeds, and therefore more fuel should be distributed to Chamber. Generally, high load operation points at high and low speeds

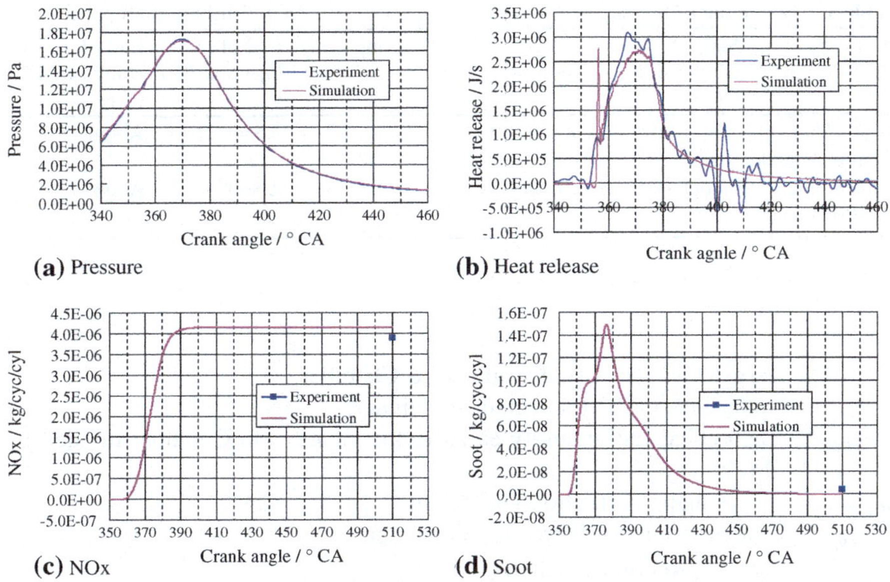

Fig. 6 The result comparison of CFD and engine performance test

Fig. 7 History of zoning controls of fuel distribution

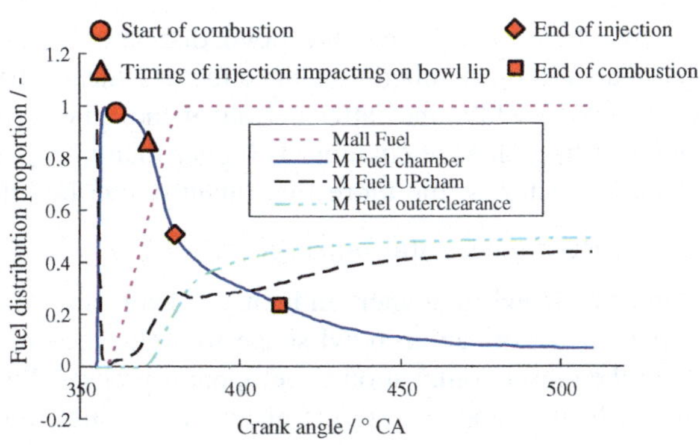

should be taken into account simultaneously to optimize the configuration parameters of combustion system. Based on tests and simulations, the optimal zoning control history of fuel distribution of high-pressure common-rail direct-injection diesel engine was obtained and could be evaluated as following:

$$Q_{Chamber} = \alpha_{Chamber} * \beta_{injection} * \omega_{swirl} * (M_{Chamber} * \theta_{O_2})$$

Usually, $Q_{cham}\,\%$ was also selected as the target control parameter.

$$Q_{Cham}\,\% = Q_{Chamber}/Q$$

Table 1 The sample of the target values of zoning controls of fuel distribution proportion

Operating points	Start of combustion (%)	Timing of injection impacting on bowl lip (%)	End of injection (%)	End of combustion (%)
1,000 r/min, 100 % load	About 100	About 90	About 52	About 18
2,100 r/min, 100 % load	About 100	About 85	About 42	About 12

Where,

$Q_{Chamber}$ target fuel quantity distributed to Chamber;
Q total fuel quantity injected to cylinder;
$M_{Chamber}$ theoretical fuel quantity equal to the air mass in Chamber divided by 14.7;
$\alpha_{Chamber}$ bowl shape correction coefficient;
$\beta_{injection}$ injection energy correction coefficient;
ω_{swirl} swirl correction coefficient;

Bowl shape correction coefficient $\alpha_{chamber} = f(\gamma,\tau,\delta)$, where bowl diameter influence factor $\gamma = f(\frac{D}{B})$, D means the maximum diameter of bowl, B means cylinder bore. The bigger the bowl diameter is, the more fuel quantity can be distributed into Chamber. Guidance factor τ means the capability of guiding fuel to UPcham at middle and latter period of fuel injection, for example, the guidance factor of lip guidance chamber is bigger than no lip guidance one. The bigger the guidance factor is, the more fuel quantity can be injected into Chamber at early period. Reentrance ratio influence factor is $\delta = f(\frac{d}{D})$, where d means reentrance diameter. Bowl diameter influence factor and the guidance factor are two important factors about bowl shape to high-pressure common-rail diesel engine.

In the diesel combustion development process, by controlling Q_{cham} % at these four critical timings, the target history of zoning control of fuel vapor distribution can be achieved. For example, the control targets of two critical operating points of a heavy commercial vehicle diesel engine, 1,000 r/min, 100 % load and the maximum power conditions, at critical timings are shown in Table 1.

3.2 Spatial and Temporal Control Approaches

Several common engineering approaches that can change the fuel distribution are listed as follow:

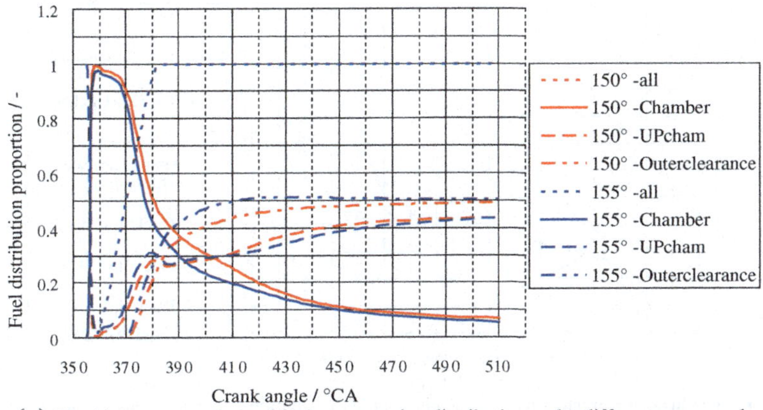

(a) The history comparison of fuel vapor zoning distribution under different spray angles

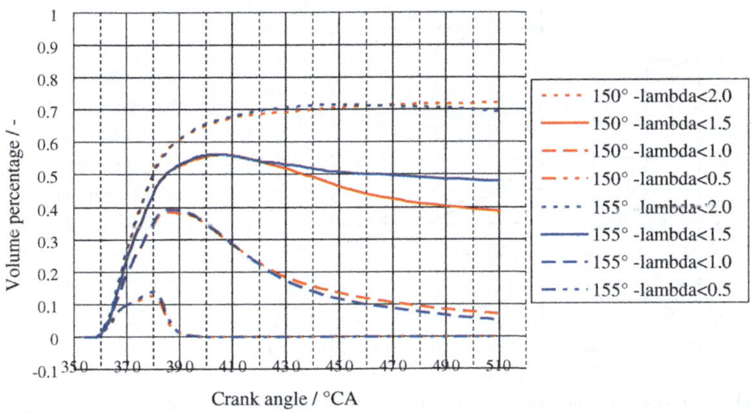

(b) The history comparison of air utilization ratio under different spray angles

Fig. 8 Effects of different spray angles on fuel distribution and mixture concentration

3.2.1 Spray Angle

Spray angle is an effective approach to change fuel distribution. The advantage of this approach is so effective that increasing (or reducing) spray angle can remarkably reduce (or increase) fuel quantity distributed to Chamber. But its disadvantage is that when this quantity increases it increases both at early and latter period. The fuel quantity in Chamber should be more at early period but less at latter period, which is the best status. So in generally speaking, the choice of spray angle can only be a tradeoff.

As shown in Fig. 8, when the spray angle was chosen as 150°, the fuel quantity distributed to Chamber was more than 155°. Heat release was more at early period but combustion was worse at latter period, and the region of lambda <1 was larger and the region of 1<lambda<1.5 was smaller, which is not good for soot oxidation.

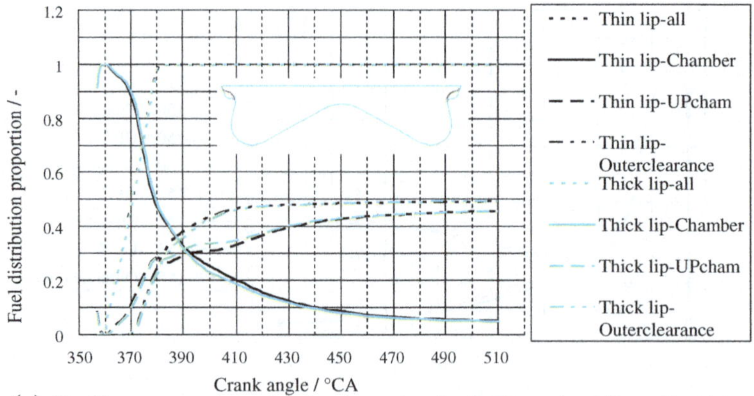

(a) The history comparison of fuel vapor zoning distribution under different lip thickness

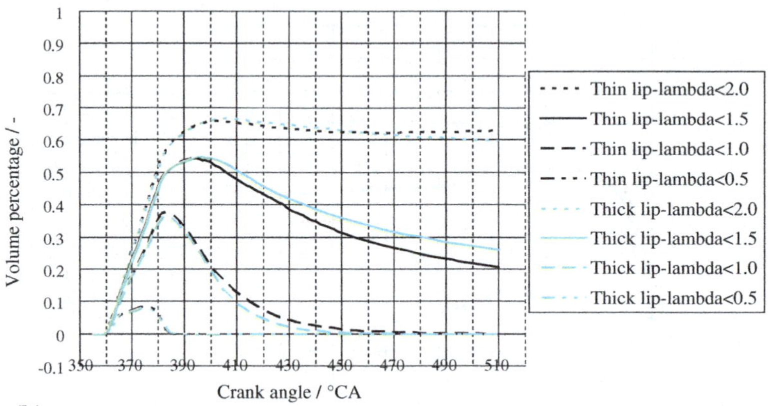

(b) The history comparison of air utilization ratio under different lip thickness

Fig. 9 Effects of lip guidance on fuel distribution and mixture concentration

3.2.2 Bowl Shape

Fuel distribution guidance:

The key strategy of fuel distribution is reducing the fuel quantity distributed to Chamber, increasing that to UPcham and restricting that to Outerclearance at latter period of injection. Therefore the bowl shape can be designed to have this guidance effect [3], for example, optimizing the lip shape. As shown in Fig. 9, keeping the lip slope and increasing the lip thickness could guide the fuel distributed in Chamber to UPcham evidently. The advantage of this approach is that more fuel quantity can be distributed to Chamber at early period, because this quantity will decrease at middle and latter period due to the fuel distribution guidance of bowl shape. Its disadvantage is that the effect is not as remarkable as spray angle.

Bowl combustion capacity:

Besides guidance, another main function of bowl shape is combustion capacity, that is, the ability of burning fuel quantity efficiently and cleanly in the period of

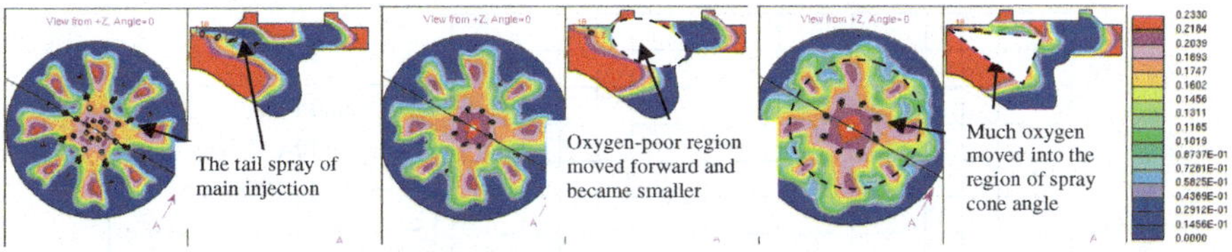

(a) 381° CA **(b)** 383° CA **(c)** 385° CA

Fig. 10 Oxygen concentration field with post injection (mass concentration/-)

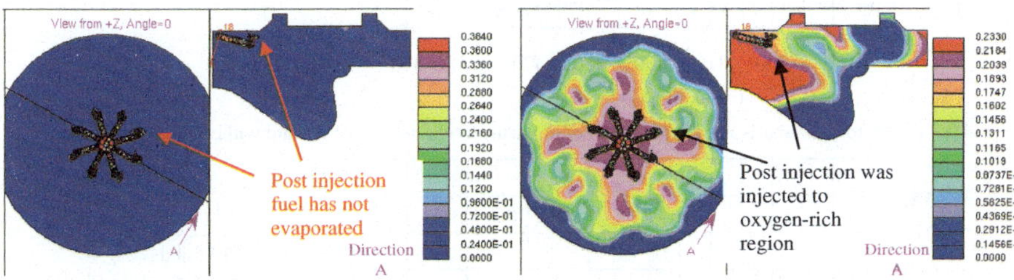

(a) Fuel vapor concentration field at 386ºCA **(b)** Oxygen concentration field at 386ºCA

Fig. 11 Fuel vapor and oxygen concentration field with post injection (mass concentration/-)

main fuel consumption in bowl. In general, if the bowl diameter increases, the penetration distance of the fuel vapor increases before it impacts on the wall. This leads to more entrainment of air, which improves the combustion [4]. So increasing the bowl diameter can promote the upper limit of fuel quantity that can be distributed to Chamber, therefore its combustion capacity becomes stronger.

3.2.3 Multiple-injection

Obviously, multiple-injection can change directly and flexibly the region that fuel is injected into, which can not only change the fuel zoning distribution, but also can inject fuel into oxygen-rich region to improve mixing. As shown in Figs. 10, 11, during the interval of main and post injection, oxygen-poor region moved away and much oxygen moved into the region of spray cone angle, which made post injection injected to oxygen-rich region, not poor region, so combustion was improved. Post injection (PI) made the fuel quantity distributed to Chamber decrease near 390 °CA evidently, and its fuel was injected to UPcham directly as shown in Fig. 12. The region of lambda <0.5 decreased evidently, in a certain range of crank angle lambda<1 region decreased too, and 1<lambda<1.5 region increased, which are beneficial to reduce the soot formation or strengthen the soot oxidation.

The emissions can be improved by optimizing quantity and interval of post injection to obtain a reasonable concentration field. As shown in Fig. 13, under the condition of constant injection timing, when 20.5 mg post injection and 5 °CA

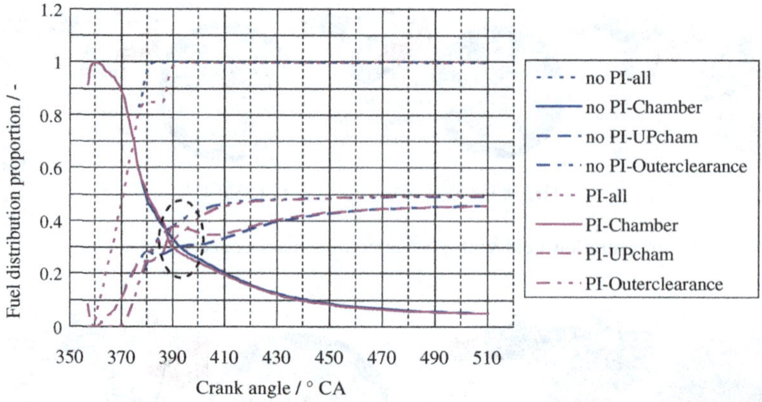

(a) The history comparison of fuel vapor zoning distribution between with and without post injection

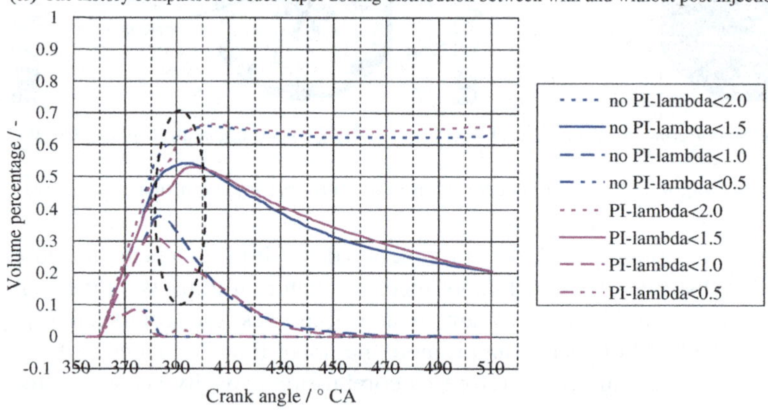

(b) The history comparison of air utilization ratio between with and without post injection

Fig. 12 Effects of post injection (PI) on fuel distribution and mixture concentration

interval were used, NO emissions decreased by about 6.5 % soot emissions decreased by 30 %, while the thermal efficiency decreased very little.

3.2.4 Swirl Ratio

Studies showed that the swirl ratio (Rs) had a remarkable influence on fuel distribution. Increasing swirl ratio made the spray marching distance along radius reduced and the timing of impacting on wall delayed, so more fuel located in UPcham, which was more evident after the end of the injection. As shown in Figs. 14, 15, when swirl ratio increased, fuel in Chamber decreased remarkably and moved to UPcham. Correspondingly, the region of lambda <0.5 and lambda <1 significantly decreased, while that of lambda <1.5 and lambda <2 increased significantly as shown in Fig. 15b, which were very favorable changes.

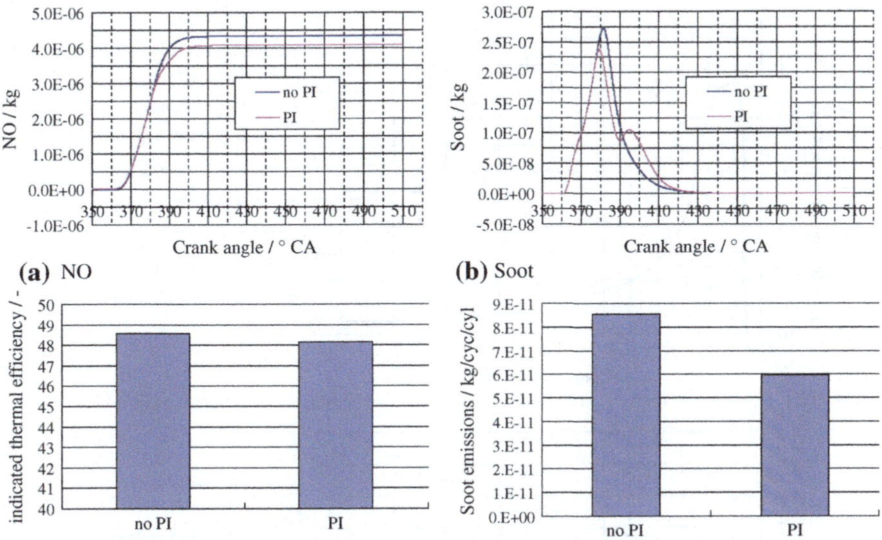

(a) NO (b) Soot

(c) Indicated thermal efficiency (d) Soot emissions

Fig. 13 Effects of post injection (PI) on indicated thermal efficiency and emissions

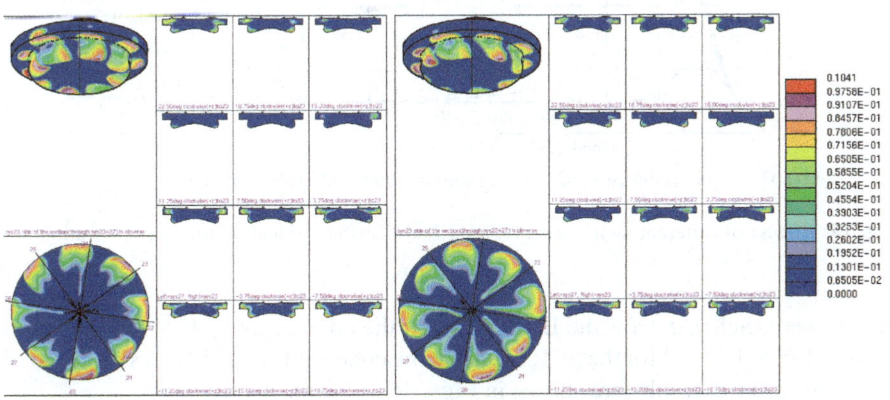

(a) Swirl ratio 1.1 (b) Swirl ratio 2.2

Fig. 14 The comparison of fuel vapor concentration filed under different swirl ratios (mass concentration/-)

4 The Application of Distributed Combustion Design in the Diesel Engine Development

Two combustion chambers were designed for the EGR and SCR solution of Euro IV of a 9 L diesel engine respectively. Combustion system optimization was done based on the concept of spatial and temporal zoning controls. The final optimization

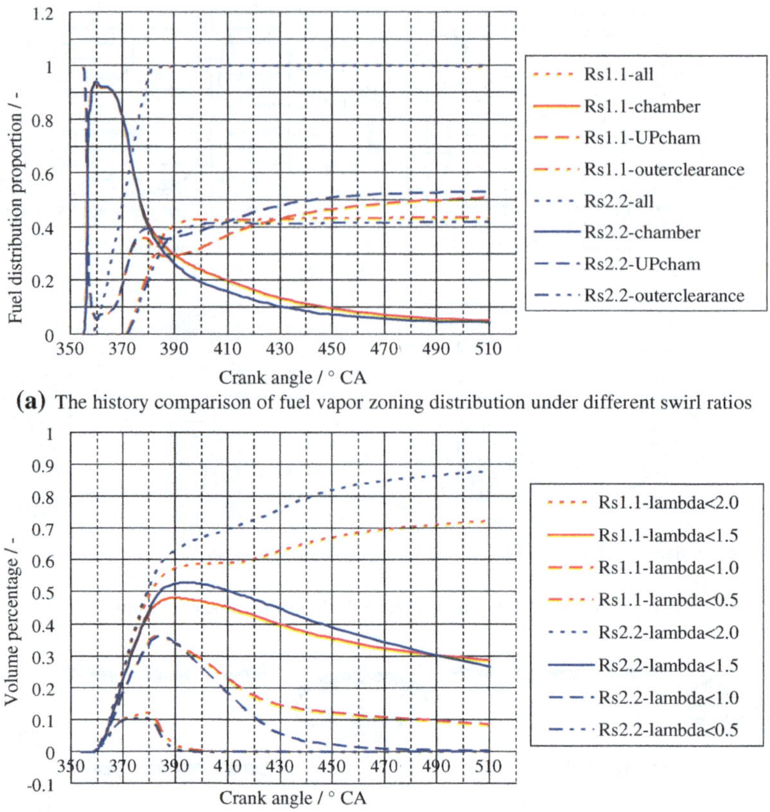

(a) The history comparison of fuel vapor zoning distribution under different swirl ratios

(b) The history comparison of air utilization ratio under different swirl ratios

Fig. 15 Effects of different swirl ratios (Rs) on fuel distribution and mixture concentration

results were: Scheme 1 for the EGR solution: the cone angle of 155° and the swirl ratio of 1.6; Scheme 2 for the SCR solution: the cone angle of 155° and swirl ratio of 2.2. Optimization results are shown in Figs. 16, 17 respectively.

The two schemes have both achieved their objectives of performance and emissions. Scheme 2 realized 2 % reduction of fuel consumption than the original engine, and its thermal efficiency was higher than scheme 1, because the bowl combustion capacity of scheme 2 was strong than that of scheme 1 and more fuel quantity could be distributed to Chamber. But NOx emissions were more too.

Based on the DCD, high efficiency and low emissions combustion was achieved on a 9 L diesel engine of FAW.

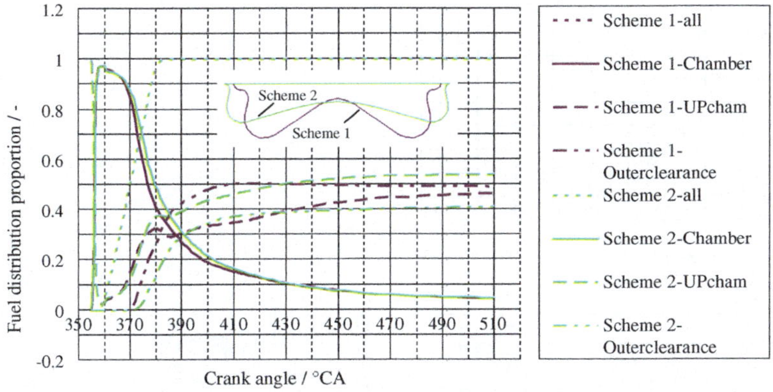

(a) The history of fuel vapor zoning distribution

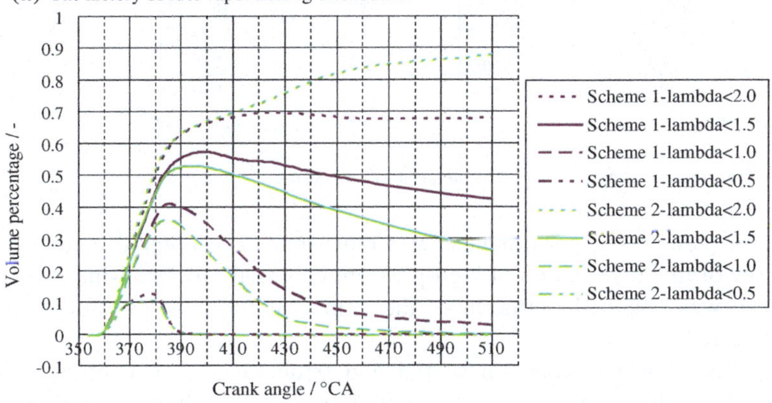

(b) The history of air utilization ratio

Fig. 16 Optimization results of fuel distribution and mixing

5 Summary and Conclusions

1. In order to achieve precise control of combustion, the whole space in cylinder was divided into three zones, Chamber, UPcham and Outerclearance, and the start of combustion, the timing of injection impacting on bowl lip, the end of injection and the end of combustion were selected as critical timings for temporal zoning controls.
2. Studies have shown that the fuel quantity distributed to Chamber was the most important and sensitive factor to control the mixture concentration distribution, combustion and emissions. Several common engineering approaches can change the fuel distribution such as spray angle, bowl shape, multiple-injection, swirl ratio and so on.
3. Based on the engine bench tests and CFD analysis, the results of spatial and temporal zoning controls were associated with engine structural parameters,

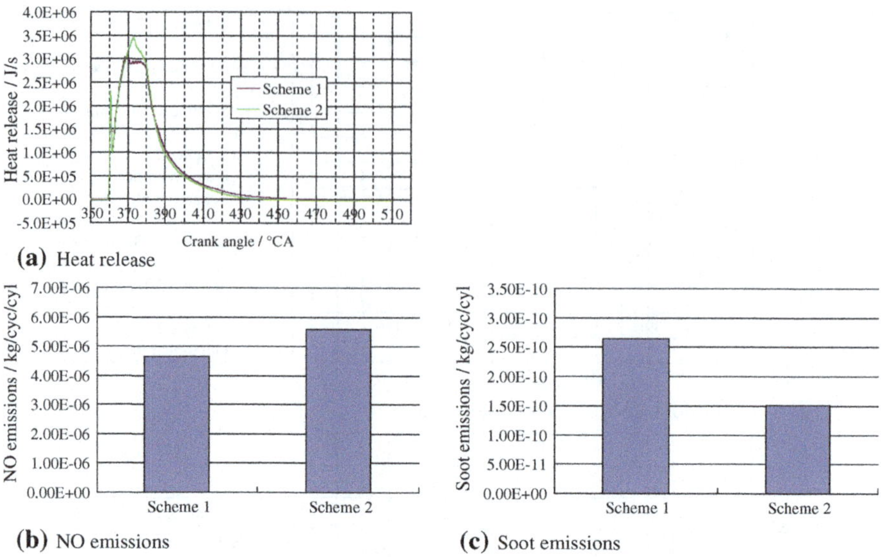

Fig. 17 Optimization results of performance and emissions

control parameters, performance, and emissions. The optimal zoning control history of fuel distribution of high-pressure common-rail direct-injection diesel engine was obtained, furthermore the empirical formulas of zoning controls were concluded.

4. The DCD concept based on spatial and temporal zoning controls helps to understand the combustion organization in cylinder. It has strong physical meanings and good maneuverability. The diesel engine development practice showed that the DCD met the engineering applications of product development and was a successful method.

References

1. Pischinger F (1995) The diesel engine for cars is there a future. ASME Fall Conference
2. Akihama K, Takatori Y, Inagaki K (2001) Mechanism of the smokeless rich diesel combustion by reducing temperature. SAE Paper 2001-01-0655
3. Takashi Kaminaga, Jin Kusaka Improvement of combustion and exhaust gas emissions in a passenger car diesel engine by modification of combustion chamber design. SAE Paper 2006-01-3435
4. Bianchi GM, Pelloni P, Corcione FE, Mattarelli E, Luppino Bertoni F Numerical study of the combustion chamber shape for common rail h.s.d.i. diesel engines. SAE Paper 2000-01-1179

Part VII
Heat Transfer and Waste Heat Reutilization

Numerical Investigation into the Cooling Process of Conventional Engine Oil and Nano-Oil Inside the Piston Gallery

Peng Wang, Jizu Lv, Minli Bai, Chengzhi Hu, Liang Zhang and Hao Liu

Abstract A numerical simulation model has been developed for the transient flow and heat transfer problems of oil inside the piston gallery using a coupled VOF/Level Set method. Detailed cooling processes of Cu-oil nanofluids, with the nanoparticle size of 50 nm and the volume fractions of 1, 2 and 3 %, have been investigated. The oil fill ratio (OFR) and heat transfer coefficients (HTC) variations at different crank angles have been examined as well. The results have demonstrated that the nano-oil is able to improve the heat transfer capacity by a large margin. Compared with the conventional engine oil, the overall average heat transfer coefficients of the nano-oil, with the volume fractions of 1, 2 and 3 %, increase by 5.80, 14.51 and 28.11 % respectively.

Keywords Piston cooling gallery · Nanofluids · Multiphase flow · VOF · Level set

1 Introduction

Over the years in the past, the overall heat loading of internal combustion engines increases rapidly with the special power rising. In a variety of high-temperature components, the working condition of piston is the worst due to prolonged exposure to the combustion chamber. The piston suffers periodic mechanical and thermal loadings. The cooling gallery inside the piston head is commonly adopted to optimize the heat extraction. The heat flux removed by the cooling gallery can

F2012-A07-002

P. Wang (✉) · J. Lv · M. Bai · C. Hu · L. Zhang · H. Liu
Dalian University of Technology, Dalian, China
e-mail: dlutwp@126.com

SAE-China and FISITA (eds.), *Proceedings of the FISITA 2012 World Automotive Congress*, Lecture Notes in Electrical Engineering 190,
DOI: 10.1007/978-3-642-33750-5_26,

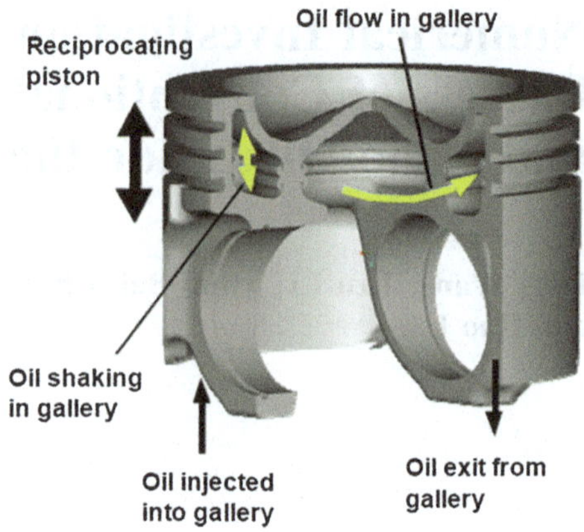

Fig. 1 Schematics of piston gallery oil flow

occupy as much as 45 ∼ 75 % of the total heat passed to the piston from the combustion gases, so heat transfer coefficient of this region is of great importance to the accurate calculation of piston temperature.

However, the structure of cooling gallery limits the piston strength to some extent. With the peak cylinder pressure in diesel engines steadily increased, simply extending or modifying the cooling gallery configuration may be insufficient when power rating is above a certain level and additional cooling measures are quite necessary. In 1995, Choi [1] proposed the concept of nanofluids, which is a new type of heat transfer medium. It contains nanoparticles of metal or metal oxides with dimensions smaller than 100 nm, suspended uniformly and stably in base fluid. Inclusion of solid particles in liquid will greatly enhance the overall average thermal conductivity due to high thermal conductivity and large specific surface area of solid particles. Therefore, if nanofluids can be used in the heat transfer process of engine piston, especially as the cooling medium of piston gallery, the heat transfer efficiency will be effectively improved.

Due to small space and complicated structure of the piston gallery, it is quite difficult through experiments to gain accurate visualization process of oil flow and heat transfer characteristics against crank angles. So far, what's widely accepted in predicting HTC is an empirical formula suggested by Bush [2]. Only an overall averaged heat transfer coefficient result can be given, without considering the temporal and spatial distributions. With the development of computer technology, CFD numerical simulations have become an alternate tool to accurately understand the transient flow and heat transfer mechanism inside the piston gallery. It can provide precise thermal boundary conditions for finite element analysis on the piston temperature and further analysis for the heat flux distribution in the combustion process.

In the current paper, the nanofluids is treated as a single-phase fluid and the Level Set function is introduced on the basis of the traditional VOF model. The two models are coupled together to simulate the two-phase flow formed by nanofluids and air. In addition, the traditional engine oil and Cu/oil nanofluids with

Fig. 2 Geometry configuration of the gallery

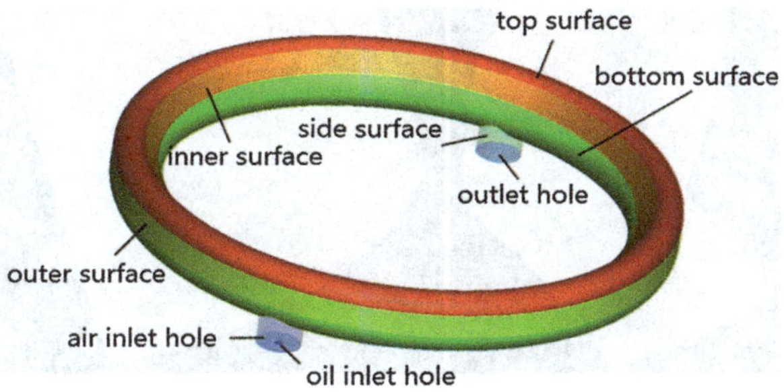

various concentrations are employed as the cooling medium to investigate the transient flow and heat transfer process numerically inside the piston gallery. The oil fill rate and heat transfer coefficient distributions are discussed under different crank angles and the flow and heat transfer mechanism are further explored.

2 Geometry Model

As schematically described in Fig. 1, the cooling oil is injected into the gallery through the entrance hole, flowing around in circumferential direction, exiting via the outlet orifice. The gallery is extracted from the entire piston and simplified inlet and outlet are added in order to simplify the numerical simulation. The geometry configuration is shown in Fig. 2, in which the entity elements are the piston gallery of actual design and the translucent parts are the added inlet and outlet. The geometry depicted in Fig. 2 is symmetric, so only half of the gallery is modeled. The computational domain and mesh is displayed in Fig. 3. The oil and air are injected into the cooling gallery together and form a periodic two-phase flow. Two concentric circles entrances are introduced. The cooling gallery has two openings separated by 180°, each having 8 mm in diameter. The oil jet of 1.6 mm diameter hits the center of the inlet opening with $V = 15$ m/s and $T = 373$ k. Pressure boundaries are assigned to oil jet surrounding area of the inlet and outlet orifices. The gallery surface is mapped with a spatially varying temperature as thermal boundary condition, recommended by Zhang Weizheng [3] et al. At the initial moment, the region is filled with air, with the temperature of 373 K. The reciprocating motion of the piston gallery is simulated by two sets of coordinate systems. All parameters will reach a periodic steady state after about 20 operation cycles.

3 Math Model

The VOF and Level Set are the most common numerical strategies used to predict interface motion. This paper will combine the advantages and overcome the disadvantages of the VOF and the Level Set to achieve a new model, which is

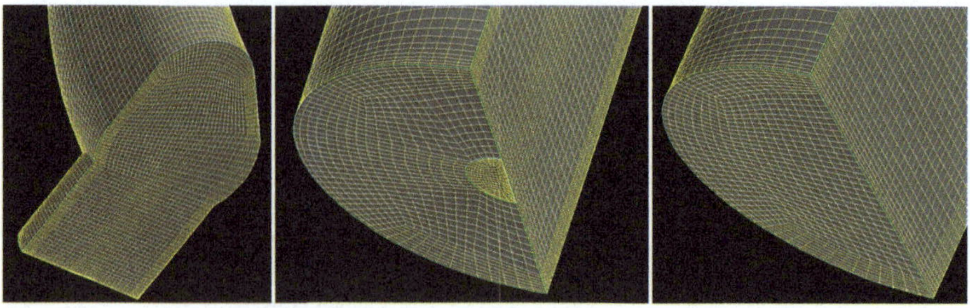

Fig. 3 Mesh of the gallery model

generally superior to either method alone. The VOF method is used to capture interfaces, which can abandon the disadvantage of mass non-conservation in Level Set. While the Level Set function is applied to compute the accurate curvature and sooth the discontinuous physical quantities near interfaces.

4 Thermophysical Properties Model for Nanofluids

4.1 Density and Specific Heat Models

Density model:

$$\rho_{nf} = (1-\phi)\rho_f + \phi\rho_p \tag{1}$$

Specific heat model:

$$\rho_{nf}C_{nf} = (1-\phi)\rho_f C_f + \phi\rho_p C_p \tag{2}$$

where ϕ is particle volume fraction, ρ_{f}, ρ_{p}, ρ_{nf} are the base fluid, particle and nanofluids density, $\mathrm{C_f}$, $\mathrm{C_p}$, $\mathrm{C_{nf}}$ are the base fluid, particle and nanofluids specific heat.

4.2 Thermal Conductivity Model

Murshed [4] et al. developed an effective thermal conductivity model:

$$k_{nf} = \frac{(k_p - k_{lr})\phi k_{lr}[2\beta_1^3 - \beta^3 + 1] + (k_p + 2k_{lr})\beta_1^3[\phi\beta^3(k_{lr} - k_f) + k_f]}{\beta_1^3(k_p + 2k_{lr}) - (k_p - k_{lr})\phi[\beta_1^3 + \beta^3 - 1]} \tag{3}$$

where subscript p, lr, f, nf represent the particle, interface layer, base fluid and nanofluids.

4.3 Viscosity Model

Yu Xiaoli [5] et al. proposed a forecast model for nanofluids viscosity:

$$\mu_{nf} = \mu_f[1 + 2.5\phi(1 + \frac{8.868}{r})] \tag{4}$$

where μ_f is base fluid viscosity, μ_{nf} is nanofluids viscosity, ϕ is nanoparticle volume fraction, r is nanoparticle radius.

5 Results and Analysis

The flow and heat transfer processes of engine oil inside the piston gallery take on periodical features under the action of reciprocating inertia force. This study focuses on oil distribution over gallery surfaces, as heat transfer occurs mainly near the walls during the cooling process. A huge advantage is observed in the nanofluids heat exchange by comparing the flow and heat removal process of the traditional oil and nano-oil at the same moment and location.

5.1 Qualitative Analysis of the Convective Heat Transfer Coefficient and the Oil Fill Rate with the Crank Angle Variations

The relative velocity between the piston and injected oil gradually increases when the piston moves from the TDC to BDC. The maximum HTC near the impingement position rises as well under the exertion of opposite accelerated motion. Most of the cooling oil concentrates in the upper section of the piston gallery due to the existence of the inertia force and travels downward. There is almost no oil flowing out from the inlet and outlet openings, as shown in Fig. 4a, which increases the oil fill ratio. The HTC at the upper part of the gallery reaches its maximum within a cycle because of high oil fill ratio and low oil temperature. The top and inner surfaces are completely covered by oil, while the outer surface is only partly attached.

The thermal conductivity of nanoparticles is three orders of magnitude larger than engine oil Therefore even if the volume fraction of Cu nanoparticles is very low, it also can make the thermal conductivity of nano-oil improve greatly. Compared with the traditional engine oil, the HTC of nano-oil enlarges, thus enhancing the heat transfer capability effectively. It can be found that the nano-oil viscosity grows as well, which subsequently raises the boundary layer thickness contacted with the wall and lowers the heat transfer effects. The nano-oil overall specific heat fell as a result of the addition of nanoparticles. Thus, the nano-oil

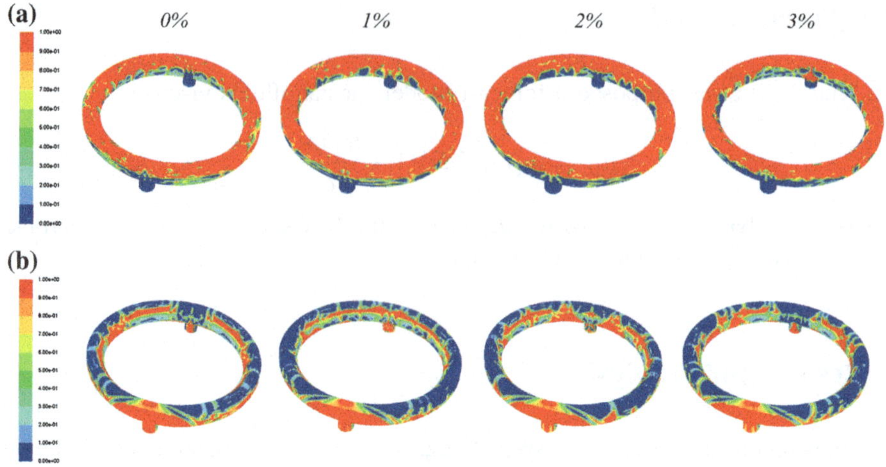

Fig. 4 Spatial distribution of traditional oil and nano-oil inside piston gallery. **a** 0°CA. **b** 180°CA

temperature rise increases and the temperature discrepancy between the gallery surface decreases under the same conditions, which weakens the heat transfer effects to some extent. As discussed above, the transformations in the physical properties of the nanofluids do not all conduce to the heat transfer. After careful contrasts of nanofluids with three different volume fractions, it's clearly indicated that the average viscosity increases by 5.9 %, the average specific heat drops by 16 % and the average thermal conductivity increases by 30.5 %. The enhancement of thermal conductivity still plays a decisive role in the actual cooling process. The nano-oil does improve the heat transfer capacity from the comprehensive aspect, but the increase rate is bound to less than that of thermal conductivity. At the TDC, the heat transfer coefficients of nano-oil with the volume fractions of 1, 2, 3 % improve by 5.61, 17.62, 24.79 % respectively in relation to the engine oil, as depicted in Fig. 5a. The heat transfer effectiveness is considerable on the upper half of the surfaces. The enhancement of heat transfer capacity increases with the increase of nanoparticle volume fraction in nano-oil. The main reason is attributed to the large improvement in thermal conductivity and few variations in viscosity and specific heat with the increase of nanoparticle volume fraction. The HTC on bottom part of the gallery reaches its minimum within a cycle, as it's almost covered with air whose thermal conductivity is much smaller than that of traditional oil or nano-oil. As shown in Figs. 5a and 6c, there are few discrepancies on the HTC between the traditional oil and nano-oil as for almost no oil film coverage over the bottom wall.

When the piston approaches the vicinity of 90 °CA, the relative velocity between the cooling gallery and injected engine oil reaches peak, as displayed in Fig. 7. The HTC also reaches its maximum near the impingement point throughout the piston motion cycle, as illustrated in Fig. 5b. The engine oil starts to fall off the

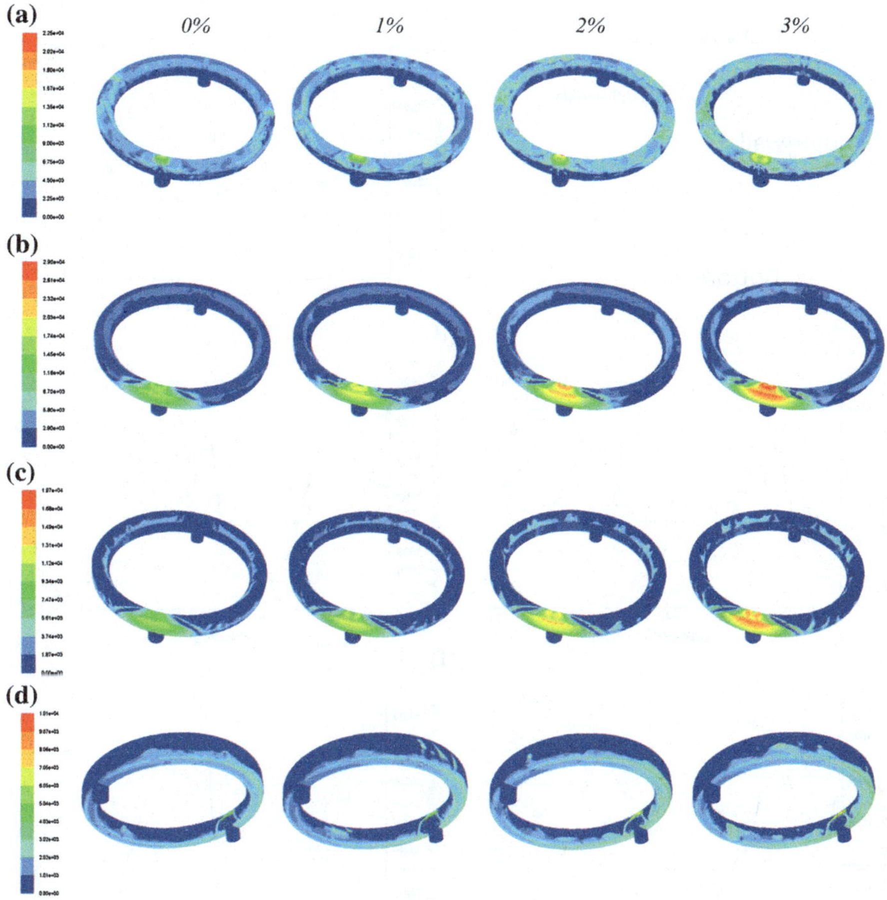

Fig. 5 HTC distribution of traditional engine oil and nano-oil on gallery surfaces. **a** 0°CA. **b** 90°CA. **c** 180°CA. **d** 270°CA

upper half of the surfaces due to the inertia force as the piston continues to move downward and decelerates. Part of the oil exits the gallery through the inlet and outlet holes on the bottom. But the oil fill ratio is still increasing, as depicted in Fig. 8. The rest is thrown to the bottom surface with a relative high speed, generating a sloshing effect. The turbulence intensity greatly improves compared with the oil flow in circumferential direction. The wall boundary layer thickness gets thin instantly and therefore the heat transfer effect is enhanced immediately. As the piston continues to travel to about 120°CA during the down stroke, the oil fill ratio reaches its peak value within a cycle, and then begins to fall, as displayed in Fig. 8. Before the piston arrives at BDC, major engine oil has accumulated on the bottom wall. The HTC on that surface is the maximum over a cycle and is the largest of the four surfaces as it is almost full of air in the upper portion of the cooling

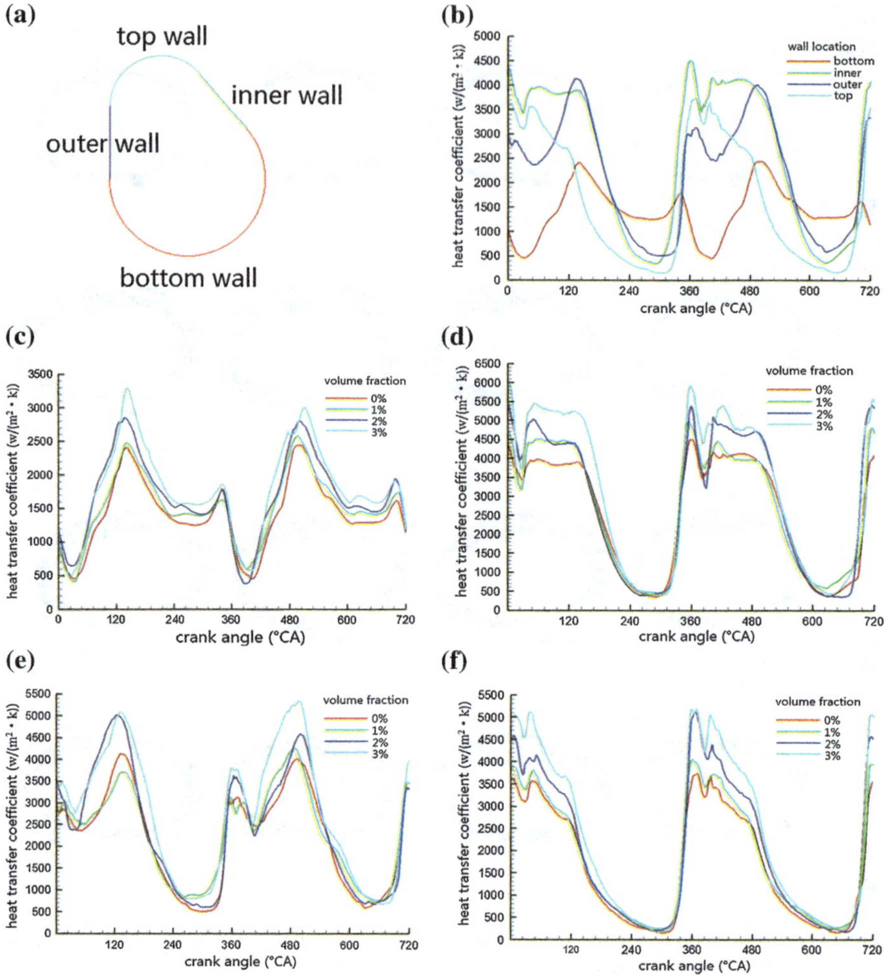

Fig. 6 Variations of gallery surface heat transfer coefficients over piston motion cycle **a** Surface locations inside the gallery, **b** HTC distributions of engine oil on different surfaces, **c** *Bottom wall*, **d** *Inner wall*, **e** *Outer wall*, **f** *Top wall*

gallery, as shown in Figs. 5c and 6b. After the heat exchange with the upper parts of the gallery, the oil temperature rises, making the temperature difference between the engine oil and the bottom surface decrease. Future more, the bottom surface temperature is much lower than those of upper surfaces. Hence, the overall average heat transfer coefficient on the bottom wall is the smallest inside the gallery during a cycle, as plotted in Fig. 6b.

The nano-oil density rises relative to the conventional engine oil, which enhances the inertia force and the impingement interaction. Therefore, these effects decrease the boundary layer thickness and further offset the boundary layer

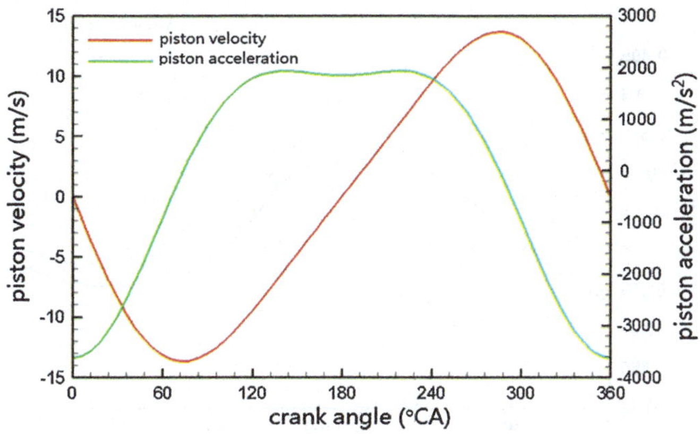

Fig. 7 Variation of piston velocity and acceleration

increase by higher viscosity of nano-oil. The nano-oil heat transfer coefficient is larger than that of conventional oil as a result of thermal conductivity increase. At BDC, the nano-oil heat transfer coefficients with the volume fractions of 1, 2 and 3 % increase by 3.90, 12.95 and 19.08 % respectively, compared with the engine oil. The heat transfer enhancement on the bottom surface is evident, as shown in Fig. 5c, and is proportional to the nanoparticle volume fraction in nano-oil. This is linked to the large improvements in thermal conductivity and density, and small deterioration in viscosity and specific heat with the increase of nanoparticle volume fraction. All other surfaces inside the gallery, such as the top, inner and bottom walls have the same trend. It is also important to notice that there is more attached nano-oil over the upper walls of the gallery because of higher viscosity in relation to the engine oil, when the piston moves towards to the nozzle. Although most of the upper domain is filled with air, more heat flux can still be removed at the same moment, compared with the conventional oil. Therefore, the heat transfer coefficient is relative huge and increases with the increase of nanoparticles volume fraction, just as illustrated in Fig. 6f.

The piston begins to accelerate in the opposite direction just shortly after the bottom dead center. The heat transfer coefficient near the impingement position and relative speed between the piston and the injected oil gradually decrease due to the motion in the same direction. Major oil is oppressed upon the bottom surface because of the inertia effect. Lots of oil drains out of the inlet and outlet openings after heat exchange with the four surfaces. The oil fill ratio starts to decline significantly, as plotted in Fig. 8. However, the heat transfer coefficient on the bottom surface is still the largest of the entire cooling gallery. When the piston travels to the vicinity of 270°CA, its upward speed reaches the maximum, while the relative velocity between the oil and the gallery reaches the minimum. The heat transfer coefficient near the impingement position has its least value within a cycle as well, as shown in Fig. 5d. As the piston moves to about 300°CA, the oil fill ratio reaches minimum value over a cycle. The piston then starts to decelerate

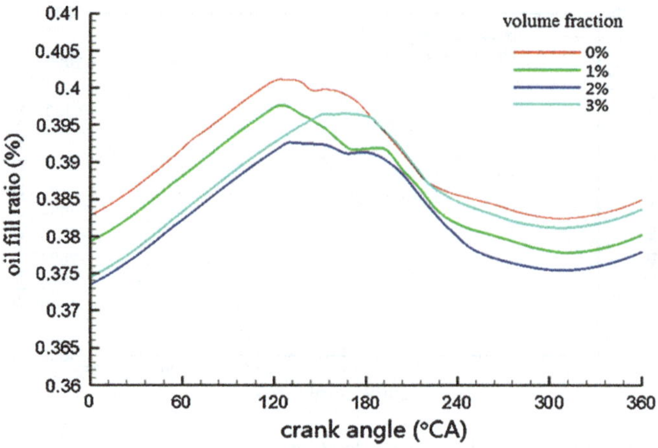

Fig. 8 Variation of gallery oil fill ratio

when continuing the upward motion. Due to the inertia effect, the oil gradually falls off the bottom surface instead of exiting the inlet and outlet holes. The relative velocity of between the gallery and the injected oil increases, causing the oil fill ratio rising again. The oil flushes the top wall with a high speed, as displayed in Fig. 8. When the piston arrives at the top dead center, a large amount of injected oil with low temperature concentrates at the upper portion of the gallery. The heat transfer coefficients on the top, inner and outer surfaces rise significantly, whereas that on the bottom surface drastically reduces. So far, a periodic motion is accomplished, and then another operation cycle proceeds. The heat transfer enhancement mechanism of nano-oil during the return stroke follows similar trend to that from TDC to BDC and will not be discussed repeatedly.

In the circumferential direction, the previous oil spreads over the surfaces and is pushed away by the injected oil that followed to leave the outlet opening. The oil temperature rises after the heat exchange with the surfaces, under the shaking effect, leading to decreased temperature difference between the oil and surfaces. Therefore, the heat transfer coefficients on all gallery surfaces are continuously decreasing along the circumferential direction, for both the traditional oil and the nano-oil. It's obvious to find out that the heat transfer coefficient near the inlet opening is relative large, while that around the outlet hole is a little small. But at the same moment and location, the heat transfer coefficient of nano-oil is still higher and increases in the volume fraction, as illustrated in Fig. 5a–d.

Figure 7 describes the variations of the acceleration and speed of piston against crank angles over an operation cycle. The piston peak velocity is well less than the injected oil (V = 15 m/s), which is quite desirable to ensure the cooling oil is always able to rush into the gallery throughout the whole periodic motion. The average oil fill ratio over a reciprocating cycle is around 38.5 %, which is in acceptable range for piston of actual design. Assume that the top dead center is at

Table 1 Average heat transfer coefficients on different surfaces over a cyle

Increase rate/ %	Top (%)	Bottom (%)	Outer (%)	Inner (%)	Average (%)
1 nano-oil	8.05	7.72	4.54	4.28	5.80
2 nano-oil	20.49	18.11	11.08	11.37	14.51
3 nano-oil	36.36	26.11	26.70	24.88	28.11

the 0 °CA. The piston dynamics naturally produces higher inertial force at TDC than at BDC. It can be seen that the acceleration at TDC is about twice that at BDC, leading to a much larger inertia force and sloshing effect. This characteristics, combined with the gallery shape, results in much higher peak values of heat transfer coefficient at TDC than at BDC. In fact, this higher heat transfer coefficient at upper part of the gallery is quite desirable for effective piston cooling in terms of controlling temperatures at critical locations, such as bowl rim and top ring groove. The heat transfer coefficients of nano-oil with the volume fractions of 1, 2 and 3 % increase by 5.61, 17.62 and 24.79 % respectively at TDC, while at BDC, they are only 3.90, 12.95 and 19.08 % respectively. The aforementioned analysis has indicated that the temperature discrepancy between oil and upper surfaces is larger than bottom surface, which is also one of the reasons why the heat transfer coefficients on upper surfaces are higher than bottom surface.

The heat transfer coefficients on the surfaces take on a periodic transformation when the piston works into a stable state. The overall mean heat transfer coefficient is large during the down stroke and relative small in the return stroke which seems to be consistent with the oil mass distribution, in the case of both the traditional and nano-oil. The spatial distributions are almost the same for both the traditional oil and nano-oil with peak HTC on the impingement position, whereas the only difference exists in the HTC magnitude. It's demonstrated that the heat transfer enhancement by the nano-oil does not rely on raising the oil fill ratio. It's obvious to find out that the heat transfer coefficients of nano-oil have greatly improved on all surfaces by adding the nanoparticles, as listed in Table 1.

5.2 *Quantitative Comparison and Analysis of the Numerical Simulation Results with the Empirical Formula Data*

The numerical simulation results are compared with the traditional formula so as to quantitatively validate the accuracy of the model. Bush [2] et al. proposed an empirical correlation for the convective heat transfer process inside the cooling gallery, based on the similarity criteria through a series of experiments:

$$Nu = 0.495(\mathrm{Re})^{0.57}(De/H)^{0.24}(\mathrm{Pr})^{0.29} \tag{5}$$

where $\mathrm{Re_B} = \omega_\mathrm{B}\mathrm{De}/v$, $\omega_\mathrm{B} = \mathrm{nH}/30$, De is gallery equivalent diameter, H is average height of cross section, $\mathrm{Re_B}$ is reciprocating Reynolds number, ω_B is

Table 2 Comparisons of the numerical simulation results and empirical formula data on HTC

Volume fraction/ %	Numerical simulation/w/(m^2 k)	Empirical formula/w/(m^2 k)	Error/%
0	1,964	2,254	14.77
1	2,078	2,322	11.74
2	2,249	2,586	14.98
3	2,516	2,834	12.64

reciprocating velocity, v is oil dynamic viscosity, n is rotation speed. The overall mean convective heat transfer coefficients given by the numerical simulation and empirical formula are listed in Table 2:

After careful contrasts, it's found that the maximum error of the overall averaged HTC between numerical simulation and empirical formula is no more than 15 %, which is in an acceptable range. In the numerical simulation, only the thermal conductivity variation with temperature is taken into account, without considering the change in viscosity in relation to temperature. Inevitably, there is kind of error to some extent, which leads to lower convective heat transfer coefficient. As the temperature increases, the viscosity and boundary layer thickness decrease, which subsequently will improve the heat transfer capacity. The numerical simulation results will more agree with the empirical formula under the consideration of viscosity changes. However, the traditional empirical formula purely gives a uniform heat transfer coefficient, while the numerical simulation in present paper is able to accurately provide HTC with spatial and temporal distributions as thermal boundary conditions for further finite element analysis of the piston. This is much more realistic and will greatly improve the precision of the FEA predictions.

6 Conclusions

1. The physical properties of engine oil are changed by addition of Cu nanoparticles. Large improvements in thermal conductivity and density, and small deteriorations in viscosity and specific heat with the increase of nanoparticle volume fraction enhance heat transfer process. The overall average heat transfer coefficient of nanofluids with volume fractions of 1, 2 and 3 % increase by 5.80, 14.51 and 28.11 % respectively, compared with the traditional engine oil, which can effectively reduce the piston temperature in practical applications.
2. The numerical simulation results correspond well with the empirical formula. But the traditional empirical formula can purely give overall averaged heat transfer coefficient, while numerical simulation in this paper is able to accurately provide thermal boundary conditions with spatial and temporal distributions for further finite element analysis of the piston.

3. The addition of nanoparticles does not significantly change the oil distribution, which demonstrates that the heat transfer enhancement by the nano-oil does not rely on raising the oil fill ratio.

References

1. Choi S (1995) U S. enhancing thermal conductivity of fluids with nanoparticles. Dev Appl Non-Newtonian Flows 231:99–105. ASME FED, New York
2. Bush JE, London AL (1965) Design data for cocktail shaker cooled pistons and valves. SAE paper 650727
3. Zhang W, Cao Y, Yuan Y, Yang Z (2010) Simulation study of flow and heat transfer in oscillating cooling pistons based on CFD. Trans CSICE 28(1):74–78
4. Murshed SMS, Leong KC, Yang C (2006) Thermal conductivity of nanoparticle suspensions. IEEE Conf 99(8):084308–084308-6
5. Peng X (2007) Study of nanofluids heat transfer performance in high temperature condition based on vehicular cooler. Zhejiang University, Zhejiang

Applying Design of Experiments to Develop a Fuel Independent Heat Transfer Model for Spark Ignition Engines

Joachim Demuynck, Kam Chana, Michel De Paepe, Louis Sileghem, Jeroen Vancoillie and Sebastian Verhelst

Abstract Models for the convective heat transfer from the combustion gases to the walls inside a spark ignition engine are an important keystone in the simulation tools which are being developed to aid engine optimization. The existing models are, however, inaccurate for an alternative fuel like hydrogen. This could be caused by an inaccurate prediction of the effect of the gas properties. These have never been varied in a wide range before, because air and 'classical' fossil fuels have similar values, but they are significantly different in the case of hydrogen. We have investigated the effect of the gas properties on the heat flux in a spark ignition engine by injecting different inert gases under motored operation and by using different alternative fuels under fired operation. Design of Experiments has been applied to vary the engine factors in a systematic way over the entire parameter space and to determine the minimum required amount of combinations. This paper presents the validation of a heat transfer model based on the Reynolds analogy for the entire experimental dataset. The paper demonstrates that the heat flux can be accurately predicted during the compression stroke with a constant characteristic length and velocity, if appropriate polynomials for the gas properties are used. Furthermore, it shows that a two-zone combustion model needs to be used as a first step to capture the effect of the flame propagation. However, alternative characteristic lengths and velocities will need to be investigated to capture the intensified convective heat transfer caused by the propagating flame front and to accurately predict the heat flux during the expansion stroke.

F2012-A07-004

J. Demuynck (✉) · M. De Paepe · L. Sileghem · J. Vancoillie · S. Verhelst
Ghent University, Ghent, Belgium
e-mail: joachim.demuynck@ugent.be

K. Chana
University of Oxford, Oxford, UK

SAE-China and FISITA (eds.), *Proceedings of the FISITA 2012 World Automotive Congress*, Lecture Notes in Electrical Engineering 190,
DOI: 10.1007/978-3-642-33750-5_27,

Keywords Heat transfer · Internal combustion engine · Design of experiments · Model · Alternative fuel

1 Introduction

A lot of research efforts focus on further improving the efficiency, emissions and power output of internal combustion engines (ICE) to propel ever heavier vehicles with lower fuel consumption and emissions. To enhance this development, computer tools are being developed to simulate the engine cycle. An important part in these models is the prediction of the heat transfer inside the engine, since it has an effect on all the improvement targets given above. In particular, the convective and radiant heat transfer from the combustion gases to the inner cylinder walls is needed at every calculation step to solve the conservation equations of mass and energy. Therefore, in-cylinder heat transfer has been the subject of many research programs since the 1950s [1, 2]. In spark ignition engines, only convection is considered since radiation is negligible [1]. Although the heat transfer process inside an engine is transient, it is assumed to be quasi-steady. As a consequence, the convective component of the heat flux (q) can be described by a convection coefficient (h), defined in Eq. 1.

$$h = \frac{G}{(T_g - T_w)} \tag{1}$$

where T_w is the wall temperature and T_g the bulk gas temperature. The underlying assumption of most of the models in literature, e.g. Refs. [3–5], is the Reynolds analogy [6], which describes the analogous behaviour of heat and momentum transfer. The analogy results in a relation between the Nusselt (=h.L/k), Reynolds (=V.L/ν) and Prandtl (=α/ν) number, given in Eq. 2. The models differ in the choice of the parameters (a, b and c) and the characteristic length (L) and velocity (V).

$$Nu = a \cdot Re^b \cdot Pr^c \tag{2}$$

Although there has been a lot of research on the heat transfer in combustion engines, the mechanisms are still not well characterized, especially not for new combustion types (e.g. HCCI) and alternative fuels (e.g. hydrogen [7, 8]). The most obvious possible reason for the models being inaccurate for an alternative fuel like hydrogen, is that the effect of the gas properties is not well captured. These gas properties have never been significantly varied before, since only air and fossil fuels with comparable properties have been used. Next, the effect of the Prandtl number has often been neglected. Furthermore, assumptions on the temperature dependency of the gas properties were made instead of using proper polynomials. The resulting error for fossil fuels could be small, but the thermal conductivity and heat capacity of hydrogen, on the other hand, differ significantly compared to air.

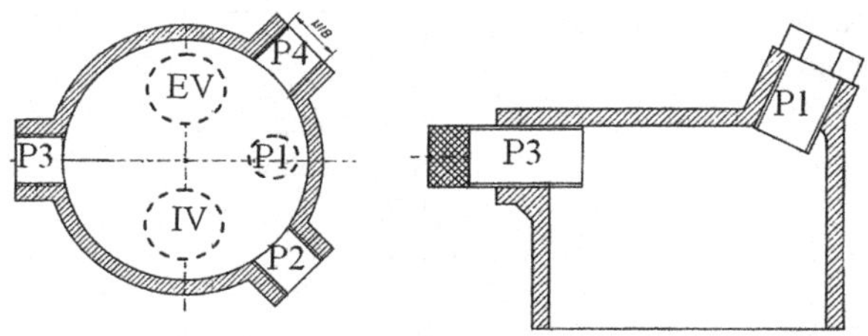

Fig. 1 Cross section of the CFR engine, *P1* spark plug position, *P2–P4* sensor positions, *IV* intake valve, *EV* exhaust valve

Solving the unsteady, one-dimensional energy equation for the boundary layer has been proposed as an alternative [9] for the quasi-steady assumption presented above. However, additional assumptions have to be made and it has not yet been demonstrated that this approach results in more accurate predictions. Therefore, this paper will investigate the applicability of the Reynolds analogy to model the convective heat transfer inside a spark ignition engine. It will mainly focus on the effect of the thermal conductivity (k), viscosity (ν), heat capacity (c_p) and density on the heat flux in an SI-ICE in an attempt to develop a model that is capable of simulating the difference in the heat flux between different fuels. The gas properties are varied in a wide range by injecting different inert gases (helium, argon, carbon dioxide) under motored operation an by using different alternative fuels (hydrogen, methane, methanol) under fired operation. In addition to the gas properties, the effect of the following engine factors is investigated: the throttle position (TP), the compression ratio (CR), the ignition timing (IGN) and the air-to-fuel equivalence ratio (λ). The experiments are designed and analysed with methods of Design of Experiments to investigate the variable engine factors in a consistent way over the entire parameter space.

2 Experimental Method

2.1 Equipment

The research is conducted on a CFR (Cooperative Fuel Research) engine, which is a four-stroke single-cylinder spark ignition engine. It has a variable compression ratio, but can only be operated at a constant speed of 600 rpm. The engine is equipped with a port fuel injection system. Two types of injectors are available, one for gaseous fuels and one for liquid fuels. The injection and ignition is controlled by a MoTeC M4Pro electronic control unit. Figure 1 shows the possible mounting positions for sensors in the engine. Table 1 gives an overview of the relevant engine properties. The CFR engine has recently been revised, resulting in a larger bore diameter and different valve timings compared to older publications.

Table 1 CFR-engine properties

Bore	83.06 mm
Stroke	114.2 mm
Connecting rod length	254 mm
Swept volume	618.8 cm^3
IVO	10 °CA ATDC
IVC	29 °CA ABDC
EVO	39 °CA BBDC
EVC	12 °CA ATDC

The heat flux and wall temperature are measured at positions P2, P3 and P4 with a Vatell HFM-7 sensor. The measurement positions are at the same height in the cylinder wall and are equally distributed around the circumference of the cylinder. The spark plug was mounted in position P1. The compression ratio has to be kept below 10, because the moving piston would otherwise damage the heat flux sensor. More detailed information on the test bench can be found in previous publications [7, 10].

2.2 Experimental Design

Both the experiments under motored and fired operation were designed according to DoE methods (Design of Experiments) described in Ref. [11]. The purpose of the DoE is to vary several factors in a systematic way and to define the minimum required number of combinations to investigate the heat flux in the entire parameter space, because it was not possible to run all the combinations. The statistical analysis of each experiment results in an experimental surface of the heat flux over the entire parameter space. This experimental surface will be compared with the surface obtained by the model prediction to test the model's capability of predicting the effect of engine factors in addition to variations within a certain engine cycle.

A detailed description of the experimental design of the measurements under motored operation can be found in Ref. [10]. In summary, three inert gases were injected in the intake manifold with the injector which is normally used to inject a gaseous fuel: helium, argon and carbon dioxide. In addition to the type and quantity of the gas, the compression ratio and throttle position were varied. The throttle position can be varied between 0 and 87°, but it only affects the ingoing flow rate when varied between 63 and 87°, because of the low engine speed. These two values will be labelled as wide open throttle (WOT) and fully closed throttle (FCT). The heat flux was only measured at P2, as we observed from previous measurements that there is no significant spatial variation in the heat flux under motored operation.

Table 2 Overview of the factors' extreme levels

Factor	Fuel	Level	Lecvel code
TP	All	Fully closed throttle (FCT)	−2
		Wide open throttle (WOT)	2
CR	All	8	−2
		10	2
IGN	CH_4	38 °ca BTDC	−2
		10 °ca BTDC	2
	H_2	5 °ca BTDC	−2
		−15 °ca BTDC	2
	CH_3OH	25 °ca BTDC	−2
		5°ca BTDC	2
λ	CH_4	1.2	−2
		0.8	2
	H_2	2.2	−2
		1.4	2
	CH_3OH	1.3	−2
		0.7	2

For the measurements under fired operation, a Response Surface Method (RSM), more specifically a Central Composite Design (CCD), was used. Consequently, each factor was tested at 5 levels (coded as −2, −1, 0, 1 and 2). Since the fuel and measurement position are categorical factors, they were not included in the DoE design and the RSM was repeated for all the levels of these factors (three times for the different measurement positions and three times for the different fuels). The levels of IGN and λ are fuel dependent to cover the optimal and widest possible range for each fuel. A combined wide range of IGN and λ was not possible for hydrogen, due to the occurrence of abnormal combustion at the extreme combinations (e.g. an early ignition for rich mixtures). Consequently, λ has been varied at the lean side for hydrogen to cover the widest possible range for IGN. An overview of the extreme levels of the continuous factors is given in Table 2.

It was not possible to run the full factorial CCD (testing all the possible combinations) for all the measurement positions for a certain fuel, so a fractional factorial design was selected. No information was available in literature on which interactions could be neglected. Consequently, the full factorial experiment was run in one measurement position (P2) on methane to provide this information. The ANOVA (analysis of variance) results of that experiment showed that only interactions of IGN with the other factors were significant. To have a clean estimation of all the main effects and the interactions of IGN with CR, TP and λ, a fractional factorial design of half of all the possible cubical points (combinations of the factors at the levels −1 and 1) could be run. This fractional factorial design enabled the completion of an entire measurement set for a certain fuel on one day.

2.3 Data Reduction

The Nusselt, Reynolds and Prandtl number were calculated for a total of 18 points between 290 and 420 °ca (most important part of the cycle) for each of the cases to build the database for the model validation. Several properties are needed to determine those dimensionless numbers, which are described here. First, the bulk gas temperature needs to be calculated, in addition to the measurement of the heat flux and wall temperature, to determine a convection coefficient out of Eq. 1. That bulk gas temperature is derived from the measured in-cylinder pressure with a Three-Pressure-Analysis (TPA) in GT-Power [12]. For motored measurements, a uniform temperature is assumed throughout the combustion chamber. For fired operation, a two-zone gas temperature approach is used, where the in-cylinder volume is divided in an unburned and a burned zone to capture the effect of the propagating flame front. Furthermore, the gas properties need to be calculated for each zone. Polynomials for those gas properties of the pure components as a function of temperature are taken from the DIPPR database [13]. The properties of the unburned and burned mixtures during the engine cycle are calculated from the ones of the pure components with the proper mixing laws described in Ref. [14]. At a certain measurement position, the gas properties and gas temperature are switched from unburned to burned at the instant the propagating flame front arrives at the measurement position. In this paper, this is determined by the instant the derivative of the heat flux trace exceeds a certain threshold value.

3 Results and Discussion

The entire database was divided in six different flow regimes to be able to study the effect of the engine factors and gases in a systematic way without and with the presence of a propagating flame front:

1. compression stroke before ignition
2. compression stroke after ignition, in contact with the unburned zone
3. compression stroke after ignition, in contact with the burned zone
4. expansion stroke, in contact with the unburned zone
5. expansion stroke, in contact with the burned zone
6. expansion stroke after the end of combustion.

The measurements under motored operation belong to part 1 or 6. A regression analysis will be used to determine the parameters a, b and c in Eq. 2. The Prandtl number only differs significantly for a limited part of the dataset. Consequently, the ability of the Reynolds analogy to predict the heat flux for the different parts can be visualised by plotting the Nusselt number against the Reynolds number on a logarithmic scale. This is shown in Fig. 2 for the reference case of a constant characteristic length (engine bore) and velocity (mean piston speed), as chosen by

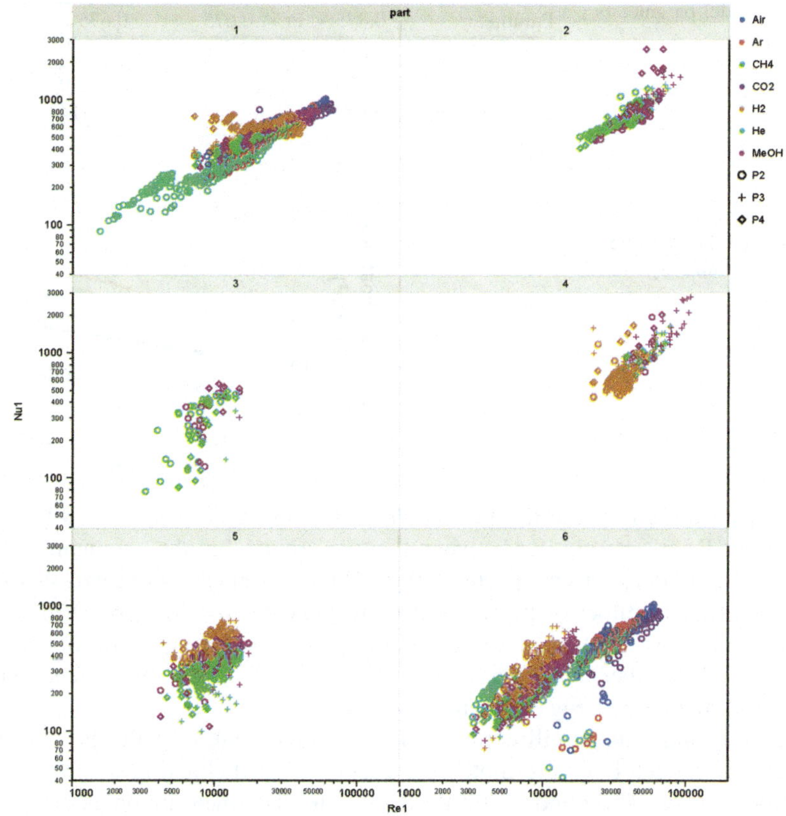

Fig. 2 The relation between the Nusselt and Reynolds number on a logarithmic scale for the six flow regimes

Annand [3]. The different gases and fuels are plotted in a different colour and different markers are used for the measurement positions. The graph shows that there is a strong linear relation between the two dimensionless numbers in the first and second part, indicating that the Reynolds analogy could be appropriate to model the heat flux within a spark ignition engine. However, the relation for regimes 3–5 is not linear at all in the reference case. Part 6 again shows a linear trend, but each case has a different offset.

First, it will be demonstrated that the Reynolds analogy accurately predicts the differences caused by the gas properties and engine effects for part 1 if a constant L and V are used (will be referred to as model 1). The usage of proper polynomials and mixing laws for the gas properties is a necessity. This is, however, not the case for some models in literature as mentioned above, e.g. Ref. [5], so these models even fail to predict the differences for flow regime 1. Furthermore, the regression analysis of the data showed that the effect of the Prandtl number differed depending on its level. Therefore, the model parameters have been determined for

Table 3 Model parameters for part 1

Parameter	0.475 < Pr < 0.675	0.675 < Pr < 0.74
a	2.67	1.71
b	0.573	0.549
c	0.872	−0.614

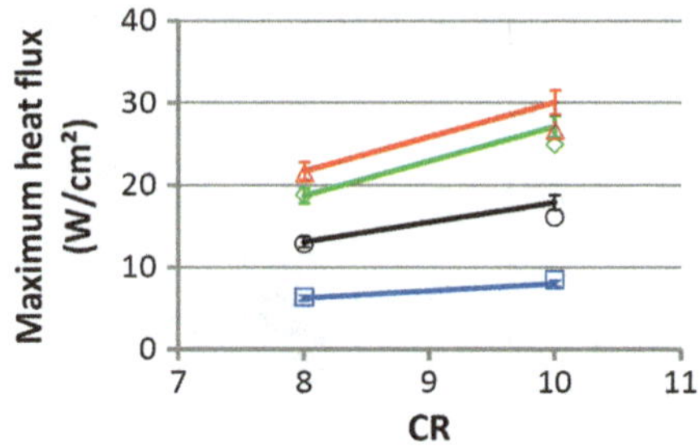

Fig. 3 The model does not accurately predict the effect of CR for helium, argon and air

two groups according to the level of the Prandtl number, as shown in Table 3. The exponent of Re is very similar between the two groups, but the exponent of Pr of the second group has an unexpected value. The data actually indicated that there was no significant effect of the Prandtl number between 0.675 and 0.74, so the regression analysis resulted in an artificial exponent which needs further investigation. The Pr number did show a clear effect for helium and hydrogen, causing the offset in the data in Fig. 2 for these gases.

The following figures will compare the measured peak heat fluxes with those predicted by model 1 for the motored measurements (belonging to part 1). The effect for each gas is plotted in a different colour. The measurements are plotted with a filled marker (connected with trend lines) and the model predictions with an unfilled marker. Error bars are added for the experimental uncertainty on the measured heat flux to indicate the accuracy of the model predictions. For the measurements with the inert gases, Figs. 3–5 show that only the effect of the compression ratio (Fig. 3) is not accurately predicted by model 1 (for air, argon and helium). The model does accurately predict the effect of the throttle position (Fig. 4) and the effect of an increasing addition of helium in the ingoing air flow (Fig. 5). However, it is noteworthy that the plotted error bars only show the experimental uncertainty in the measured heat flux (±5 %). The resulting error in the derived convection coefficient which is being modelled is significantly higher (±11 %). Consequently, if this is taken into account, the model accurately predicts all the effects within the experimental uncertainty. The same results are obtained for part 1 of the measurements under fired operation. Figure 6 compares the effect of TP and CR predicted by the model with that obtained out of the measurements at 320 °ca, being the last instant in the cycle belonging to part 1 for all the cases (earliest ignition for CH4 is 38°ca BTDC). The simulation results are slightly outside the experimental uncertainty on the measured heat flux, but this is negligible taking into account the remark about the actual uncertainty on the convection coefficient given above.

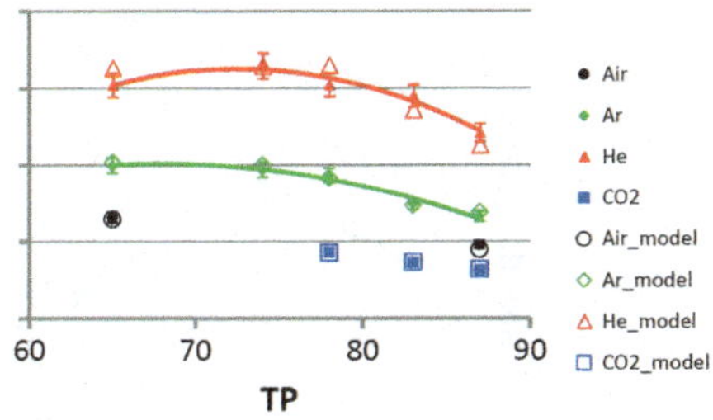

Fig. 4 The model accurately predicts the effect of the throttle position (65 = WOT, 87 = FCT)

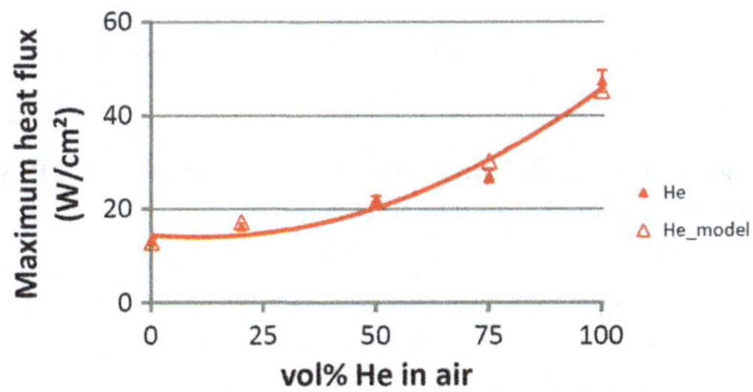

Fig. 5 The model is capable of predicting the increasing heat flux when the amount of helium is augmented

Figure 7 demonstrates that model 1 also accurately predicts the instantaneous heat flux within the cycle during the compression stroke for the motored measurements. However, the model fails to predict the heat flux during the expansion stroke (flow regime 6). The discontinuity between compression and expansion is caused by a different model fit for part 6 which is not presented here. Alternative lengths (2) and velocities (1) proposed in the literature have been explored and some improved the model fit. The first alternative length was suggested by Hohenberg et al. [12], being the diameter of a sphere with the same volume as the in-cylinder volume. The second one was the height above the piston, which was suggested by Chang et al. [15]. None of these lengths improved the data fit. The alternative velocity was suggested by Morel et al. [4] and included the effect of the turbulent kinetic energy. This model is referred to as model 2 and its predictions are added in Fig. 7. The results show that the fit is improved compared to model 1. However, the results are still not accurately enough.

The most important heat loss occurs during combustion and model 1 also fails in the simulation of the peak heat flux under fired operation for hydrogen and methanol. Figure 8 shows that satisfactory predictions are obtained for methane and that the effect of the engine factors (CR and TP are not shown here) is well predicted, except the effect of λ for hydrogen. However, model 1 predicts an average heat flux level for hydrogen and methanol which is too low. The alternative lengths and velocities presented above did not result in more accurate results and Fig. 9 indicates that this is caused because of the effect of the flame propagation which is not yet fully captured. Using a two-zone combustion model can only partially explain the initial increase in the heat flux at the instant the propagating flame front arrives at the measurement position. The second peak in the measured heat flux trace is being captured by the model because it is caused by

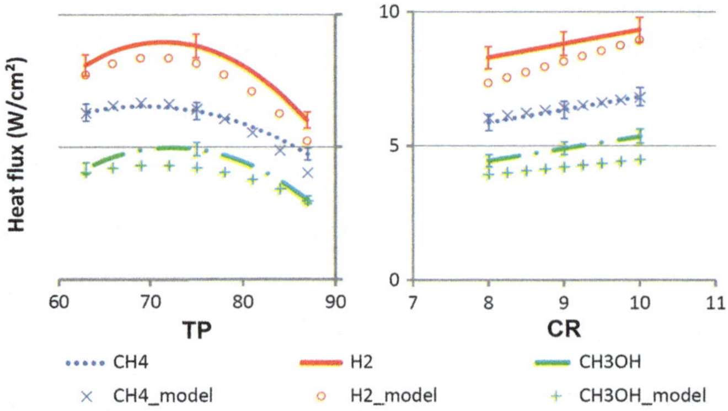

Fig. 6 The model predicts the correct effect of TP and CR during compression, but slightly outside the measurement uncertainty

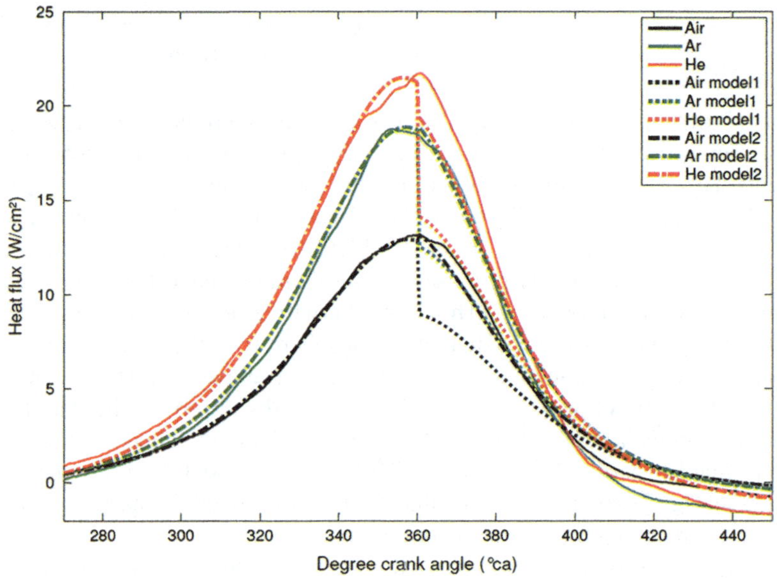

Fig. 7 Measured heat flux vs. model predictions under motored operation at P2 (CR = 8, WOT)

the peak in the burned gas temperature. The first peak must occur because the propagating flame front results in an intensified convective heat transfer or a momentarily larger gas temperature. The alternative lengths and velocities did not improve the simulation results, because they only change the level of the peak, but do not induce a second peak. Consequently, extra lengths and velocities will have to be investigated to improve the simulation results. Figure 9 further demonstrates that model 1 accurately simulates the heat flux during the compression stroke before and after ignition (part 1 and 2), but that no good agreement is obtained during the expansion stroke.

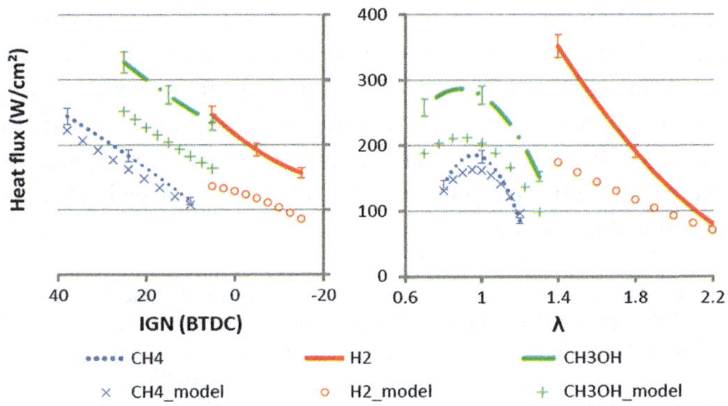

Fig. 8 The model fails to predict the correct peak heat flux under fired operation for hydrogen and methanol

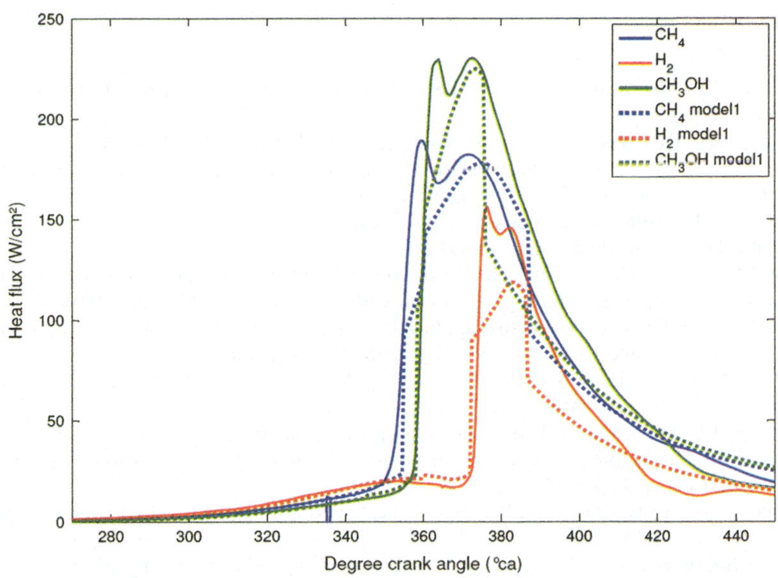

Fig. 9 Measured heat flux vs. model predictions for all the factors at level 0 at P2

4 Conclusion

This paper has presented the validation of a heat transfer model for spark ignition engines, based on the Reynolds analogy, with an experimental database that focuses on the effect of the gas properties. It was shown that the model accurately predicts the heat flux during the compression stroke, both for the measurements under motored and fired operation, if a constant characteristic length and velocity is used. Correct polynomials and mixing laws for the gas properties are indispensible keystone in the accurate results, which is not always the case in models in literature. The simulation results were not accurate in the other parts of the engine cycle. Alternative lengths and velocities proposed in literature also did not result in satisfactory results. Therefore, extra characteristic lengths and velocities will have to be explored, e.g. to better capture the intensified convective heat transfer caused by the propagating flame front.

Acknowledgments The authors of this paper like to acknowledge the technical assistance of Koen Chielens and Patrick De Pue. The research is carried out in the framework of a Ph.D. which is funded by a grant (SB-81139) of the Institute for the Promotion of Innovation through Science and Technology in Flanders (IWT-Vlaanderen). The experimental equipment is funded by a Research Grant (1.5.147.10 N) of the Research Foundation—Flanders (FWO). These financial supports are gratefully acknowledged.

References

1. Borman GL, Nishiwaki K (1987) Internal-combustion engine heat-transfer. Prog Energy Combust Sci 13(1):1–46
2. Demuynck J, De Paepe M, Sierens R, Vancoillie J, Verhelst S (2011) Literature review on the convective heat transfer measurements in spark ignition engines. In: 13th European automotive congress (EAEC 2011), paper EAEC2011_A46. Valencia, Spain
3. Annand WJD (1963) Heat transfer in the cylinders of reciprocating internal combustion engines. Proc Instn Mech Engrs 177(36):973–996
4. Morel T, Rackmil CI, Keribar R, Jennings MJ (1988) Model for heat transfer and combustion in spark ignited engines and its comparison with experiments. SAE paper 880198
5. Woschni G (1967) A universally applicable equation for the instantaneous heat transfer coefficient in the internal combustion engine. SAE paper 670931
6. Welty JR, Wicks CE, Wilson RE, Rorrer GL (2001) Fundamentals of momentum, heat and mass transfer, 4th edn. Wiley, New York
7. Demuynck J, De Paepe M, Huisseune H, Sierens R, Vancoillie J, Verhelst S (2011) On the applicability of empirical heat transfer models for hydrogen combustion engines. Int J Hydrogen Energy 36(1):975–984
8. Shudo T, Suzuki H (2002) Applicability of heat transfer equations to hydrogen combustion. JSAE Rev 23(3):303–308
9. Kleinschmidt W (1999) Instationäre Wärmeübertragung in Verbrennungsmotoren: Theorie, Berechnung und Vergleich mit Versuchsergebnissen. Fortschr.-Ber. VDI Reihe 12 nr. 383. Düsseldorf: VDI Verlag
10. Demuynck J, De Paepe M, Vancoillie J, Sileghem L, Verhelst S (2012) Applying design of experiments to determine the effect of gas properties on in-cylinder heat flux in a motored SI engine. SAE Int J Engines 5(3)
11. Berger PD (2002) Experimental design with applications in management, engineering, and the sciences. Duxbury, Belmont. ISBN 0-534-35822-5
12. GT-Power (2009) Engine performance application manual. Gamma Technologies
13. Rowley RL, Wilding WV, Oscarson JL, Yang Y, Zundel NA, Daubert TE, Danner RP (2003) DIPPR® data compilation of pure compound properties. Design Institute for Physical Properties, AIChE, New York

14. Reid RC, Prausnitz JM, Poling BE (1988) The properties of Gases and Liquids, 0–07-100284-7. McGraw-Hill Book Co, Singapore
15. Chang J, Guralp O, Filipi Z, Assanis D (2004) New heat transfer correlation for an HCCI engine derived from measurements of instantaneous surface heat flux. SAE paper 2004-01-2996

Performance Analysis of a Thermoelectric Generator Through Component in the Loop Simulation

Guangyu Dong, Richard Stobart, Anusha Wijewardane and Jing Li

Abstract As a low maintenance solid state device, the thermo-electric generator (TEG) provides an opportunity to recover energy from the exhaust gas directly. In this paper, dynamic behaviour of a TEG applied to the EGR path of a non-road diesel engine has been analysed. Through the component in the loop (CIL) method, the proposed TEG was simulated by a virtual model in real time. A Nonlinear Auto-regressive exogenous (NLARX) model was operated in an xPC as the virtual model. Based on calculated results of the model, the position of EGR valve and the coolant valve were adjusted to simulate the effect of TEG installation on the gas flow in EGR path. Analysis of the TEG performance under Non-road Transient Cycle (NRTC) was conducted base on above method. With above analysis, following conclusions can be drawn. First, the component in the loop simulation method demonstrates a better performance to predict the TEG dynamic behaviour comparing to the software based methods. Using component in the loop methodology, transient performance of the TEG device can be predicted with a satisfied error range. Besides of the TEG performance prediction, the effect of the installation of the TEG was also analysed. The temperature variation and the pressure drop which affected by TEG system were predicted, and a more accurate TEG performance prediction can be achieved accordingly.

Keywords Thermoelectric generator · Energy recovery · Exhaust gas recirculation · Component in the loop · Non-road transient cycle

F2012-A07-006

G. Dong (✉) · R. Stobart · A. Wijewardane · J. Li
Department of Aeronautical and Automotive Engineering, Loughborough University, Loughborough, LE11 3TU, UK
e-mail: G.Dong@lboro.ac.uk

SAE-China and FISITA (eds.), *Proceedings of the FISITA 2012 World Automotive Congress*, Lecture Notes in Electrical Engineering 190,
DOI: 10.1007/978-3-642-33750-5_28, © Springer-Verlag Berlin Heidelberg 2013

1 Introduction

The pursuit of improved fuel economy is becoming an increasingly important objective for automotive manufacturers. Concerning Internal Combustion Engines (ICE), approximately 30–40 % of the energy supplied by the fuel is rejected to the environment through exhaust gas [1]. Thus, there is a possibility for further significant improvement of ICE efficiency with the utilization of exhaust gas energy and its conversion to mechanical energy or electrical energy. In this context, TEGs have been identified as a reliable solid state technology for power generation. TEGs have many advantages in comparison to other thermal energy recovery methods i.e., no moving parts, produce no noise and vibration, low maintenance, environmentally friendly and low quality thermal energy directly convert into high quality electrical energy. Because of these properties, TEGs have been used in military applications and in deep space exploration missions (as RTGs -Radioisotope Thermo-electric Generators) by NASA [2].

For automotive industries, since diesel engines are the most efficient power train solution, to achieve the waste heat recovery by thermoelectric on diesel engine appears to be a promising solution to improve the efficiency. For this purpose, TEGs have been studied for years. The reports from Hi-Z projects initially demonstrated that the efficiency of TEGs tested by most of the past automotive projects were fairly low. But Saqr et al. provided a basis for an efficient thermoelectric generator in 2008 [3]. Recently the performance of TEG applied on a truck has been implemented to 1 kW, which greatly increased the possibility of applying the TEG on mass productive engines [4]. Now the TEG for automobiles constructed with skutterudites could increase the efficiency of diesel engine up to 5–10 %, depending on engine operating points [5]. Moreover, a lot of automobile manufacturers have already worked on TEG architectures design and optimization, such kind of activities should promote the application of TEG on large scale.

Based on above reports, it can be deduced that the advantages of a TEG device can only be fully explored if the device is placed in the context of its application in a working engine. The engine environment is particularly demanding, and given the experimental nature of TE devices, initial evaluation away from the engine permits a prediction of performance before robust hardware is deployed. In this paper, the performance and dynamic behaviour of a TEG applied to the exhaust gas re-circulation path of a non-road diesel engine has been analysed. To achieve such a goal, a component in the loop methodology was developed accordingly. CIL methods have been established for hybrid vehicle engineering [6], but not so far in engine sub-system research. Through the CIL method, the proposed TEG was simulated by a virtual model in real time. A nonlinear ARX model was operated in an xPC as the virtual model. Based on calculated results of the model, the position of EGR valve and the vane of the turbine were adjusted to simulate the effect of TEG installation on the gas flow in EGR path. Data developed from bench tests is used in a real time model running on a fast computer connected to the engine control system. The device outputs (the voltage and current of an electricity

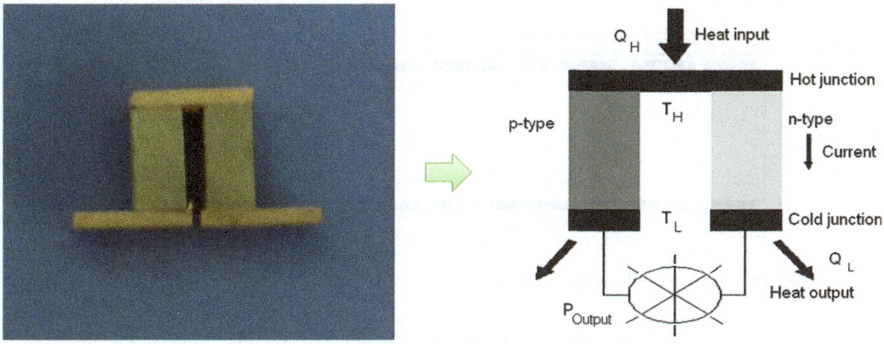

Fig. 1 Diagram of thermoelectric generation process

supply for example) are calculated and their values injected back into the experimental hardware as if the device were in situ.

In this paper, analysis of the TEG performance under Non-road Transient Cycle (NRTC) was conducted base on above method. Both availability of the exhaust gas in EGR path and power output of the TEG were predicted in real time. Besides, effect of the TEG installation on temperature, pressure and mass flow of the exhaust gas were estimated during the whole NRTC cycle. Results show that the dynamic behaviour of the TEG device can be well predicted through a component in the loop methodology, and a NLARX model of the TEG demonstrates good response ability under such a transient cycle. Comparing to the simulation results through 1-D GT-power model, CIL simulation results are more accurate under varied operation conditions and the effect of thermal mass can be reflected clearly as well. However, a small error still exists due to limitation of the trade off relationship of real time ability and modelling accuracy.

2 Fundmentals of Thermoelectric Generator

2.1 Thermoelectric Theory

TE materials fall into two categories, n and p whose behaviour differs in the form of the electrical conduction mechanism, as shown in Fig. 1. By combining these two TE materials with two junctions across a temperature difference, a potential difference can be developed. The "Seebeck Effect", named after the physicist who first noted the effect is the foundation of TE device behaviour [7].

A TE converter is a heat engine and like all other heat engines it is subject to the laws of thermodynamics. Efficiency of the TEG is given by,

$$\eta_{TEG} = \frac{P_{out}}{Q_H} \tag{1}$$

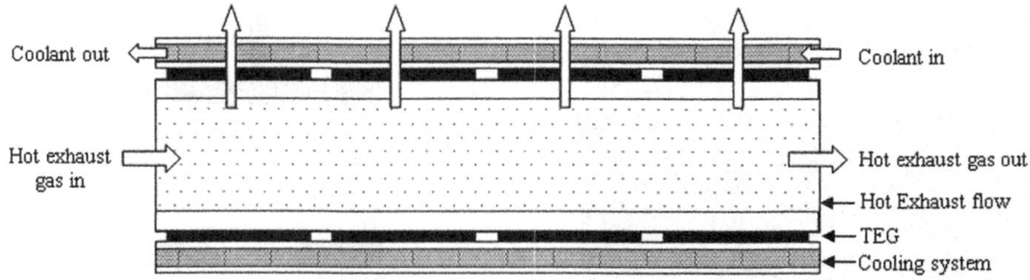

Fig. 2 Diagram of the bench-mark TEG model

where Poutput is energy supplied to the load, QH is heat absorbed at the hot junction. Conversion efficiency of a TEM depends on many parameters such as; electrical and thermal resistances of the TE materials, Seebeck and Peltier coefficient of the junctions, length and area of TE element, TH and TC and internal resistance of the TEG. The maximum conversion efficiency of a TEM is expressed as,

$$\eta_{conversion} = \left[\frac{\sqrt{1+ZT}-1}{\sqrt{1+ZT}+\left(\frac{T_C}{T_H}\right)}\right]\left[\frac{T_h-T_c}{T_h}\right] \tag{2}$$

where ZT is thermo-electric figure of merit, Tc and TH are cold and hot side temperatures of a TEM respectively. The conversion efficiency increases with the figure-of-merit of the material. The power, on the other hand, always increases with ΔT; therefore, to maximize the power production, the hot junction temperatures should be as high as possible and the cold junction temperature as low as possible, subjected to material limitations [8]. Hence, the performance of a TE device for a given material and heat exchanger will mainly decided by the gas mass flow rate and a temperature of exhaust gases on diesel engines.

2.2 Heat Exchanger Design

Besides of the thermoelectric materials, the heat exchanger design also plays a vital role in thermoelectric generation because of the electricity generation is directly related to the heat transfer rate between hot and cold sides of the TEG. Capturing thermal energy from a low quality heat source, such as exhaust waste heat of an automobile engine is really challenging. Therefore it's a necessity to design a high efficient and high performance heat exchanger to recover more energy out of the exhaust flow. In this section, the heat transfer performance of the heat exchanger which is used for the validation of CIL method is analyzed. The diagram of a proposed bench-mark TEG, which has corrugated open pipe architecture, was shown in Fig. 2. From the figure it can be seen that 8 TE modules can be sandwiched between the exhaust pipe and the TEG coolant pipe. Thus, heat

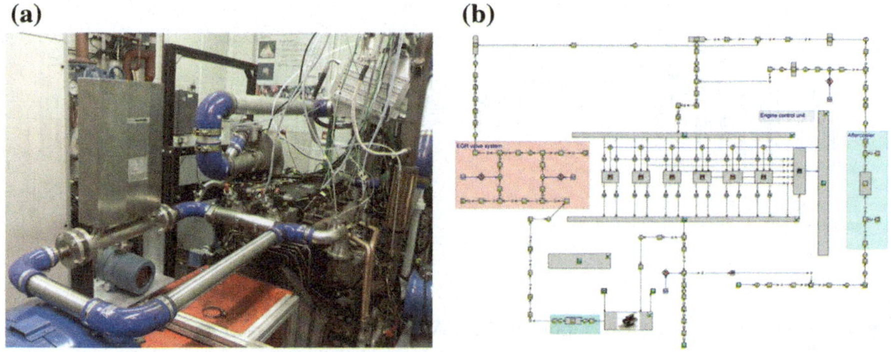

Fig. 3 Test bench setup and 1-D caterpillar engine model. **a** Engine test bench. **b** Engine 1-D GT-Power model

transfer in the TEG involves, "convection" in hot & cold fluids and the "conduction" across the separating walls. The ceramic wafers which have high heat transfer coefficient are used in both sides of the TEMs to electrically insulate the TEG system. Then the heat transfer across the TEG and the temperature difference across the TEMs, entirely depend on the thermo-physical properties of the TEG material and the hot and cold fluids.

3 Engine Test Bench and Modeling Works

In this paper a 6 cylinder Caterpillar heavy-duty off-highway engine is used for the component in the loop simulation of TEG system. Also, a 1-D TEG model which achieved in GT-Power environment was built up to achieve an initial analysis of the TEG dynamic behaviours. For the validation of the simulation results, the bench-mark TE generator was installed in the EGR path of the engine as shown in Fig. 3a. The engine calibration used in this work produces up to 159 kW at rated speed (2300 rpm) with peak torque of 920 Nm at 1400 rpm. A high pressure loop EGR system and a variable geometry turbine system were calibrated for experimental purposes.

Figure 3b is the schematic diagram of the engine model programmed in GT-Power environment. As a one dimensional model, the configuration parameters of the structure are all measured from the real engine or from the engine datasheet. For achieving the analysis of the TEG performance installed in the EGR path, a detailed TEG model, including 4 TE modules, heat exchanger and cooling system was built up, and then was integrated with the engine model. With above modelling work, the dynamic characteristics and the transient behaviour of the engine can be simulated. And the analysis of the operating conditions and performance of TEG can be achieved as well.

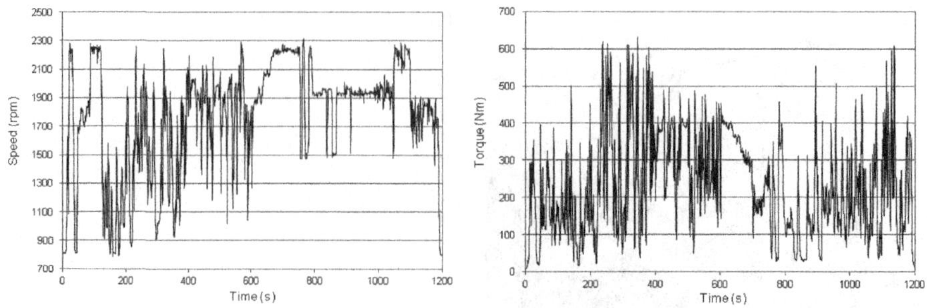

Fig. 4 Speed and Torque distribution of the CAT C6.6 L engine over the non-road transient cycle (NRTC)

4 Cil Simulation of the Thermoelectric Generator

4.1 Limitations of Current Software Based TEG Simulation Methods

As mentioned above, the performance of a proposed TEG system is strongly depend on a serial of design issues like heat exchanger design, TE material selection and the integration of the TEM into its working environment. However, due to the development progress it is hardly to test all of these aspects at the same stage. Also it is a difficult task to test each design of the TEG subsystem through physical experiments. Hence, several methods were developed to predict the performance of TEG which applied for the engine exhaust heat recovery [9, 10]. Among of these methods, computational fluid dynamics (CFD) proved to be a solution which can predict the heat transfer and power regenerating process accurately [11]. However, the time consumption of the CFD analysis makes it is hardly to be use for dealing with the TEG behaviours under unsteady states. To predict the transient performance of TE generator, one dimensional analysis provides an opportunity without the huge time consumption. GT-Power is a 1-D engine simulation tool which can be used for this purpose. However, the accuracy of the TEG power output prediction is stilled limited under unsteady operation conditions. Figure 4 shows the operating points of the test engine under NRTC cycle. The NRTC test is a transient driving cycle for mobile non-road diesel engines developed by the US EPA in cooperation with the authorities in the European Union (EU). Through the load and speed varying profile it can be deduced that the NRTC cycle is a highly transient test cycle which represents the typical off-road operating conditions of the heavy-duty engines. For Caterpillar C6.6 engine, the operating speed is in the range of 800–2,300 r/min, and the load distribution covers a wide range from 0 (idle stage) to 700 N.m under the test cycle.

Figure 5a shows the 1-D modelling results of the bench-marked TE generator in GT-Power environment. From the figure it can be seen that the the variation

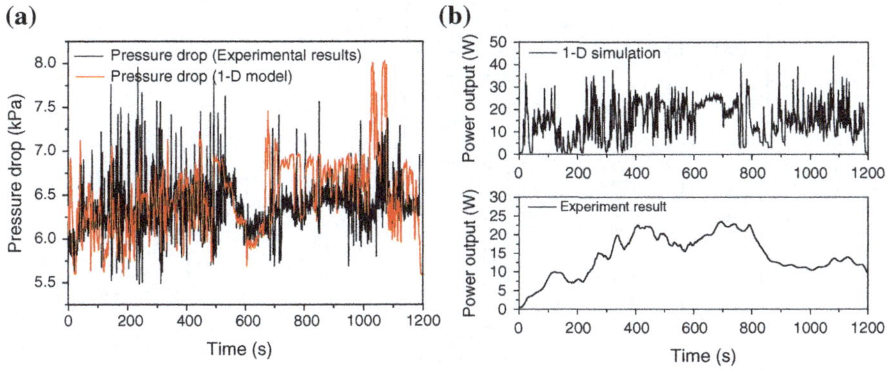

Fig. 5 Comparison between simulation and experimental results. **a** Pressure drop of the TEG. **b** Power output

tendency of calculated pressure drop correlated well with the experimental results. The root mean square error is less than 1.17 kPa. So the GT-Power model exhibits a good performance when it is applied to predict the dynamic behavior of the exhaust gas flow in EGR path. However, the modeling result about the TEG power output is very limited. Figure 5b shows both modelling result and experimental results of the TEG power output, and an obvious difference does exist between them. Such a difference is mainly because the effect of the TEG's thermal mass cannot be fully simulated by the model. Based on above analysis, the software based simulation methods are limited for the prediction of TEG performance under transient operating conditions. And then the component in the loop simulation method which supported by the experimental data will be studied in this paper.

4.2 Concept of Component in the Loop Simulation

As a rapid prototyping methodology, CIL provides a good opportunity for TEG system optimization and will reduce development time. Figure 6 illustrates the CIL experimental system for TEG applied to the EGR path of the C6.6 engine. A real time TEG model was constructed using XPC® technology. The model can be used to predict the power output, the outlet temperature and pressure of a proposed TEG. On the other hand, the effect of the TEG system on the EGR path can be simulated through the control of the temperature, pressure and mass flow of the EGR path. When the pressure and temperature drop caused by TEG was predicted by the real time model, the position of EGR valve and the coolant valve were adjusted based on calculated results of the model. Then the effect of TEG installation on the gas flow in EGR path was simulated.

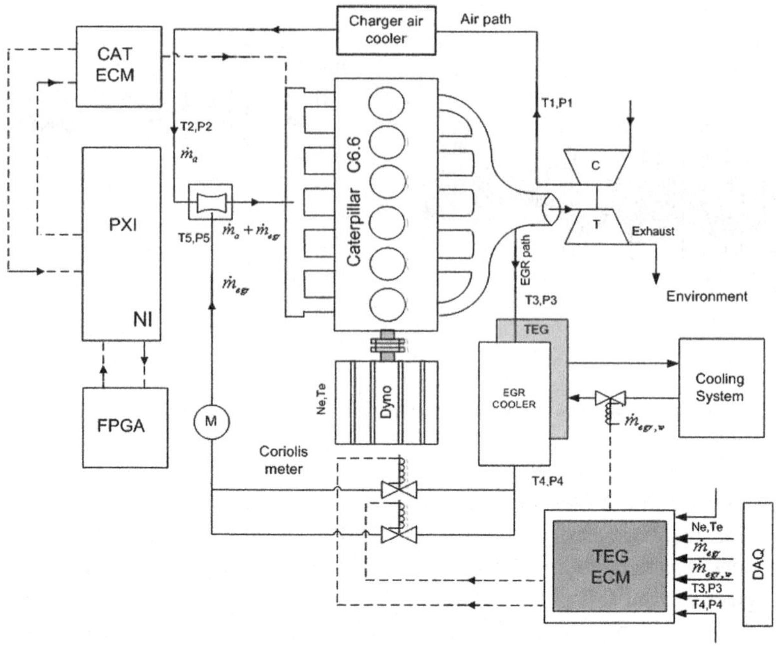

Fig. 6 Component in the loop design for TEG system applying in EGR path

4.3 Structure of the Real Time TEG Model

Figure 7 shows the structure of the real time TEG model. The engine operating parameters like torque, speed, exhaust temperature and pressure, EGR mass flow and VGT position, were collected by the NI PCI-6024 DAQ card and used as the model inputs. Through the normalization process these parameters were transferred into the non-linear ARX neuron network based TEG model. The predicted TEG voltage and power output were recorded by the xPC system and transferred to the host machine. Also, the predicted pressure and temperature drop which caused by the installation of TEG were used as the reference values to adjust the EGR valve, VGT vane and coolant valve position. Based on such values, the commands will sent to these actuators through the CAN communication system.

The real time neural network (NN) model of the TEG was shown in Fig. 8. As mentioned above, the aim of the model is to predict the TEG performance and adjust the EGR cooler to perform as a TE device. NLARX neural network model proved to be the best candidate for predict real time simulation of the TEG in terms of accuracy and execution time for the model. The figure shows the NLARX structure. It was selected for the current work because of its means of representing dynamic systems [12]. The output of the model is fed back to the input layer and current input data is applied with a comparison of previous input. To train the NN

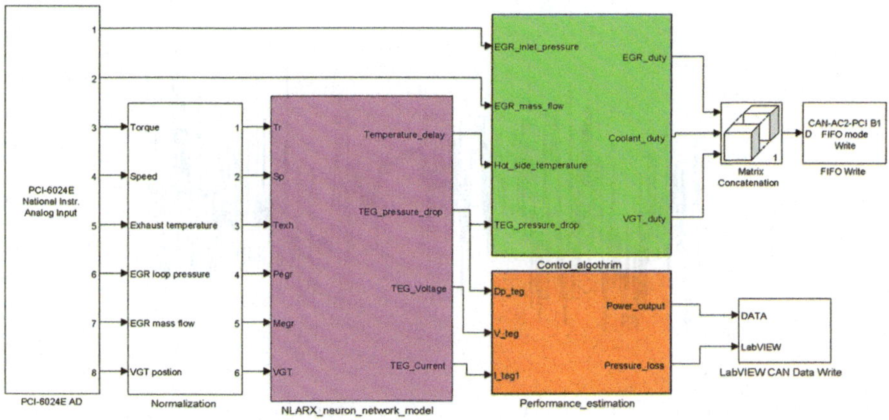

Fig. 7 Structure of the real time TEG model

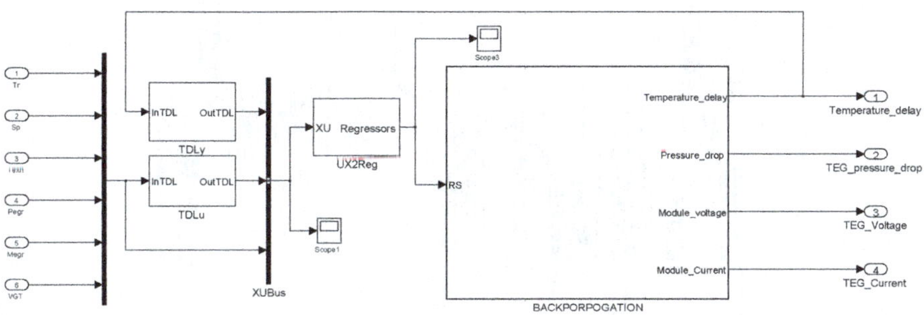

Fig. 8 The structure of non-linear ARX model

model, a series of tests of the bench-mark TE generator were carried out to provide the real boundary conditions and the validation data of the model.

5 Results and Discussions of Cil Based Teg Simulation

Under the NRTC cycle, the variation of the temperature and mass flow were shown in Fig. 9a and b. Figure 9a is the measured temperature at the inlet of the EGR cooler. Here it can be seen that the temperature varied within a range of 400–800 K during the whole cycle. Meanwhile the coolant temperature was also varied according to the operation conditions. Figure 9b illustrates the variation of mass flow rate during the whole test cycle. Such a variation of the mass flow is mainly dominated by the EGR valve control algorithm. The temperature and mass flow of EGR path are dominating factors which decide the availability of the energy in the EGR flow.

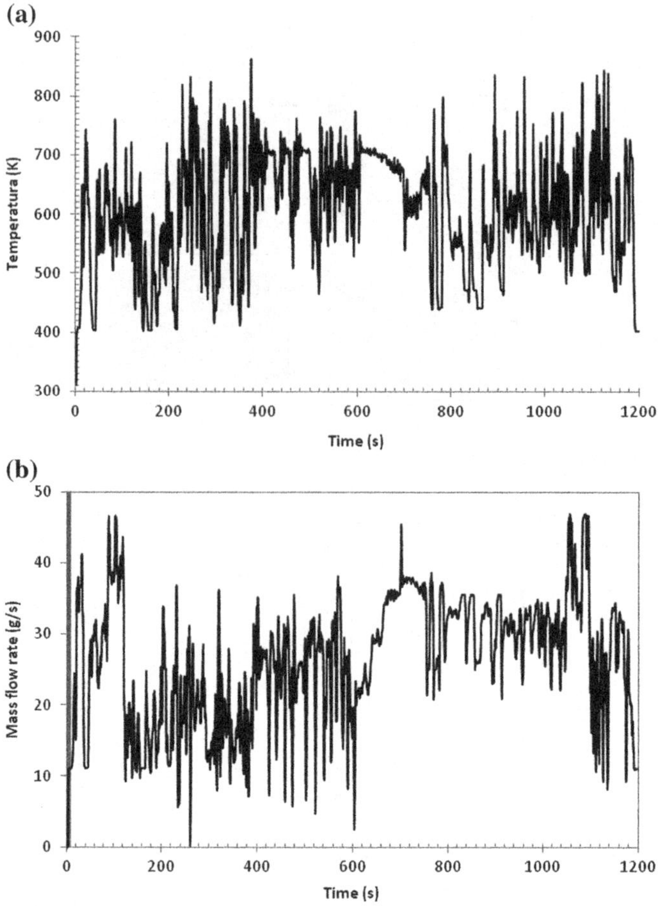

Fig. 9 Temperature and mass flow varying history under NRTC cycle

The modelling results about the TEG power output under NRTC cycle was shown in Fig. 10. Here 4 TE modules and 4 dummy modules were installed into the TE generator. As mentioned above, for achieving accurate predicting results, the effect of the TEG's thermal mass should be carefully predicted by the model. For the NLARX model, the linear input delay can be used for this purpose. On the other hand, as a discrete system, the reducing of the sampling rate will also increase the right of the previous state in the whole input values. However, the sampling rate cannot be too low because of the consideration of the system response. From Fig. 10a and b, it can be seen that when the input delay layer increased from 1 to 3, and the sampling rate reduced from 5–1 HZ, the deviation between the model predicting result and experimental result was contained and a better power output estimation was achieved.

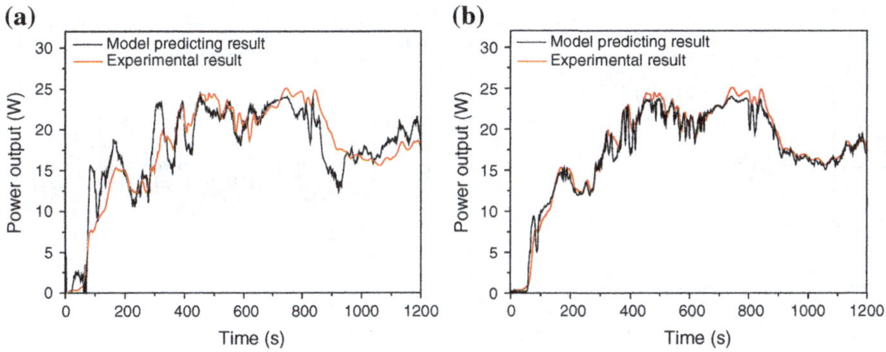

Fig. 10 Effect of NN model structure and sampling rate on simulation results. **a** Input delay layer:1 Sampling rate: 5 HZ. **b** Input delay layer:3 Sampling rate: 1 HZ

Besides of the TEG performance prediction, the effect of the installation of the TEG was also need to be simulated. When the TEG is installed into the EGR path, the temperature and pressure drop of the EGR path were changed, such changes should be evaluated and they will effect on the accuracy of the predicted TEG power output. Thus, for the component in the loop simulation process, the outlet temperature of the EGR cooler and the pressure drop of the EGR path should be controlled as a real TEG installed there. Figure 11 shows the xPC command based on the predicted TEG behaviours. The cooler outlet temperature control can be seen in Fig. 11a, a closed loop control was applied here since the response of the coolant system is not fast. Figure 11b demonstrates the control results of the pressure drop through the adjust EGR valve duty. It can be seen that based on the predicted pressure drop caused by the TEG, the pressure drop of the EGR path was adjusted higher than the case which control by original engine ECU. Thus, through the CIL simulation, the TEG's effect on the EGR path can be studied. And, based on the predicted parameters of the EGR path, a more accurate TEG performance prediction can be achieved as well.

Figure 12 shows the predicted TEG power outputs which achieved by CIL simulation. Figure 12a shows the performance of the TEG with total 8 TE modules under a 100 % NRTC cycle. It can be seen that the model predicted value correlates well with the experimental result. The averaged power out of the total 8 TEG modules is 32.7 W during the test cycle. The root mean square error (RMSE) of TEG power output was 2.14 W, which represented an error of less than 7 %. Moreover, the same TEG was also tested under a 60 % NRTC cycle (the load was decreased to 60 % of its original value). As shown in Fig. 12b, the predicted TEG power also correlates well with the test value, the average power output is 17.7 W, and the RMSE error is 0.87 W which represented an error of less than 5 %.

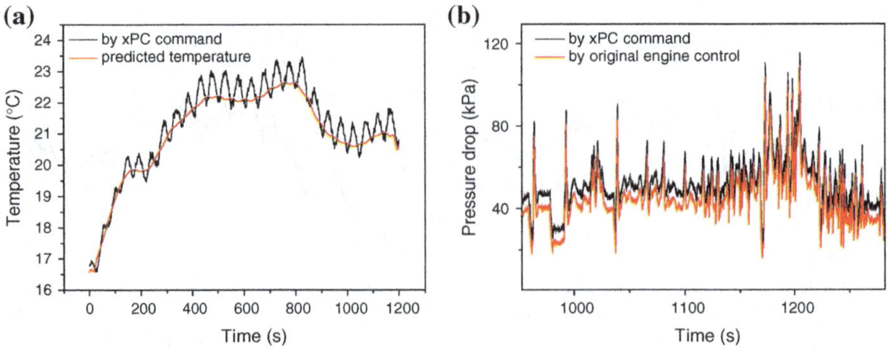

Fig. 11 xPC command based on the predicted TEG behaviours. **a** Coolant valve control. **b** EGR valve control

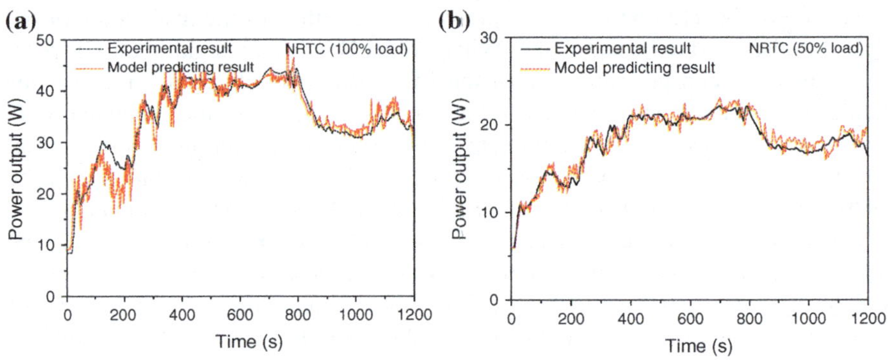

Fig. 12 Power output of TEG under different transient conditions

6 Conclusions

In this paper, dynamic behaviour of a TEG has been analysed through the component in the loop (CIL) method. The proposed TEG was simulated by a virtual model in real time, and the model was validated by the experimental data. Through the analysis of the TEG transient behaviours and the model structure, following conclusions can be drawn:

Comparing to the software based simulation; the component in the loop simulation method demonstrates a better performance to predict the dynamic behaviour of the TE generator.

The power output of the bench-mark TEG was well simulated by a NLARX neuron network model which was operated in an xPC. Increase input delay layer or reducing the sampling rate proved to be efficient to overcome the effect of the TEG's thermal mass and improve the model prediction accuracy.

Comparing to the experimental results, the error of the predicted power output was less than 7 % under the 100 % NRTC drive, and less than 5 % under 60 % NRTC cycle. Thus both the transient and averaged TEG power output can be accurately predicted by the model.

Besides of the TEG performance prediction, the effect of the installation of the TEG was also analysed. The temperature variation and the pressure drop which affected by TEG system were predicted and a more accurate TEG performance prediction can be achieved accordingly.

References

1. Richard S, Milner D (2009) The potential for thermoelectric regeneration of energy in vehicles. SAE paper 2009-01-1333
2. Robert DA (2006) Space missions and applications. Jet Propulsion Laboratory, California Institute of Technology
3. Saqr KM, Mansour MK, Musa MN (2008) Thermal design of automobile exhaust based thermoelectric generators: Objectives and challenges. Int J Automot Technol 9:155
4. Bass JC, Elsner NB, Leavitt FA (1994) In: Proceedings of the 13th international conference on thermoelectrics. AIP Kansas City
5. Hussain Q, Brigham D, Maranville C (2009) Thermoelectric exhaust heat recovery for hybrid vehicles SAE technical paper. doi:10.4271/2009-01-1327
6. Filipi Z et al. (2006) Engine-in-the-loop testing for evaluating hybrid propulsion concepts and transient emissions—HMMWV case study. SAE paper 2006-01-0443
7. Goldsmid HJ (1960) Principles of thermo-electric devices. Br J Appl Phys 11:209–217
8. Dresselhaus MS et al (2007) New direction for low-dimensional thermo-electric materials. Adv Mater 19:1–12
9. Stobart R, Wijewardane A (2010) The potential for thermoelectric devices in passenger vehicle applications. SAE paper, 2010-01-0833
10. Stobart R, Wijewardane A (2011) Exhaust system heat exchanger design for thermal energy recovery in passenger vehicle applications. VTMS
11. Baker C, Shi L (2012) Experimental and modelling study of a heat exchanger concept for thermoelectric waste heat recovery from diesel exhaust. SAE paper 2012-01-0411
12. Deng J, Maass B, Stobart R, Winward E, Yang Z (2011) Accurate and continuous fuel flow rate measurement prediction for real time application. SAE Int J Engines 4(1):1724–1737

Analysis and Simulation of Hybrid Electric Turbocharger and Application on ICE and HEV

Feng Tian, Guofeng Ren, Shumei Zhang and Lin Yang

Abstract Hybrid electric turbocharger (HET) has been proven to be an effective pathway to eliminate turbo-lag and improve ICE fuel economy. But HET is not appropriate for traditional vehicles. It is because that the low voltage electrical system (12 V) of these vehicles is not capable of handling the electrical current in excess of 150 A. While HEV has achieved considerable market share in recent years, this provide more flexibility for the application of HET, particularly as while energy recovered by HET can effectively applied on HEVs. The improvement of fuel economy by the combined HET and HEV is investigated in this chapter. For the purpose of exhaust energy recovery, the relevant exhaust gas temperature and availability from a Higer hybrid bus were analysed. An exhaust energy recover model based on the availability balance method is developed to evaluate the potential energy which can be recovered in a practical hybrid implementation. Experimental ICE data are used as inputs for the model. The ICE with HET was modelled and validated by experimental data. And the steady-state fuel benefits of the HET were studied. Finally the engine model was implemented to HEV model in GT-DRIVE to predict the potential of the HET benefit for fuel economy benefits in UDDS driving cycle. The exhaust availability calculation results show that 10 ~ 15 % of the exhaust energy could be recovered, which is slightly higher than the prototype vehicle. This may be caused by the constant load conditions for the ICE in the HEV, and this is a potential advantage for the implementation of HET on HEV. The low-speed steady-state characteristic of ICE is improved by increasing volumetric efficiency. The transient simulation shows

F2012-A07-008

F. Tian (✉) · G. Ren · S. Zhang · L. Yang
School of Mechanical Engineering, Shanghai Jiaotong University, 800 Dongchuan Road, Shanghai, People's Republic of China
e-mail: toyota_benny@sjtu.edu.cn

SAE-China and FISITA (eds.), *Proceedings of the FISITA 2012 World Automotive Congress*, Lecture Notes in Electrical Engineering 190,
DOI: 10.1007/978-3-642-33750-5_29,

that HET could effectively improve ICE turbo-lag. The simulation under driving cycle indicates that combined HET/HEV technology could reduce fuel consumption by 25 %, which is slightly higher than middle-hybrid vehicle. Since the ultra-high speed motor is not commercialized, combined HEV and HET could only be applied on commercial vehicles because of the relatively low turbocharger speed required. Because of the high exhaust temperature, cooling system design of the high speed motor is very important and is now under investigation. Combined HET/HEV technology could significantly improve fuel economy without extra weight and cost. This novel configuration also provides new flexible energy utilization method. The HEV diesel engine's exhaust temperature and availability characteristics were studied. HET is a more applicable technology on HEVs than traditional vehicles due to the high voltage of the on-board electric system and HEV engine working characteristic. Simulated results show that HET could effectively balance engine working by improving engine low speed torque output and high speed exhaust heat recovery. Combined HET/HEV technology could significantly improve ICE fuel consumption. Simulation results show that this novel configuration could improve ICE fuel consumption by 25 % during UDDS driving cycle.

Keywords Exhaust energy · Hybrid electric turbocharger · Downsize · Mild-HEV · Fuel economy

1 Introduction

Downsizing was approved to be an effective approach to achieve better fuel economy and exhaust emission performance for internal combustion engines [1, 2]. Turbo charging and electric hybrid are the most common technical methods for engine downsizing. Using turbocharger could take partial use of the exhaust energy to simultaneously increase intake air pressure and decrease engine displace volume (downsizing usually by reduction of cylinder numbers). However, because of the working characteristics of the turbocharger, turbo-downsized engine could not achieve satisfactory performance under low speed and transient conditions, called 'turbo-lag'. Hybrid electric turbocharger (HET) has been proven to be an effective approach to eliminate 'turbo-lag' of the turbo-downsized engine during low speed and transient conditions [3–6].

Analysis of hybrid technology is used to show that much of the fuel economy benefit derived by hybrid electric vehicles (HEV) is due to engine downsizing [7]. Other HEV control strategies, like start/stop and fuel cut-off during coasting/deceleration, also contribute to fuel economy improvement. For hybrid with higher degree of hybridization, in order to compensate for engine torque deficit due to downsizing, bigger electric motor and traction battery must be adopt. However, research results by Shahed [8] show that the recovered brake energy is almost used up by the extra weight of traction motor and battery pack.

It is proposed that combined mild-HEV and HET technology could achieve equivalent fuel economy and emission benefits without extra weight and cost. Engine torque curving and engine downsizing could be achieved by HET configuration. And the mild-HEV configuration could be used for fuel cut-off during idling, coasting and decelerating conditions and auto start/stop.

In the following sections, the exhaust energy characteristics of diesel engine will be firstly analysed using the first- and second-law of thermodynamics. And the application of HET on diesel engine will be discussed. Moreover, the cost benefit of the combination of mild-HEV and HET is compared with a middle-hybrid configuration.

2 Exhaust Energy Analysis

The first law of thermodynamics has been used to model the engine processes for many years. But in order to get better insight into engine operation, the availability analysis using the second law of thermodynamics must be applied [8].

The engine is selected as a two-inlet, single-exit control volume. And the combustion and the exhaust gas each forms ideal gas mixtures. The potential energy of the intake air and exhaust gas flow and kinetic energy of the intake air flow are ignored. According to the first law of thermodynamics, the energy rate balance for the control volume at steady state can be written as:

$$\dot{E}_{ex} = \dot{Q}_{cv} - P_{brake} - P_f + \dot{m}_{fuel}LHV + \dot{m}_{air}h_{air} \tag{1}$$

where $\dot{E}_{ex}$ is the exhaust power, $\dot{Q}_{cv}$ is the heat loss rate from engine cylinders, P_{brake} is the brake power, P_f is the friction power, $\dot{m}_{fuel}$ is the fuel injection flow rate, *LHV* is the low heat value of the diesel, $\dot{m}_{air}$ is the intake air flow rate, h_{air} is the specific enthalpy of the intake air flow.

According to the second law of thermodynamics, the availability of exhaust at steady state can be expressed as:

$$\dot{A}_{ex} = \dot{m}_{ex}\left(h(T) - h(T_0) - T_0\left(s^0(T) - s_0^0(T_0) - R_g ln\frac{p}{p_0}\right)\right) \tag{2}$$

where $\dot{A}_{ex}$ is the exhaust availability, $\dot{m}_{ex}$ is the exhaust gas flow rate, h_i is the specific enthalpy, s^0 is the specific entropy of exhaust at datum state, p_0 is the ambient pressure, p is the exhaust pressure.

By using the above mentioned energy and availability analysis methods and experimental data of ICE, the specific energy and availability characteristics of the diesel engine at full load were analysed. The results are shown in Fig. 1. The specific availability of the exhaust heat increases with its temperature. The specific

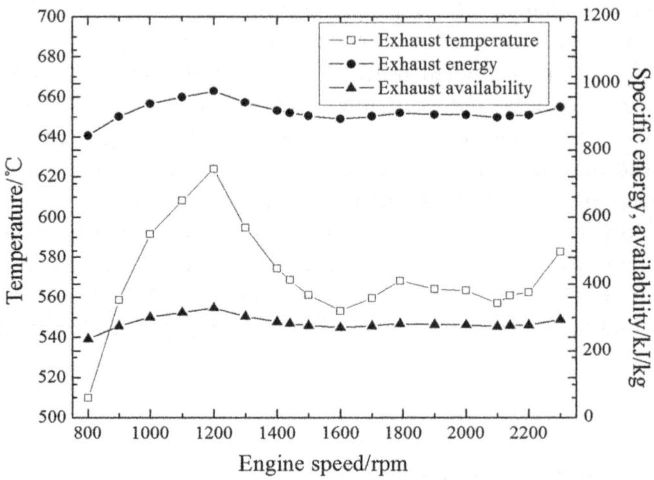

Fig. 1 Temperature, specific energy and availability of the diesel engine at full load

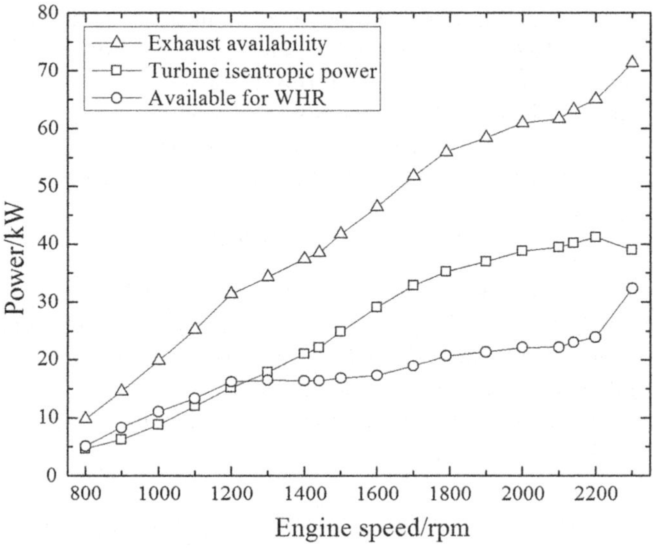

Fig. 2 Availability for waste heat recovery at full load

availability of the exhaust if about 30 % of its energy value at exhaust temperature above 600 °C. The power consumed by the turbine must be taken into consideration for the sake of minimize the modification on the base engine operation. Figure 2 shows the characteristics of the exhaust availability, turbine isentropic

Table 1 Base engine specifications

Engine type	In-line 4 cylinders, common rail diesel
Aspiration	Turbocharger
Displacement	5.13 L
Bore × Stroke	110 × 135 mm
Compression ratio	17.24
Rated power	150 kW@2300 rpm
Rated torque	780 Nm@1300 ~ 1700 rpm

power and surplus power for waste heat recovery (WHR) for various speeds at full load. If 50 % of the surplus power in the exhaust can be recovered, the overall engine efficiency can be improved by 5 % at least.

3 Application of HET on Diesel ICE

A four cylinder diesel ICE from FAW and HET were modelled using GT-Suite software. The target engine is a mid-duty diesel engine with a common rail fuel injection system and turbocharger. The main features of the target engine are listed in Table 1. The system architecture of the combined HET and HEV powertrain is presented in Fig. 3. In this configuration, an electric motor/generator is incorporated on the same shaft of the turbocharger. At low engine speeds and tip-in conditions, the electric motor/generator can work as a motor to speed up the turbocharger. So the torque deficit, always called 'turbo-lag', can be overcome. At high engine speeds and tip-out conditions, when there is surplus exhaust energy, power is generated through the motor/generator and can be used to supplement the integrated starter generator (ISG) or charge the battery.

The engine model was calibrated using steady-state experimental data collected on an eddy current dynamometer at various speed-load operating points. The fuel injection quantity and timing were determined using look-up tables based on engine speeds and loads. The target values and controlling parameters of PID controller for fuelling and boost pressure were also determined by look-up tables. The unpredicted direct-injection Wiebe model was used as the combustion model. The parameters of Wiebe model were calibrated using cumulative heat release profiles obtained by in-cylinder pressure measurements.

The steady-state simulation results of the HET technology, which was applied on the base diesel engine, are shown in Fig. 4. As can be seen from this figure, the steady-state power generated/consumed by the HET is mostly affected by engine speed. The ESC weighted average steady-state power is 2.84 kW. The ESC weighted steady-state BSFC improvement is 3.2 %. The BSFC improvement was calculated using the following equation:

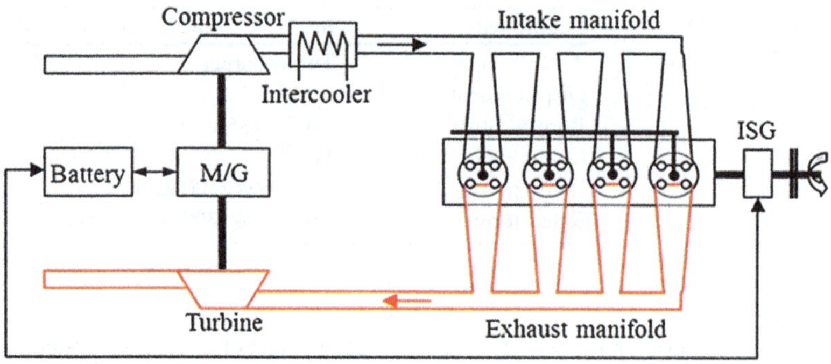

Fig. 3 System architecture of the combined HET and HEV technology

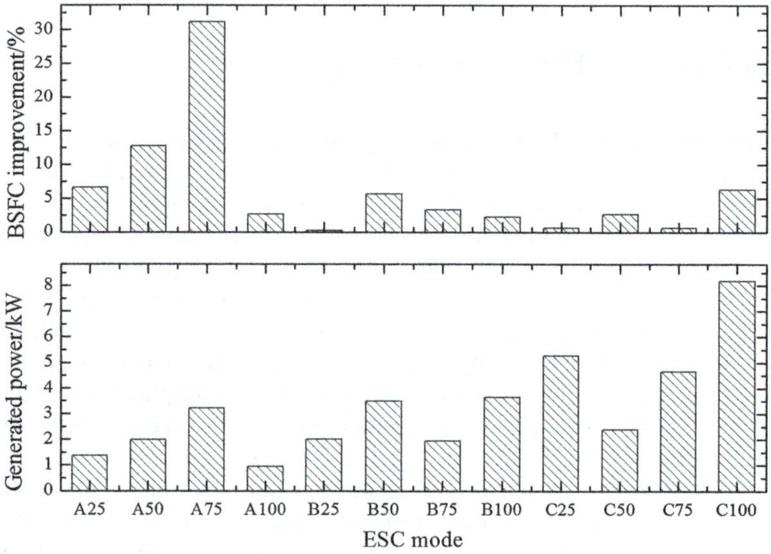

Fig. 4 Optimized BSFC improvement and exhaust heat recovery for the configuration at different ESC modes

$$\Delta_{BSFC} = \frac{\left(P'_{eng} + P_{mtr}\right) - P_{eng}}{P_{eng}} \times 100\,\% \tag{3}$$

where P'_{eng} is the brake power of ICE with HET, P_{mtr} is the generated power of motor/generator, P_{eng} is the brake power of the prototype ICE.

It is shown in Fig. 4 that the BSFC improvement of the diesel engine is proportional to engine loads at constant engine speed. The generator/motor needs to consume electric power from battery to speed up the compressor at low speeds. It can be explained that the A/F ratio is comparatively small because of the high fuel

Table 2 Main characteristics of the prototype bus

Dimensions (m)	12 × 2.55 × 3.18
Front face area (m^2)	6.89
Drag coefficient	0.58
Coefficient of rolling friction	0.006
Mass (t)	12.4
Max velocity (km/h)	85
Wheel base (m)	6.1

injection rate at high loads. And the ratio of the recoverable exhaust energy to engine power is fewer than that in low loads. In contrary, the A/F ratio is high at low loads. The ratio of the recoverable exhaust energy to engine power is higher than that in high loads. But in fact, the low-load, high speed conditions could rarely present because of the existence of transmission.

It is not practical for the application of HET system on a conventional vehicle. It is because that the 12 V low voltage system of these vehicles could not deal with the instantaneous heavy electrical current in excess of 200 A. And the capacity of the battery on conventional vehicle is too small to provide enough energy storage before battery protection is engaged and the electrical energy generated from exhaust heat by the HET is wasted in form of heat again in the charging process.

4 Combination of HET and Mild-HEV

The main characteristics of the prototype bus are shown in Table 2. The original engine is a six-cylinder 7.13 L diesel engine. The fuel consumption of the prototype bus in China typical urban drive cycle is 36.5 L/100 km. The test fuel consumption of the middle-hybrid bus is about 28 L/100 km. The reduction of fuel consumption is 23.3 %. The cost of this achievement is 10 % extra weight and 30 % added cost of the prototype bus. It was shown that the lost energy of middle-hybrid is mainly during transient conditions. And the extra weight and cost of middle-hybrid system mitigate its fuel economy benefits. It is proposed that a mild-hybrid system combined with HET can get the same benefits without extra weight and cost.

The aforementioned diesel engine model with HET was implemented to the mild-HEV model in GT-DRIVE to evaluate the economy benefit of combine HET and mild-HEV technology in China typical urban drive cycle. The traction motor and battery pack were downsized from 100 kW and 120 Ah to 60 kW and 80 Ah, respectively. The simulation results show that with the application of HET, the fuel economy benefit could be achieved by 25 % without extra weight and cost.

5 Conclusion

An energy and availability analyses were applied on a turbocharged four cylinder diesel engine. The results show that 10 ~ 15 % exhaust energy could be recovered.

HET is a promising technology to improve fuel economy and eliminate turbolag effect of ICE. Because of the low voltage adapted on conventional vehicles, the application of HET on conventional vehicles is limited. But the proposed combined HET and HEV technology could not only recover the exhaust energy but also downsize the weight and cost of ISG and battery pack. With appropriate control strategy of the combined HET and HEV, the downsized ICE of HEV powertrain could work on more economical points to improve HEV fuel benefits furthermore.

References

1. Gheorghiu V (2011) Ultra-downsizing of internal combustion engines. The Automotive Research Association of India, Pune
2. Shahed SM, Shahed KH (2009) Parametric studies of the impact of turbocharging on gasoline engine downsizing. SAE Int J Engines 2(1):1347–1358
3. Arnold S, Balis C, Barthelet P et al (2005) Garrett electric boosting systems (EBS) program: final report. Honeywell Turbo Technologies
4. Kolmanovsky I, Stefanopoulous AG, Powell BK (1999) Improving turbocharged diesel engine operation with turbo power assist system. Proceedings of the 1999 IEEE International conference on control applications
5. Yamashita Y, Ibaraki S, Sumida K et al (2010) Development of electric supercharger to facilitate the downsizing of automobile engines. Mitsubishi Heavy Ind Tech Rev 47(4)
6. Panting J, Pullen KR, Martinez-Botas RF (2001) Turbocharger motor–generator for improvement of transient performance in an internal combustion engine. Proc Inst Mech Eng: J Automobile Eng
7. Katrašnik T (2007) Hybridization of powertrain and downsizing of IC engine—A way to reduce fuel consumption and pollutant emissions—Part 1. Energy Convers Manage
8. Moran MJ (1982) Availability analysis: a guide to efficient energy use. Englewood Cliffs, Prentice-Hall, NJ

New Compact and Fuel Economy Cooling System "SLIM"

Junichiro Hara, Mitsuru Iwasaki and Yuichi Meguriya

Abstract A new engine cooling system is proposed that alters sub-radiator function according to driving conditions without valves and actuators. This system enables substantial fuel consumption improvement by: (1) Simplified heat exchanger configuration located in the front end of a vehicle with only two layers instead of three, which leads to 30 % reduction in pressure drop of air flow. (2) Cooling fan power and air-conditioning compressor load reduction associated with above mentioned air flow improvement. This performance improvement mecha nism of the new system called SLIM (Single Layer Integrated cooling Module) is studied in various aspects such as air pressure drop in comparison with full water-cooled condenser systems. Furthermore, by extending the concept to CAC (Charge Air Cooler) with a focus on the use of optimum heat transfer efficiency, a new 2-way CAC structure is proposed. 30 % improvement in heat exchange performance over conventional water cooled CAC is confirmed.

Keywords Engine · Cooling · Radiator · Condenser · Charge air cooler

1 Introduction

New generation powertrain such as hybrid, clean diesel or downsizing turbo requires an additional heat exchanger. Accordingly pressure drop of heat exchangers is likely to increase, which further leads to increased cooling fan load

F2012-A07-013

J. Hara (✉) · M. Iwasaki · Y. Meguriya
Calsonic Kansei Corporation, Chiyoda, Japan
e-mail: junichiro_hara@ck-mail.com

SAE-China and FISITA (eds.), *Proceedings of the FISITA 2012 World Automotive Congress*, Lecture Notes in Electrical Engineering 190,
DOI: 10.1007/978-3-642-33750-5_30,

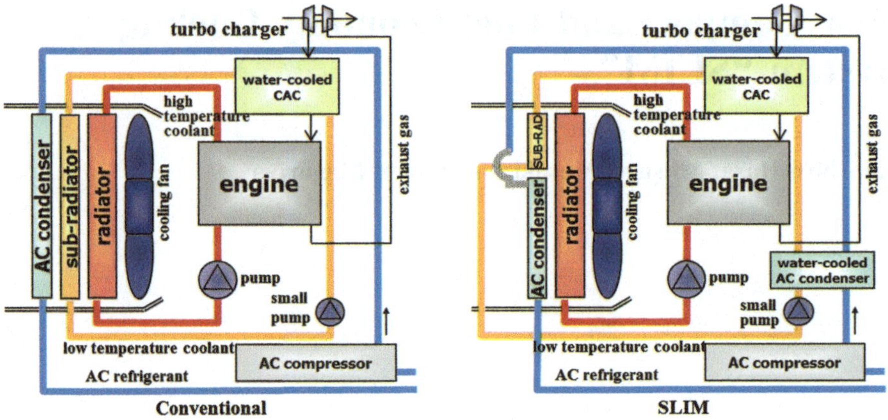

Fig. 1 Comparison of conventional cooling system and SLIM

and worsened fuel consumption. The proposed system focuses on integration of various cooling functions to meet various driving conditions. By analyzing load situations of condenser for air conditioning system and sub-radiator for CAC or "motor and inverter" for hybrid vehicle, it was found that the loads for each of the systems are complementary. Thus, it is conceivable to reduce the number of heat exchanger layers if the function of sub-radiator can vary depending on the driving conditions. Consequently cooling fan power can also be reduced. A compact cooling system based on this concept is called SLIM (Single Layer Integrated cooling Module) and has been studied in a wide range of vehicle operation cases. Additionally it is worth noting that emission regulation trend will require larger capacity CAC, although it might become more difficult to increase its capacity without enlarging the surface area of heat exchangers [1–3]. The SLIM concept is found to be also effective to CAC performance improvement.

2 Structure

Figure 1 shows a schematic of a SLIM system compared with a conventional system. As indicated in the figure, AC (Air Conditioning) condenser is divided into water-cooled and air-cooled ones. The function of sub-radiator is cooling the water-cooled condenser and a CAC. The water-cooled condenser is placed in the sub-radiator tank. The coolant of sub-radiator is below 60 °C.

Refrigerant of the AC system that flows out of the AC compressor flows into the water-cooled condenser. Then the AC refrigerant goes to the air-cooled condenser. In the condenser part in gas area, the AC refrigerant is cooled by water flow which has higher heat transfer coefficient than air flow condenser part. And by using an air-cooled condenser, the AC refrigerant can be cooled to near the temperature of air flow. The volume of water-cooled condenser is small (less than 150 cm^3 in this case) but the heat dissipation capacity can be as large as 2 kW (Fig. 2).

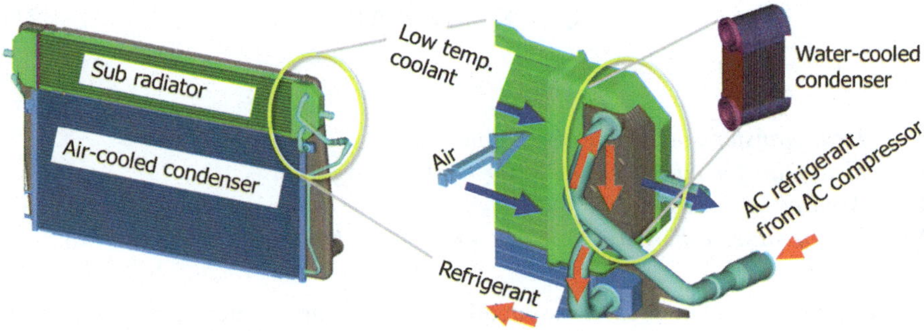

Fig. 2 Water-cooled condenser

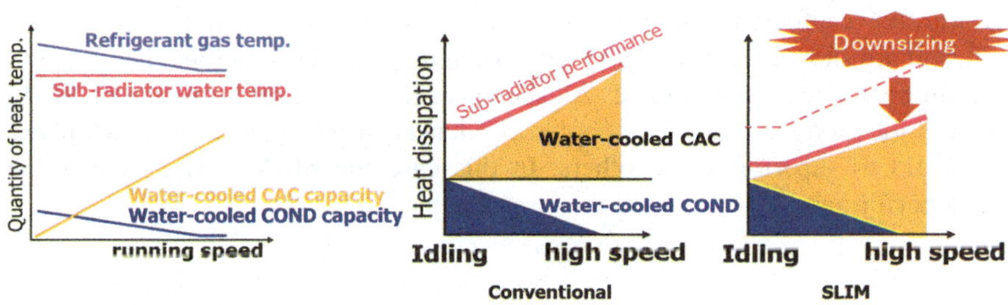

Fig. 3 Heat dissipation in various driving scenes

3 Principle in Operation

The thermal loads of water-cooled CAC and water-cooled condenser changes in various driving scenes (see Fig. 3). In idling condition, engine speed is low and the heat dissipation of the CAC is minimal. The vehicle speed is zero and the load of AC condenser is very high. In case of high speed driving and hill-climbing condition, the load to CAC is high while the load to AC condenser is low. System operations in such various driving scenes have been studied with 1-D simulation tool and actual test drives. Based on the simulation and the vehicle experimental data, it was confirmed that SLIM can cope with superimposed condition of the heat loads of CAC and AC condenser.

4 Improvement of Heat Transfer

In SLIM system, two condensers are used: (1) a small water cooled condenser is placed in a low temperature sub-radiator; and (2) a conventional air cooled condenser.

$$Q = K \times A \times \Delta T \tag{1}$$

Where $\frac{1}{K} = \frac{1}{\alpha_1} + \frac{1}{\alpha_2}$

α_1 *Heat transfer coefficient of air flow*
α_2 *Heat transfer coefficient of water or AC refrigerant*
A *Area of heat dissipation*
ΔT *Temperature difference = Water temp. – Air flow temp.*

The water cooled condenser essentially cools the superheated refrigerant gas to saturated conditions; and the second condenser (air cooled condenser) further cools it through two-phase condensation process. In both heat exchangers, heat transfer coefficient has been optimized as indicated in the following Table 1.

According to Eq. 1, heat dissipation is decided by temperature difference and heat transfer coefficient. In Table 1, temperature difference and heat transfer coefficient are shown for different system configurations. Data in italic letter shows optimum heat transfer coefficient. As shown in this table, performance of SLIM system is the same as full water-cooled condenser in gas area; and in two phase area, SLIM is superior to the others. In this case the SLIM system shows the highest performance.

5 Primary Factors of Improvement of Fuel Consumption

In SLIM, the reduction of pressure drop or air flow is essential. By this effect, the followings are achieved.

1) Reduction of cooling fan power due to the lower air flow resistance
2) Power reduction of AC compressor by enhancing heat transfer performance of condenser

The reduction of power of the cooling fan and AC compressor consequently leads to improved fuel mileage.

6 Application to Vehicles

SLIM system was applied to several prototype vehicles. The examples are shown as follows (Fig. 4) (Table 2):

In each vehicle, the following advantages of the SLIM system were confirmed.

(1) Improved fuel consumption
(2) Weight reduction
(3) Compactness of heat exchangers

Table 1 Heat transfer for various cooling systems

		Conventional + sub RAD K × ΔT	Full water-cooled condenser K × ΔT	SLIM K × ΔT			
Refrigerant gas area	Temperature difference ΔT (K)	34.5	1.92	24	2.38	24	2.38
	K value[b]	0.06		0.10		0 10	
	Heat transfer coefficient α1 (kW/(m^{2}K))[c]	0.077 fin to air		[a]*0.11 fin to air*		[a]*0.11 fin to air*	
	Heat transfer coefficient α2 (kW/(m^{2}K))	0.2 gas to fin		[a]*1.0 water to air*		[a]*1.0 water to air*	
Refrigerant two phase area	Temperature difference ΔT (K)	25	1.90	15	1.49	25	2.69
	K value[b]	0.08		0.10		0.11	
	Heat transfer coefficient α1 (kW/(m^{2}K))[b]	0.077 fin to air		[a]*0.11 fin to air*		[a]*0.11 fin to air*	
	Heat transfer coefficient α2 (kW/(m^{2}K))	[a]*5.0 ref to fin*		1.0 water to fin		[a]*5.0 ref to fin*	

[a] *Italic style* shows optimum choice
[b] In the calculations, the are ratios have been omitted
[c] Air side heat transfer coefficient is dependent on the airside face velocities (Temperatures of fluid used for above calculations)
Outside temperature 35 °C refrigerant temperature inlet 69.5 °C
Sub RAD water temp. inlet 1 59 °C refrigerant saturation temperature 60 °C
Sub RAD water temp. inlet 2 50 °C (gas area uses inlet 1, refrigerant two phase area uses inlet 2)

Fig. 4 Reduction of fan load

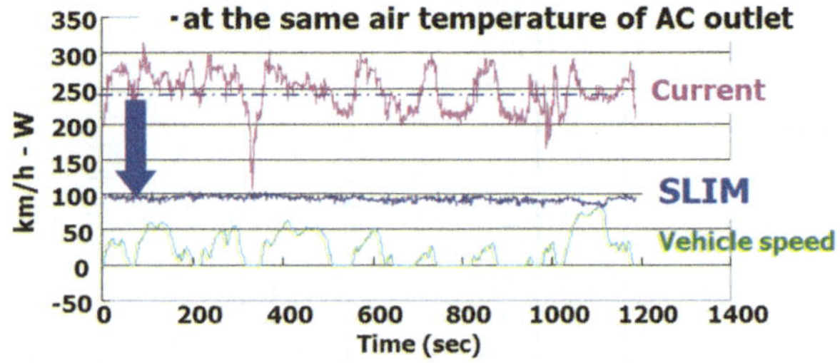

7 Analysis of Improvement of Fuel Consumption

The two contribution factors in fuel mileage have been studied as follows:

(1) Reduction of cooling fan load

As shown in Fig. 4, when the discharge pressure of AC system is the same as a conventional case, the cooling fan load of SLIM can be reduced by about 150 W.

(2) Thermal storage effect of water

During the acceleration, the flow rate of refrigerant in the AC system and accordingly discharge pressure increase. Larger thermal storage effect of water than air, however, is utilized in SLIM to keep the temperature change moderate in such conditions. A part of the AC condenser is cooled by water in the SLIM system, which lowers the refrigerant temperature than conventional systems. Thus, the AC compressor load is reduced and engine can avoid large BSFC (Brake Specific Fuel Consumption) area (Fig. 5).

8 Analysis of Optimal Ratio of Sub-Radiator

Because air-cooled condenser and sub-radiator are placed in vertical position, the optimal ratio of the height of sub-radiator to total height is necessary to be considered. This ratio is constrained by the minimum compression ratio of refrigerating cycle, which is in various cases below 40 %. It is necessary to cool CAC even in the highest speed in this ideally proportional sub-radiator.

In our vehicle experiments, we can confirm that sub-radiator cools CAC enough for compact cars.

Figure 6 shows three cooling types of AC condenser.

- Conventional air-cooled condenser (at left hand side)
- SLIM (combined air-cooled and water-cooled condenser)
- Full water-cooled condenser (at right hand side) [4, 5].

SLIM can choose the optimal heat transfer mechanism, and then the discharge pressure of AC system is the lowest. The lower pressure brings the vehicle fuel

Table 2 Application for vehicles

		VW Golf TSI 1.4L	2L Diesel	BMW 530d Diesel
Engine Cooling System		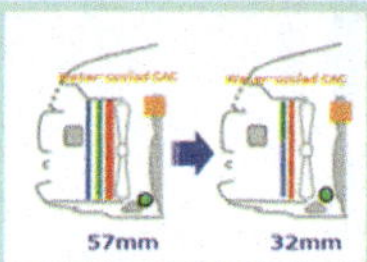	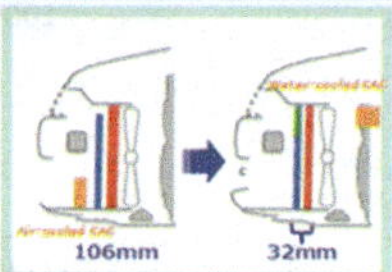	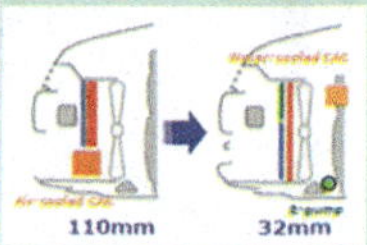
Performance	Load	• Cooling fan load 150W reduced	• Cooling fan load 210W reduced	• AC Discharge press 20% reduced • Cooling fan load 100W reduced
	Fuel Use	4% improved (at AC ON ,ambient 35℃,JC08)	3.5% improved (at AC ON ,ambient 35℃,40km/h)	5.3% improved (at AC ON ,ambient 40℃,40Km/h)
	Weight, Space	• Weight 30% reduced • Core thickness 44% reduced	• Weight 30% reduced • Core thickness 70% reduced	• Weight 30% reduced • Core thickness 71% reduced

AC means Air-conditioning. "JC08" is Japanese regulation on fuel consumption

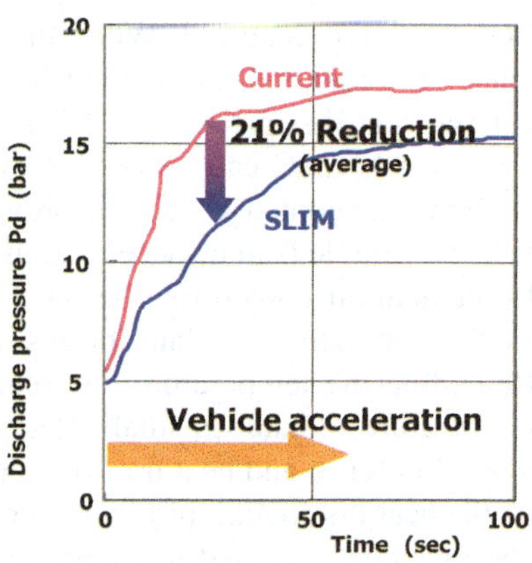

Fig. 5 Moderation of pressure rise at vehicle acceleration

mileage to be better. AMESim which was developed by LMS Corporation as 1-D simulation for analyzing a refrigerant cycle to reduce the number of prototypes and experiments [6, 7].

9 Enlargement of Heat Dissipation of CAC

In Φ-T map (equivalence ratio—temperature diagram), to reduce NO_x, low temperature combustion is necessary. EGR (Exhaust Gas Recirculation) including CO_2 has larger specific heat than N_2 and O_2 molecules. Then enlarged capacity of

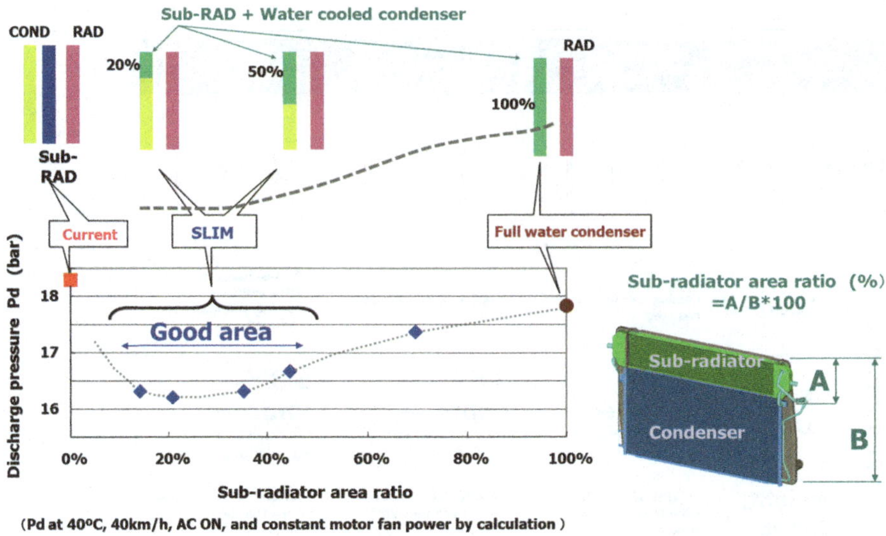

Fig. 6 Optimum ratio of sub-radiator height

EGR cooler is required. When the ratio of EGR is increased, a lot of unburned matter is generated. So, cooled flesh air with oxygen is necessary. Therefore, to enlarge heat dissipation of CAC is necessary.

In usual, CAC can be cooled by air flow, engine coolant or low temperature coolant. Air-cooled CAC can provide lower gas temperature of outlet of CAC. When a vehicle is in an acceleration state, gas temperature of water-cooled CAC is lower than air-cooled one because of thermal storage effect of water as coolant. Low temperature coolant brings the gas temperature of CAC to be lower. Regarding the temperature difference of coolant and air, heat dissipation of low temperature radiator is small. Because heat dissipation is determined by temperature difference and heat transfer coefficient, the limiting factor of CAC is found to be the heat dissipation of low temperature radiator. Therefore, the CAC is divided, and one side is cooled by engine coolant and the other is cooled by low temperature radiator, that is a sub-radiator. In this divided CAC system, called 2-way CAC system, the total heat dissipation of is increased compared with a conventional CAC system by above 30 % (Figs. 7 and 8).

10 Comparison with Full Water-Cooled Condenser System

Comparison with full water-cooled condenser system with water-cooled CAC has been studied. The heat dissipation of AC condenser and CAC were estimated for vehicle idling, constant speed 40 km/h and hill climbing conditions. In each case,

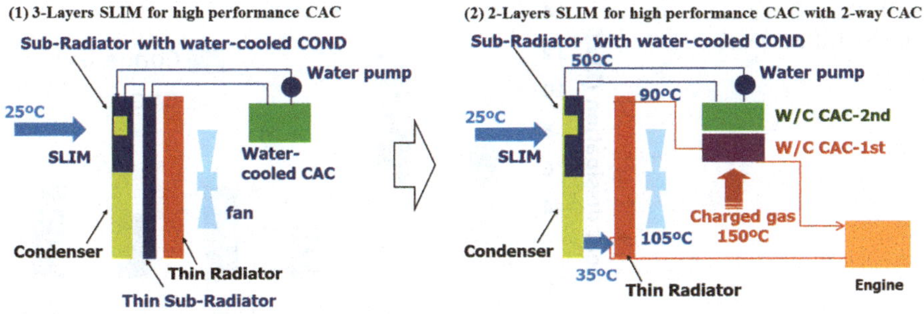

Fig. 7 2-way CAC system

Fig. 8 Performance of radiators

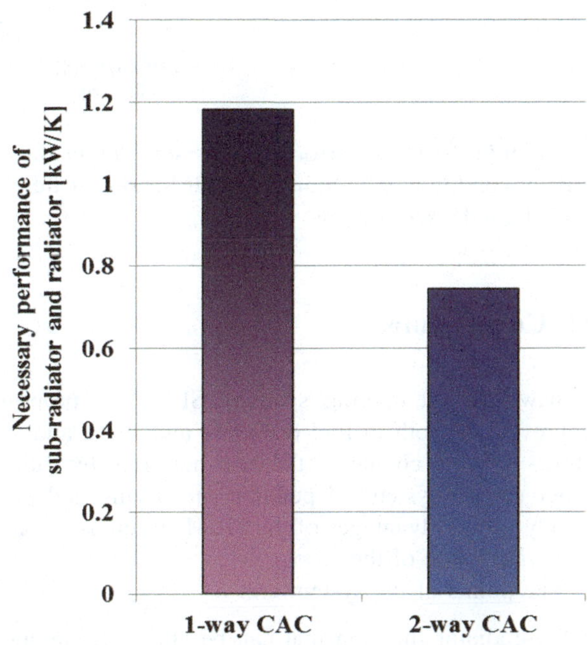

the heat dissipation of CAC is assumed to be the same, the amount of which is based on experimental data with actual vehicle with 1.6 L turbo-charged petrol engine. The results are shown in Fig. 9.

In the figure the bars on the left hand side graph are for a hill-climbing driving condition. Full water-cooled condenser system shows lower performance than SLIM in this condition. This is due to high temperature of sub-radiator outlet water temperature that cools CAC. The center pair of the bars shows an idling condition case. Since temperature of coolant of a sub-radiator is high, in the case of full water-cooled condenser system, its heat dissipation of AC condenser is inferior to SLIM. The right hand side bars corresponds to a constant speed driving condition

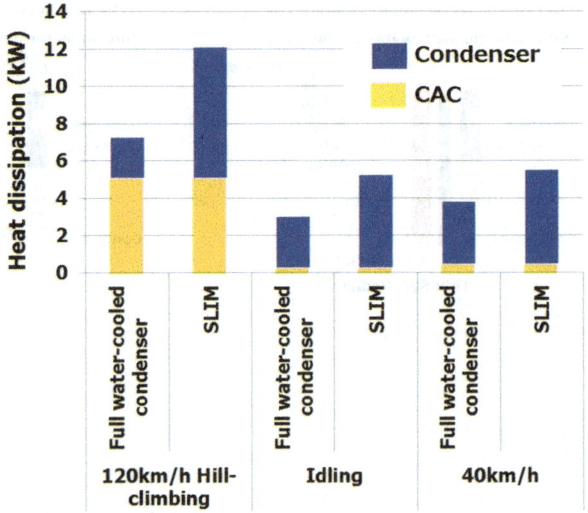

Fig. 9 Comparison with full water-cooled condenser system

at 40 km/h. SLIM's performance is superior in this condition as well. In every driving condition, consequently, SLIM shows better performance than full water-cooled condenser systems.

11 Conclusions

A new concept cooling system "SLIM" is proposed that enables substantial improvement both in fuel economy and space and weight savings. This concept allows optimal choice of the best heat transfer efficiency within air-cooled and water cooled systems depending on driving and environmental conditions that widely vary. Advantages of the SLIM system is extensively studied and confirmed in a wide range of the cases.

The points of the system are:

- Sub-radiator function that can be altered by temperature difference of coolant and target heat loads,
- (Above-mentioned) switching performed automatically without valves and actuators,
- Fuel consumption improvement by 3–5 % by load reduction of cooling fan power and AC compressor power, and
- Improvement in spatial efficiency of engine cooling module by 40 % (for turbo-charging petrol and diesel engine vehicles).

In addition, a new 2-way CAC system is conceived along the SLIM concept to achieve 30 % improvement in heat exchange performance. Considering future emission regulation for diesel engines etc., this 2-way CAC system has a potential to cope with various requirements in a flexible way.

References

1. Iwasaki M (2010) Development of new energy saving cooling system SLIM. Automotive Summit, Nov. 2010
2. Peuvrier O (2011) Development of compact cooling system 'SLIM'. VTMS 2011
3. Mathur GD (2012) Development of an innovative energy efficient compact cooling system "SLIM". SAE Apr 2012
4. AP NS (2008) Ultimate cooling™ system application for R134a and R744 refrigerant. SAE July 2008
5. Malvicino C (2011) Dual level vehicle heat rejection system. VTMS10, Mar 2011
6. Yagisawa K (2008) From AC to complete vehicle thermal management using LMS imagine lab. AMESim International LMS engineering simulation conference, Oct. 2008
7. Rozier T (2008) From the modeling of AC system to the complete vehicle thermal management using one single platform. Haus der Technik 2008

Influence of Operating Parameters on the Thermal Behavior and Energy Balance of an Automotive Diesel Engine

Christian Donn, Daniel Ghebru, Wolfgang Zulehner, Uwe Wagner, Ulrich Spicher and Matthias Honzen

Abstract The paper gives a comprehensive overview regarding all relevant operation parameters mainly influencing the energy balance of a state-of-the-art automotive Diesel engine. A detailed experimental analysis of the influence of injection timing and pressure, EGR rate, boost pressure, charge air temperature, coolant temperature and swirl flap position is presented. The investigation includes combustion chamber energy balances based on pressure trace analyses (internal energy balance) as well as outer energy balances that take into account all out going and incoming energy flows of the engine based on extensive temperature and flow measurements of the relevant fluids. Measurements of the engine structure temperatures are used to complete the analysis of the thermal engine behaviour depending on the operating parameters. The aim of the paper is to give an overview which operating parameters are crucial for thermal engine application and simulation.

Keywords Diesel engine · Thermal management · Heat flow analysis · Energy balance

Nomenclature

Air	Intake air
ATDC	After top dead centre
Aux	Auxiliaries

F2012-A07-014

C. Donn (✉) · D. Ghebru · W. Zulehner · U. Wagner · U. Spicher
Karlsruhe Institute of Technology, Institut fuer Kolbenmaschinen, Karlsruhe, Germany
e-mail: christian.donn@kit.edu

M. Honzen
Audi AG, Neckarsulm, Germany

SAE-China and FISITA (eds.), *Proceedings of the FISITA 2012 World Automotive Congress*, Lecture Notes in Electrical Engineering 190,
DOI: 10.1007/978-3-642-33750-5_31, © Springer-Verlag Berlin Heidelberg 2013

BTDC	Before top dead centre
CA	Crank angle
CAC	Charge air cooler
CC	Combustion chamber
CO	Carbon monoxide
EGR	Exhaust gas recirculation
HC	Hydrocarbon
$\mathrm{IMEP_n}$	Net indicated mean effective pressure
$\mathrm{MFB_{50\ \%}}$	Mass fraction burned 50 %
MI	Main injection
PiI	Pilot injection
rpm	Revolutions per minute
SOI	Start of injection
TDC	Top dead centre
VTG	Variable turbine geometry
α_{Gas}	Gas side heat transfer coefficient (W/(m^{2}K))
$\dot{H}_{Exh}$	Exhaust gas enthalpy flow (kW)
m	Injected mass per cycle (mg)
n	Engine speed (rpm)
$\dot{Q}$	Heat flow (kW)
$\dot{Q}_{in}$	Input energy flow (kW)
P	Power (kW)
p_{cyl}	Cylinder pressure (bar)
T_{gas}	Gas temperature (°C)

1 Introduction

Modern passenger car Diesel engines are controlled by the engine control unit through several independent and interdependent operating parameters. Operating parameters can be adapted within the same operating point to optimize fuel consumption, emissions and noise. Depending on ambient conditions and especially during engine warm-up, the operating parameters change significantly and affect the thermal behavior of the engine. Since experimental analyses of the engine energy balance are time consuming and cost intensive, simulation models that are able to predict the thermal engine behaviour under various ambient conditions and operating strategies get more and more important. Such models offer a great potential for application strategy pre-tests and concept analyses. The aim of this work is to quantify the influence of the operating parameters described above and give an overview about the importance of these operating parameters regarding the engine energy balance and their implementation into thermal engine simulation.

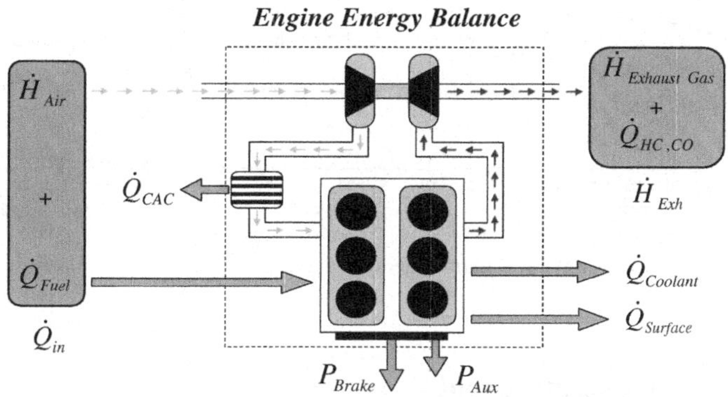

Fig. 1 System boundary for engine energy balance

2 Experimental Setup and Heat Flow Calculation

Within this work an engine energy balance by applying the First Law of Thermodynamics for open systems to a control volume that surrounds the engine is used to investigate the thermal engine behaviour. The system boundary is characterized by transfer of mass as well as heat and work. Figure 1 shows the different energy flows to be considered as described in detail in [1]. In this work the authors made two additions: On the one hand the mass flow of unburned HC and CO is calculated based on measured emissions to determine the energy content of the unburned components. This portion of energy is added to the exhaust gas enthalpy flow. On the other hand the surface heat flow caused by convection and radiation is splitted into two parts to account for the different heat sources. One part is assigned to the engine itself with coolant and oil as heat source, the other part is assigned to the exhaust system with exhaust gas being the heat source. The input energy consists of intake air enthalpy and fuel energy flow and is used in the following as reference for the engine energy balance result values.

Moreover, pressure trace analyses for the high-pressure cycle are used to evaluate the accuracy of the cylinder wall heat flow calculation on the basis of an inner energy balance. The approaches suggested by Woschni, Woschni/Huber and Hohenberg were compared. Since the Hohenberg approach [2] provides very good values for the inner energy balance within a wide range of operating points, all pressure trace analysis results presented in the following are based on this approach. In Addition, a modified approach from Ghebru [3], based on the originally published approach of Zapf [4] is used to quantify heat transfer to the coolant at the intake and exhaust ports. Mass flow rates and gas temperatures out of a 1D GT-Power model were used for the port wall heat flow calculation. Test engine is a 2010 Audi 3.0 L V6 TDI with high pressure EGR and common rail injection system [5]. The engine is equipped with in-cylinder as well as intake and exhaust manifold pressure sensors and with a total of 140 temperature sensors as well as coolant flow measurement devices [1].

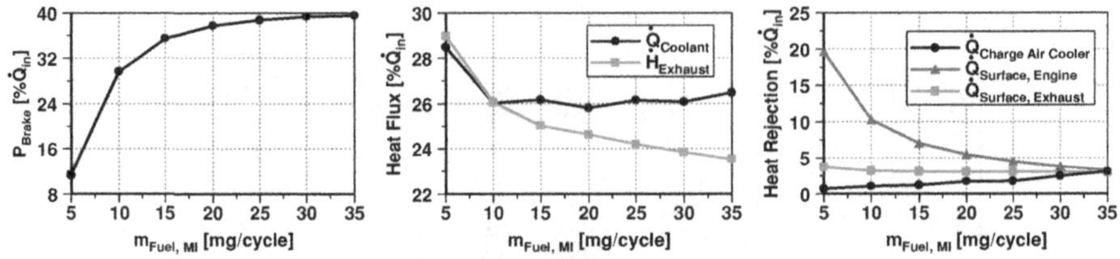

Fig. 2 Energy balance results for main injection quantity variation at 1500 rpm (1.5–13 bar $IMEP_n$)

3 Influence of Engine Load

Besides speed and load, the operating point of an engine mainly defines the amount of exhaust gas enthalpy and heat rejection to the ambient via the radiator and the engine surface [6, 7]. However, the energy balance is an engine specific matter and also depends on operating parameters [1]. At constant speed and combustion phasing the relative exhaust gas enthalpy flow of the test engine decreases with increasing load. The relative coolant heat flow is higher at low loads and stays on an almost constant level within the investigated load range of 1.5–13 bar $IMEP_n$ (see Fig. 2).

Since the coolant temperature was constant for the entire load sweep, the engine surface temperature only slightly increases. This is the reason why the relative engine surface heat flow decreases with higher loads while heat rejection at the exhaust system is almost constant for different loads because the exhaust gas temperature rises with the engine load. However, the absolute values of coolant heat and exhaust gas enthalpy flow strongly depend on speed and load. Hence, a correct load calculation within a thermal engine model has the highest priority and requires a precise combustion and friction model.

4 Influence of Operating Parameters

All parameter sweeps shown in the following were performed with constant fuel mass, without EGR and with fully opened swirl flap. Only for the EGR and swirl flap position sweeps these two parameters were varied. For better comparability all values of the engine energy balance are specified in percent of the total input energy (compare Fig. 1).

4.1 Main Injection/Combustion Phasing

Main injection timing plays a special role among the Diesel engine operating parameters, because it directly defines combustion phasing. A start of injection (SOI) variation for the main injection simultaneously implies a $MFB_{50\,\%}$ shift and

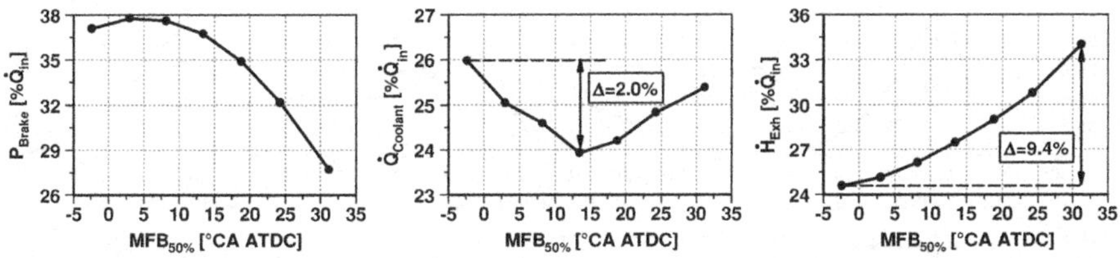

Fig. 3 Engine energy balance values for variation of start of main injection at 2000 rpm (start: 7 bar $IMEP_n$)

an increasing burn duration with later injection timing. The effect is a rising exhaust gas temperature which leads to an increasing exhaust gas enthalpy flow of 9.4 % of the input energy (see Fig. 3).

The presented sweep was done with constant air mass flow and shows, that coolant heat flow decreases up to a point of $MFB_{50\ \%}$ at 15 °CA after top dead centre. Once this point is reached, coolant heat flow starts to increase again. This interesting behaviour is not completely explainable with the results of a pressure trace analysis (see Fig. 4).

The upper part of Fig. 4 shows the measured cylinder pressure as well as the mean gas temperature as a result of a pressure trace analysis using the Hohenberg approach to calculate cylinder wall heat flow. Figure 4 also shows the calculated heat transfer coefficients, combustion chamber wall heat flows, measured wall temperatures and a sweep overview of wall heat flow changes in the combustion chamber and exhaust port. Since the charge air temperature and the air mass flow is constant, the wall heat flow in the intake ports is also constant and not relevant for the change in total wall heat flow. It can be seen, that the total wall heat flow as a combination of heat transferred in the combustion chambers and exhaust ports drops by approximately 1.5 % of the input energy within the range of $MFB_{50\ \%}$ from 2 °CA before to 10 °CA after top dead centre. For later main injections and thus combustion phasing the wall heat flow stays at a constant level and does not match the coolant heat flow behaviour measured in the engine energy balance (compare Fig. 3). Since the friction level only changes within a range of 0.5 % of the input energy, the Hohenberg approach seems to reach its limits of accuracy with late combustion phasing. The values of the inner energy balance and the measured wall temperatures presented in Fig. 4 support this assumption. While wall temperatures in the cylinder head and liner persist at the same level after $MFB_{50\ \%}$ of about 15 °CA is reached, the calculated wall heat flow further drops in an almost linear way. However, the authors found that the approach of Hohenberg even with late combustion works better than the other ones. In summary can be stated, that the combustion phasing strongly influences the exhaust gas enthalpy but also has an effect on coolant heat flow. For all further parameter sweeps combustion phasing was held constant to evaluate the influence of these operating parameters on the engine energy balance independently from combustion phasing.

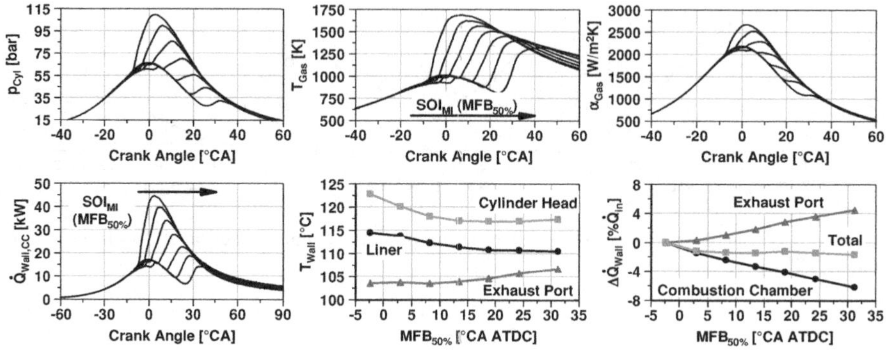

Fig. 4 Pressure trace and coolant heat input analysis results for variation of start of main injection

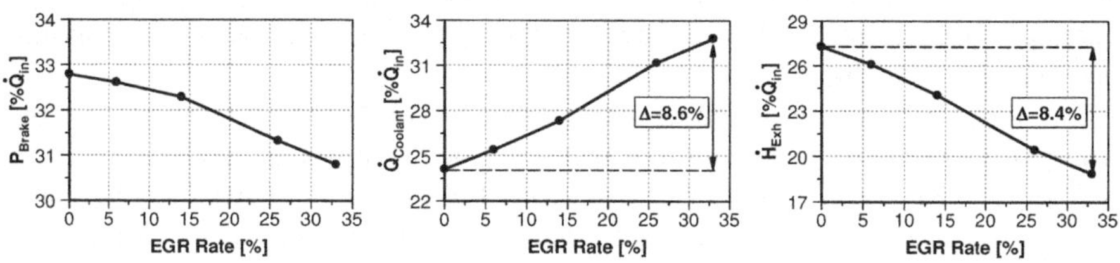

Fig. 5 Engine energy balance values for EGR Rate variation at 1500 rpm (start: 4 bar IMEP)

4.2 EGR Rate

EGR rate is the most important operating parameter regarding the Diesel engine energy balance as the authors also presented within a coolant heat input analysis in previous work [1]. Figure 5 shows results of an EGR rate variation from 0 to 33 % at constant boost pressure.

Due to a lower indicated efficiency resulting from EGR in combination with increasing intake manifold temperatures, relative break power is reduced by 2 % within the EGR rate variation from 0 to 33 %. Coolant heat input increases by 8.6 % while exhaust enthalpy flow is reduced by 8.4 %. Since the exhaust gas recirculation is water-cooled in the test engine, an increased EGR-Rate implicates an increased gas temperature in the intake manifold (in this case 30 K) which leads to a higher wall heat flow due to a higher mean gas temperature in the cylinder and ports during the entire cycle. However, the main part of the additional coolant heat input resulting from operation with EGR is transferred to the coolant in the EGR cooler.

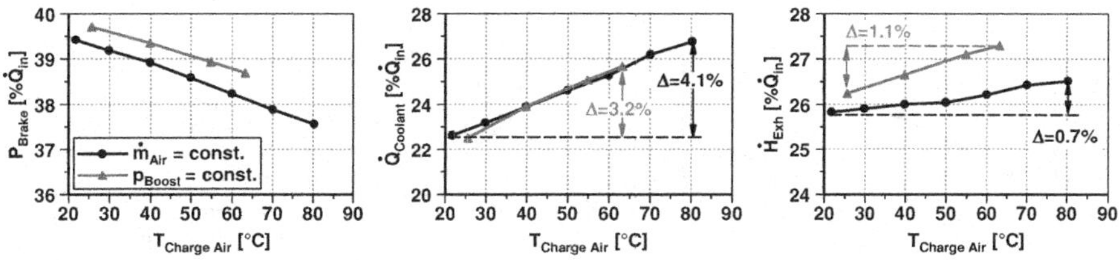

Fig. 6 Engine energy balance values for charge air temperature variation at 2000 rpm (start: 9 bar $IMEP_n$)

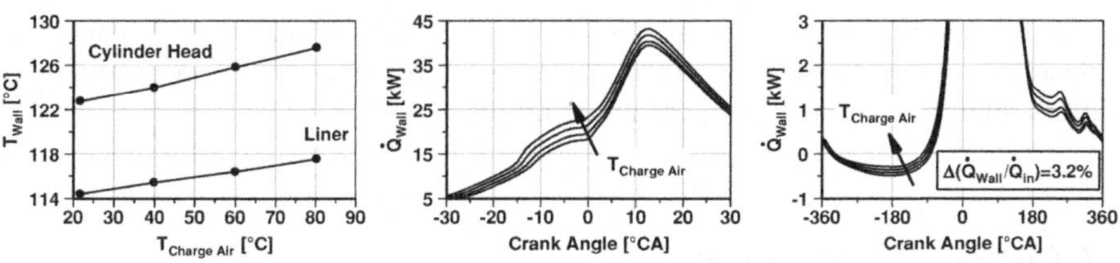

Fig. 7 Pressure trace analysis results for charge air temperature variation at constant air mass flow

4.3 Charge Air Temperature

For the investigation of the influence of charge air temperature two measurement series were made: one with constant boost pressure at constant VTG position of the turbo and another with adjusted boost pressure to keep air mass flow constant with rising charge air temperature. Figure 6 shows the measurement results.

Charge air temperature mainly influences coolant heat input due to a higher temperature level in the intake and exhaust ports and the combustion chamber. Almost independent on whether boost pressure or air mass flow is kept constant, relative coolant heat flow rises by approximately 0.75 % per 10 K charge air temperature. Coolant heat flow increases by 4.1 and 3.2 % while charge air temperature rises by 58 and 37 K respectively. Exhaust gas enthalpy flow for the engine energy balance is measured downstream the turbocharger (see Fig. 1). This is the reason why its value depends on whether VTG position is changed to adjust boost pressure or not. In case of boost pressure adjustment for constant air mass flow and thus constant air to fuel ratio, a greater portion of the exhaust gas energy is used at the turbine. However, in both cases exhaust gas enthalpy flow is increased due to the higher gas temperature level by 0.7 and 1.1 % respectively. Figure 7 shows some results of the cylinder pressure trace analysis for the sweep with constant air mass flow.

Cylinder wall heat flow increases by 3.2 % among the charge air temperature sweep from 22 to 80 °C while cylinder wall temperature rises by 4 °C at the liner and by 6 °C at the cylinder head. The percentage of heat transferred to the wall in

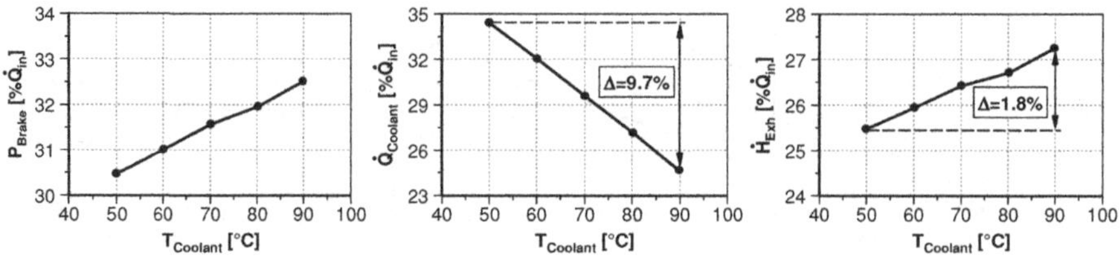

Fig. 8 Engine energy balance values for coolant temperature variation at 1500 rpm (start: 4 bar $IMEP_n$)

the high pressure cycle decreases from 90.3 to 87.9 % within the charge air temperature sweep from 20 to 80 °C, because the relative change in gas temperature is higher during the gas exchange period.

4.4 Coolant Temperature

A coolant temperature sweep implicates a changed engine surface temperature and thus heat rejection caused by radiation and convection is different. Since the coolant heat flow for the engine energy balance is measured at the radiator, it also includes the change in surface heat flow of the engine block and cylinder head. The change in surface temperature and thus in surface heat flow is negligible for a constant coolant temperature like in all other operating parameter sweeps shown in this paper. Within a coolant temperature variation from 50 to 90 °C, the surface heat rejection increases by 5.5 % of the input Energy. Figure 8 shows the results of the engine energy balance.

In total an increased coolant temperature of 40 K leads to a reduced relative coolant heat flow of 9.7 % at the radiator. Taking into account the rising heat losses through surface heat rejection, an effective decrease in coolant heat input of 4.2 % results. This decrease in coolant heat input comes from reduced heat flows at the port and cylinder walls caused by reduced temperature differences between gas and wall as well as from a reduced overall friction heat input. The pressure trace and port heat flow analysis show a decrease in relative wall heat flow of 0.9 % of the input energy in the cylinder, 0.8 % in the exhaust ports and 1.2 % in the intake ports. Friction power is reduced by 1.8 % of the input energy. Summing up these values leads to a total decrease in coolant heat input of 4.7 %. Taking into account the achievable accuracy of measurement, this shows the validity of the approaches from Hohenberg [2] and Zapf [4] in this case. In summary it can be stated that a coolant temperature reduction in a conventional engine without encapsulation leads not only to a significant increase in coolant heat input (in this case plus 4.2 % of the input energy) but also to a redistribution of heat rejection via the radiator and engine surface that has to be taken into account for an overall vehicle thermal management optimization.

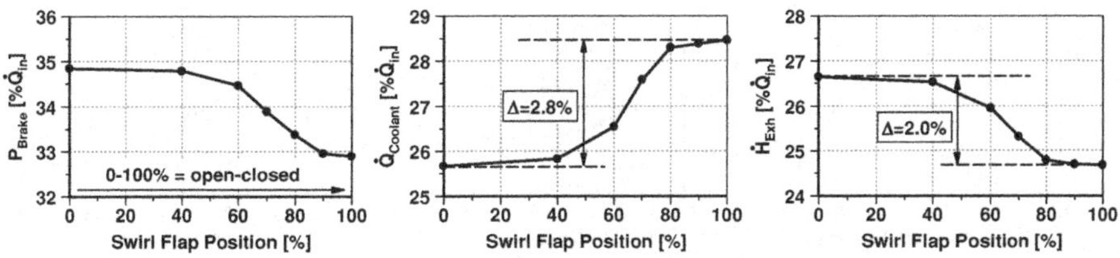

Fig. 9 Engine energy balance values for swirl flap position variation at 1500 rpm (start: 5 bar $IMEP_n$)

4.5 Swirl Flap Position

Swirl flaps in the intake system of Diesel engines are applied to create a higher in-cylinder charge motion. They are mainly used to reduce soot and CO emissions at low speed and loads. However, higher in-cylinder turbulence also causes increased heat transfer to the wall. Figure 9 shows the influence of swirl flap position on the engine energy balance of the test engine.

Boost pressure was adjusted by the VTG position to equalize higher pressure losses in the intake caused by the swirl flap. Thus, the air mass flow for the sweep was constant. The coolant heat flow increases by 2.8 % of the input energy when the swirl flap is closed. Brake power is reduced by almost 2 % due to higher throttling and wall heat losses. Exhaust gas enthalpy flow is reduced by approximately 2 % because of lower exhaust gas temperatures after the turbocharger. The effect of swirl flap position on charge motion and thus on coolant heat flow is an engine specific issue that has to be considered for accurate wall heat flow calculation. Coolant heat flow measurements at different swirl flap positions offer the possibility to modify existing heat transfer coefficient correlations in an appropriate way.

4.6 Boost Pressure

To analyze boost pressure influence on the engine energy balance, a boost pressure sweep from 1.15 to 1.75 bar was performed at constant charge air temperature. To quantify and compare the influence of different operating parameters, the authors decided to introduce ranges that are relevant for application (grey areas in the following figures) if parameters were varied outside a reasonable application range for the test engine. In this case the engine efficiency decreases significantly for boost pressures greater than 1.45 bar. Thus this value was chosen to be the limit of the application relevant range. The pressure sweep was realized by changing the turbocharger rack position and leads to an increased air mass flow of approximately 50 % and thus to higher flow velocities in the intake and exhaust system. Figure 10 shows the results of the engine heat balance regarding brake power, coolant heat flow and exhaust gas enthalpy flow.

Because of the greater mass flow rate and an increasing power consumption of the turbine, exhaust gas temperature is reduced significantly. The overall exhaust

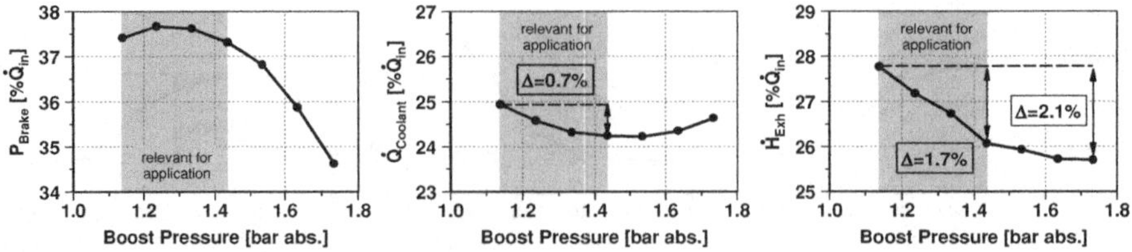

Fig. 10 Engine energy balance values for boost pressure variation at 2000 rpm (start: 7 bar $IMEP_n$)

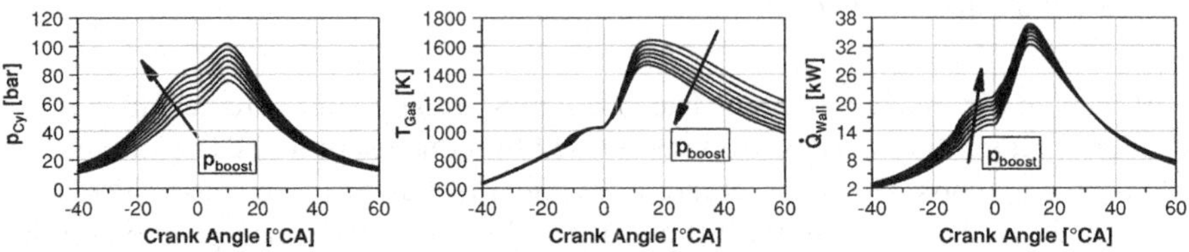

Fig. 11 Cylinder pressure trace analysis results for boost pressure variation at 2000 rpm (start: 7 bar $IMEP_n$)

gas enthalpy flow decreases by 2.1 % of the input energy, because the lower exhaust gas temperature predominates the higher exhaust gas mass flow rate. While the exhaust gas enthalpy flow decreases continuously, the coolant heat flow shows a minimum at a boost pressure of approximately 1.5 bar. The increasing coolant heat flow at boost pressures over 1.5 bar cannot be explained with the cylinder wall heat flow. A pressure trace analysis shows that higher boost pressure leads to significantly lower mean gas temperatures during combustion and expansion (see Fig. 11).

Both, higher cylinder pressure and lower gas temperature, lead to continuously higher in-cylinder heat transfer coefficients. In total, the combination of both effects leads to a higher wall heat flow during compression and the first part of combustion. About 30 °CA ATDC the lower gas temperatures predominate the higher heat transfer coefficients and the trend changes. The accumulated wall heat flow for the entire cycle shows a continuously increasing cylinder wall heat flow with increasing boost pressure up to 2.1 % of the input energy (see Fig. 12). The coolant heat flow minimum at a boost pressure of 1.5 bar as described above results from the heat flow behaviour in the intake and exhaust ports.

Coolant heat input in the exhaust ports decreases and coolant heat removal in the intake ports increases with rising boost pressure. In the intake ports a combination of constant charge air temperature and higher heat transfer coefficients due to increased mass flow leads to a constantly increasing amount of heat transferred from the coolant to the intake air. The situation in the exhaust ports is different, because the exhaust gas temperature is not constant. The exhaust gas temperature decreases on a diminishing scale with rising boost pressure, because rising boost pressure also results in a higher exhaust gas back pressure. In total the

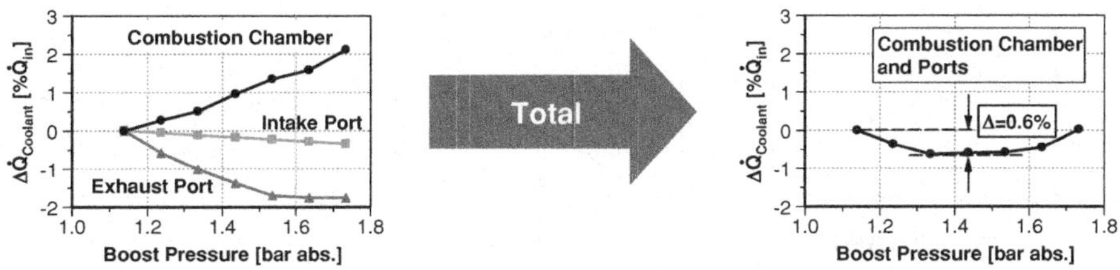

Fig. 12 Coolant heat input analysis results for boost pressure variation at 2000 rpm (start: 7 bar $IMEP_n$)

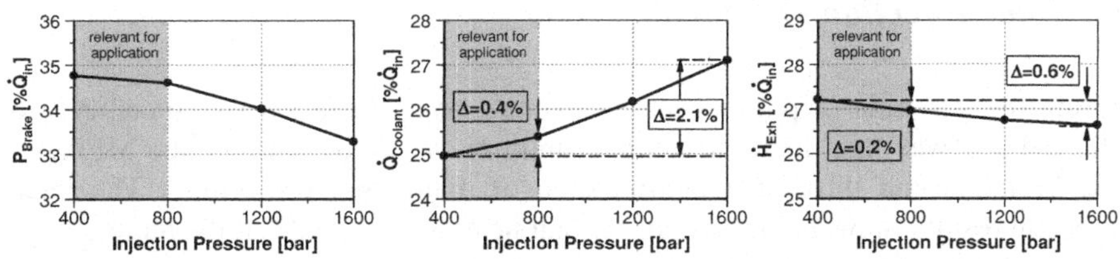

Fig. 13 Engine energy balance values for injection pressure variation at 2000 rpm (start: 5 bar $IMEP_n$)

result of the pressure trace analysis and the heat transfer analysis in the intake and exhaust ports leads to a result close to the measurement of the overall coolant heat input shown above (compare Figs. 10 and 12).

4.7 Injection Pressure

The investigation of the influence of the injection pressure on the engine energy balance showed that a high injection pressure under low load conditions lead to an increasing coolant heat input of up to 2.1 % of the input energy (see Fig. 13).

The influence of injection pressure is very low for the application relevant range (0.4 % of the input energy) and changes within the measurement uncertainty. However, the effect is remarkable and was confirmed with a pressure trace analysis. The inner energy balance of the combustion chamber during the high pressure cycle shows a reduction of about 4 % including a higher amount of unburned HC and CO due to high injection pressure at low load. Since all other relevant operating parameters were constant for the measurements (only main injection timing was adjusted slightly to keep $MFB_{50\,\%}$ constant), the too low calculated wall heat flow is the reason for this change in the in-cylinder energy balance. This behaviour could not be investigated in more detail within this work, but high pressure fuel injection leads to a more intense charge motion and combustion that proceeds closer to the cylinder wall. Rising heat transfer coefficients due to the higher turbulence level are the consequence and the most feasible explanation for the effect.

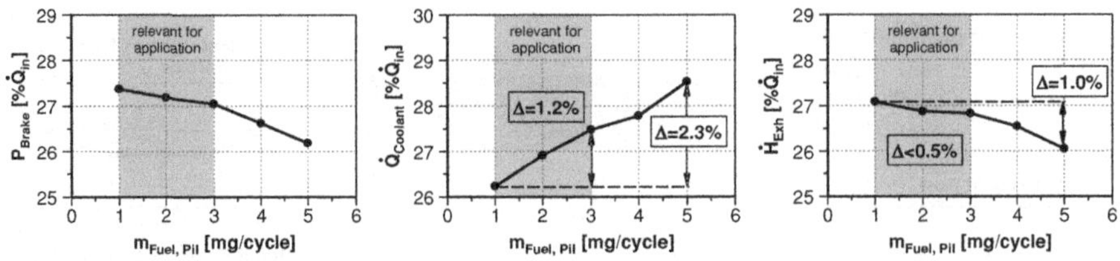

Fig. 14 Engine energy balance values for pilot injection quantity variation at 1500 rpm (start: 3 bar $IMEP_n$)

4.8 Pilot Injection Quantity

Within the variation of pilot injection quantity, main injection fuel quantity was adjusted in a way that the total fuel consumption was kept constant and $MFB_{50\ \%}$ was kept constant through a modification of main injection timing. The measurement results show an increase in coolant heat flow of 2.3 % of the input energy and a decrease of exhaust gas enthalpy flow of 1.0 % for a pilot injection quantity sweep from 1 to 5 mg/cycle (see Fig. 14).

The pressure trace analysis results show, that significant differences in-cylinder pressure only occur around firing top dead centre in the region of the pilot injection combustion (see Fig. 15). The average cylinder wall heat flow for the entire cycle only increases by 0.8 %.

Taking into account decreasing wall heat flow in the exhaust ports due to slightly lower exhaust gas temperatures, the overall increase of coolant heat input only adds up to 0.4 % of the input energy. Although friction increases by 0.6 % of the input energy due to higher maximum pressure rises with increasing pilot injection quantity from 2.7 to 7.3 bar/°CA, the calculated overall coolant heat input is less than the measured value. This leads to the assumption that the existing heat transfer correlations have room for improvement in case of low loads and pronounced pre-combustions in the top dead centre region. Research results published by Bargende [8] deal with the question how strong combustion turbulence influences wall heat transfer. It is shown that especially wall heat flow during pre-combustion in Diesel engines is higher than predicted by the existing models. This effect would explain the measured increase in coolant heat flow with an intensification of pre-combustion through a higher pilot injection quantity. However, the overall influence of pilot injection quantity on the engine energy balance within the application relevant range is comparatively low.

4.9 Pilot Injection Timing

Pilot injection timing also has a little effect on the engine energy balance. Figure 16 shows the results of a pilot injection timing variation with constant main injection timing and a pilot injection quantity of 2 mg/cycle. The variations in

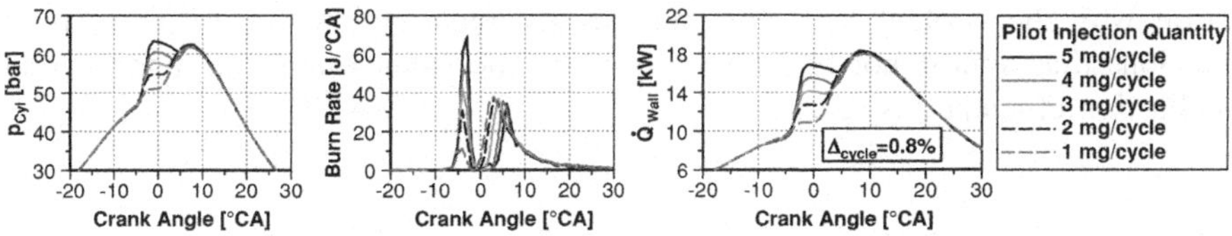

Fig. 15 Pressure trace analysis results for pilot injection quantity variation at 1500 rpm (start: 3 bar $IMEP_n$)

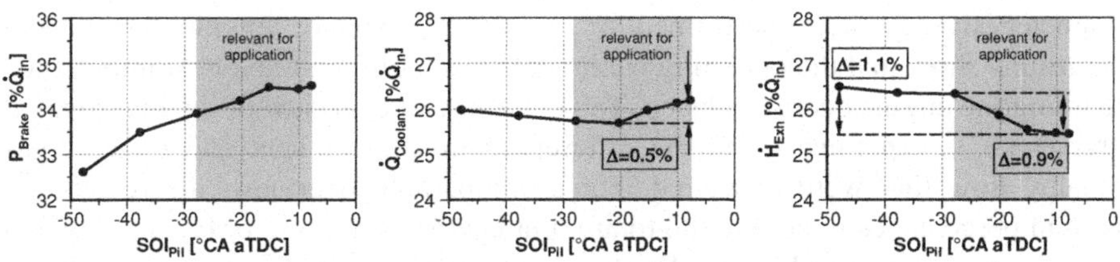

Fig. 16 Engine energy balance values for start of pilot injection at 1500 rpm (start: 5 bar $IMEP_n$)

Table 1 Influence of operating parameters on the engine energy balance, overview

Operating parameter	Investigated sweep	$\Delta\dot{Q}_{coolant}[\%\dot{Q}_{in}]$	$\Delta\dot{H}_{exhaust}[\%\dot{Q}_{in}]$	Importance
EGR rate	0–33 %	8.6	8.4	Very high
Main injuction timing	$MFB_{50\ \%}$:2°CA BTDC-32 °CA ATDC	2.0	9.4	High
Coolant temperature	50–90 °C	9.7	1.8	
Intake air temperature	22–80 °C	4.1	0.7	Medium
Swirl flap position	0–100 %	2.8	2.0	
Boost pressure	1.14–1.74 bar	0.7	2.1 (1.7)	
Pilot injection quantity	1–5 mg/cycle	2.3 (1.2)	1.0 (0.3)	Low
Pilot injection timing	48–8 °CA BTDC	0.5	1.1(0.9)	
Injection pressure	400–1600 bar	2.1 (0.4)	0.6 (0.2)	

coolant heat flow with a maximum difference of 0.5 % of the input energy stay within the range of accuracy of measurement. Exhaust gas enthalpy flow differences are a little higher, because unburned HC and CO-emissions are imputed to the exhaust gas enthalpy flow as described above. Early pilot injections only burn partly which results in increasing HC and CO emissions and in an increasing ignition delay. This results in a greater premixed portion and thus harder combustion of the fuel injected in the main injection event. This seems to be the reason why the coolant heat flow reaches its maximum with very early pilot injections that almost not burn and late pilot injections where the highest total combustion efficiency is reached.

5 Summary and Conclusion

The paper gives an overview of the influence of operating parameters on the thermal behaviour and energy balance of an automotive Diesel engine. Operating parameter sweeps were performed on an energy balance engine test bench. The results show that the operating parameters reach different levels of importance regarding the engine energy balance. Table 1 gives an overview of the investigated parameters, sweep ranges and the differences in coolant heat and exhaust gas enthalpy flow that were measured. For a comparable overview, the results are given in percentages related to the total input energy consisting of fuel energy and intake air enthalpy (see Fig. 1). The bracketed values show the maximum differences within the range that is relevant for application.

EGR rate, main injection timing and coolant temperature are the most important parameters which can shift energy flows by more than 8 % of the input energy. The absolute changes in coolant heat and exhaust enthalpy flow can reach values of up to 40 % depending on the operating point. While main injection timing mainly affects exhaust gas enthalpy flow and coolant temperature mainly affects coolant heat flow, the EGR rate has a strong influence on both values. Intake air temperature, swirl flap position and boost pressure also have a significant influence on the energy balance, while pilot injections and injection pressure only have a minor impact. The parameters with low importance do not offer much scope for a thermally optimized application and are not crucial for thermal engine simulation. Parameters with medium and high importance should be considered carefully for overall vehicle thermal management optimization. The importance classification of operating parameter influence can be seen as generally applicable for the energy balance of this size of light duty Diesel engines.

References

1. Donn C, Zulehner W, Ghebru D, Spicher U, Honzen M (2011) Experimental heat flux analysis of an automotive diesel engine in steady-state operation and during warm-up, SAE technical paper no. 2011-24-0067
2. Hohenberg G (1979) Advanced approaches for heat transfer calculations, SAE technical paper no. 790825
3. Ghebru D, Donn C, Zulehner W, Kubach H, Wagner U, Puntigam W, Strasser K (2012) Vehicle warm-up analysis with experimental and co-simulation methods, FISITA world automotive congress 2012, Beijing
4. Zapf H (1968) Untersuchung des Wärmeübergangs in einem Viertakt-Dieselmotor während der Ansaug- und Ausschubphase. PhD-thesis, Technische Universitaet Muenchen

5. Bauder R, Bach M, Fröhlich A, Hatz W, Helbig J, Kahrstedt J (2010) The new generation of the audi 3.0 TDI engine—low emissions, powerful, fuel-efficient and lightweight, 31. International vienna motor symposium, Austria
6. Pischinger R, Klell M, Sams T (2002) Thermodynamik der Verbrennungskraftmaschine. Springer, Wien, ISBN 3-211-83679-9
7. Smith LA, Preston WH, Dowd G, Taylor O, Wilkinson KM (2009) Application of a first law heat balance method to a turbocharged automotive diesel engine, SAE technical paper 2009-01-2744
8. Bargende M, Heinle M, Berner H-J (2011) Some useful additions to calculate the wall heat losses in real cycle simulations, 13th congress. D Graz, Austria

Development of Electric Engine Cooling Water Pump

Atsushi Saito and Motohisa Ishiguro

Abstract Aisin has developed the electric water pump for engine cooling as a pioneer in Japan. It has been necessary to downsize the pump and reduce cost to install the electronically controlled components into the engine. But Aisin has accomplished it with sufficient reliability for engine installation by developing various ways as follows.

Keywords (Standardized) EV and HV systems · Electrical accessories · Cooling and temperature management [A3]

1 Preface

Now, heightened interest in the environment is ever increasing the need for improvement of vehicle fuel efficiency.

Further improvement of fuel efficiency is also required for hybrid engines and all the more necessary for gasoline engines.

Against this background, one of the items that contribute to improving fuel efficiency for hybrid engines would be the optimal control over the flow rate of engine cooling water [1]. But, controlling the flow rate of cooling water was difficult because the pumping rates for the conventional mechanical water pumps depended on the engine rpm.

F2012-A07 016

A. Saito (✉) · M. Ishiguro
Aisin Seiki Co. Ltd, 2-1 Asahi-machi, Kariya, Aichi 448-8650, Japan
e-mail: a-saito@engin-t.aisin.co.jp

SAE-China and FISITA (eds.), *Proceedings of the FISITA 2012 World Automotive Congress*, Lecture Notes in Electrical Engineering 190,
DOI: 10.1007/978-3-642-33750-5_32,

Fig. 1 Structure of electric pump

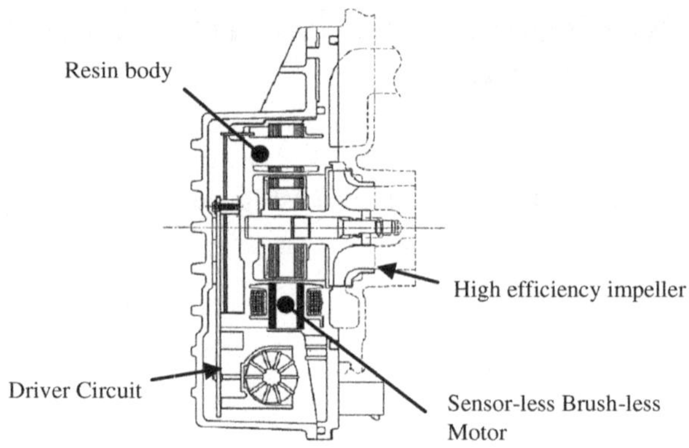

Our company originally developed electric inverter cooling water pumps; thinking that it was possible to implement an optimal control over the flow rate of cooling water by eliminating such dependency on the engine rpm through electrically-driven operation by adapting those pumps for use with engines, we started to tackle the development of an electric engine cooling water pump.

But, if we had tried to secure the maximum discharge flow rate equivalent to those of the conventional mechanical water pumps when developing this electric engine pump, we would have resulted in producing a large frame pump.

It would have been inevitable that the result not only would affect its mountability to the limited space within an engine room, but also would bring forth a high-cost electric engine cooling water pump.

For this reason, we participated in the engine cooling system working group to consider and implement elements, in cooperation with the other members in the group, such as the reduction of pressure loss in an entire cooling system and optimization of flow rates required for engines and devices, and thereby we could bring forth an electric engine cooling water pump that was low-cost and suitable for controlling hybrid engines.

2 Overview of Electric Engine Cooling Water Pump

Figure 1 shows the overview of the electric engine cooling water pump we have developed recently.

The structure generally consists of the pump, motor, and driver sections, and the number of parts in the pump and motor sections has been reduced by integrating the water room and stator, and the impeller and rotor, etc.

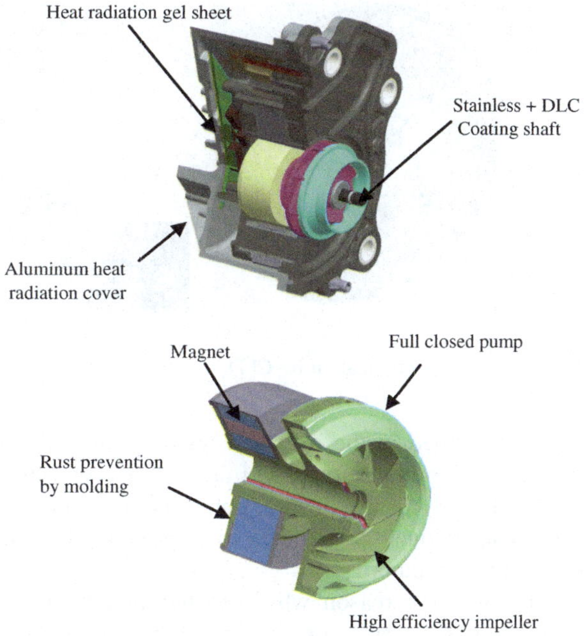

Fig. 2 Efficient technology in electric pump

3 Development Concept

The development concept for the electric water pump is saving power and reducing friction aimed at the improvement of fuel efficiency for single engines.

As electric engine cooling water pumps are intended to replace mechanical counterparts, the former should be able to be mounted to positions similar to those for the latter in an engine.

For this reason, we set it out as the development object that we would consider first mounting an electrical pump directly to an engine, and downsize it to the size similar to that of its corresponding mechanical pump.

4 Development Case Examples

4.1 Making Electric Cooling Water Pump More Efficient

There are many ways to downsize electric cooling water pumps, and if we adopted any means easily, it would surely result in a high-cost pump.

We thought that we could downsize the electric cooling water pump by improving its efficiency.

Figure 2 shows an example of the high-efficiency technology used for this pump.

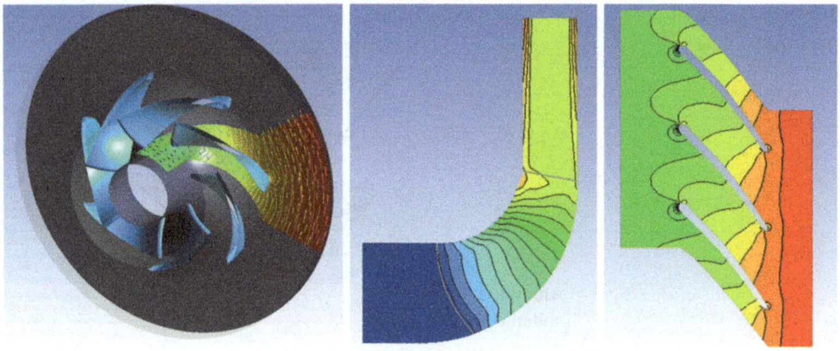

Fig. 3 Impeller optimization by CFD

As described in the overview of the pump, this electric cooling water pump consists of a pump, motor, and driver, each of which has its own efficiency.

As the efficiency of the pump is the lowest among the three elements, improving it most contributes to improving the efficiency of the entire electric cooling water pump.

This was the reason why we attempted to improve the pump efficiency by optimizing the shape of the impeller. Figures 3 and 4 show an example of optimizing the impeller shape, using CFD.

We were successful in optimizing the impeller shape so that we could get the maximum efficiency at the working point of the pump.

In addition, we adopted a centrifugal pump for this pump; as it frequently happened that the working point at which efficiency was high in the centrifugal pumps was normally different from that at which motor efficiency was high, we attempted to achieve high efficiency by performing the fitness of characteristics or getting the two high-efficiency points closer to each other.

4.2 Securing Rust Prevention Strength for Rare Earth Magnet

Next, we describe a case example of attempting to make the pump smaller and more efficient by reducing the parts used for it.

The mechanical cooling water pump gains the driving force by conveying the crank revolution output of an engine to the pump pulley through the belt.

The pump is structured such that the pulley is connected to it through the shaft and the cooling water is sealed by applying a mechanical seal to the shaft because the impeller is in the cooling water and the pulley is outside the engine.

As the electric cooling water pump has an advantage that it has driving force given by the motor, we thought that we could eliminate this mechanical seal.

The elimination of the mechanical seal would contribute not only to lowering cost due to part reduction, but also to downsizing the motor due to reduced friction by removing the sliding portion.

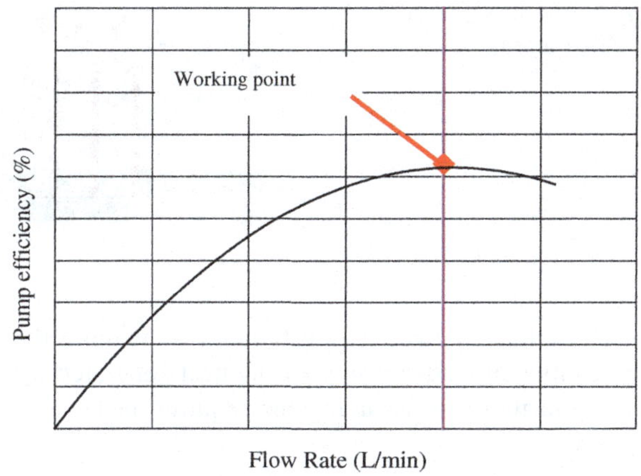

Fig. 4 Pump flow rate and efficiency

For this reason, we could think of a structure in which the yoke and magnet part of the motor were made to be immersed in cooling water, but both the yoke and magnet had a property of getting rusty against water.

This fact made us adopt a structure of wrapping the yoke and magnet in cooling water with resin to secure rust prevention for use of them in cooling water.

Wc adopted PPS resin with little water absorbency as the resin material, and have prevented water from penetrating into the magnet part through forming the resins in two colors.

In fact, the structure is such that the boundary face between two resins is sealed by forming a circular protrusion through the first color molding in advance in the boundary surface section between them that will be subjected to two-color molding, and by melting this protrusion with heat generated by the resins formed at the time of the second color molding.

4.3 Heat-Resistance Reliability by Low-Cost Electronic Substrate

This section introduces a case example in which we considered adopting a low-cost printed circuit board to develop a low-cost electric cooling water pump.

An issue when adopting a printed circuit board under an environment in which the motor is mounted on an engine is to secure heat resistance of its electronic parts.

If we tried to secure heat resistance of the electronic elements themselves, it would be inevitable that the elements would be high-cost.

For this reason, we secured a heat release path with an aluminum chassis as the heat sink for heat generated within the circuit as shown in Fig. 5, and thereby we succeeded in lowering the heat resistance level required for the printed circuit board to be used even in an environment where the board is mounted on an engine.

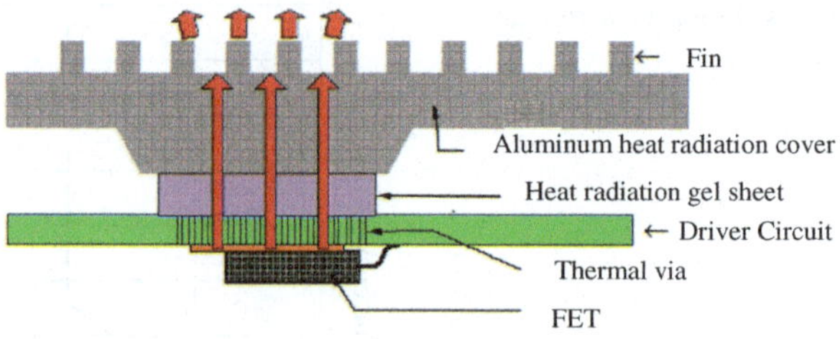

Fig. 5 Increase of heat release capacity

In addition, power saving brought by making the electric cooling water pump more efficient leads to suppressing heat generated by the elements themselves, and also contributes to reducing the required heat resistance level for printed circuit boards.

5 Conclusion

The developments described above have enabled us to develop a low-cost electric cooling water pump that can be mounted to an engine interchangeably with an existing mechanical cooling water pump, and making the engine cooling water pump electrically-driven has enabled us to contribute to vehicle fuel efficiency.

We have received reports that this electric engine cooling water motor has brought in other benefits, including 2 % improvement of fuel efficiency due to the following effects:

- Reduced friction
- Water pump power saving
- Improved heating performance
- Improved fuel efficiency in the range of low and medium loads and knocking control in the range of high loads
- Early warming-up
- Maintenance free

Acknowledgments Finally, we extend our gratitude to all involved both inside and outside the company who helped us greatly when developing and commercializing this electric engine cooling water pump.

Reference

1. Kiyoshi N, Toshihiro K, Nobuki K, Naoki T, Kenji H, (2009) Development of new 1.8-liter engine for hybrid vehicle [J]. A collection of preprints for the academic lecture of the Society of Automotive Engineers of Japan, Inc. (JSAE), 2009, No. 79-09, p 1–6

Structure and Parameters Optimization of Organic Rankine Cycle System for Natural Gas Compressor Exhaust Gas Energy Recovery

Yongqiang Han, Zhongchang Liu, Yun Xu and Jing Tian

Abstract In the paper, the structure and working principle of free piston based organic rankine cycle (ORC) exhaust gas energy recovery system have been presented in detail. A GT-Power software based prototype model has been established in order to find out the effects of different parameters. Comparisons under different thermo parameters of heated and cold work material have been analyzed thoroughly. Simulation results show that the thermo parameters of heated work material are very sensitive to net heat efficiency.

Keywords Organic rankine cycle · Waste energy recovery · Natural gas compressor · Free piston · Simulation

1 Introduction

In this paper, a free piston based organic rankine cycle exhaust gas energy recovery system has been presented for the overall compressor of natural gas recovery [1]. The system uses organic rankine cycle to recycle exhaust gas energy and transforms into mechanical energy by similar sterling cycle. In addition, combining with the structure of free piston, the system can realize kinds of lifting by coupling the ratio of the motive power piston area to the compression piston area and the Thermal state of working medium in Power cavity.

F2012-A07-023

Y. Han (✉) · Z. Liu · Y. Xu · J. Tian
State Key Laboratory of Automotive Simulation and Control, Jilin University, Jilin, China
e-mail: hanyq@jlu.edu.cn

SAE-China and FISITA (eds.), *Proceedings of the FISITA 2012 World Automotive Congress*, Lecture Notes in Electrical Engineering 190,
DOI: 10.1007/978-3-642-33750-5_33,

The working principle can be described simply as follows: High temperature and pressure rankine working medium in the tank make free piston do linear reciprocating motion in cylinder by produce super-saturation steam and compress low-pressure natural gas when the connecting rod is driving compression piston. Liquid medium will absorb the heat of working medium which has finished doing work and make it liquid after liquid medium has been injected into work cavity when circulation work has been completed. By this way, the power waste heat and the latent heat of vaporization of working medium can be used fully so that the purpose of saving energy and environmental protection will be realized.

The system has six characteristics which can be described as follows: (1) Using organic rankine cycle to recycle exhaust gas energy and similar Sterling cycle to transforms into mechanical energy; (2) The structure of free piston, which can realize kinds of lifting by coupling the ratio of the motive power piston area to the compression piston area and the Thermal state of working medium in Power cavity; (3) The injection of High temperature and pressure organic liquid working medium, flash urgent boiling to medium pressure and superheated steam, keeping working medium expanding power always in the superheated steam state; (4) Achieving the function of too low back pressure and heat recovery; (5) Using time constant of heat transfer of space multiphase flow to determine the limit working frequency and the injection of working medium frequency to adjust lifting rate; (6) Considering that the natural gas lifting is a heat release process, a cooler is set up to maintain a suitable thermodynamic state for rankine cycle access. The follows are the basis of selecting this system:

For one hand, A lot of researches show that organic rankine Cycle has a good prospect in the low-grade exhaust gas energy recovery [2–6]. One study shows that ORC cycle thermal efficiency of the waste gas of a miniature gas turbine in 260 °C which uses R123 as working medium can reach up to about 18 % [7]. Because that a overall natural gas compressor is a stationary power machinery and its site is compared capacious commonly, the device of exhaust gas energy recovery of organic rankine cycle is very feasibility.

Compared with sterling cycle, rankine cycle's thermal efficiency is low because of involving phase change and the latent heat of vaporization loss which account for more than 50 % of the total energy of the working medium. Besides that, rankine cycle also exist the problem of not fully expansion for the reason that working medium's degree of superheat changes when it expands. Therefore, recycling the latent heat of vaporization and keeping the degree of superheat are basic ways of improving rankine cycle's thermal efficiency. The utilization of the latent heat of vaporization of rankine cycle can be realized through the way of set tint up a pressurization regenerator which is similar to steam engine cycle or rankine cycle uses.

On the other hand, free piston system has advantages of seldom moving parts, simple structure and few mechanical losses. In addition, the system saves power consumption of lubrication and cooling system because of making rankine cycle working medium as cooling medium and lubrication medium, and it has efficient potential.

In conclusion, a free piston based organic rankine cycle exhaust gas energy recovery system can advance rankine cycle's efficiency by making the latent heat of vaporization be used, reduce the cost and power consumption of system by using free piston which has a simple structure and have a optimistic application prospect.

2 System Structure and Work Principle

A free piston based organic rankine cycle exhaust gas energy recovery system is made up by the follows roughly: 2. Low-pressure gas pipeline 4. High-pressure natural gas pipeline 8.22. Comparison cylinder 9.21. Comparison piston, 15. Power piston, 28. Working medium tank, 31.41. Working medium collecting tank, 35. Cooler 43,48. Working medium mixer 51. Heating boiler, etc. Both ends of low-pressure gas pipeline is joined to the comparison cylinder's intake respectively; Both ends of High-pressure natural gas pipeline is joined to the comparison cylinder's outlet respectively; Two comparison piston links with the power piston through the connecting rod; The ratio of power piston's size to comparison piston's depends on low-pressure natural gas's pressure value and high-pressure natural gas's by the way of formula computing; The power piston is placed in the working cylinder; The Low-pressure gas pipeline is connected to the heating boiler; The working medium tank links with the cooler; The two work material collecting tanks are place on the two sides of power cylinder respectively and link with Heating boiler. Figure 1 shows the system structure.

The working process of the system can be described simply as follows: high temperature and pressure working medium enter into power cavity after multilevel heating (compressor's cooling fluid, compressor's lubricant, lift high temperature natural gas, compressor's high temperature exhaust). We could make the initial power piston when it is located on the right dead point as an example: high temperature and pressure rankine working medium is injected into right of the power piston (a large Middle area of the cylinder) and boils rapidly in it. Then the working medium will produce 0.6–0.8 MPa of super-saturation steam and it make the piston move to left. Natural gas will be compressed to 1.5–1.6 MPa by Comparison piston and pass into high-pressure natural gas pipeline through one-way valve. When power piston close to the left dead point mass under cooling working medium in the working medium tank is sprayed into right of the power cylinder and exchanges space of multiphase flow heat such as evaporation and liquefaction, and make working medium liquid, so that the pressure to right power cylinder falls sharply and under cooling working medium absorbs all the latent heat of vaporization. At the same time high temperature and pressure working medium is injected into left power cylinder after it has mixed in the working medium mixer and boils rapidly to high-pressure super-saturation steam. This super-saturation steam make the power piston and comparison piston move and make natural gas compressed and lift. When the power piston near to the right head point under cooling working medium is sprayed into the left power cylinder

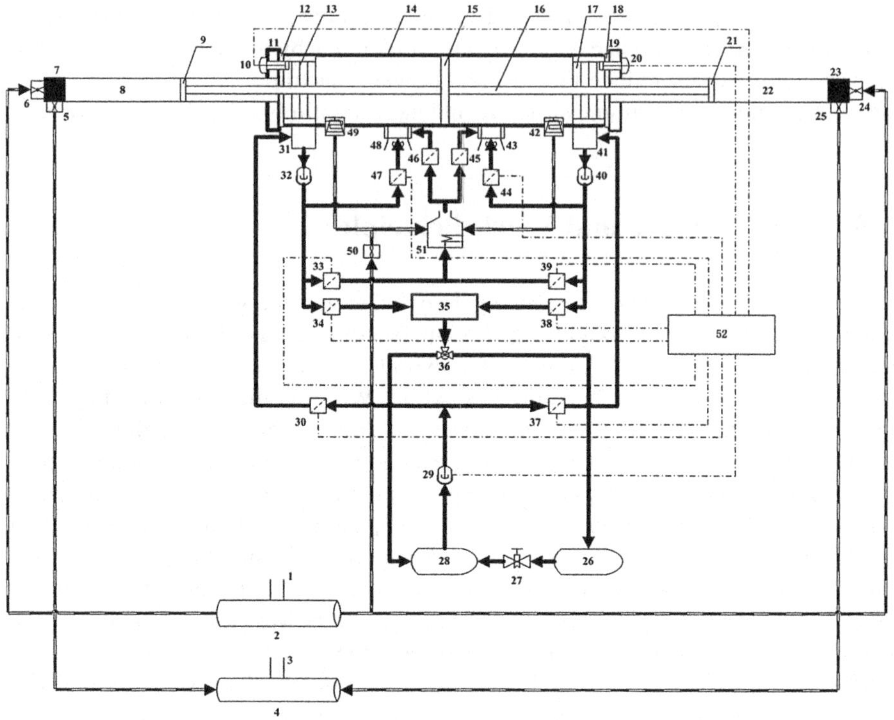

Fig. 1 Structure of a free piston based organic rankine cycle exhaust gas energy recovery system

and is cooled urgently in the cylinder. This leads pressure of the left power cylinder to fall. And high temperature and pressure working medium is injected into the right power cylinder and boils rapidly. This makes pressure of the right power cylinder rise, then the power piston move left again. So go round and round to make high-pressure gas produce. After the working medium completes doing work its pressure reduces and the working medium flows into working medium tank. The control unit monitors thermal state of rankine cycle's pipeline currently and control the working medium flow of the cooler, so as to ensure the suitable thermal state of rankine cycle's pipeline.

3 Simulation and Discussion

3.1 Model Specification

There are few kinds of commercial software include ORC reference and library. Several papers have discussed the performance of traditional ORC waste energy recovery system by using home-made Simulink codes. The GT-Suite based model

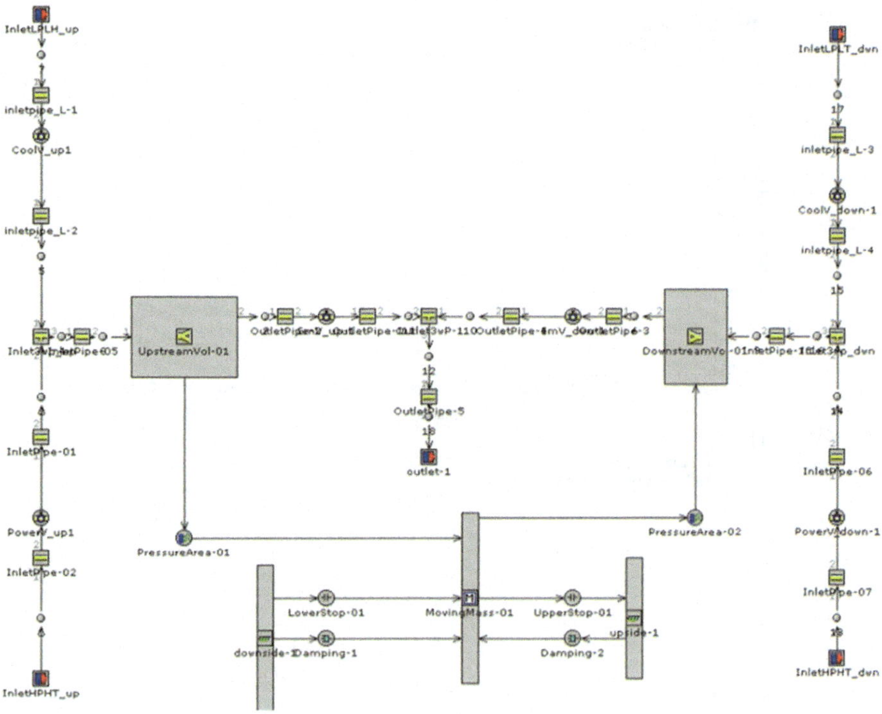

Fig. 2 Model diagram

has been established as Fig. 2. The model mainly composed power piston, up-down cylinders, power and cooling work material injectors, cooled work material outlet port subsystems. In order to calculate the total work, a constant resistance is attached to the power piston.

3.2 Simulation Results

Figure 3 shows the power piston displacement, the pressure of down- cylinder and the volume of up-cylinder histories, which are gained by adopting the parameters listed in Table 1. Each work material is sprayed for 0.5 s in a toothed curve. The piston displacement fluctuates from–90.1 mm to 92.5 mm and the down-cylinder pressure differs from 4.0 bar (outlet P) to 9.0 bar (power P) and accord to the prediction. Therefore, the above mentioned model has provided a convenience for the free piston based ORC waste energy recovery system study.

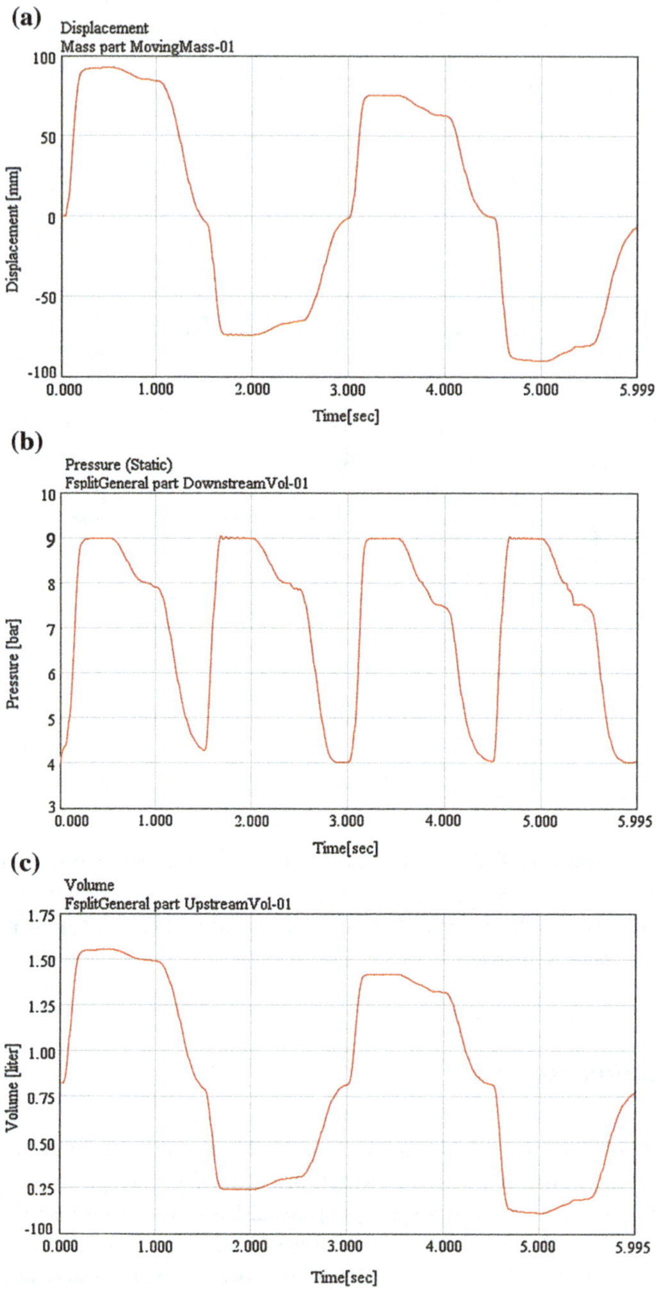

Fig. 3 Typical results of simulation. **a** Piston displacement. **b** Down-cylinder pressure. **c** Up-cylinder volume

Table 1 Model key parameters of case 1

Work material	R245ca	Power work material Temperature/K	430
Damp force/(N-s/m)	1,000	Cooling work material T/K	320
Up-cylinder D/mm	100	Outlet work material T/K	320
Down-cylinder D/mm	100	Power work material Pressure/bar	9.0
Piston Stroke/mm	200	Cooling work material P/bar	9.0
Piston mass/kg	1.0	Outlet work material P/bar	4.0
Piston Diameter/mm	100	Cycle duration/s	3.0
Simulation duration/s	6.0	Cooling work material injector D/mm	7.5

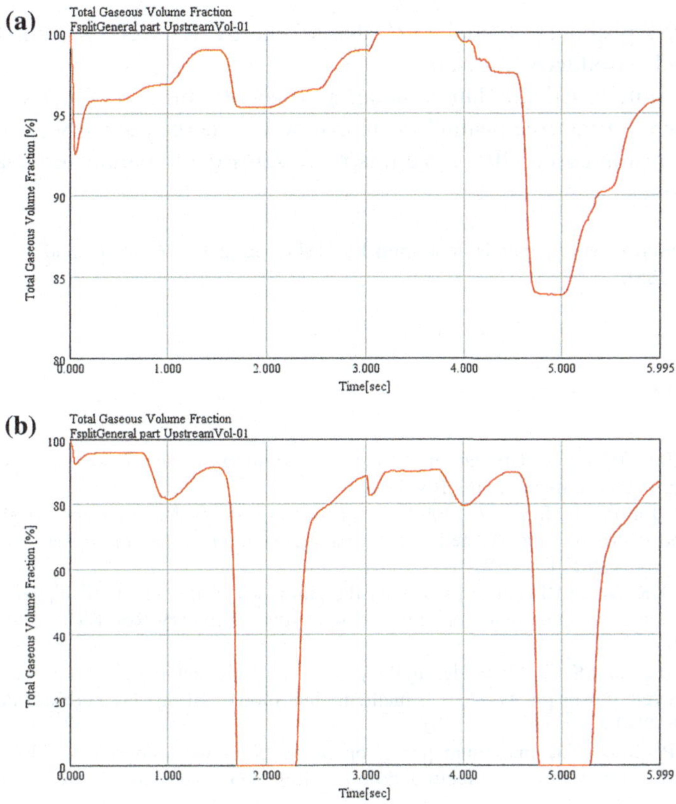

Fig. 4 Comparison of gaseous volume fraction under different cases. **a** Case 1. **b** Case 2

3.3 Sensitive Parameter Analysis

Figure 4 shows the Total Gaseous Volume Fraction comparison between case 1 and case 2 (only enlarges the cooling work material injector diameter from 7.5 to 10.5 mm, case 2).

The gas volume fraction indicates the absorption completeness of the vaporization latent heat in free piston ORC system. By changing the sprayed quantity of cooling work material can make the power cylinder remained material from over heat vapor to over cold liquid, which means more latent heat has been reused. Therefore, because of the complexity and sensitivity in the free piston based ORC system, more attentions should be paid in parameters optimization.

4 Summary

A novel free piston structured ORC based energy recovery system has been designed and introduced in detail.

The GT-Suite based simulation model has been established, which can provide a convenience for the free piston based ORC waste energy recovery system study.

The free piston based ORC system is more sensitive to parameters than that of traditional ones.

Acknowledgments This paper is supported by NSFC project (50976046) and NCET project (NCET-2011-185).

References

1. Han Y, Zhao H, et al. Utilization device of natural gas compressors energy [P]. Jilin Universitiy, Patent number: 2,011,104,339,64.1
2. James AM, Jon RJ, Jr, Jiming C (2009) Experimental testing of gerotor and scroll expanders used in and energetic and exergetic modeling of, an organic rankine cycle. J Energy Res Technol Trans ASME 131:012201
3. Srinivasan KK, Mago PJ (2008) Improving the efficiency of the advanced injection low pilot ignited natural gas engine using organic rankine cycles. J Energy Res Technol Trans ASME 130:022201
4. Danov SN, Gupta AK (2004) Modeling the performance characteristics of diesel engine based combined-cycle power plants—Part I: mathematical model. J Eng Gas Turbines Power Trans ASME 126:28–34
5. Richard BP (2007) The maximum power operating point for a combustion-driven thermoelectric converter with heat recirculation. J Eng Gas Turbines Power Trans ASME 129:1106–1113
6. Korakianitis T, Grantstrom J, Wassingbo P (2005) Parametric performance of combined-cogeneration power plants with various power and efficiency enhancements. J Eng Gas Turbines Power Trans ASME 127:65–72
7. Zhao W, Du J, Xu J (2009) The design and analysis of combination cycle of miniature gas turbine and organic rankine cycle. Chin Electr Eng J 29(29):19–24

Study on Exhaust Heat Recovery Utilizing Organic Rankine Cycle for Diesel Engine at Full-load Conditions

Yan Chen, Yanqin Zhang, Hongguang Zhang, Bin Liu, Kai Yang and Jian Zhang

Abstract In this chapter, organic rankine cycle (ORC) system was utilized to recover diesel engine exhaust heat. The performance of ORC was discussed under the full-load conditions of diesel engine. The results indicate that when the turbine inlet temperature is 450 K, the max network of ORC is 15.09 kW at the speed of 1,599 r/min, and the diesel power increased by 6.17 %. At the same speed, the network of ORC reduces as increasing as the turbine inlet temperature when the exhaust heat changes. The proper range of the exhaust temperature leaving the evaporator is 373–383 K. At last, the diesel engine power increases by about 6 % under the full-load conditions of the diesel engine, and the relationship between the network of ORC and the diesel engine power was given in this chapter.

Keywords Organic rankine cycle · Diesel engine · Exhaust heat recovery · Full-load conditions · Performance analysis

According to vehicle engine heat balance, heat for the power output is only 30–40 % of the total fuel burning energy while the waste heat is 60–70 %. The common method cannot effectively use this part of the low-grade energy. Therefore the low-grade energy utilization technologies are needed. At present, the low-grade energy technologies include the organic rankine cycle (ORC), micro-organic rankine cycle, the Stirling machine, cogeneration, and reverse Brayton cycle etc. Bianchi et al. [1] analyzed the bottoming cycle for power generation technology.

F2012-A07-024

Y. Chen (✉) · Y. Zhang · H. Zhang · B. Liu · K. Yang · J. Zhang
College of Environmental and Energy Engineering, Beijing University of Technology, 100124, Beijing, China
e-mail: chenyan8769@hotmail.com

SAE-China and FISITA (eds.), *Proceedings of the FISITA 2012 World Automotive Congress*, Lecture Notes in Electrical Engineering 190,
DOI: 10.1007/978-3-642-33750-5_34,

Table 1 Properties of R123

M	152.93	kg/kmol
$T_{critical}$	456.83	K
$P_{critical}$	3.6618	MPa
$\rho_{critical}$	550.0	kg/m^3
T_{min}	166.0	K
T_{max}	600.0	K
P_{max}	40.0	MPa

The result showed ORC displayed the best performance in the recovery of low-grade heat source.

Currently, there are some scholars using the ORC system for the engine exhaust heat recovery. Srinivasan et al. [2], Vaja et al. [3], Weerasinghe et al. [4] and Ringler et al. [5] have designed ORC system to recover the engine exhaust heat in foreign. The research showed that the ORC system can not only improve the combustion efficiency, but also effectively reduce emissions of NO_X, CO etc. Fang Jinli et al. [6] and He Maogang etc. [7, 8] analyzed the performance of the ORC system recovering vehicle engine exhaust heat. But the engine exhaust was treated as a constant temperature heat source in all the above study. So at full-load conditions of a diesel engine, the performance of the ORC system was studied in this chapter.

Because R123 shows good performance in the low-grade ORC system, non-toxic and environmentally friendly [6, 9–15], R123 was selected as the working fluid in this chapter.

1 Thermodynamic Analysis of ORC

Basic ORC is utilized to recover the exhaust heat of a 6-cylinder diesel engine for heavy trucks in this chapter. The performance of the basic ORC was evaluated at the conditions of various turbine inlet temperature, various exhaust gas temperature leaving the evaporator and the full-load conditions of the diesel engine.

The properties of R123 are calculated by REFPROP 8.0 [25]. Table 1 shows main properties of R123

Calculation conditions are assumed that the ORC cycle is an ideal cycle; the condenser outlet temperature is 310 K; isentropic efficiencies of the turbine and pump are constant 65 and 80 %; the ambient temperature is 298 K. Firstly, a proper speed was set at the full-load conditions of the diesel engine, at which a speed corresponds to one exhaust gas temperature. The performance of the basic ORC was calculated at that temperature. From the calculation, the working parameters at best system performance were found. Then, this chapter evaluated how the amount of the heat transfer affected the system performance by changing

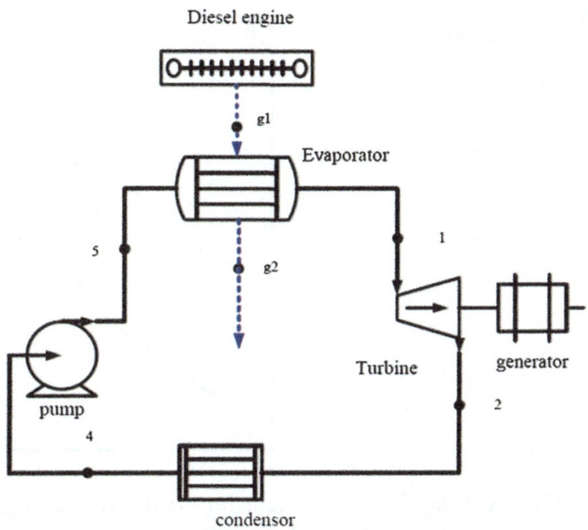

Fig. 1 Basic components of ORC assumed for diesel engine

the exhaust temperature leaving the evaporator. At last, the various characteristics of the basic ORC were analyzed at the full-load conditions of the diesel engine.

The basic components of ORC assumed for diesel engine are shown in Fig. 1. There are four different processes of the system as observed in Fig. 2: process 1, 2(s) (expansion process) that process 1, 2 is an actual expansion process and process 1, 2 s is an ideal expansion process, process 2(s)–4 (constant-pressure condensing process), process 4, 5 (pumping process), process 5–1 (constant-pressure evaporating process).

(1) Process 1, 2(s)

Heat is converted into work in this process. At ideal process, work is produced by the turbine using the saturated vapor of R123 during process 1–2 s. However, the actual process is the process 1–2 as it is not an isentropic process in the turbine. The shaft power of the turbine can be expressed as

$$W_t = \dot{m}(h_1 - h_2) \tag{1}$$

where h_2 can be considered as

$$h_2 = h_{2s} + (1 - \eta_{oi})(h_1 - h_{2s}) \tag{2}$$

where η_{oi} is the isentropic efficiency of the turbine.

The turbine irreversibility rate can be expressed as

$$\dot{I}_t = \dot{m}T_0(s_2 - s_1) \tag{3}$$

where T_0 is the ambient temperature.

(2) Process 2(s)–4

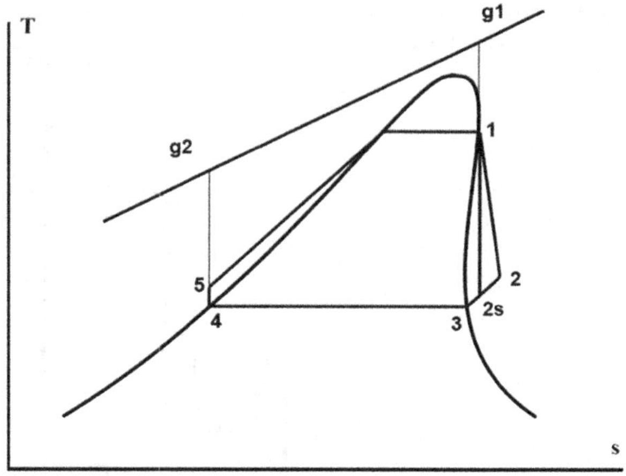

Fig. 2 T-s diagram of the system using R123

The working fluid is condensing to the saturated liquid at a constant-pressure condition in this process. Heat exchanged in the condenser can be expressed as

$$Q_c = \dot{m}(h_2 - h_4) \tag{4}$$

The condenser irreversibility rate can be expressed as

$$\dot{I}_c = \dot{m}T_0\left[(s_4 - s_2) - \frac{h_4 - h_2}{T_L}\right] \tag{5}$$

where T_L is the temperature of the low-temperature reservoir, which is considered to $T_L = T_4 - \Delta T_{pp}$.

(3) Process 4–5

The pump work can be expressed as

$$W_p = \frac{W_{p,ideal}}{\eta_p} = \frac{\dot{m}(h_5 - h_4)}{\eta_p} \tag{6}$$

Since the pump irreversibility is relatively small, it can be considered as isentropic, i.e. $\dot{I}_p \approx 0$. So there is another expression of the pump work as

$$W_p = \frac{1000\dot{m}(p_5 - p_4)}{\rho_c \eta_p} \tag{7}$$

(4) Process 5–1

The working fluid absorbs heat from the diesel engine exhaust gas, then transforms into saturated vapor with high enthalpy. This is a constant-pressure process. Heat exchanged in the evaporator can be expressed as

$$Q_e = q_m(h_{g1} - h_{g2}) = \dot{m}(h_1 - h_5) \tag{8}$$

The evaporator irreversibility rate can be expressed as

$$\dot{I}_e = \dot{m}T_0\left[(s_1 - s_5) - \frac{h_1 - h_5}{T_H}\right] \tag{9}$$

where T_H is the temperature of the high-temperature reservoir, which is considered to $T_H = T_1 + \Delta T_{pp}$.

(5) Total Cycle

The net power output of the system can be expressed as

$$W_{net} = W_t - W_p \tag{10}$$

The thermal efficiency of the system can be expressed as

$$\eta_{sys} = \frac{W_{net}}{Q_e} \tag{11}$$

The system second law efficiency can be expressed as

$$\eta_{\|} = \frac{W_{net}}{Q_e\left(1 - \frac{T_L}{T_H}\right)} = \frac{\eta_{\text{sys}}}{1 - \frac{T_L}{T_H}} \tag{12}$$

The total irreversibility rate of the system can be expressed adding Eqs. (3), (5) and (9) as following

$$\dot{I} = \dot{I}_c + \dot{I}_p + \dot{I}_e + \dot{I}_t = \dot{m}T_0\left(-\frac{h_1 - h_5}{T_H} - \frac{h_4 - h_2}{T_L}\right) \tag{13}$$

2 Results and Analysis

2.1 At 1,599 r/min of Full-Load Conditions

2.1.1 At Various Turbine Inlet Temperature Conditions

Table 2 shows the engine parameters. The speed at 1,599 r/min of the diesel engine is selected in this section, assuming the temperature of the exhaust gas is 383 K leaving the evaporator, and the range of the turbine inlet temperature is in the range of 330–456 K. Noticed, if the turbine inlet temperature exceeds the critical temperature of the working fluid, the properties of the working fluid may be unstable. It may be decomposed, which leads to the bad working conditions, even the system can't work and the equipment may be damaged.

Table 2 Engine characteristics at 1,599 r/min

Speed	1,599	r/min
Torque	1,460	N·m
Power	244.4	kW
Exhaust gas temperature	694	K
Exhaust gas mass flow	0.313	kg/s

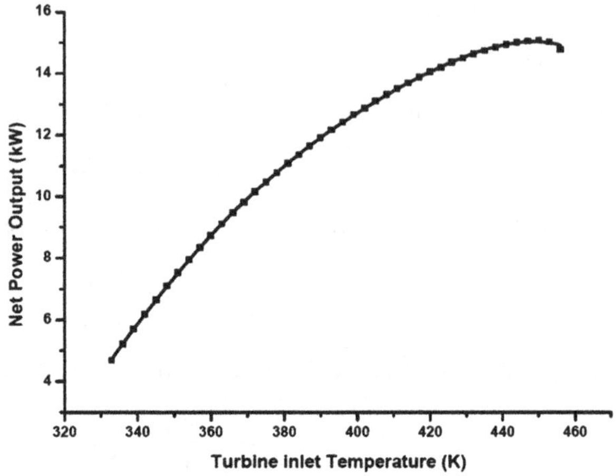

Fig. 3 ORC network versus turbine inlet temperature

The calculating results are shown in Figs. 3, 5. Figure 3 illustrates that the system net power output increases with the increment of the turbine inlet temperature, but if the temperature is reaching the critical value, the output will decrease. Due to Eq. (11), the system net power output is only related to the saturated vapor enthalpy of R123. Figure 4 shows the saturated vapour enthalpy of R123 has a peak value when reaching the critical value, so the system net power output has a peak value when reaching the critical point. The peak net power output is 15.09 kW at 450 K, and the engine power increases by 6.17 %.

Figure 5 illustrates the increment of the irreversibility rates of the key components of ORC system with increasing the turbine inlet temperature or pressure. The total irreversibility rate increases with the increment of turbine inlet temperature, in which the evaporator and turbine irreversibility rate account for high proportion. Because the condenser outlet temperature is assumed as a constant value, the condenser irreversibility rate is at a lower level in the total, and keeps essentially steady. The turbine irreversibility rate shows a smooth variation of growth, which is larger than the evaporator irreversibility rate at the condition of a low turbine inlet temperature. The evaporator irreversibility rate exceeds the turbine at 420 K. Therefore, it can be considered that the total irreversibility rate comes from the turbine and evaporator mainly as the condenser outlet temperature assuming as a constant value.

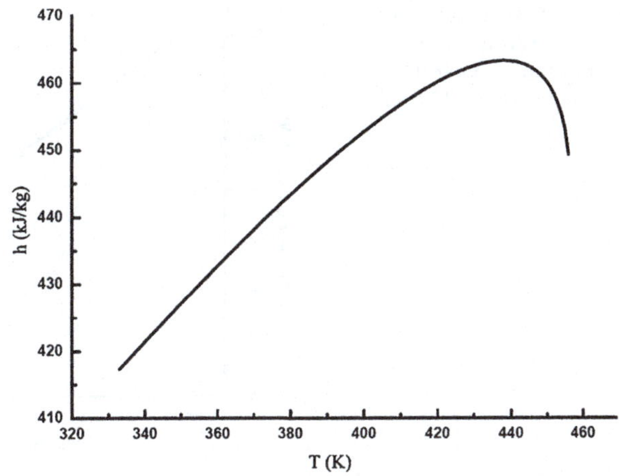

Fig. 4 T-h diagram of R123 at saturated condition

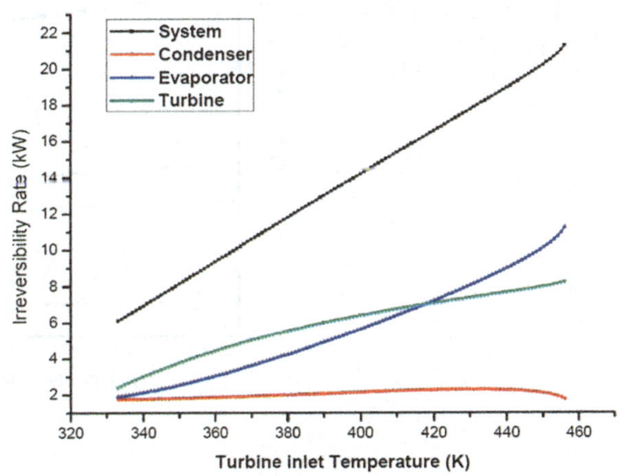

Fig. 5 The irreversibility rate of the key components versus turbine inlet temperature

2.1.2 At Various Exhaust Gas Temperature Leaving the Evaporator

In this section, the exhaust gas temperature leaving the evaporator is considered from 323 to 383 K. The speed of the diesel engine is also selected at 1,599 r/min, and the range of the turbine inlet temperature is assumed as 450 K.

Figure 6 demonstrates the variation of the system net power output with the increment of the exhaust outlet temperature leaving the evaporator when the turbine inlet temperature is constant at 450 K. It is observed that there is a linear relationship between the net power output and the exhaust gas temperature. The results show that reducing the exhaust gas temperature leaving the evaporator as much as possible can improve the net power output highly. However, drops may appear in the exhaust pipe when the exhaust gas temperature reduces below 373 K,

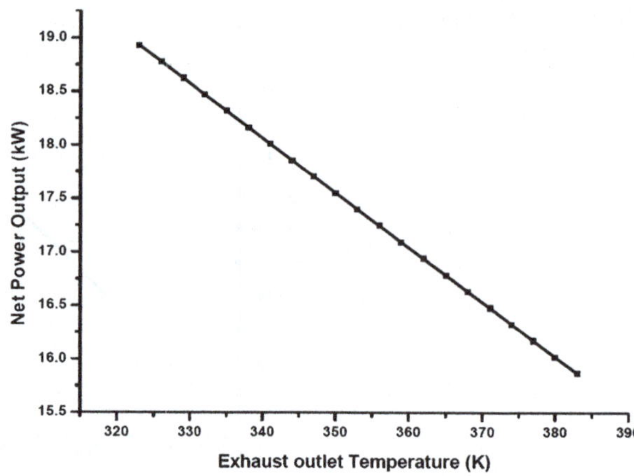

Fig. 6 System net power output versus exhaust outlet temperature leaving the evaporator (T_1 = 450 K)

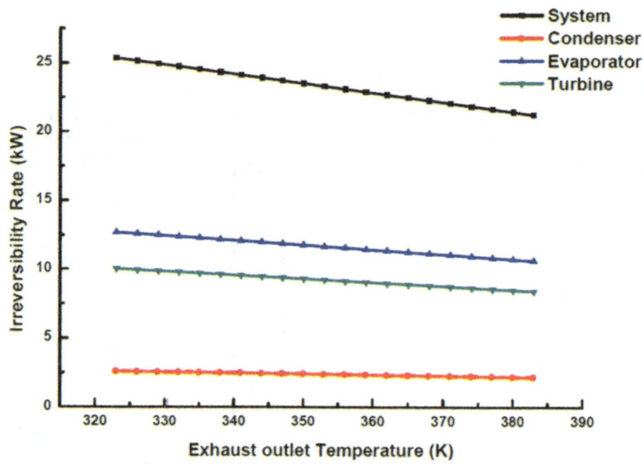

Fig. 7 The irreversibility rate in the key components of the cycle versus exhaust outlet temperature leaving the evaporator (T_1 = 450 K)

which may react with sulfide or nitride in the exhaust gas leading to damage the exhaust pipe. Therefore, selecting a proper exhaust gas temperature leaving the evaporator is needed.

Figure 7 demonstrates the irreversibility rate in the key components of the cycle varies according to exhaust outlet temperature leaving the evaporator at the turbine inlet temperature of 450 K. It can be observed that all the irreversibility rate reduce with increasing the exhaust outlet temperature leaving the evaporator. So the higher exhaust outlet temperature leaving the evaporator is needed. However, the higher exhaust outlet temperature leaving the evaporator is selected, the lower the net power output appears in Fig. 6. So the proper temperature range is considered between 373 and 383 K.

Table 3 ORC max network at full-load conditions of diesel engine

Speed (r/min)	Power (kW)	Net power output (kW)	The increment of the engine power %
799	80.4	4.97	6.18
897	99.2	5.96	6.00
1,003	118.7	7.15	6.02
1,101	147.5	9.15	6.20
1,199	175.2	10.52	6.00
1,302	199.8	12.07	6.04
1,402	220.3	13.30	6.04
1,500	235.3	14.82	6.30
1,599	244.4	15.09	6.17
1,695	245.6	14.82	6.04
1,796	246.4	14.93	6.06
1,904	246.7	14.43	5.85
1,978	173.7	8.90	5.12

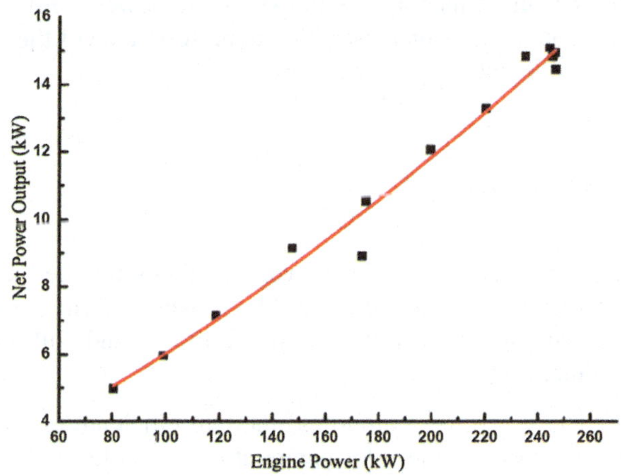

Fig. 8 ORC network versus the diesel engine power

2.2 At Various Speeds of Full-Load Conditions

The performance of the ORC is calculated at the full-load conditions of the diesel engine in this section. It is also assumed that the turbine inlet temperature is selected from 330 to 456 K, and the exhaust gas temperature leaving the evaporator is 383 K.

Table 3 shows the best net power output of the ORC system at every speed of full-load conditions. The result is the power of diesel engine increased by about 6 % with the ORC system, and the maximum increment is 6.30 %.

Figure 8 shows the relationship between the system net power output and the engine power. It can be observed that the system net power output becomes larger when the engine power is larger. Figure 9 demonstrates the irreversibility rate in the key components of the cycle varies at full-load conditions. The system net

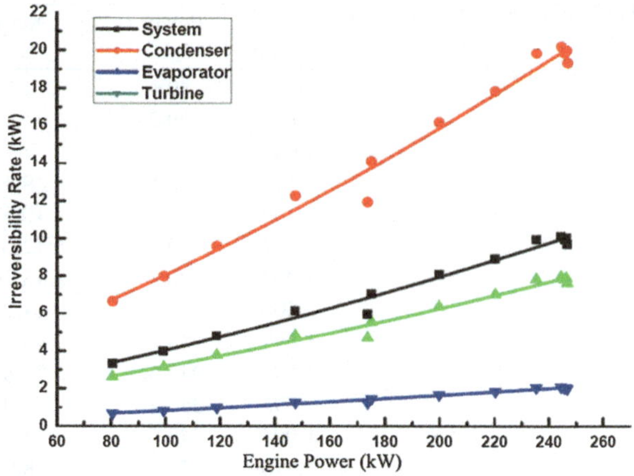

Fig. 9 The irreversibility rate of the key components versus the diesel engine power

power output increases with the engine power, while the irreversibility rate in the key components increases. So the performance of the ORC system becomes worse when the engine power is larger.

3 Conclusion

This chapter used R123 as working fluids for the basic organic rankine cycle to recover a diesel engine exhaust heat. The performance of ORC was analyzed at the constant speed of 1,599 r/min conditions and full-load conditions of the diesel engine, respectively.

1. The net power output and thermal efficiency of ORC increase with the increment of turbine inlet temperature at 1,599 r/min, but once turbine inlet temperature is reaching the critical value, the net power output and thermal efficiency decrease. The peak net power output and thermal efficiency are 15.087 kW and 13.38 %. The diesel engine power increases by 6.17 %. The system irreversibility rate increases as rising as turbine inlet temperature, in which the evaporator and turbine irreversibility rate make up large proportion.
2. At the condition of various exhaust gas temperature leaving the evaporator, the system net power output decreases with increasing the exhaust gas temperature leaving the evaporator. However, the total irreversibility rate is also decreases. So the best range of proper exhaust gas temperature leaving the evaporator is between 373 and 383 K.
3. At the full-load conditions of the diesel engine, the engine power increases by about 6 % with the ORC system. The performance of the ORC system becomes worse when the engine power is larger.

Acknowledgments This chapter is finally support by the National Basic Research Program of China (973 Program) (Grant No. 2,011CB707202, No. 2011CB710704) and the National High-Tech Research and Development Program of China (863 Program) (Grant No.2009AA05Z206).

References

1. Bianchi M, de Pascale A (2011) Bottoming cycles for electric energy generation: parametric investigation of available and innovative solutions for the exploitation of low and medium temperature heat sources. Appl Energy 8(5):1500–1509
2. Srinivasan KK, Mago PJ, Krishnan SR (2010) Analysis of exhaust waste heat recovery from a dual fuel low temperature combustion engine using an organic rankine cycle. Energy 35(6):2387–2399
3. Vaja I, Gambarotta A (2010) Internal combustion engine (ICE) bottoming with organic rankine cycles (ORCs). Energy 35(2):1084–1093
4. Weerasinghe WMSR, Stobart RK, Hounsham SM (2010) Thermal efficiency improvement in high output diesel engines a comparison of a rankine cycle with turbo-compounding. Appl Therm Eng 30(14–15):2253–2256
5. Ringler J, Seifert M, Guyotot V et al. Rankine cycle for waste heat recovery of IC engines. SAE 2009-01-0174
6. Fang J, Wei M, Wang R et al (2010) Simulation of waste heat recovery from a heavy-duty diesel engine with a medium temperature ORC System. Trans Csice 28(4):362–367
7. He M, Zhang X, Zeng K (2009) A new combined thermodynamic cycle for waste heat recovery of vehicle engine. J Xian Jiaotong Univ 43(11):1–5 91
8. Zhang X, He M, Zeng K et al (2010) Selection of working fluid used in vapor power cycle for waste heat recovery of vehicle engine. J Eng Thermophys 31(1):15–18
9. Chen Y, Lundqvist P, Johansson A et al (2006) A comparative study of the carbon dioxide transcritical power cycle compared with an organic rankine cycle with R123 as working fluid in waste heat recovery. Appl Therm Eng 26(17–18):2142–2147
10. Xu R-J, He Y-L (2011) A vapor injector-based novel regenerative organic rankine cycle. Appl Therm Eng 31(6–7):1238–1243
11. Hung TC (2001) Waste heat recovery of organic rankine cycle using dry fluids. Energy Convers Manage 42(5):539–553
12. Liu BT, Chien KH, Wang CC (2004) Effect of working fluids on organic rankine cycle for waste heat recovery. Energy 29(8):1207–1217
13. Hettiarachchia HDM, Golubovica M, Worek WM et al (2007) Optimum design criteria for an organic rankine cycle using low-temperature geothermal heat sources. Energy 32(9):1698–1706
14. Roy JP, Mishra MK, Misra A (2010) Parametric optimization and performance analysis of a waste heat recovery system using organic rankine cycle. Energy 35(12):5049–5062
15. Yamamoto T, Furuhata T, Arai N et al (2001) Design and testing of the organic rankine cycle. Energy 26(3):239–251

The Impact of Vehicle Heating Systems on the Energy Consumption Determined Based on the Vehicle Exhaust Emission Tests Under Actual Operating Conditions

Jerzy Merkisz, Maciej Bajerlein, Łukasz Rymaniak, Andrzej ZióŁkowski and Dariusz Michalak

Abstract Global policies aim at a reduction of the greenhouse gas emissions (particularly CO_2) and tightening the emission limits for vehicles. As a result, automotive manufactures have been forced to optimize the propulsion systems as well as other units such as air conditioning or heating systems. Currently, a great deal of research has been carried out on reducing energy consumption in vehicles [1–3]. Institute of Combustion Engines and Transport of Poznan University of Technology, conducted a study on the impact of heating systems on the energy consumption in a variety of vehicles. Two parking heaters were compared in terms of their emission characteristics and the temperature inside the vehicle measured while the engine was off. Based on these results and the heating rate inside the vehicle were determined. At the same time the temperatures inside the vehicle were measured. The study was conducted under the same conditions. In order to determine the impact of these systems on the vehicle energy consumption the emission measurements were carried out with all systems activated. The obtained results were used to determine both the and gas mileage as well as the specific emission of CO, HC, NO_x, CO_2. The temperatures were used to determine the effect of the heating system design on the travel comfort. A portable gas analyzer (PEMS) SEMTECH DS (made by Sensors Inc.) was used to measure CO, HC, NO_x, CO_2, O_2 and the exhaust gas mass flow rate. Temperatures were measured with 6 PERSONAL IOTECH DAQ 3000 thermocouple transmitters.

F2012-A07-027

J. Merkisz (✉) · M. Bajerlein · Ł. Rymaniak · A. ZióŁkowski
Poznan University of Technology, Poznan, Poland
e-mail: maciej.bajerlein@put.poznan.pl

D. Michalak
Solaris Bus and Coach S.A, Bolechowo, Poland

SAE-China and FISITA (eds.), *Proceedings of the FISITA 2012 World Automotive Congress*, Lecture Notes in Electrical Engineering 190,
DOI: 10.1007/978-3-642-33750-5_35,

Keywords Energy consumption of vehicles · Exhaust emissions · Exhaust emission tests under actual traffic conditions

1 Introduction

The main purpose of HVAC systems (Heating, Ventilation, Air Conditioning) in vehicles is to ensure not only the traveling comfort (climate control, air cleanliness inside the vehicle) but also safety (sufficient window visibility) under any weather conditions [4]. The realization of these aims significantly contributes to the improvement of the traveling comfort. The traveling comfort is of paramount importance in the case of city transit, as the number of passengers of varied social background using this means of transport is growing (students, construction workers, office workers etc.). The HVAC thus, must provide the driver with the possibility of adjusting the three basic values [5]:

1. Air temperature,
2. Air exchange rate,
3. Air distribution (blowing direction).

It is thus assumed that the value of the air temperature inside the vehicle on the level of the passenger's head should be lower than that on the level of his legs irrespective of the time of year. Such a distribution of temperatures reduces the feeling of sleepiness and fatigue of the passengers and the driver. This also has a great impact on the driver's reflexes and his capability of evaluating road situations. It is hard to determine the traveling comfort in an absolute way as different users have different preferences in this respect. Nevertheless, the system must quickly and effectively chill the interior in the summer and heat it in the winter.

2 Research Methodology

The exhaust emission tests performed on the parking heaters were realized under stationary conditions (Fig. 1). In order to determine the influence of the heaters on the vehicle energy consumption the measurements were carried out at the same temperature for all tested devices (approximately 10 °C). In order to objectively compare the emissions of the tested heaters the interiors of the vehicles were conditioned beforehand to obtain identical temperature during the tests.

The objects of the investigations were two parking heaters widely used in city transit vehicles. The heaters were fitted into vehicles of identical configuration (the same length and location of the convectors and heat exchangers). The technical specifications have been shown in Table 1.

Fig. 1 The emission tests performed on the heaters under actual operating conditions

Table 1 Technical specifications of the heaters

	Heater A	Heater B
Heating power (kW)	30	30
Power consumption (W)	90	110
Fuel consumption (kg/h)	3,0	3,3
Rated voltage (V)	24	24
Dimensions (mm)	600 × 240 × 220	610 × 246 × 220
Weight (kg)	18	19

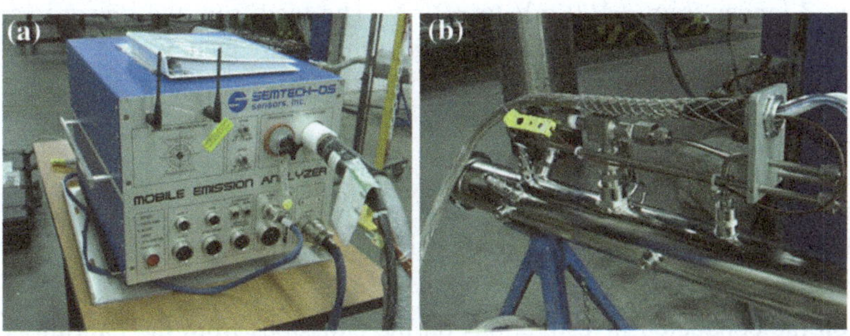

Fig. 2 Portable exhaust emission analyzer SEMTECH DS (**a**) exhaust mass flow meter (**b**)

For the tests the authors used SEMTECH DS a portable exhaust emission analyzer (PEMS) manufactured by Sensors Inc. (Fig. 2) The exhaust sample is taken through a probe and supplied by a heated line maintaining the temperature of 191 °C. Subsequently, the exhaust gases are filtered out of the particulate matter (diesel engine only). Then the measurement of the concentration of hydrocarbons

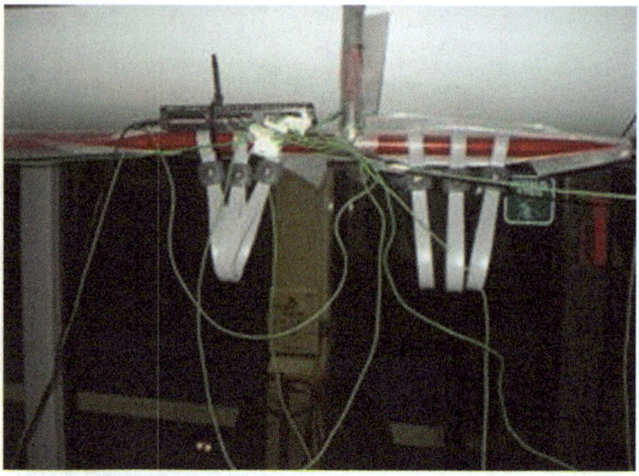

Fig. 3 The view of the temperature-measuring module

is carried out in the FID (*Flame Ionization Detector*). The sample is chilled to the temperature of 4 °C and in the NDUV (*Non-Dispersive Ultraviolet*) analyzer the measurement of the nitric monoxides and nitric dioxides is performed and in the NDIR (*Non-Dispersive Infrared*) the concentration of carbon monoxide is measured. The measurement of oxygen is carried out through an electrochemical sensor. Semtech also records parameters taken from the vehicle OBD and GPS (*Global Positioning System*). The analyzer can be controlled by a personal computer and through a wired or wireless connection [6].

The measurement of temperature was performed with the use of 6 thermocouples using the IOTECH PERSONAL DAQ 3000 signal converter (Fig. 3). The measuring module was fitted with a USB interface and a fast A/C converter (1 MHz/16 bit). The used converter had 16 analog single-ended inputs (8 differential inputs), 4 analog outputs, 24 digital in/out lines. The device was programmable in 7 ranges from ±100 mV to ±10 V. The feed from the sensor is sent to the computer recording the data according to the set frequency.

The figure below presents the location of the measuring points and the location of the thermocouples (Figs. 4 and 5).

3 Results of the Parking Heater Emission Tests

Based on the conducted investigations on the emissions from the parking heaters using the portable exhaust emissions analyzer (SEMTECH DS) the authors obtained the courses of the concentrations of CO, HC, NO_x and CO_2 in the exhaust gases. The tests were conducted under the same thermal conditions—conditioned interior before the measurements and similar ambient temperature. The test for heater A lasted longer by approximately 100 due to its lower thermal efficiency. The tests were continued until the temperature of 22 °C was obtained inside the vehicle.

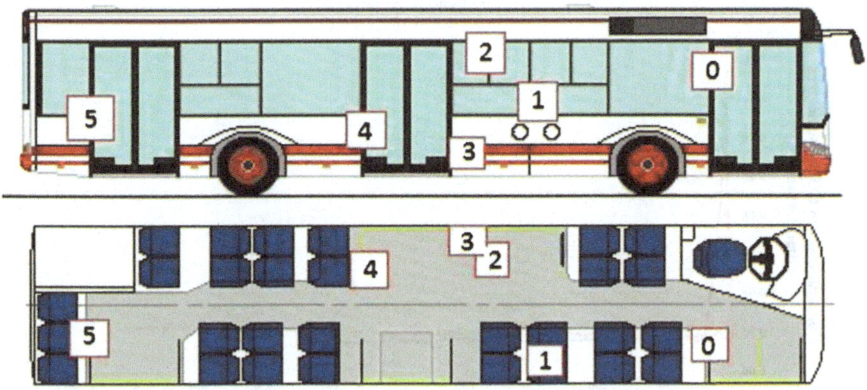

Fig. 4 Location of the measuring points inside the bus

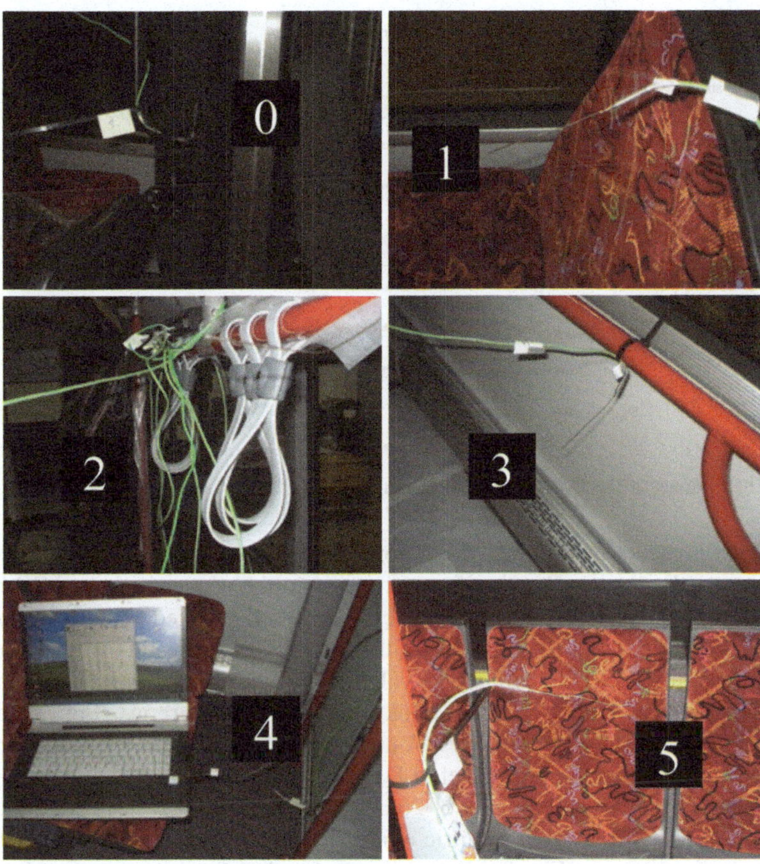

Fig. 5 Images of the individual measuring points inside the bus

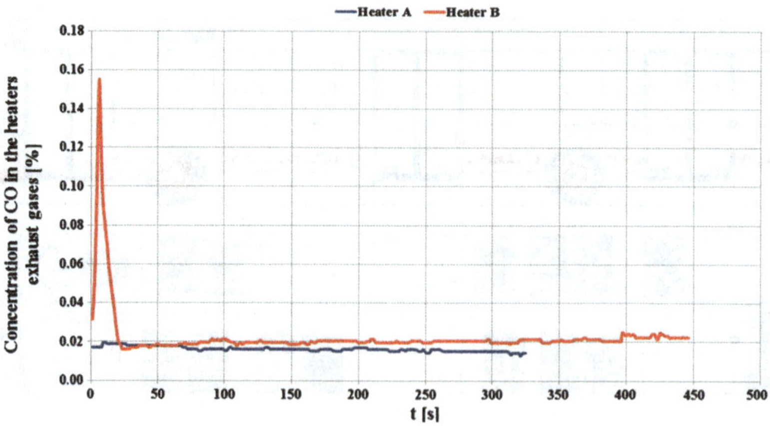

Fig. 6 The courses of the carbon monoxide concentration for the parking heaters

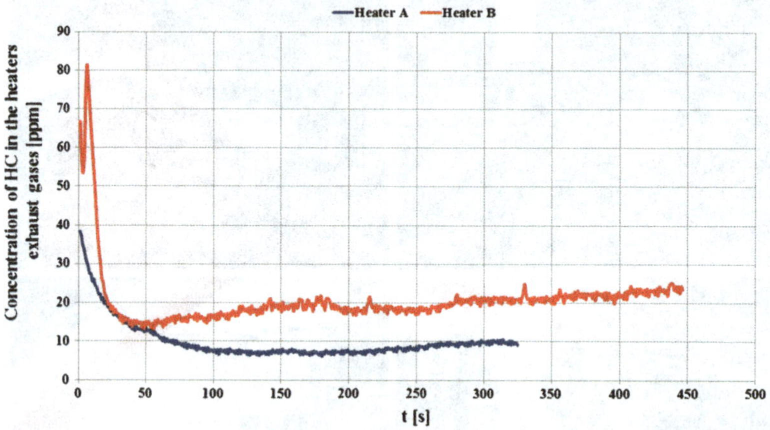

Fig. 7 The courses of the hydrocarbons concentration for the parking heaters

Figure 6 presents the course of the concentration of carbon monoxide for the tested heaters. For heater B a significant increase in the concentration of CO in the initial phase of the test was observed, which was directly related to the activation of the heating system. After the first 30 s of the test the heater would stabilize and the CO concentration oscillated around the value of 0.02 %. This level was higher than that of heater A. A similar scenario took place for hydrocarbons (Fig. 7). In the beginning of the test for both heaters (A and B) the concentration of hydrocarbons grew drastically to a maximum level of 80 ppm (for heater B). Next, the HC level dropped and remained on the level of 20 and 10 ppm (A and B respectively).

The greatest concentration of nitric oxides occurred for heater B and reached a maximum of 120 ppm (Fig. 8). It is noteworthy that for both heaters the concentration of NO_x in the exhaust gases grew with time, which resulted from the

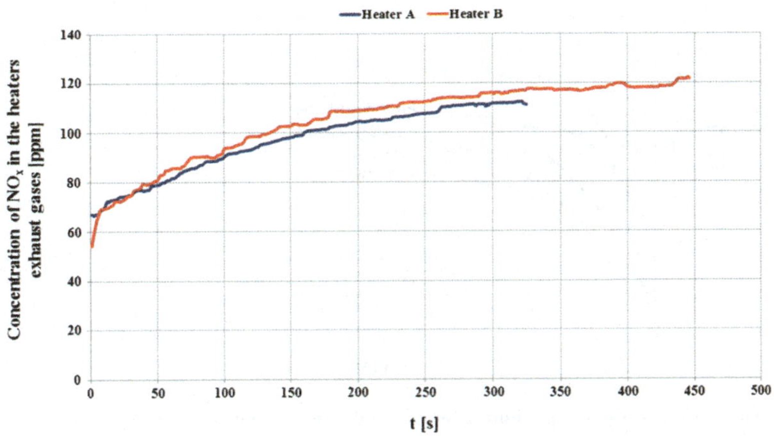

Fig. 8 The courses of the nitric oxides concentration for the parking heaters

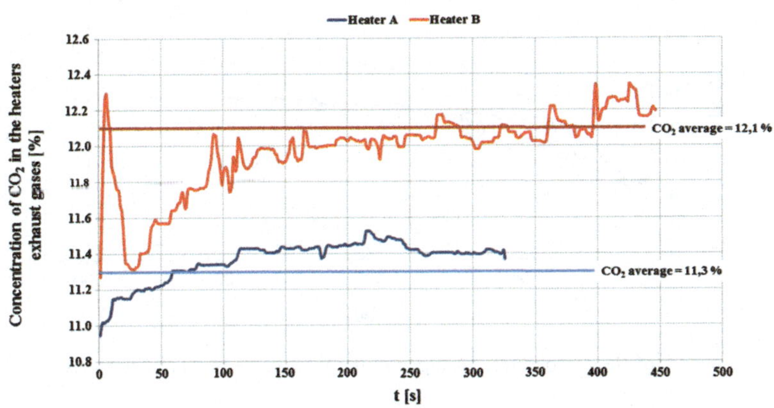

Fig. 9 The courses of the carbon dioxide concentration for the parking heaters

thermal state of the heaters. As the time passed the temperature inside the combustor grew, which greatly contributed to the increase in the concentration of nitric oxides. The differences in the NO_x concentrations for heater B and A oscillated around 10 ppm.

Figure 9 presents the courses of the concentration of carbon dioxide. For heater A the lowest concentration of CO_2 occurred in the initial phase of the test. Then, as the time passed this concentration grew. For heater B the situation was quite contrary. The heater initially had high CO_2 concentration (approximately 12.3 %) then it dropped to a level of 11.3 % and then grew again until the value of 12.4 % was obtained. The average concentration of CO_2 for heater A was 11.3 % and was lower by 0.8 % than the average value for heater B. This fact directly resulted from the higher overall efficiency of heater A.

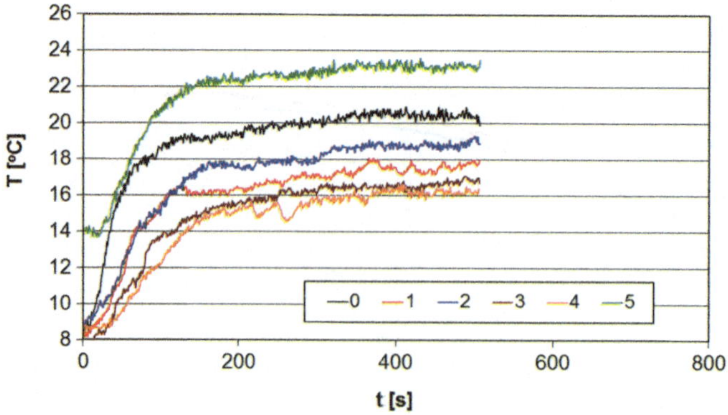

Fig. 10 The course of the temperature changes in the measuring points for heater A

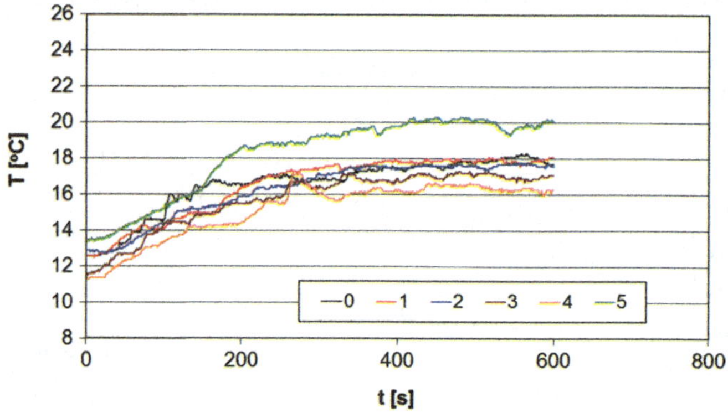

Fig. 11 The course of the temperature changes in the measuring points of heater B

4 Temperature Distribution Inside the Vehicle

Figures 10 and 11 present the courses of the changes of the temperature at the measuring points inside the vehicle. For heater A the highest temperature was observed at points 5 and 0. Its highest level was in the range of 20–24 °C. The lowest temperature value was for points 3 and 4 (they were similar—approximately 16 °C). The interior warm-up time was approximately 500 s. In comparison to heater B it was shorter by 100 s. This resulted from a higher thermal efficiency of heater A.

The interior of the vehicle heated by heater B had the highest temperature—20 °C that was measured at the measuring point 5. At this point the temperature was lower by approximately 4 °C than in the same measuring point of the interior heated with heater A. For heater B similar temperature level was observed for measuring points 0–4 (in the time 300–600 s), which oscillated around 16–18 °C.

Based on the determined temperature distributions, relative temperature increments were calculated inside the buses heated by the tested devices. The

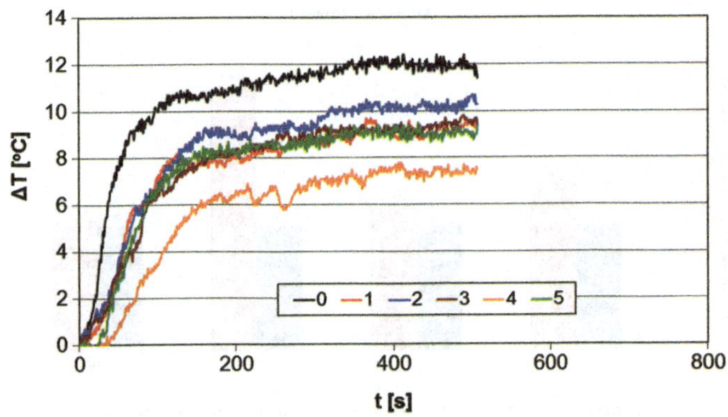

Fig. 12 Temperature increment in the measuring points inside the vehicle heated with heater A

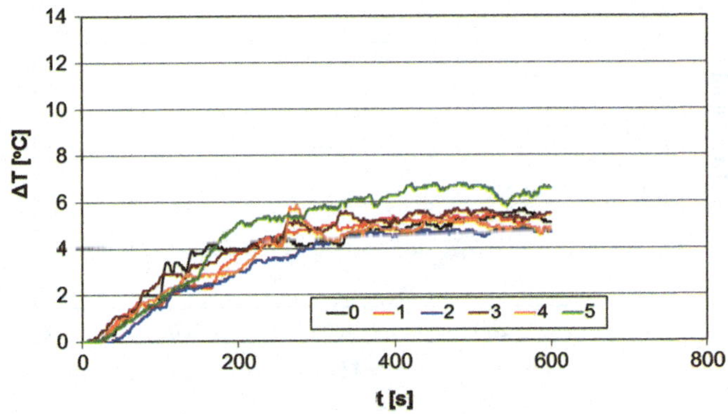

Fig. 13 Temperature increment in the measuring points inside the vehicle heated with heater B

greatest increment for heater A occurred at the measuring point 0 and was 12 °C (Fig. 12). The smallest increment occurred at the measuring point 4 and was approximately 8 °C. The interior of the vehicle heated with heart B had a temperature increment up to 7 °C, which, in comparison to heater A, was lower by 75 % (Fig. 13). The authors observed that the greatest increment occurred for measuring point 5. Measuring points 0–4 had a similar relative temperature increment that oscillated around 4–5 °C.

5 Conclusions

The greatest relative concentration of hydrocarbons for heater B was twice as high as for heater A (Fig. 14). The authors also observed that the average value of the concentration of CO_2 was higher by 5.6 % for heater B. This fact not only

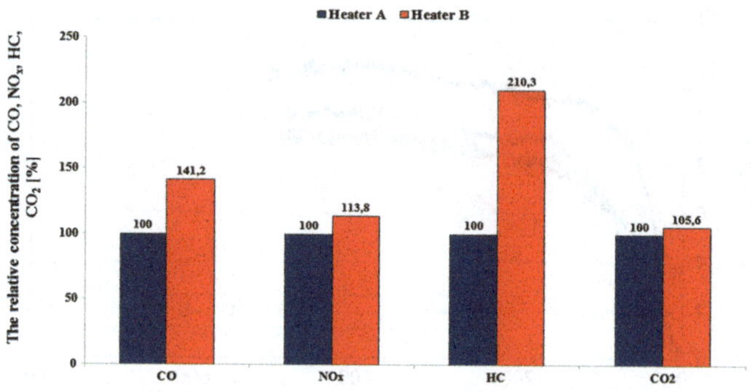

Fig. 14 Relative concentrations obtained during the heater emission tests

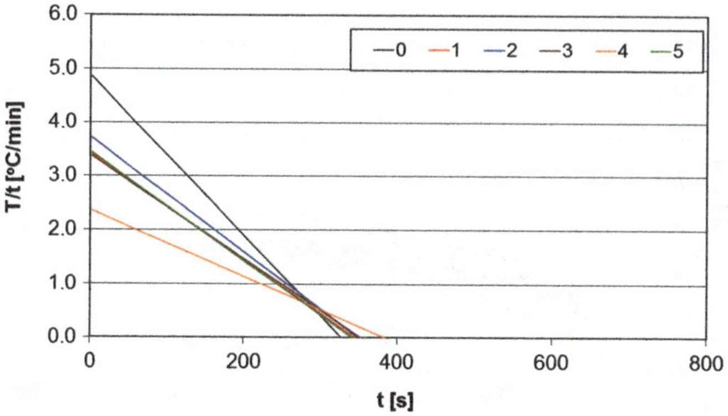

Fig. 15 Temperature growth rate in the vehicle heated with heater A for the measuring points (approximated courses)

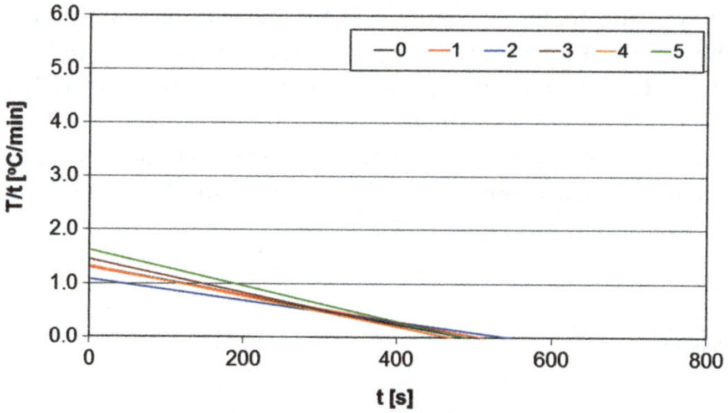

Fig. 16 Temperature growth rate in the vehicle heated with heater B for the measuring points (approximated courses)

confirms a lower overall efficiency of heater B against heater A but also higher hourly fuel consumption (CO_2 concentration is equivalent to fuel consumption).

The interior of the vehicle heated with heater A had the greatest temperature increment that was, on average, approximately 3 °C/min (Fig. 15). As for the interior of the vehicle heated with heater B, the increment was greater by 1.5 °C/min (Fig. 16). The authors also observed that heater A had a better heating efficiency as it generated the highest average temperature of the interior in a shorter time at a lower hourly fuel consumption.

The above conclusions confirm that heater A positively influences the energy consumption of the vehicle and it is purposeful to apply such systems in all types of city buses: conventional buses, hybrid buses and trolleybuses.

References

1. Merkisz J, Michalak D, Bajerlein M, Nowak M, Rymaniak Ł, Ziółkowski A (2012) Wpływ zastosowania różnych konstrukcji piecy ogrzewania postojowego w autobusie miejskim na zmniejszenie energochłonności. Autobusy—Technika, Eksploatacja, Systemy Transportowe nr 3/2012
2. Merkisz J, Fuc P (2010) The exhaust emission from light duty vehicles in road test in urban traffic. In: International powertrains, fuels & lubricants meeting, May 2010, Rio De Janeiro, Brazil, SAE Paper 2010-01-1558
3. Merkisz J, Lijewski P, Fuc P, Pielecha J (2010) Exhaust emission tests from agricultural machinery under real operating conditions. In: SAE 2010 commercial vehicle engineering congress, Oct 2010, Chicago, USA, SAE Paper 2010-01-1949
4. Deh U (2008) Klimatyzacja w samochodzie, WKiŁ, Warszawa
5. Prochowski L, Żuchowski A (2011) Samochody ciężarowe i autobusy. WKiŁ, Warszawa
6. SEMTECH®-DS On Board, In-Use Emissions Analyzer, Manual, Michigan 2007

Part VIII
Emission Standard and International Regulations

Study on a Test Procedure for the NO_x Emission of Heavy Duty Vehicle

Mengliang Li, Yueyun Xu, Hui Guo, Maodong Fang and Long Sun

Abstract The NO_X emission of 19 heavy duty vehicles compliance with China stage IV emission standard had tested by Portable Emission Measurement System in the real world and gained the second-by-second data. These test data is processed according to NTE. Because the local driving cycle is also much difference from that in the United States, over 70 % data over the NTE zone, the number of the NTE events is fewer and it is hard to estimate these vehicles emission level. Extending the NTE zone and shorten the NTE event time to adapt to actual cycle conditions can make more the cycle points falling into NTE zone, and the effective NTE events also increased. This paper reports a work on developing a new NTE zone and NTE event time beseemed the local actual driving cycle. And the new procedure can clearly distinguish that vehicles emission is lower or higher than emission standard.

Keywords Diesel engine · NO_X emission · NTE · Driving cycle · HDV

1 Preface

The current heavy-duty vehicle emissions certification mainly uses the engine bench test methods, and the test cycles are the European Steady Cycle (ESC) and the European Transient Cycle (ETC.). Due to significant differences between the ETC. and on-road cycle, a large number of high-emission test points are not

---

F2012-A08-004

M. Li (✉) · Y. Xu · H. Guo · M. Fang · L. Sun
China Automotive Technology and Research Centre, Tianjin, China
e-mail: xuyy2007@163.com

SAE-China and FISITA (eds.), *Proceedings of the FISITA 2012 World Automotive Congress*, Lecture Notes in Electrical Engineering 190,
DOI: 10.1007/978-3-642-33750-5_36, © Springer-Verlag Berlin Heidelberg 2013

controlled, so the actual emissions of on-road vehicles are much higher than the regulatory limit values. Domestic and foreign researches on emissions of Stage IV heavy-duty vehicle show that due to the influence of cycle conditions, the emissions performance of heavy-duty diesel vehicle of China stage IV and Stage V may be worse than that of Stage III. For effective monitoring the status of actual emissions and making up for the deficiencies of the existing regulations, we should develop and set up test methods based on heavy-duty vehicle emissions in order to strengthen the control and management of actual on-road emissions of heavy-duty vehicle and improve scientific and effective supervision of heavy-duty vehicle emissions. This report referred to the Not-to-Exceed (NTE) test method enacted by Environmental Protection Agency (EPA) in the United States, revised the time requirements of NTE events and NTE region, and acquired the test and evaluation methods in accordance with actual driving conditions of bus in Beijing, provided technical support for the government to supervise emissions of heavy-duty vehicle on the roads.

2 Overview of NTE

2.1 Test Requirements

The numbers of the test refer to the measurement requirements of the NTE (U.S.) test procedures. The test does not specify a particular test driving cycle, but follows the daily use of the test vehicle to drive on the real road under routine driving conditions with non-idling time for at least 3 h. The test duration should be no less than one working day. The test will be run again if the non-idling time is less than 3 h during the first time test [1, 2].

2.2 Test Equipments

Using the experiences of U.S. and Europe for reference on emissions control measures of heavy-duty diesel engine vehicles in actual road driving, considering the technology level of portable emission measurement system (PEMS) and the feasibility of using it for vehicle emissions test, we determine to use PEMS to test the heavy-duty vehicle emissions. PEMS mainly consists of three major components, NO_X emissions testing equipment, engine ECU data reading equipment and GPS for recording the driving track and speed.

OBS-2200 system is a set of equipment for measuring related parameters, such as exhaust emissions and fuel consumption and etc., under actual road conditions. It consists of anti-shock gas analyzer unit, main controlling computer, related software, subsidiary sensor and the sampling probe of Pitot tube flow meter connected to exhaust pipe.

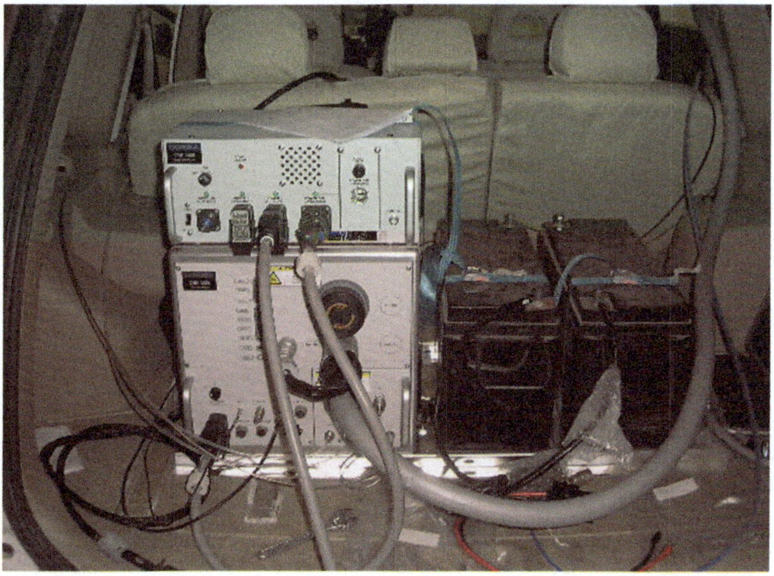

Fig. 1 OBS2200 test sketch

The data in the analyzer and affiliated sensor store in the computer through the control software. External signal also input into the computer through affiliated port of OBS-2200. Figure 1 shows the OBS-2200. Speed data records via GPS during the test [3, 4].

ECU information is read by Canalyzer through the particular port module during the test. At present, the main parameters that can be read are as following: engine speed, load, reference torque and ratio of engine torque under present working status. The communication protocol of this equipment is J1939 which is widely used by heavy-duty vehicle.

2.3 NTE Region and Computing Method

The NTE region is defined as the engine speed and torque area where engine speed is greater than the n_{15} speed and engine torque is greater than or equal to 30 % of the maximum engine torque or greater than or equal to 30 % of maximum engine power, whichever is greater. The speed n_{15} is calculated from Eq. 1 [5].

$$n_{15} = n_{lo} + 0.15 \times (n_{hi} - n_{lo}) \tag{1}$$

n_{hi} engine speed above rated speed at which full load power is 70 % of maximum power of the engine.

Fig. 2 NTE region

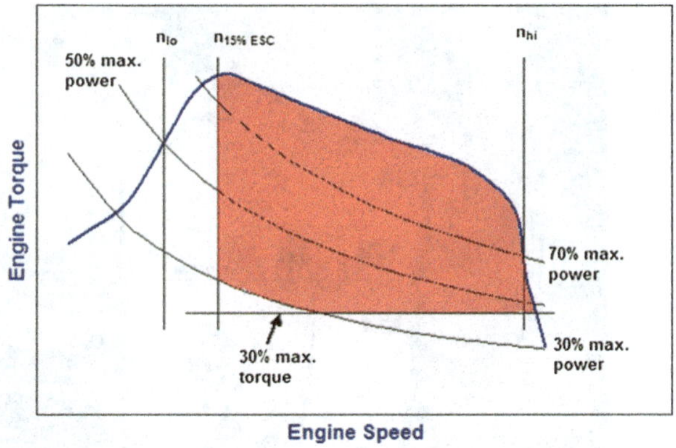

n_{lo} engine speed above rated speed at which full load power is 50 % of maximum power of the engine [6] (Fig. 2).

NTE event is that at least continuous 30 s of engine work points fall into the NTE region. The emissions of NTE events are the accumulative emission quality of a NTE event in divided by the work of the NTE event. The maximum time of NTE event is determined by the minimum value of the following three conditions.

1. The actual continuous time of the NTE event
2. 10 times of the shortest time of all NTE events
3. 600 s

The judgment of NTE method is to compare the brake NO_X emissions of every NTE event to the NTE limits, then, sum the weighted time of NTE events whose results are lower than NTE limits (recorded as a); sum the weighted time of all NTE events (recorded as b); calculate the ratio of both (recorded as c): $c = (a/b) * 100\ \%$. If $c > 90\ \%$, and if the largest emission results of all NTE events must be less than two times NTE limits or 2.00 g/BHP-hr, then the vehicle emissions are qualified.

3 Analysis of Results

3.1 Result of Vehicle Emission

In order to provide evidence for vehicle emissions evaluation method and limits determination, verify the feasibility of the testing procedure, we started the verification test according to the test procedures. So far, we collected emissions and engine ECU data from 19 vehicles to verify the method validation. All of the 19 vehicles are Stage IV heavy-duty buses. The after-treatments of the tested vehicles

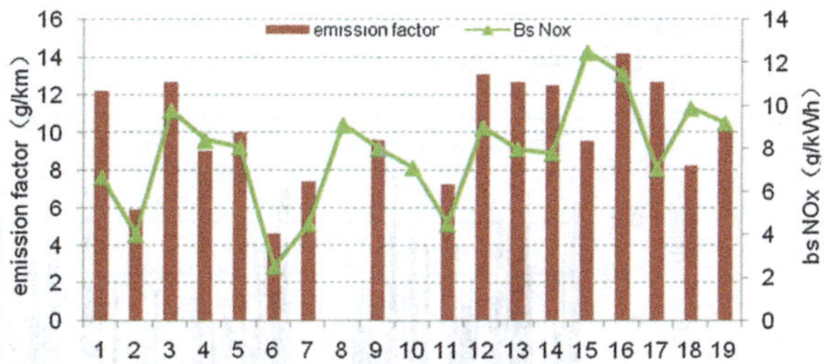

Fig. 3 The results of the tested vehicles

are all using the SCR (Selective Catalytic Reduction) system. All vehicles fill stage IV light diesel from the fixed oil station in daily use. We checked the tested vehicles before testing by special equipments to ensure that they are in good conditions, and found all of the tested vehicles are met the test requirements. Figure 3 shows the emission factors (g/km) and $bsNO_X$ (g/kWh) results of the 19 vehicles. There are no results of emission factors for the NO. 7 and NO. 9 vehicles since their GPS abnormal signals during the testing. From the results, the vehicle emissions levels are not exactly the same, emission factor results are basically higher than 7 g/km, and $bsNO_X$ emissions are higher than 6 g/kWh, much higher than 3.5 g/kWh which is the limit of Stage IV heavy-duty engines specified in GB17691-2005. Figure 3 also tells emission factors and $bsNO_X$ have the same tendency, if the vehicle emission factor is higher, the $bsNO_X$ is also higher.

3.2 Analysis of NTE

We choose five results of the tested vehicles (NO. 6–10 in Fig. 4) to analysis since the sample size of the test is large, The engine cycle points of every vehicle which fall into NTE region distribution are shown in Fig. 4a–d. There is no results of the NO. 7 vehicle in Fig. 4 because it has no NTE event throughout the whole test process.

In Figure 4, the blue line is the NO_X transient emission rate (g/s) in the test process, the pink line is the transient emission rate of NTE events. We can find from the figure that the percentage of the cycle points from the NTE events is very small, lower than 10 %, because the test vehicles are all public transport vehicles, they run in the urban of Beijing, the characteristics of these buses driving conditions is that vehicle speed and engine speed are both low, and the load of engine is also low, but the NTE region mainly concentrated in this region where the engine speed is higher than medium speed and engine load is much bigger, so there were a few points fallen into the NTE region. Figure 5 shows the results of 5 vehicles according to NTE method.

In Figure 5, the NTE events are very few for 5 vehicles, the most is the NO. 6 vehicle, but there are only 23 events happened. In addition we found that the $bsNO_X$ results of NO. 9 and NO. 10 vehicles are very low, they are lower than 5 g/kWh, but their emissions are very high (in Fig. 1). Therefore, using the America

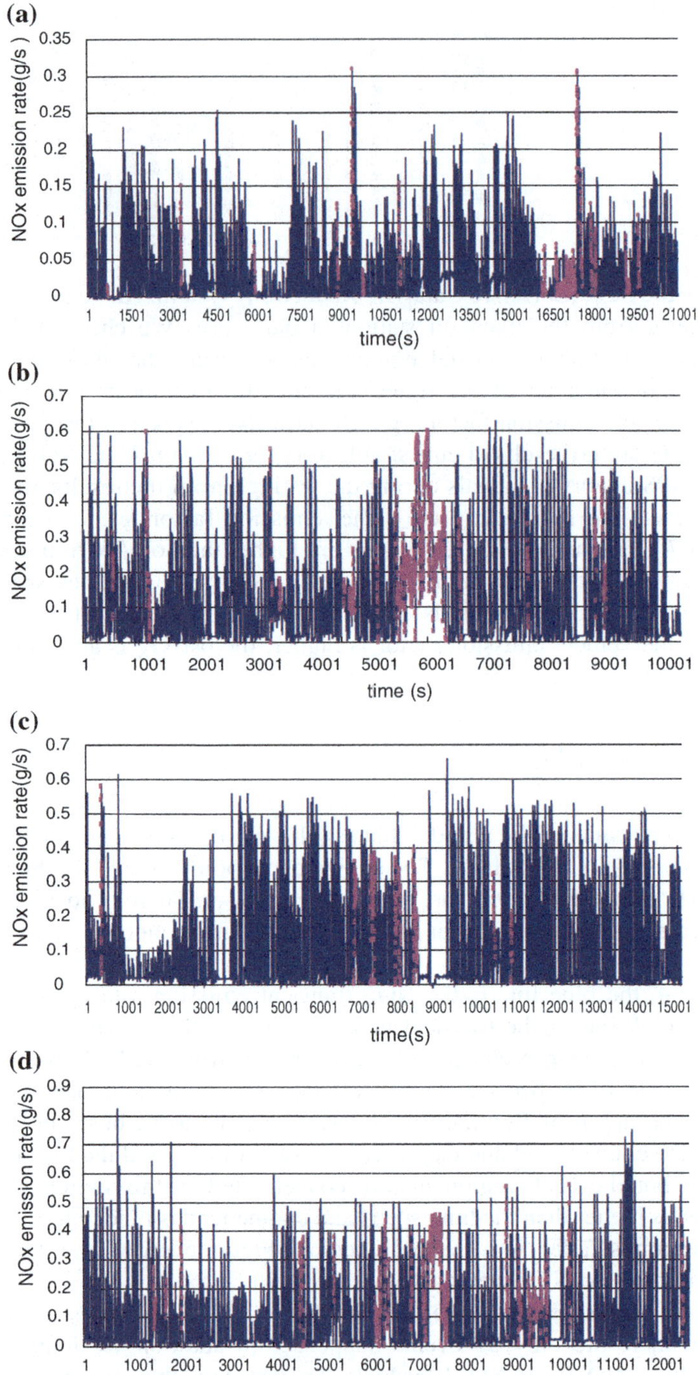

Fig. 4 The statistics of every NTE event points fall into the NTE region

Fig. 5 The results of NTE method

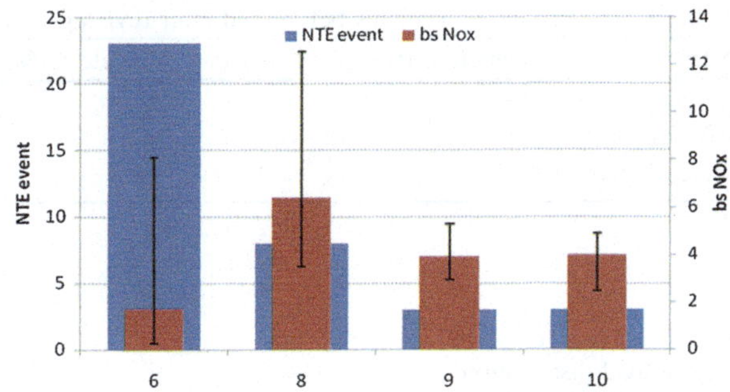

Fig. 6 Results of different equivalent power curves

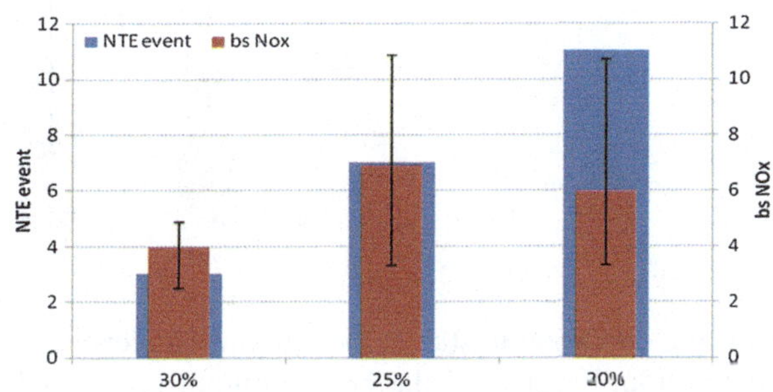

NTE directly to judge the vehicle emission level is not reasonable, the NTE region and the time requirement of NTE events limit the number of the NTE events. So we should not only refer the NTE method, but also revise the NTE method to adapt to the Beijing conditions.

3.3 NTE Revised Method

According to the engine cycle of heavy-duty vehicles on-road driving cycle in Beijing, the American NTE control region is revised to make it reflect the heavy-duty vehicles cycle in Beijing area and more reasonable evaluation of vehicle on emissions. By lowering the engine speed and reducing the 30 % maximum power to enlarging the NTE region. The engine speed of NTE reduced from n_{15} to n_{lo}; and the motivation of the power curve border was adjusted properly. Figure 6 shows the number of NTE events and the bsNO_X emission results of the 30, 25 and 20 % of the maximum power, respectively.

The Figure 6 shows, when choosing 30 % of the maximum power curve as border, there were only 3 NTE events, and with the curve move down, the NTE events also were increasing, when choosing 25 %, the NTE events increased to 7, and when the choice is 20 %, the NTE events increased to 11. In order to make more cycle points drop into the NTE region, the equivalent power curve is the lower the better, but when the engine power is too low, it can lead to engine

Table 1 The change of vehicles before and after revising

NTE region	Vehicle 6 (%)	Vehicle 7 (%)	Vehicle 8 (%)	Vehicle 9 (%)	Vehicle 10 (%)
Before revised	33	21	20	14	16
After revised	42	35	26	21	22
Increasingly(%)	9	14	6	8	6

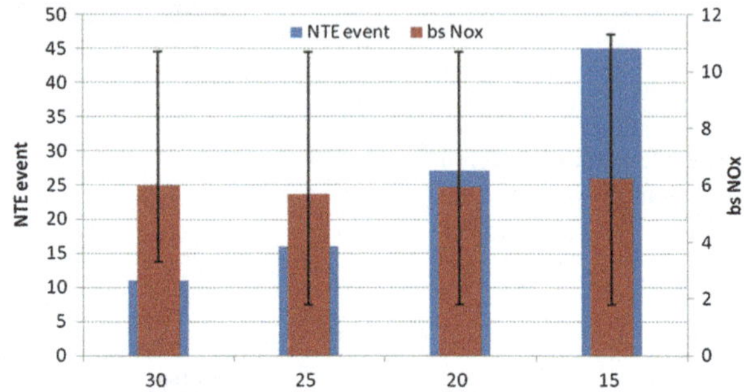

Fig. 7 NTE regions and engine load distribution before and after NTE revision

working condition unstable, the emission under these conditions is hard to control. In addition, because of the engine output power is large, exhaust temperature of the vehicle is higher, the reduction effect of SCR system is also better; if the power curve is too low, the SCR system reduction effect is bad, it will appear wrongly convicted phenomenon. Comprehensive consideration, the 20 % of maximum power curve will not neither cause the temperature of the exhaust too low, nor lead the state of engine to unstable status. In sum, 20 % of maximum power curve is a better choice.

After revising according to NTE methods, the number of vehicles cycle points falling into NTE region is different from that before. Table 1 shows the change of NO. 6 to NO. 10 vehicles before and after revising. After revising, the points of NO. 7 vehicle fall into the NTE region increased by 14 %, and make up about 35 % of all running points.

The time length of NTE events will also affect the number of the NTE events. Figure 7 shows the number and the $bsNO_X$ of NTE events results of one vehicle in different time length of NTE events. The number of NTE events is increasing when the time length of NTE events is shortened. When the time length shorten from 30 to 15 s, the number of NTE events increased by 34. Comparing the results between 20 and 15 s of the NTE events minimum duration, we find that the number of events is different, but the $bsNO_X$ emission results, not only the values of $bsNO_X$ emissions but also the maximum and minimum values of the emissions are very close. This shows that when choosing 20 s for NTE event minimum duration, the results of the test vehicles tend to be stable, even if the events to time is shortened, the results won't fluctuate. Therefore, the duration of NTE events minimum defined as 20 s is a reasonable choice.

Table 2 Results of before and after revising NTE method

Vehicle NO	Number of NTE events		Percentage		Qualified or not	
	After	Before	After (%)	Before (%)	After	Before
6	114	23	94	100.00	Yes	Yes
7	25	0	91	–	Yes	–
8	21	8	31	80.65	No	No
9	14	3	26	100.00	No	Yes
10	27	3	55	100.00	No	Yes

Note percentage in the table is the proportion of the accumulative time of NTE events which $bsNO_X$ emission is lower than 6.00 g/kWh in the accumulative time of all NTE events

3.4 Revisional NTE Decision Method

The limits of the revised NTE method are determined by referring the America NTE method. It consists of three parts: the standard value, the wide limits of in use compliance and wide limits of new equipment.

The standard value is 1.5 times of the FTP limits;
The wide limit of new equipment is 0.45 g/HP-hr (0.60 g/KWh).
The wide limit of in use compliance is 0.10–0.20 g/HP-hr (0.13–0.27 g/kWh).

The heavy-duty vehicle emissions stage of Beijing is different from that of the U.S. it leads to different levels of heavy-duty vehicle emissions. But we can refer the principles of the constitution of the NTE limits in United States to determine the limits in Beijing as following, the standard value is the 1.5 times limits of ETC., which is specified in GB17691-2005. The wide limits of in use compliance and new equipment continue to use the values of the United States NTE limits. Therefore, NTE limits defined as: $5.25 + 0.13\ (0.27) + 0.6 = 6.0\ (6.14)$ g/kWh.

The limits of NTE method is 6.00 g/kWh. If the emission of a heavy-duty vehicle is qualified, the accumulative time of NTE events which $bsNO_X$ emission is lower than 6.00 g/kWh should be at least 90 % of the accumulative time of all NTE events.

3.5 Results Analysis Before and After Revision

Revising according to the test and limits requirements of the NTE method, we calculated and analyzed NO. 6–NO. 10 vehicles emissions results of NTE method respectively. The requirements are the same, except the NTE region and the time requirement of NTE events. Table 2 gives the results of before and after revising NTE method. From Table 2 we find that the emissions of NO. 6 and NO. 7 vehicles are better than other vehicles, they can meet the limits requirement before and after NTE method revision. But for the NO. 9 and NO. 10 vehicles, their emissions are worse than other vehicles. Their results are qualified before NTE method revising, but their results are not qualified after NTE method revising. Therefore, if we use the NTE method to judge whether the vehicle emissions are met the standard, it may lead to wrong conclusion, and the high emission vehicles are qualified by mistake. The revised NTE method can effectively discern vehicles emissions levels, providing technical support for the government to supervise emissions of heavy-duty vehicle on the roads.

4 Conclusions

The method to test the emissions of on-road heavy-duty vehicle emissions is first proposed in this report. This method improves the ability to verify the heavy-duty vehicle, and reduces the NO_X emissions of on-road heavy-duty vehicles. There are some conclusions as follows:

1. The emissions of on-road vehicles are much higher than 3.5 g/kWh which is the limit of China Stage IV heavy-duty engines. The supervision should be strengthened to improve the emission performance of on-road heavy-duty vehicles;
2. NTE method can be used to evaluate emission performance of on-road heavy-duty vehicles, but may get few NTE events if applying NTE method analysis to heavy-duty vehicle in Beijing area, such as NO. 7 vehicle. So NTE method should be revised accordingly.
3. The NTE events were increasing as the equivalent power curve moving down. But when the engine power is too low, it can lead to engine working condition unstable; the emission under these conditions is hard to control. So 20 % maximum power is a good choice. The number of NTE events may be doubled if the factor reduced from 30 to 20 %.
4. The NTE events were also increasing as the minimum duration decreasing. When the length is 20 s, the results tend to be stable. So 20 s is a good choice of minimum duration. The number of NTE events may be 1.5 times more if the factor reduced from 30 to 20 s.
5. After the treatment using NTE methods, it is easily to distinguish high emissions vehicles from low emissions vehicles.

References

1. Code of Federal Regulations (2007) Title 40, part subpart N—emission regulations for new otto-cycle and diesel heavy-duty engines; gaseous and particulate exhaust test procedures
2. Code of Federal Regulations (2007) Title 40, Part 1065—engine testing procedures
3. Kongjian Q, Junhua G (2009) Integration and validation of an on-board emission measurement system for heavy-duty vehicles. Autom Eng 31:1081–1085
4. Horiba MEXA-720 NOX (2003) Instruction manual, Horiba, Ltd., Japan

5. Benjamin CS, Daniel KC (2008) Work-based window method for calculating in-use brake-specific NOX emissions of heavy-duty diesel engines, SAE paper 2008-01-1301
6. Code of China Regulations (2005) Limits and measurement methods for exhaust pollutants from compression ignition and as fuelled positive ignition engines of vehicles (III, IV, V), ministry of environmental protection of the People's Republic of China

Application Research on SCR Post-Processing System in Non-Electronic Diesel Engine of Vehicles

Chenglin Deng, Yingfeng Zhang, Zhengfei Xu and Yan Yu

Abstract Selective Catalytic Reduction (SCR) technology as an effective method to control emission of NOx has been applied to electronic controlled diesel engine. But, how to deal with the NOx emission of non-electronic controlled diesel engine is a challenging problem. In order to resolve this, a SCR system based on exhaust temperature and exhaust back pressure is established. The relationship between exhaust back pressure and structure parameters, exhaust temperature is studied. The relationship between exhaust back pressure and structure parameters, exhaust temperature is studied, and a model of exhaust flow of engine is developed. The bench test study of SCR system is made. The trends of NOx concentration, injection rate of reductant, exhaust temperature and exhaust back pressure are analyzed. Results show that the SCR system based on exhaust temperature and exhaust back pressure is useful to reduce the concentration of NOx. This way is verified as a feasible method to control NOx emission of non-electronic controlled diesel engine.

Keywords SCR post-processing system · Non-electronic diesel engine · NOx emission · Exhaust back pressure · Exhaust temperature

---

F2012-A08-005

C. Deng (✉)
State Key Laboratory of Automotive Safety and Energy, Tsinghua University, Beijing, 100084, China
e-mail: zyfimportant@163.com

C. Deng · Y. Zhang · Z. Xu · Y. Yu
Military transportation University, Tianjin, 300161, China

SAE-China and FISITA (eds.), *Proceedings of the FISITA 2012 World Automotive Congress*, Lecture Notes in Electrical Engineering 190,
DOI: 10.1007/978-3-642-33750-5_37,

1 Introduction

With the strictness of national emission law, the NOx emission is an urgent problem that should be solved in exhaust emission field [1]. SCR technology is a method to reduce NOx emission [2, 3].

Recent years, many scientific research institutions and corporations do much work in SCR technology study and SCR system has been applied gradually to electronic controlled diesel engine [4, 5]. But, SCR system is not suitable for non-electronic controlled diesel engine. In our country, most of in-use vehicles are equipped non-electronic controlled diesel engine, which is a restriction to control the NOx emission level of in-use vehicles. With much test research, we find that there is a positive connection between exhaust temperature, exhaust back pressure and speed, load. According to this theory, a SCR system with a novel control strategy based on exhaust temperature and exhaust back pressure is established. And experimental study of SCR system with non-electronic controlled diesel engine is made and a good result is gotten to control NOx emission.

2 Foundation of SCR System Based on Exhaust Temperature and Exhaust Back Pressure

2.1 Layout Scheme of SCR

The layout scheme of SCR based on exhaust temperature and exhaust back pressure is shown as Fig. 1. The SCR system consist of air reservoir, electronic assembly of IUT (Intelligent Urea Tank), dosing control unit (DCU), supply system of reductant, SCR catalytic converter, pressure difference sensor, exhaust temperature sensor, NOx sensor and so on. Compressed air is stored in Air reservoir which offers power of drive, injection and cleaning for SCR system. IUT is an integrated system that is made up of several valves, dosing unit and so on. The addition, storage and transportation of reductant are made by the supply system. With a built-in program, DCU controls the work of IUT.

2.2 Exhaust Back Pressure

SCR catalytic converter is installed in the exhaust pipe of engine. Catalytic converter is a wall-flow filter and catalysts are attached on the surface of filter. Under the action of catalysts, a reduction reaction is occurred between exhaust gas and reductant. It is found out that a pressure loss will be made when exhaust gas flowing through the filter. The influence factors of pressure loss are not only parameters of catalytic converter but also gas flow, viscosity and so on. For a

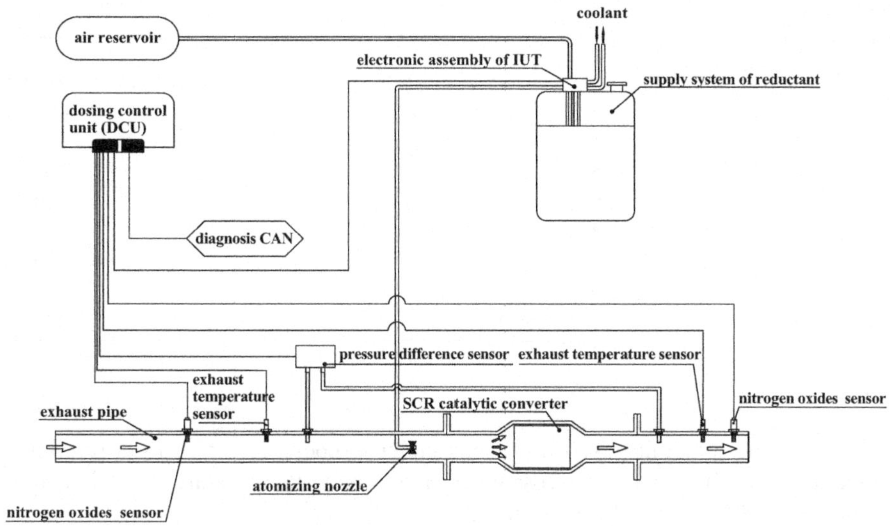

Fig. 1 Schematic diagram of SCR system

certain catalytic converter, the head loss is approximately equal to pressure loss. And the head loss can be expressed as following [6–8]:

$$\Delta p = \frac{1}{2} R_e \lambda \frac{\mu L Q}{d^2 \rho A} \tag{1}$$

where, Δp is head loss, μ is dynamic viscosity of gas, Q is volume flow of gas, A is the sectional area of filter, λ is loss coefficient, L is effective length of filter, d is hydraulic diameter of channel, ρ is density of gas.

For a certain catalytic converter, the parameters of it are determined. So, some parameters in Eq. (1) can be replaced by k:

$$k = \frac{1}{2} \frac{\lambda L}{d^2 A} \tag{2}$$

Therefore, the Eq. (1) can be rewritten as following:

$$\Delta p = k R_e \mu \frac{Q}{\rho} \tag{3}$$

2.3 Dynamic Viscosity of Gas

According to the satran formula, the dynamic viscosity of gas can be expressed as following [9]:

$$\mu = \mu_0 (T/T_0)^{\frac{3}{2}} (T_0 + B)/(T + B) \tag{4}$$

Table 1 Values of k in different conditions

Condition	Value
1	308.52
2	319.92
3	323.16
4	334.85
5	319.43
6	329.14
7	328.96
8	336.78
9	317.42
Average	324.24

where, μ is the dynamic viscosity of gas under temperature of T, T_0 is reference temperature, μ_0 is dynamic viscosity under reference temperature T_0, T is actual temperature, B is constant related to gas kind.

2.4 Determination of the Parameter k

In test research, it is no necessary to know detailed structure parameters of catalytic converter. It is enough to get the parameter k in Eq. (3). A testing method is applied to get the value of k. 9 running conditions with different speeds and load are simulated at a flow test rig. The values of 9 running conditions are shown in Table 1. The results shown in the table indicates the parameter k is slightly different. In order to simple calculation, the average value of k is used.

2.5 Velocity Ratio of SCR Catalytic Converter

There are two main influence factors of efficiency of catalytic converter: exhaust temperature and velocity ratio. The definition of velocity ratio is volumes of gases that are treated by per unit volume catalyst in unit time. The velocity ratio of catalytic converter can be calculated by following equation:

$$SV = \frac{Q_m - C}{1.293V_C} \times 1000 \tag{5}$$

where, SV is velocity ratio, Q_m is exhaust mass flow, C is fuel consumption, V_c is volume of catalytic converter.

The exhaust mass flow, fuel consumption, and volume of catalytic converter at 9 running conditions can be gotten, according to Eq. (5), velocity ratios are also obtained, as shown in Table 2.

Table 2 Values of velocity ratio in different conditions

Condition	Velocity ratio (h^{-1})
1	12,738
2	15,559
3	19,198
4	15,468
5	25,477
6	31,391
7	19,380
8	33,483
9	37,669

In order to get efficiency of catalytic converter in different conditions, a lot of bench test is done. During the test, the addition quantity of reductant in different conditions is made by manual. Its aim is to find the optimal addition in a certain condition. Through the analysis of testing data, the relationship among the efficiency, exhaust temperature and velocity ratio is established, as is shown in the Fig. 2. At the same time, the relationship between exhaust flow and exhaust back pressure, exhaust temperature is also obtained (see Fig. 3).

3 Experimental Research

Bench test is done for the feasibility of SCR system based on exhaust back pressure and exhaust temperature. On the base of earlier test, a ESC experiment are performed.

The changes of NOx concentration values with ESC conditions are shown in Fig. 4. The solid line indicates the original emission values of NOx, and the dashed line indicates the values of NOx after the catalytic converter. The results show that there are high values of NOx in original situation and the maximum values of NOx are more than 1500 ppm. The values of NOx reduce sharply after the addition of reductant and the minimum values of NOx are about 100 ppm. The maximum conversion efficiency of NOx reaches 90 %, however, in some situation, the conversion efficiency is less than 50 %. In most of ESC conditions, the conversion efficiency is more than 70 %.

The changes of injection rate of reductant with ESC conditions are shown in Fig. 5. Because exhaust temperature and exhaust back pressure can be disturbed by the changes of speed and load, moreover, feedback control is also introduced to the SCR system, the adjustment of injection rate is real time even in the same ESC condition. Its aim is to decrease NOx emission at the utmost. As can be seen from the Fig. 5, the most stable injection rate is 3700 g/h, and the instantaneous injection rate reaches a maximum 4500 g/h.

As Fig. 6 shows that there are noticeable changes in temperature of catalytic converter under ESC conditions. And the maximum temperature is up to 400 °C,

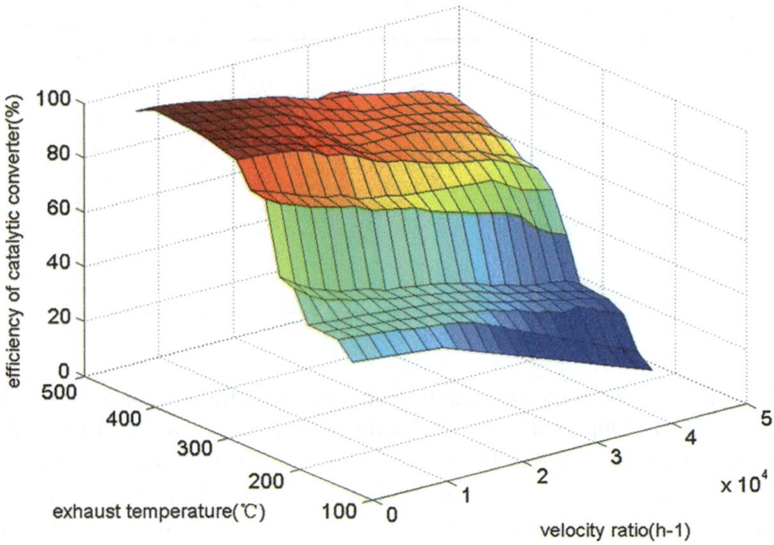

Fig. 2 Relationship between efficiency of catalytic converter and exhaust temperature, velocity ratio

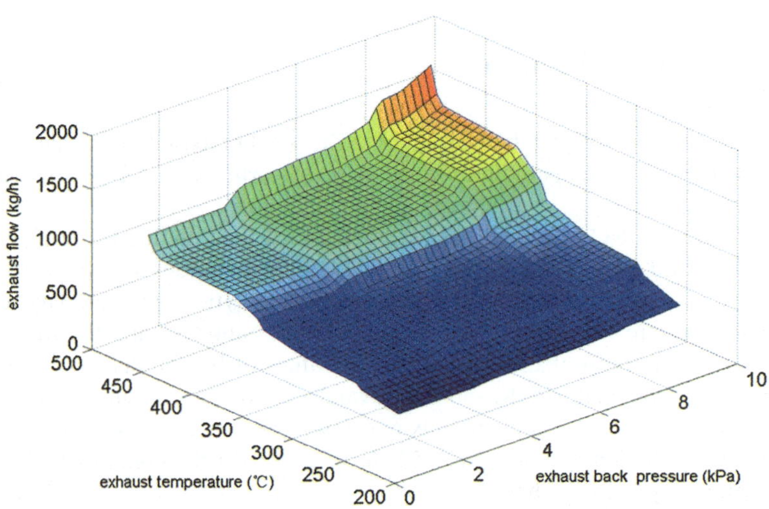

Fig. 3 Relationship between exhaust flow and exhaust back pressure, exhaust temperature

the lowest temperature is less than 200 °C. Catalyst activity is sensitive to temperature. The catalyst activity is good when the temperature is more than 300 °C. There are some operation points where the temperatures is less than 250 °C. And these points correspond to those points with lower conversion efficiency of NOx in Fig. 4. This also explains that the temperature is very important to conversion efficiency.

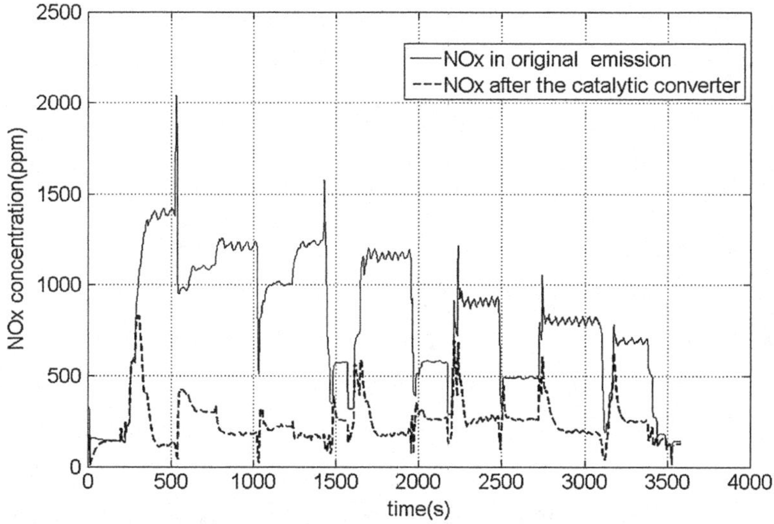

Fig. 4 NOx concentration of ESC conditions

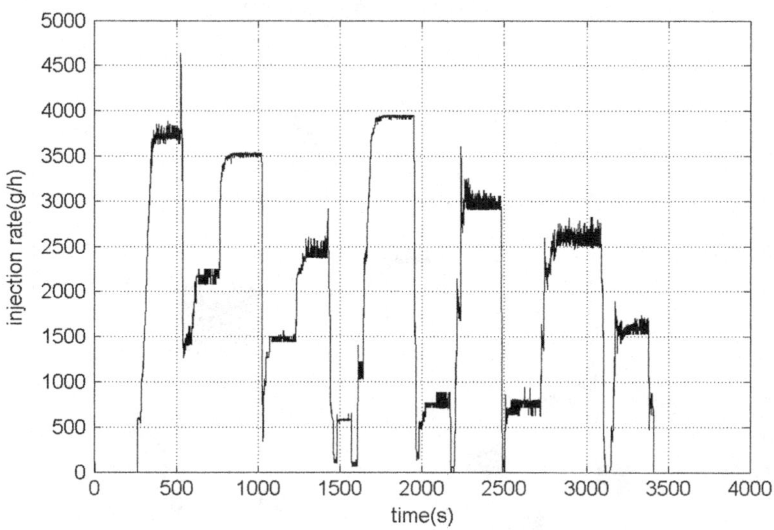

Fig. 5 Injection rate of reductant under ESC conditions

The changes of exhaust back pressure with ESC conditions is shown in Fig. 7. As can be seen from the figure, the exhaust back pressures are great different in different conditions. The minimum exhaust back pressure is about 1 kPa, and the maximum exhaust back pressure is more than 9 kPa. Through comparison between Figs. 6 and 7, the operation points with lower exhaust back pressures where the exhaust temperatures are also low and vice versa. The two parameters well correspond to each other.uctant under ESC conditions.

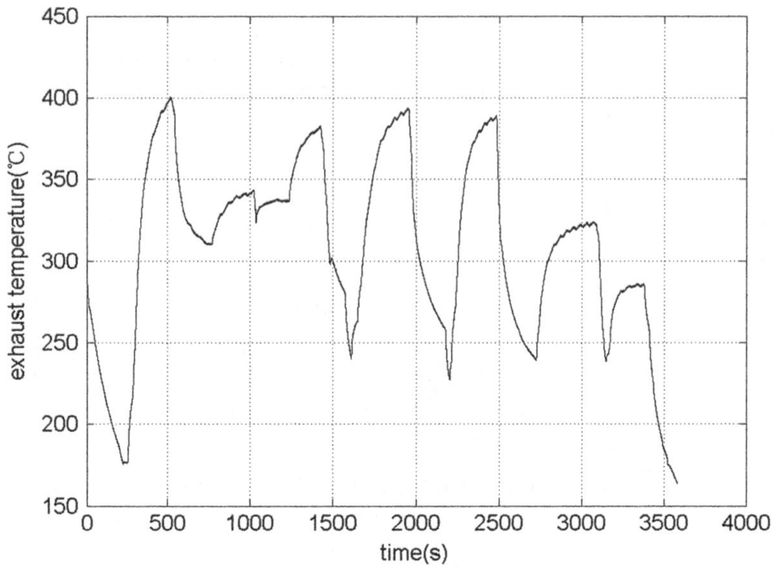

Fig. 6 Temperature of catalytic converter under ESC conditions

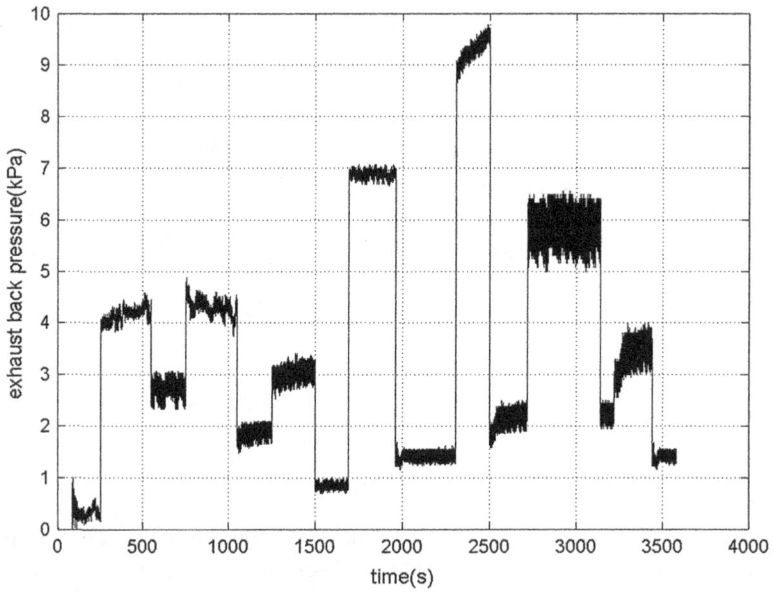

Fig. 7 Trend of exhaust back pressure under ESC conditions

From the analysis of ESC conditions, the exhaust temperature and exhaust back pressure are changed with different conditions. The exhaust back pressure and NOx concentration are associated with the exhaust flow. A relationship between exhaust back pressure and NOx concentration is established. The injection rate of reductant can be determined according to the exhaust temperature and exhaust back pressure. The signals of nitrogen oxides sensors which are located at the front and back of the catalytic converter are sent to DCU. A analysis about real conversion efficiency is done. And a comparison between the real conversion efficiency and that of Fig. 3 is also made. According to the comparison, the injection rate of reductant is adjusted in real time.

The results show that the NOx emission of non-electronic controlled diesel engine can be decreased sharply with SCR system based on exhaust temperature and exhaust back pressure. Under the control of DCU, the SCR system can run automatically.

4 Conclusion

In order to reduce the NOx emission of in-use vehicle, a SCR system based on exhaust temperature and exhaust back pressure is put forward which is aimed at non-electronic controlled diesel engine. Through experimental research, the NOx emission can be significantly reduced. This technique way is verified as a feasible method to control NOx emission of non-electronic controlled diesel engine.

Acknowledgments This work is supported by the national science and technology support plan of China (no.2011BAG06B02).

References

1. Jobson E (2004) Future challenges in automotive emission control. Top Catal 28(1–4):191–199
2. Kondratenko EV et al (2006) Influence of O2 and H2 on NO reduction by NH3 over Ag/Al2O3: a transient isotopic approach. J Catal 239:23–33
3. Kenichi S (2007) Hydrogen assisted urea-SCR and NH3-SCR with silver-alumina as highly active and SO2-tolerant de-NOx catalysis. Appl Catal B 77:202–205
4. Su L, Zhou L, Jiang D, Li W (2003) Recent development of aftertreatment of diesel emissions. Intern Combust Engines 1:1–4
5. Du B, Zhu H, Xu L (2004) The status and progress of vehicle diesel exhaust emission control. Chin Intern Combust Engine Eng 25(3):71–74
6. Yoon CS et al (2007) Measurement of soot mass and pressure drop using a single channel DPF to determine soot permeability and density in the wall flow filter. SAE 2007-01-0311, 219–238

7. Tang W et al (2007) A lumped/1-D combined approach for modeling wall-flow diesel particulate filters. SAE 2007-01-3971, 549–557
8. Zhe X, Shunxi W, Yin Z et al (2011) Pressure loss of urea-SCR converter and its influence on diesel engine performance. Trans Chin Soc Agr Eng 27(8):169–173
9. Yang D, Gao X, Wang X et al (2003) The research on regeneration occasion for electric-heated regeneration method for diesel exhaust particulate ceramic filter. Chin Intern Combust Engine Eng 24(4):42–44

11L Diesel Engine to Meet China V with Doc

Pengcheng Wang, Jiangwei Liu, Fang Hu, Zhenbo Xu, Zhong Wang, Huayu Jin, Liu Yu, Xinchun Niu and Fanchen Meng

Abstract Environmental protection is the key issue in the whole world. Emissions legislation is pushing the engine technique forward more and more. For heavy duty diesel engine the fuel economy is enhanced greatly since the introduction of after treatment system. But the basic engine performance still goes further since the appearance of the advanced components and CAE lean design. In this paper based on the combination of CFD and experimental test 11 L diesel engine meets China V standards with only DOC, which has competitive fuel efficiency compared with SCR solution.

Keywords Heavy duty · Diesel · Optical · Combustion · CFD

1 Introduction

Commercial vehicle engine platform has been established in FAW so far (15 L engine is in the conceptual design stage). Engine displacement ranges from 3 to 13 L, with power ranges from 90 to 460 Ps, and torque ranges from 240 to 2,400 Nm, adaptation for the whole products series has been completed involving light duty L series, medium duty K and M series, heavy duty P series (Fig. 1). This paper is focus on the 11 L diesel engine.

F2012-A08-007

P. Wang (✉) · J. Liu · F. Hu · Z. Xu · Z. Wang · H. Jin · L. Yu · X. Niu · F. Meng
FAW RDC, Mainland, China
e-mail: wangpengcheng@rdc.faw.com.cn

SAE-China and FISITA (eds.), *Proceedings of the FISITA 2012 World Automotive Congress*, Lecture Notes in Electrical Engineering 190,
DOI: 10.1007/978-3-642-33750-5_38, © Springer-Verlag Berlin Heidelberg 2013

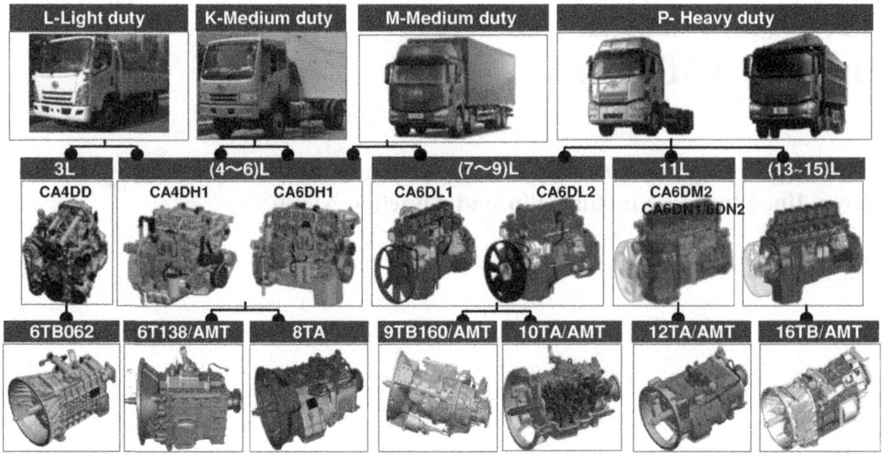

Fig. 1 FAW truck range with power trains

2 CFD Leaded Design

CFD is the cost-efficient method to run numerous combustion parameters virtually and select appropriate boundary. But the precondition is the accuracy of simulation. To calibrate the CFD model the combustion process is investigated based on the transparent single cylinder engine and endoscope. Optical engine has the function of observing the complete flame shape however is restricted in low load by PFP limit. To extend the observed load endoscope is used although the window is limited in two fuel spay. Figures 2 and 3 are the comparison of combustion flame and simulation. Based on the calibrated CFD model the optimization is carried out in terms of different fuel injection, air intake and combustion chamber with local lambda coefficient.

3 Main Features of China V EGR Engines

3.1 Injection System

The mixer energy comes from fuel injection and air flow and then influences the local air fuel ratio resulting in uneven combustion temperature and soot formation. 1,600 bar fuel injection system is used popularly in SCR market. With the introduction of EGR the mixing quality becomes poor because EGR distribution is not even in the cylinder proved by CFD simulation and smoke test in Figs. 4 and 5.

Local condition of Lambda less than 0.5 is considered to have important contribution to final soot emission. 1,800 bar system has less volume of under 0.5

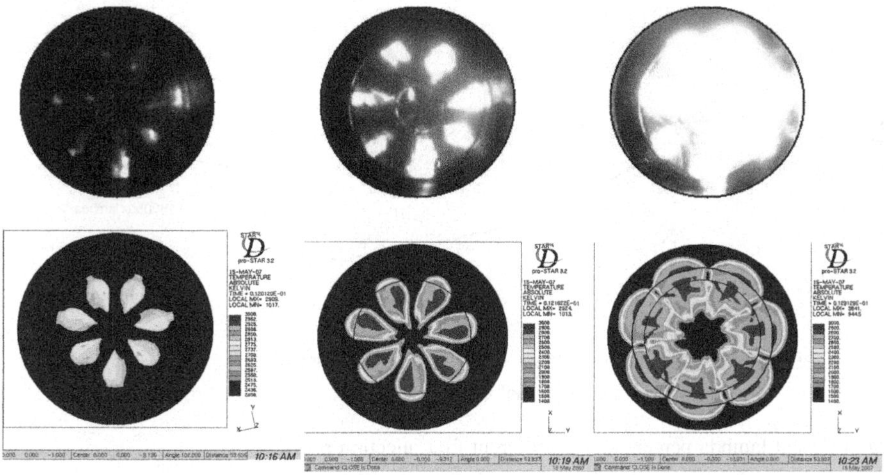

Fig. 2 Comparison of *bottom view* and simulation

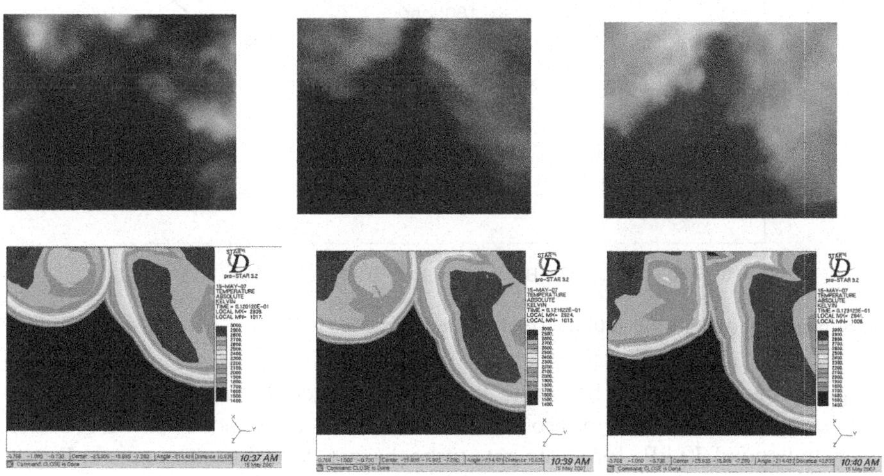

Fig. 3 Comparison of endoscope capture and simulation

than 1,600 bar system and smoke test also proves the result. Therefore 1,800 bar FIE is selected finally in Fig. 6.

3.2 Turbocharger of VGT

The application of modern diesel turbocharger takes full advantage of exhaust energy to boost inlet air. In most conditions EGR is hard to flow back to the intake manifold to affect combustion. VGT could balance well the trade off between EGR

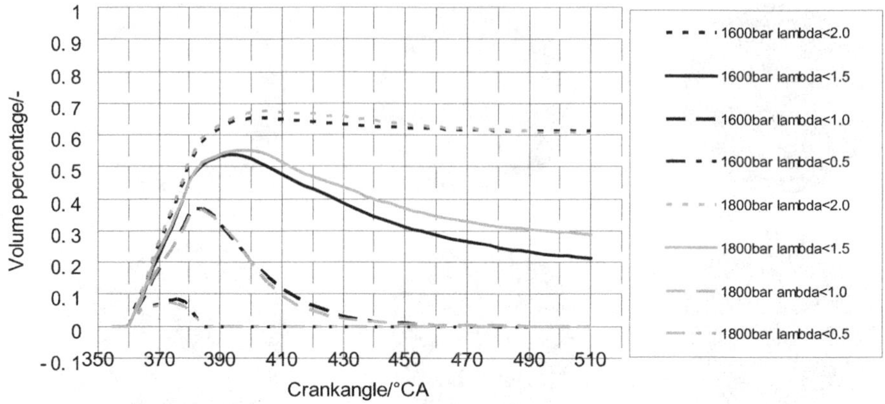

Fig. 4 Local lambda percentage for different fuel injection pressure

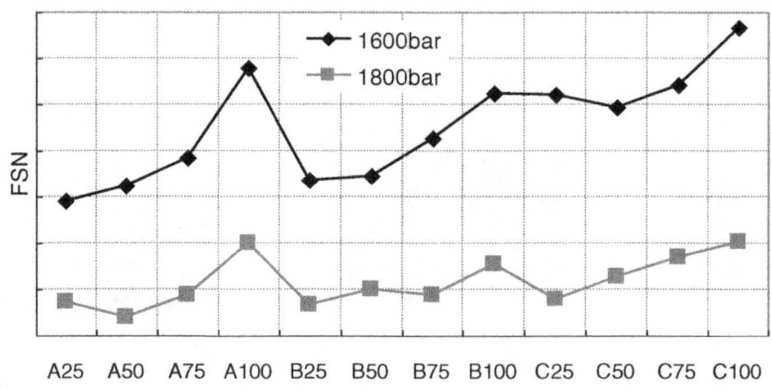

Fig. 5 Smoke test result for different fuel injection pressure

and pumping loss caused by waste gate turbocharger. Figure 7 shows the reasonable pressure difference of around 25 kPa at C100 EGR mode for VGT, which is easy to beyond 70 kPa for waste gate turbocharger. VGT installation is in Fig. 8.

3.3 External Cooled EGR

External cooled EGR is the effective solution for decreasing NOx formation in the cylinder. The negative effect is to deteriorate combustion and increase pumping loss. As mentioned above high pressure injection and AFR recovery by VGT has although balanced some deterioration EGR pipes optimization still has the potential to improve BSFC by increasing effective combined flow area. Detailed calculation has been conducted for selecting the right EGR valve, cooler and connecting pipes. The EGR layout is shown in Fig. 9.

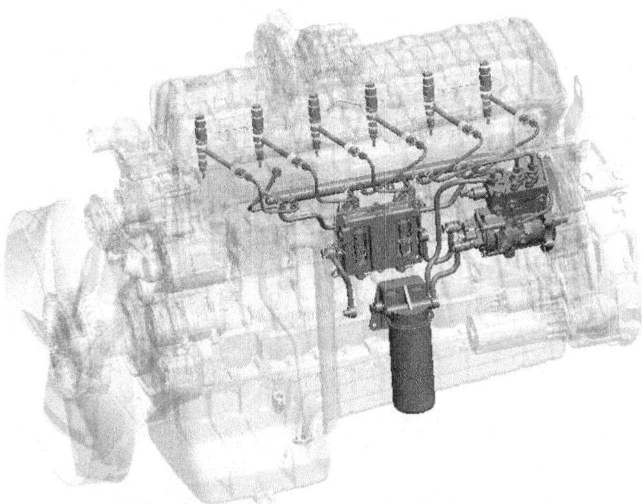

Fig. 6 Bosch common rail 1800 bar

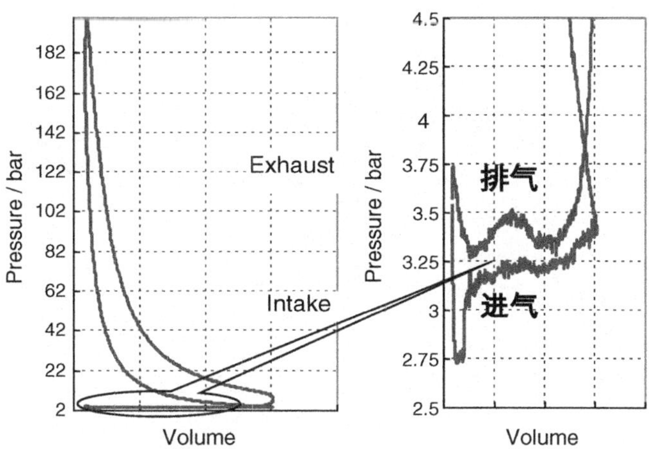

Fig. 7 Intake and exhaust pressure at C100 EGR mode

3.4 Piston

For lowering soot emission the very important feature is the fuel distribution in cylinder. In the later injection period more fuel right above the chamber and less fuel inside the cylinder and clearance will effectively decrease soot emission. Optimized chamber shape could be helpful for fuel guidance such as in Fig. 10. Applied piston bowl can be found in Fig. 11.

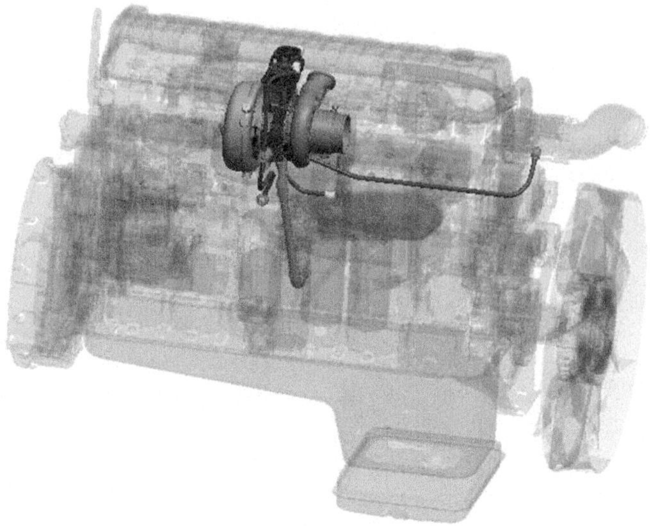

Fig. 8 VGT installation in 11 L diesel engine

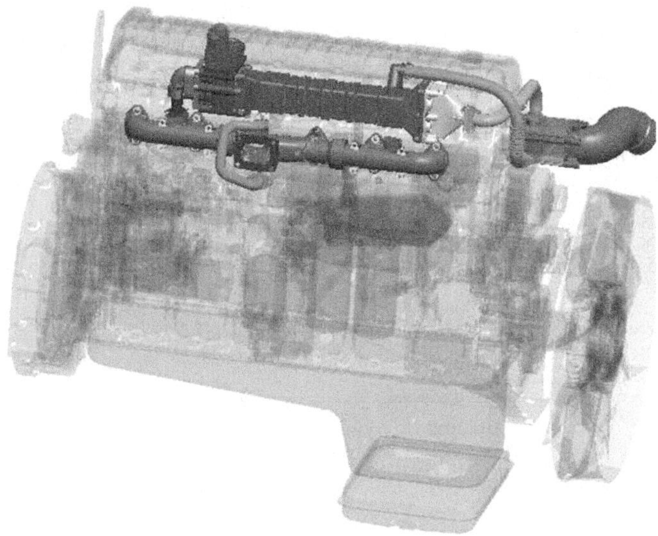

Fig. 9 EGR pipes layout in 11 L diesel engine

4 Test Results

The optimized distribution of the air and fuel decreases soot emission dramatically despite EGR introduced. Less pumping penalty is paid since the air and EGR pipe design including the application of VGT. The best brake thermal efficiency is close

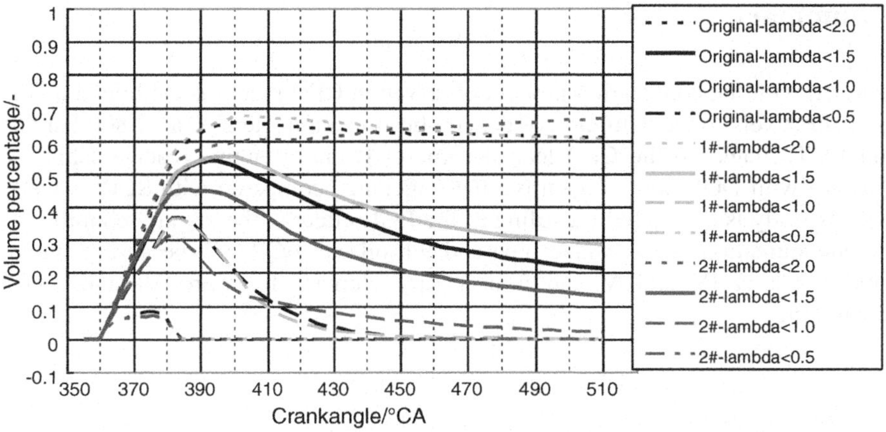

Fig. 10 Local lambda percentage for different chambers

Fig. 11 Comparison of China IV and V chamber

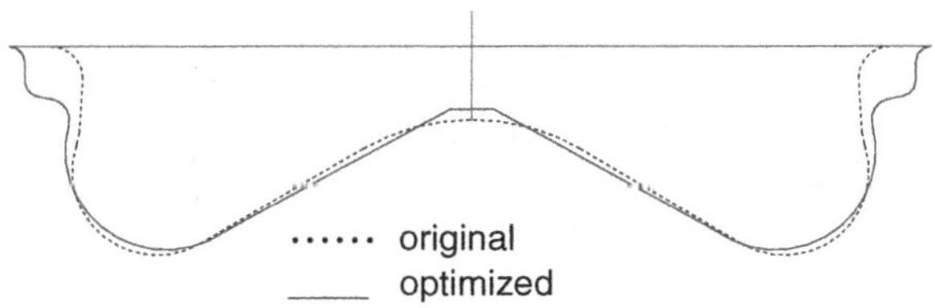

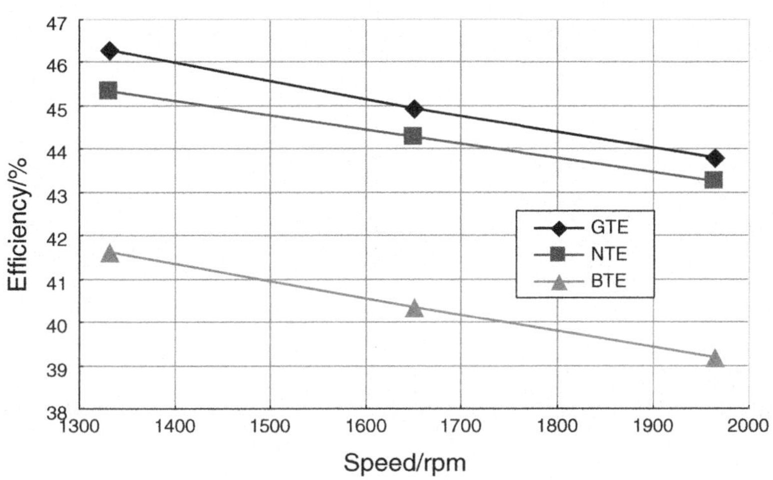

Fig. 12 Brake thermal efficiency of 11 L diesel engine

to 42 % in Fig. 12 and the competitive fuel economy is acquired compared with SCR average consumption keeping all emissions under the limits of China V.

5 Summary

Coupled with the combustion process observation CFD model is calibrated and is used effectively to the definition of the combustion parameters. With 1,800 bar CR and VGT, thanks to the CFD lean design, 11 L diesel engine reaches China V standard with DOC and competitive fuel efficiency compared with SCR as well.

Less study is conducted regarding to the EGR effect to the engine reliability as the contamination of the intake pipe and lub oil etc. EGR will be an alternative path to China IV and V and a necessary element for Euro VI and worth researching.

Part IX
Other

Study the Failure About Spring Which Used on VVA Engine in System

Ling Lin, Haizhu He, Jun Mao, Liyun Kang, Maohui Wang, Yang Qiu and Yong He

Abstract Recently a new technology called Variable Valve Action (VVA) had been used in high performance engines. The valve springs have to endure high stress and high speed etc. because of VVA system. In order to get the answer about the spring failure, the running condition were reviewed and the photo documentation focus on the fracture surface which was amplified to find the crack initiation site. And the model of VVA valve train was build up based on AVL-work space to get the detail inner information, and the prototypes were send to the test centre, so the test report document about the chemistry, microhardness and microstructure of the fractured valve springs can be got. The photo document shows that all springs fractured in approximately the same location (in the 2nd coil), and have evidence of coil-to-coil contacted, and the crack initiation site is subsurface and the crack propagates via fatigue. The test report documents show that for all springs, there is a small subsurface ductile tear from which multiple cracks are initiating and propagating via fatigue. The ductile tear is at an angle to the rest of the fracture surface, the surface of the ductile tear is not oxidized, and the chemistry is consistent with the specification for OTEVA 70 SC. The simulation results show that the sympathetic vibration and the no cam contact stress happen when the engine speed out of 6000 r/min. The experiment of valve train had completed on the test bench, the results show that the value of seating velocity is unacceptable and badly

F2012-A09-004

L. Lin (✉) · H. He · L. Kang · M. Wang · Y. Qiu · Y. He
Automobile Global Research and Development Center of Changan Automobile Co., Ltd, Mainland, China
e-mail: sunqiemail@163.com

J. Mao
Sichuan Professional Technology College, Chengdu, China

SAE-China and FISITA (eds.), *Proceedings of the FISITA 2012 World Automotive Congress*, Lecture Notes in Electrical Engineering 190,
DOI: 10.1007/978-3-642-33750-5_39, © Springer-Verlag Berlin Heidelberg 2013

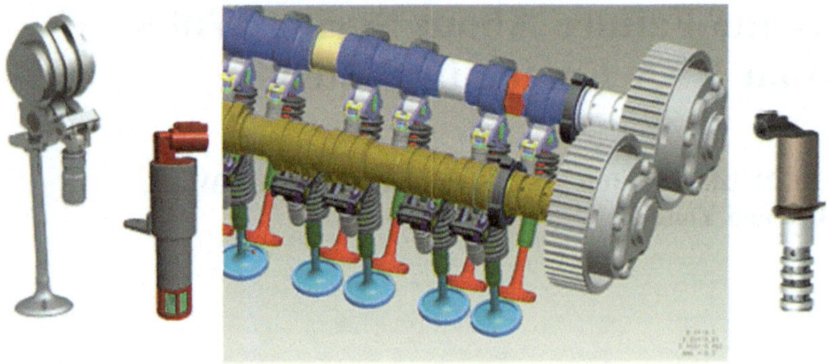

Fig. 1 Layout of VVA'S valve-train

valve bounce happened. The springs most likely fractured due to sympathetic vibration and contact-loss (no cam contact stress) when high speed, overloading (potentially from dynamic amplification of the applied stress). According to these, redesign cam profile and improve spring surface quality can solve these issues.

Keywords Valve spring · VVA · Subsurface · Fracture · Valve train

1 Preface

The new techniques of engine were given attention by human in the world because of energy sources safe and the statute of environment. The VVA engine which has excellent performance [1] was welcomed in the field of power train.

The valve spring which used on the VVA engine should have to satisfy the following requirement, such as strong stress and good reliability and high velocity and down size, so these springs face to new challenges [2–4].

2 System Description: Two Steps Valve Train

Components used to implement two steps VVA systems include a two-step roller finger (RFF) which is heavier and bulkier than the normal RFF, intake camshaft which has two high lift cams symmetrically and one low lift cam in the middle and so its structure is very complicated, two control valve for actuating the two step RFF, intake and exhaust cam phasers which were connected by the center bolt, and two control valves for actuating the cam phasers.

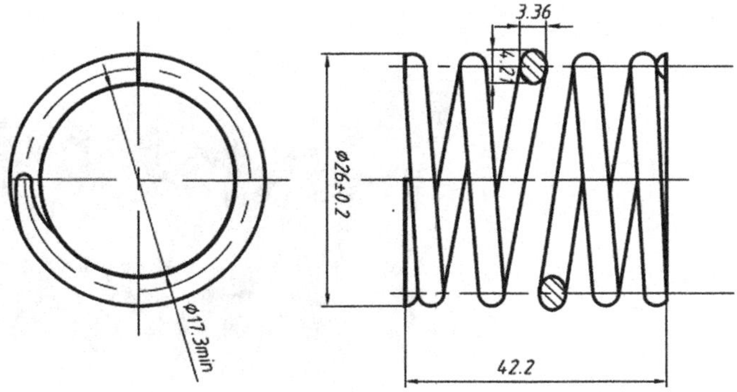

Fig. 2 Valve spring specifications

The layout is the same as normal valve train shown Fig. 1. The valve spring's specification as following the free height is 42.2 mm and the outer diameter is 26 mm, the section is ellipse. Figure 2 show more information about the spring.

3 Background Description: The Recordation of the Running Engine and the Character of Crack Spring

3.1 Working Condition of the VVA Engine

There are Three-Stages for test the engine:

(1) The mechanical check stage, the new engine running 20 h at 1000 r/min just for checking the mechanical function, everything is ok.
(2) The VVA function test stage,the engine running 40 h, and the rotate speed is from 1000 to 4000 r/min when test VVA system.
(3) The engine performance test stage, the engine running 100 h and the rotate speed is from 850 to 6500 r/min. Some abnormal voice comes from the engine.

Disassemble the valve train, then the crack traces were found at the second and the third cylinders valve springs.

3.2 Description the Fractured Spring

The Fig. 3 shows that all the springs fractured in approximately the same location (in the 2nd coil) and the Fig. 4 shows that all of springs have evidence of coil-to-coil contact.

Fig. 3 The location of fracture spring

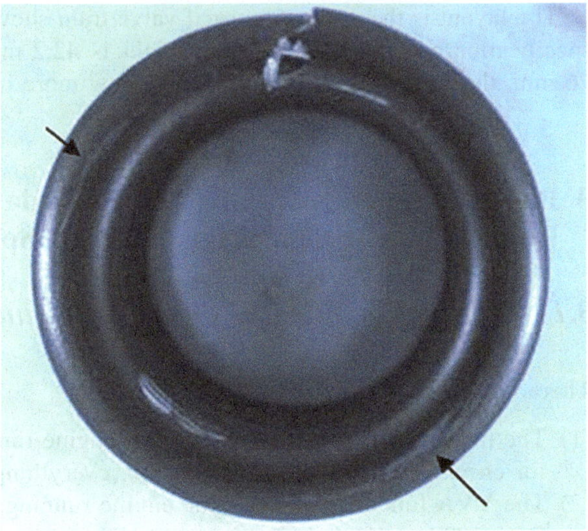

Fig. 4 The friction scrape on the surface

4 The Dynamics and Kinematics Study About Valve Train Base on Test Bench

The valve train experiment had been done on the dynamics and kinematics test bench, so the valve train's dynamics and kinematics data can be read from the velocity and displacement sensors. Please see the Fig. 5.

The experiment results show that the max speed should be lower than 6000 r/min, or else, the serious bounce and excessive closing velocity cannot be avoided. Please see the Figs. 6, 7 and 8. The closing velocity will be more than 1000 mm/s when the engine's speed at 6000 r/min, so the valve bounce happened and the bounce height of 0.16 mm. The closing velocity will be more than 2400 mm/s when engine speed more than 6500 r/min.

Fig. 5 The test bed and sensors location

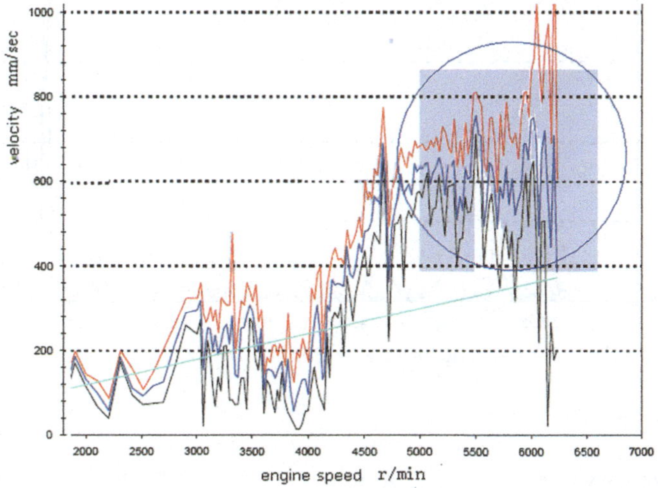

Fig. 6 Valve closing velocity

5 CAE Analyse for Valve Train

The VVA engine's 1 dimension model was set up based on AVL_tycon 5.1 software, just focus on study the kinematics and dynamic parameter, please see the Fig. 9.

The Fig. 10 show the valve profile. The parameter such as valve spring frequency and valve train's mass should be referred to in simulation.

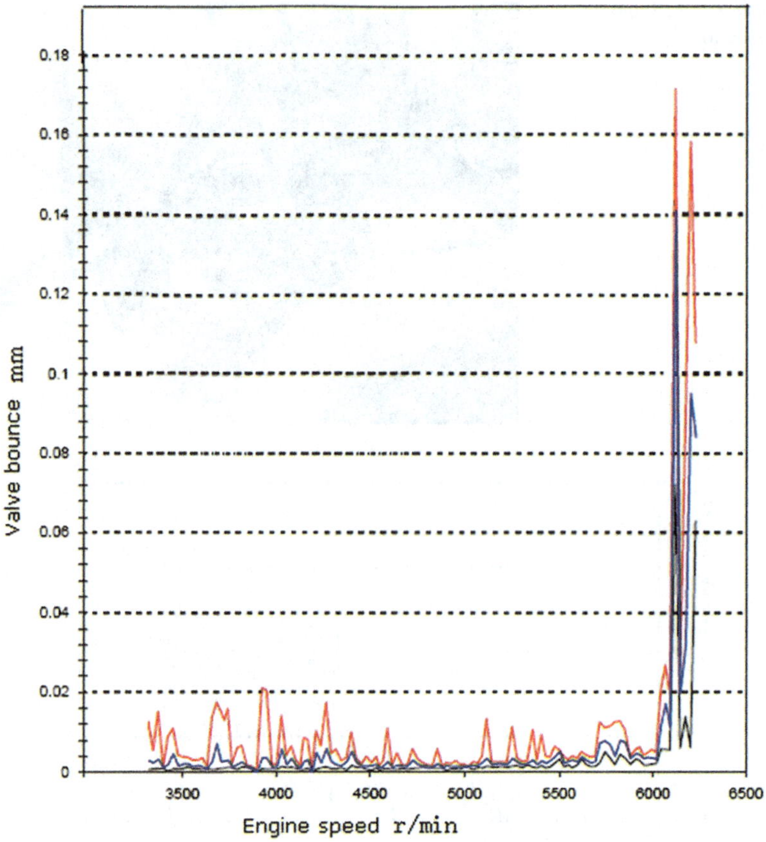

Fig. 7 Valve bounce

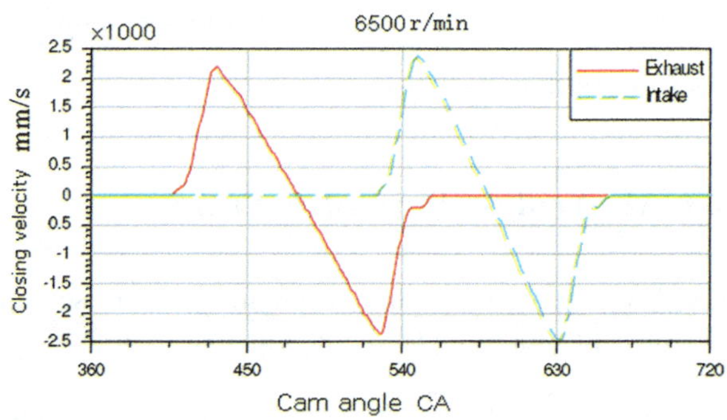

Fig. 8 Closing velocity at 6500 r/min

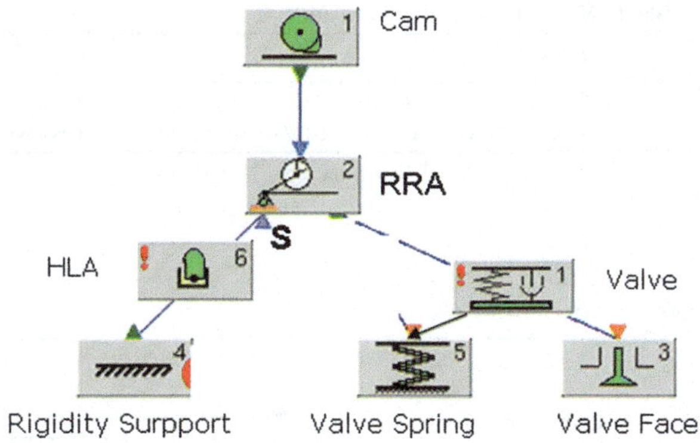

Fig. 9 Valve train model of VVA engine

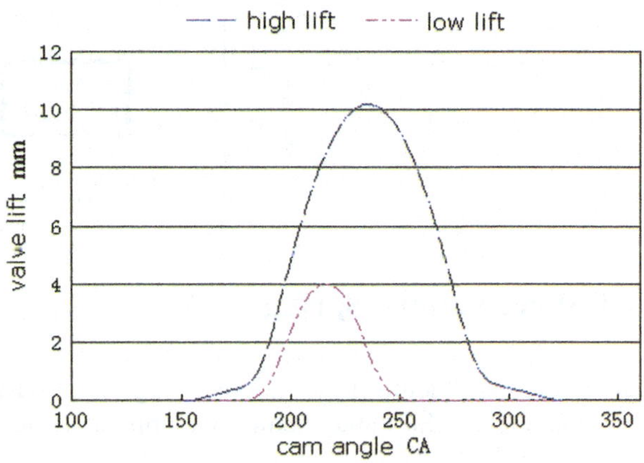

Fig. 10 Valve profile

5.1 Result of CAE Analyse

The valve train had been considered that happen contact-loss or sympathetic vibration, according to the failure cam shafts and valve springs' character. So the CAE study will focus on kinematics and dynamic. Then the CAE study results show that the cam contact stress is 0 and the valve closing velocity is quite higher when the engine speed was 6600 r/min. Please see the Table 1 and Fig. 11.

Fourier analysis of the valve lift event shows significant lift amplitude up to the 11th cam harmonic. The threshold for significant lift amplitude is 0.071 mm at any harmonic, and the valve event in this study has an amplitude of 0.068 mm at the 11th. This is large enough to be of concern if the valve spring has a natural frequency that is within the normal operating speed range of the engine. The natural frequency of the spring for this application is 564 Hz which corresponds to an engine speed of 6154 rpm. Since this speed is less than the 6400 rpm fuel cutoff speed, there is concern that harmonic excitation of the valve spring could occur within the normal operating range of the engine. Please see the Fig. 12.

Table 1 Valve-train analyses results

Speed (rpm)	High lift			Low lift		
	Contact stress MPa	Closing velocity mm/s	Acceler mm/deg^2	Contact stress MPa	Closing velocity mm/s	Acceler mm/deg^2
0	770	2400 max	0.018 max	1050	2400 max	0.012 max
3800				970		
4000	650			0∼900		
6000	630					
6600	0∼750					

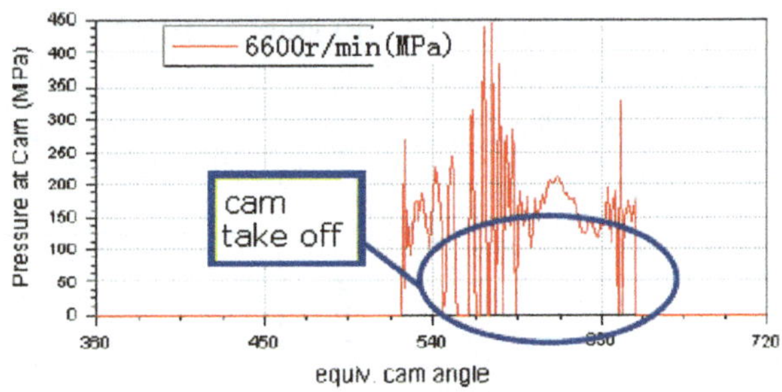

Fig. 11 Cam contact stress

6 Inspect Failure Spring

Study the inner factor of the failure spring based on check the microstructure and microhardness values and chemistry components and metallography.

6.1 The Chemistry Components and Microh-Ardness Analyse

EDX analysis of the short piece fracture surface of the failure springs. For all the springs, there is a small subsurface ductile tear from which multiple cracks are initiating and propagating via fatigue, please see the Fig. 13. The ductile tear is at an angle to the rest of the fracture surface, the surface of the ductile tear is not oxidized, and there is no evidence of inclusions at the ductile tear.

Chemistry of springs as determined via XRF. The chemistry of all the springs is consistent with the specification for OTEVA 70 SC. Please see the Figs. 14, 15, 16 and Table 2, so there is no question about the material of the springs. Table 3 show that all of the springs had similar average microhardness values, and the average microhardness is slightly lower for OTEVA 70 SC wire of this diameter. So the lower microhardness values is considered to be one of the reasons of failure.

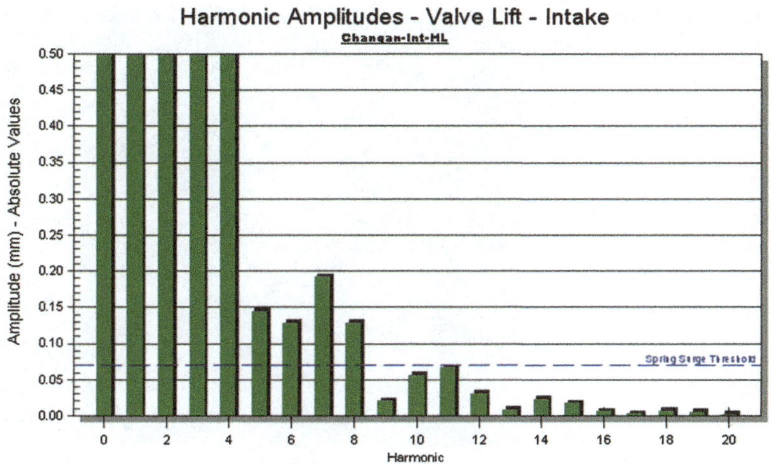

Fig. 12 Fourier analyses

Fig. 13 Fracture surface

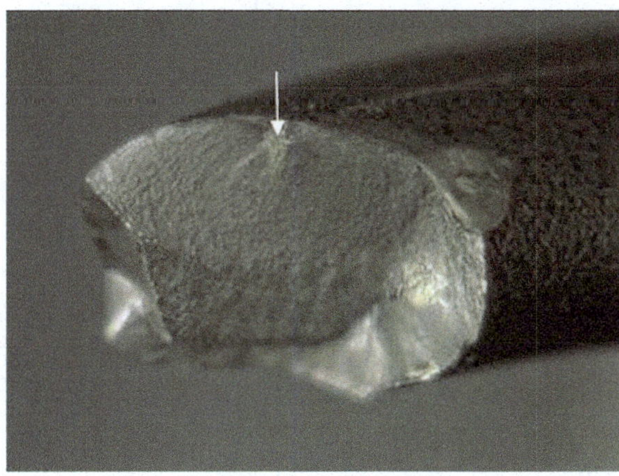

6.2 Metallography Study

Etched microstructure of longitude inner cross-section of all failure springs through the fracture surface at 500X. The microstructure of all springs consists of fine grained tempered martensite with no significant decarburization on the surface. So the metallography is ok for the spring (Figs. 17 and 18).

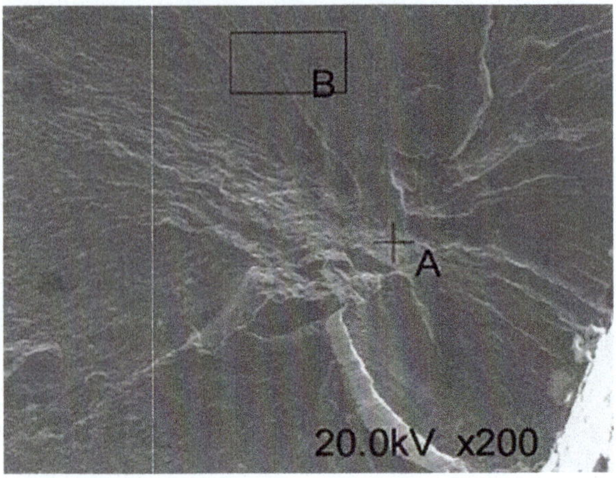

Fig. 14 Check position

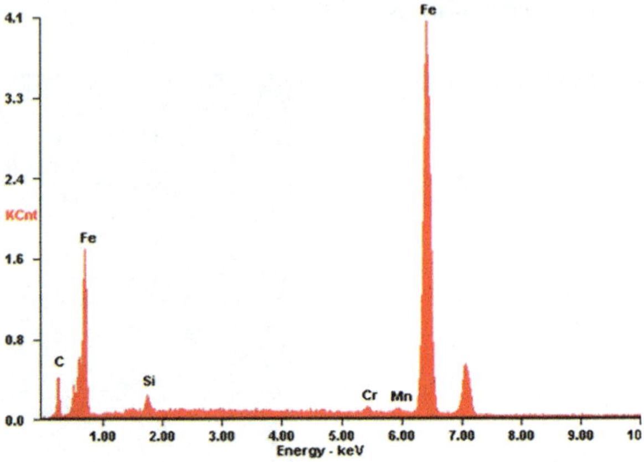

Fig. 15 EDX spectrum of box A

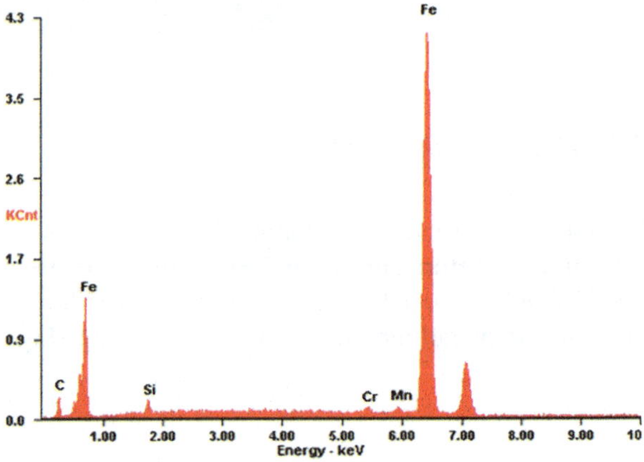

Fig. 16 EDX spectrum of box B

Table 2 Chemistry of springs

Spring	Cr	Mn	Ni	Cu	Fe
#1	0.50	0.76	0.06	0.00	Bal.
#2	0.48	0.70	0.07	0.00	Bal.
#4	0.54	0.75	0.07	0.00	Bal.
#7	0.54	0.77	0.07	0.00	Bal.

Table 3 Microhardness Values

Spring	Along fracture surface	Away from fracture surface
#1	51.3	51.7
#2	52.4	51.5

Fig. 17 Microstructure of surface

Fig. 18 Microstructure of core

7 Conclusion

1. The springs most likely fractured due to overloading, so the closing velocity and bounce and take off come up.
2. The valve train natural frequency is close to the average stiffness of valve trains from similar engines, so the sympathetic vibration comes up.
3. The crack initiation site is subsurface which has disfigurement and the crack propagates via fatigue.
4. Redesign the cam profile or choose the other frequency spring maybe solve these issues.

References

1. Lian F (2005) Material engine. Northeast University Publishing Company, ShenYang
2. Li X (2003) Study the spring of gasoline engine. Technol motorcycle 02
3. The term of internal combustion engines valve spring, GB 2785-88
4. Hu L (2011) Analysis about the failure of Valve fracture. Intern combust engine 1

The Impact of Oil-Based Diamond Nanofluids on Diesel Engine Performance

Hao Liu, Minli Bai and Yuan Qu

Abstract Experimental test were carried out to investigate the effects of adding oil-based nanodiamond lubricant both in friction tester and diesel engine. Tribology, engine performance, fuel economy and the temperature field of one piston are important aspects of this work. To this end, surface-modified nanodiamond lubricant of 0.1 wt% was used in a four-stroke diesel engine. The results indicate that adding nanodiamond particles to engine oil has a perceptible effect on engine performance, increasing the maximum engine power and the maximum torque relatively to 1.15, 1.18 % respectively, while decreasing the fuel consumption relatively to 1.27 % as compared to the engine oil.

Keywords Nanofluid · Diamond · Bench test · Engine performance · Fuel economy

1 Introduction

Machine components and friction pairs rely on high-quality lubricant oil to withstand high temperature and extreme pressure in internal combustion engine. Especially the piston and cylinder system under in the harsh environment of high temperature, overload and low speed. With high-quality lubricant oil, it can reduce engine wear, prolong engine life and improve the fuel economy. Application of nanomaterials as lubricant oil additives has become a rapidly progressing field of

F2012-A09-006

H. Liu (✉) · M. Bai · Y. Qu
Dalian University of Technology, 116024, Dalian, China
e-mail: liuyuhao1985@vip.qq.com

SAE-China and FISITA (eds.), *Proceedings of the FISITA 2012 World Automotive Congress*, Lecture Notes in Electrical Engineering 190,
DOI: 10.1007/978-3-642-33750-5_40,

research since nanomaterials are different from traditional bulk materials due to their excellent physical properties.

Tribology research shows that lubricant oils with nanoparticles (MoS_2, CuO, TiO_2, diamond, etc.) exhibit improved load carrying capacity, anti-wear and friction-reduction properties [1, 2]. Peng et al. [3] found from their experiments that the diamond nanoparticle as additive in liquid paraffin at appropriate concentration can show better tribological properties for antiwear and antifriction than the pure paraffin oil. Tao et al. [4] investigated diamond nanoparticles as an oil additive and found that under boundary lubricating conditions, the ball-bearing effect of diamond nanoparticles existed between the rubbing surfaces, the surface polishing and the increase in surface hardness effects of the diamond nanoparticles were the main reasons for the reduction in wear and friction. Friction coefficients in the range 0.04–0.20 have been reported by Hayward et al. for natural diamond-on-diamond atmospheric sliding experiments, without the necessity for additional lubrication [5]. In contrast to most other engineering surfaces, the diamond-on-diamond friction coefficient depends on the surface roughness of the films [6, 7]. In that sense, the friction coefficient for the diamond coatings tested in Hayward et al.'s work [5] gave values from 0.03 up to values considerably higher (0.50).

Experiment results shows that the diamond nanoparticles can enhance the thermal conductivity of nanofluids. Kang et al. [8] reported a 75 % thermal conductivity enhancement for ethylene glycol with 1.2 % (v/v) diamond nanoparticles between 30 and 50 nm in diameter.

The present study focus on the influence of tribology and engine performance upon the application of diamond nanofluids on AVL diesel engine bench test system supported by DEUTZ (Dalian) Engine Co. The friction and wear experiments were performed to evaluate the friction reduction and anti-wear abilities with or without the diamond nanoparticles. In addition, more investigations were conducted using diesel engine bench test system.

The nanoparticles of diamond were provided by Department of Engineering Mechanics, DUT. Dalian Diesel Engine Factory commercial engine oil was used as the base oil to prepared the nanofluids at the concentration of 0.1 wt%. At the beginning of each test, nanofluids subjected to 20 kHz ultrasonic stirring for 30 min in order to keep well suspension property.

2 Nanofluids Viscosity Measurement

Viscosity is a measure of a fluid's resistance to flow. To measure the viscosity of nanofluids, a rotation-type viscometer (Brookfield Engineering Lab, digital viscometer DV II+pro) was employed, which uses a weight-driven rotating paddle to sense the viscosity of fluid spinning at a constant angular velocity [9]. For comparative data analyses, nanofluid samples are maintained the same rotational speed at the same temperature when measuring the viscosity, our test temperature ranged from 25 to 140 °C.

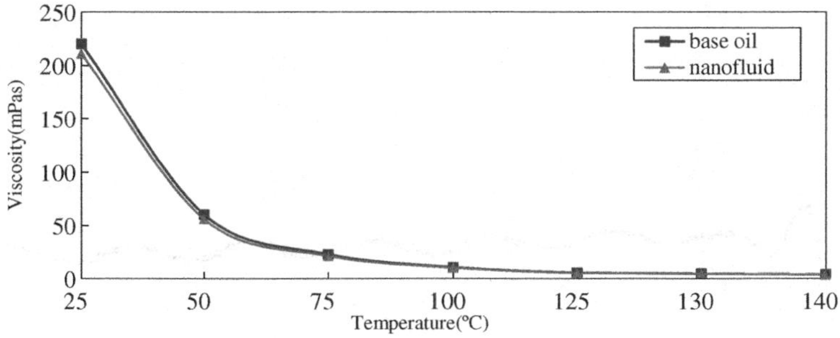

Fig. 1 Viscosity of the nanofluid and the base oil

The viscosity of nanoparticle-oil suspensions decreases in accordance with temperature increasing. At 25 °C the viscosity value achieved at 211.48 mPas, decreased by 4.07 % than the base oil; 4.44 mPas at 140 °C, decrease by 3.33 % than the base oil. The viscosity test results are shown in Fig. 1. In our tests, the viscosity decreased by adding nanodiamond additive, as the base oil were the Dalian Diesel Engine Factory commercial engine oil which is the high quality lubricate with a higher viscosity. And the nanodiamond additive were added not only the nanoparticles but also some kinds of chemical reagents. It is foreseeable that it will be enhance the mobility of the lubricants, reduce the powerconsumption, improved the efficiency and decreased in the temperature of piston.

3 Friction-Reduction Property

To study the friction-reduction properties, four ball tests were carried out according to the standard SH/T0189 by the MMW-1A friction and wear tester.

The tests performed by loading 392 N correspond to 600 rpm and last for 60 min at the temperature of 130 °C. Friction coefficient was recorded by the computer every 10 s. All tests were maintained the same condition when measuring the friction efficiency. The friction coefficients of the lubricant were obtained by tribological tests and shown in Fig. 2 that the base oil adding diamond nanoparticles can reduce the coefficient obviously. The average friction coefficient decreased by 30.25 %, indicates that diamond nanoparticles have excellent friction-reduction property. This might be attributed to the sphere like nanoparticles rolling between the rubbing surfaces, changed the situation of friction from sliding to rolling.

4 Engine Bench Tests

Our tests were carried out on AVL diesel engine bench test system supported by DEUTZ (Dalian) Engine Co. The main performance indicators and parameters are shown in Table 1.

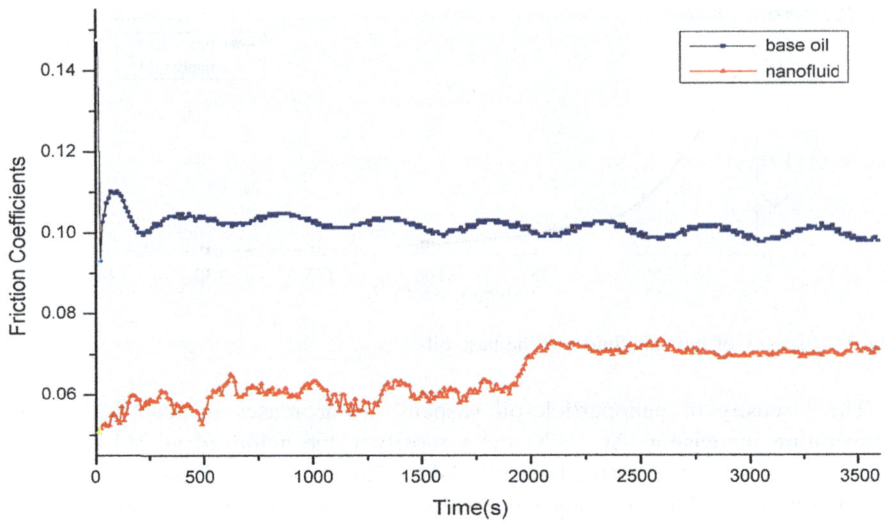

Fig. 2 Friction coefficients of the diamond nanofluid and the base oil

Table 1 Main performance indicators and parameters of the diesel engine

Engine	BF6M1013-30E3
Number of cylinders	6
Number of strokes	4
Bore × stroke	130 × 108 mm
Maximum power	220 kw
Rated power speed	2,300 RPM
Maximum torque	1,100 Nm
Dynamometer	AVL electric dynamometer

4.1 Reverse Dragging Process

Connected to the engine with the electric dynamometer and motored engine at a given speed. The power measured by electric dynamometer should be considered as the mechanical loss that are shown in Fig. 3. Based on the results, the average motor power decreased by 2.77 % associate with the additive of diamond nanofluid. It proved that nanodiamond nanofluid have a good anti-friction and lubrication performance.

4.2 External Characteristics Tests

Keep the engine fully-opened working condition, measuring the change of the torque, power and fuel consumption. The relationship between speed and fuel consumption is shown in Fig. 4. It shows that nanodiamond used as additive in

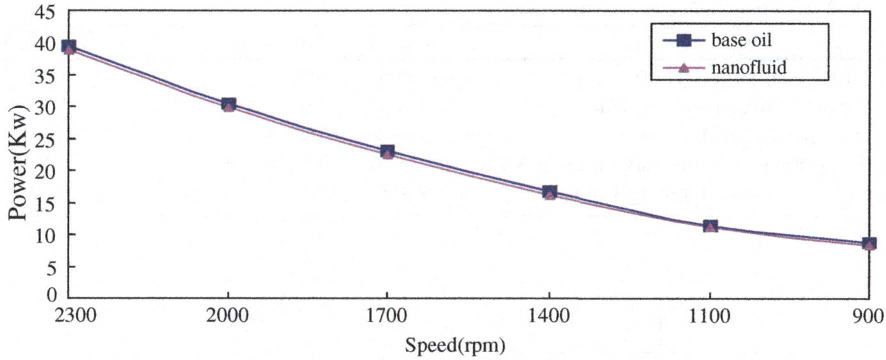

Fig. 3 Motor power in the reverse dragging process

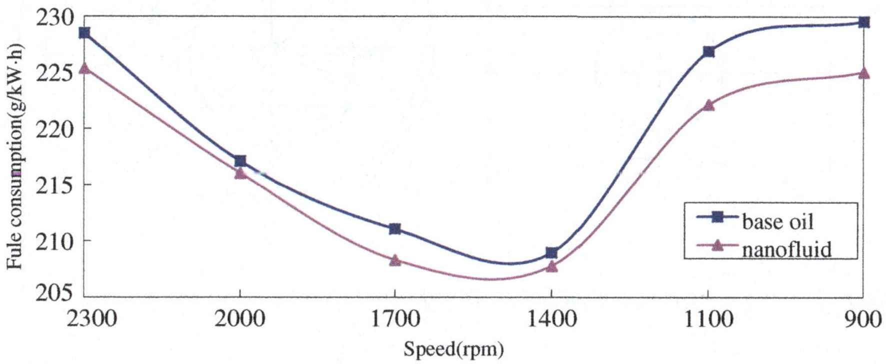

Fig. 4 The relationship between speed and fuel consumption

lubricating oil reduced the fuel consumption. As described in Table 2, the max power increased by 1.15 %, the maximum torque increased by 1.18 %, the average fuel consumption decreased by 1.27 % with the applying of nanodiamond lubrication that exhibit good friction and lubricating behavior, more output power was converted to useful work and reduced the fuel consumption.

4.3 Temperature Field Tests

The experiment of temperature measurement was made on piston by the way of hardness plug for realizing the thermal load of piston. The tests were carried out at the rated operating condition for 60 min, inlet and outlet water temperature maintained stable relatively. Figure 5 shows the measuring points of the piston surface while the results are shown in Table 3. Test points that the center of the

Table 2 External characteristics test results

	Base oil	Nano-diamond lubricant	Rate (%)
The maximum torque (N·m)	1071.8	1084.4	1.18
Power of the maximum torque (kW)	157.13	158.94	1.15
Average power (kW)	156.00	156.24	0.15
Fuel consumption of the max torque (g/kW·h)	208.94	207.76	−0.56
Average fuel consumption (g/kW·h)	220.23	217.44	−1.27

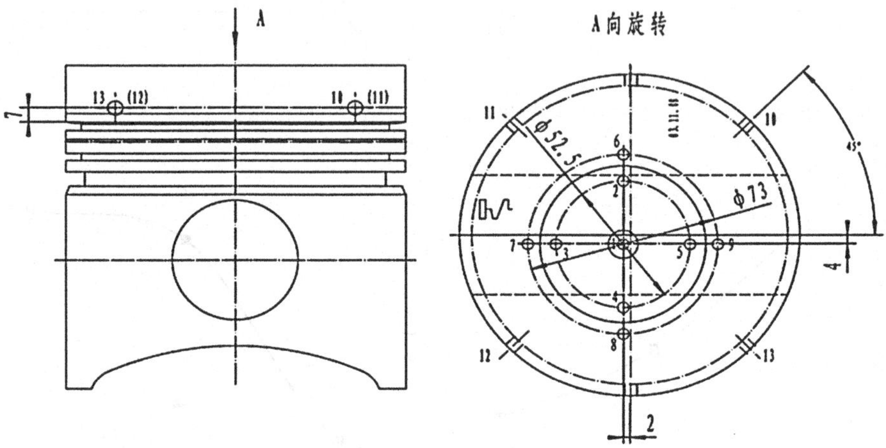

Fig. 5 The measuring points of the piston (BF1013-1097) surface

Table 3 Temperature of the measuring points

Measuring points	Base oil (°C)	Nano-diamond lubrication (°C)	Temperature change (°C)
1	306	298	8
8	330	320	10
13	273	269	4

top, the edge of the combustion chamber and the shore of first ring, it has reduced 8, 10, and 4 °C compared with the base oil. Frictional work and fuel consumption were reduced with the additive of diamond nanofluids, thus cause the lower temperature of the piston.

5 Conclusion

It can be seen through the experimental results that diamond nanofluids developed by ourselves used as additive in lubricating oil exhibit good friction and lubricating behavior, this might be attributed to the sphere like nanoparticles rolling between the rubbing surfaces, changed the situation of friction from sliding to rolling. Thus the dynamic performance of the diesel engine was improved through the reduced frictional work and fuel consumption.

References

1. Xu JF, Zhang JR, Du YW (1996) Ultrasonic velocity and attenuation in nanostructured Zn materials. Mater Lett 29 (November (1–3)):131–144
2. Verma P, Chaturvedi P, Rawat JSBS (2007) Elimination of current non-uniformity in carbon nanotube field emitters. J Mater Sci: Mater Electron 18 (June (6)):677–680
3. Peng DX, Kang Y, Chen CH, Chen SK, Shu FC (2009) The tribological behavior of modified diamond nanoparticles in liquid paraffin. Ind Lubr Tribol 61(4)
4. Tao X, Jiazheng Z, Kang X (1996) The ball-bearing effort of diamond nanoparticles as an oil additive. J Phys 2932–2937
5. Haywar IP, Singer IL, Seitzman LE (1992) Wear 157
6. Zuiker C, Krauss AR, Gruen DM, Pan X, Li JC, Csencsits R, Erdemir A, Bindal C, Fenske G (1995) Thin Solid Films 270:154
7. Miyoshi K (1999) NASA/TM-1999-208905 report
8. Kang HU, Kim SH, Oh JM (2006) Estimation of thermal conductivity of nanofluid using experimental effective particle volume. Exp Heat Transfer 19(3):181–191
9. Saeed MB, Zhan Mao-Sheng (2007) Adhesive strength of nano-size particles filled thermoplastic polyimides: Part-I. Multi-walled carbon nano-tubes (MWNT)–polyimide composite films. Int J Adhes Adhes 27(4):306–318

Citybus Microtrips Classification Using the Data Envelopment Analysis (DEA) Method Applied on Portable Emissions Measurement System (PEMS) Experimental Data

Jerzy Merkisz, Arkadiusz Barczak and Jacek Pielecha

Abstract The standard DEA models were all formulated for desired inputs and outputs. However, in the real-life applications there frequently exist also undesired inputs and/or outputs. The problem is how to treat undesirable outputs jointly produced with desirable outputs. Ignorance of the undesirable outputs is equal to the saying that they have no value in the final evaluation and may provide misleading results. It is therefore necessary to credit DMUs for their provision of desirables and penalize them for their provision of undesirables. The purpose of this chapter is to evaluate the applicability of the Data Envelopment Analysis methods in the efficiency assessment of citybus microtrips (trips of the bus between two consecutive bus stops) treated as decision making units. The aim is to help transportation authorities to not only maximize beneficial outputs, such as for example vehicle speed, but also to reduce the production of undesirable (or negative) outputs in the form of vehicle pollution emissions.

Keywords DEA · PEMS · Microtrip · Pollution · Emissions

F2012-A09-007

J. Merkisz (✉) · A. Barczak · J. Pielecha
Institute of Combustion Engines and Transportation, Poznan University of Technology, Poznań, Poland
e-mail: jerzy.merkisz@put.poznan.pl

SAE-China and FISITA (eds.), *Proceedings of the FISITA 2012 World Automotive Congress*, Lecture Notes in Electrical Engineering 190,
DOI: 10.1007/978-3-642-33750-5_41,

1 Introduction

The development of real-world driving cycle is important for traffic and transport management, vehicular pollution emissions measurement and control, energy and fuel consumption studies. The emission models involve a considerable number of factors that affect the levels of the emitted pollutants. Traditionally, the emission models are based on test-drive cycles emission data measured during the average traffic flow conditions. The driving characteristics usually include parameters: average speed, average running speed, acceleration rate, deceleration rate, fuel consumption rate, driving mode variation (idle, cruse, acceleration and deceleration), frequency of vehicle stops and others.

Existing standardized driving cycles do not reflect adequately the actual variations of the urban traffic conditions influencing the real-world emission levels.

It is necessary to reduce emission generated by transportation through careful management including the reduction of travel distance, the control of the load factors, the usage of advanced routing and scheduling systems. Thus, transportation pollution emissions may be reduced through changes in transportation policies and guidelines. As a result, authorities who establish schedules of trips and those who influence transportation infrastructure development are also responsible for the modification of the pollution emission levels and the safety of transport. It is imperative for the public transport to improve operating efficiency, safety records and sustainability. If one could benchmark the most efficient transport organization as well as scrutinize the sources of inefficiency then one would probably be capable to propose more practical strategies to ameliorate public transport. However, conventional efficiency measurements for transportation activities focus mainly on the marketable outputs and ignore side-effects of the operations (pollutions, accidents). The problem of accounting pollution emission in the transportation domain with efficiency sustainability aspects is very important nowadays [5, 7, 9, 11].

It is therefore essential to account for the impact of pollutions while measuring transport efficiency. It is also crucial to establish the relationship between desirable and undesirable outputs.

The most common used approach for identifying relative efficiency of decision making units (DMUs) when there are multiple inputs and multiple outputs is Data Envelopment Analysis (DEA) method [2]. Since the DEA method has some advantages, for instance one can handle multi-input and multi-output production technologies without the need of specifying the function form in prior, it may be successfully employed to measure efficiency in the transportation domain. When the relative efficiency of the DMU set is evaluated using DEA approach, DMUs are differentiated into two groups: efficient DMUs and inefficient DMUs. During this process the efficiency scores of inefficient DMUs are assigned values from 0 to 1 (exclusively).

The standard DEA models were all formulated for desired inputs and outputs. However, in the real-life applications there frequently exist also undesired inputs and/or outputs. The problem is how to treat undesirable outputs jointly produced with desirable ones. Ignorance of the undesirable outputs is equal to the saying that they have no value in the final evaluation and may provide misleading results. It is therefore necessary to credit DMUs for their provision of desirables and penalize them for their provision of undesirables. Different approaches have been applied to the DEA models in order to solve this problem [1, 2, 4, 7, 11, 12]. The first one treats the undesirable outputs as inputs—such a solution is easy to apply but is not consistent with the real production process. The second approach is based on the transform of the undesired outputs into new variables which then are used in the standard DEA models. The third approach makes use of the direction output function.

The purpose of this chapter is to evaluate the applicability of the Data Envelopment Analysis methods in the efficiency assessment of citybus microtrips (trips of the bus between two consecutive bus stops) treated as decision making units. The aim is to help transportation authorities to not only maximize beneficial outputs, such as for example vehicle speed, but also to reduce the production of undesirable (negative) outputs in the form of vehicle pollution emissions.

The numerical analysis conducted in this chapter was based on the data gathered during the real-world on-road driving experiment—pollution emissions were collected using the Portable Emissions Measurement System (PEMS) and complementary driving characteristics were obtained from the vehicle on-board diagnostics system (OBD).

2 DEA Models with Undesirable Outputs

In the DEA model, a production process in which a set of decision making units (DMUs) use a vector inputs to produce a vector of desirable outputs and a vector of undesirable ones may be represented by the production possibility set [1, 6]. That is, production technology T is denoted as:

$$T = \{(x, y, b) : x, \text{ can produce } y \text{ and } b\} \tag{1}$$

where:

$x = (x_1, x_2, \ldots, x_M) \in \Re_{+}^{M}$ is the vector of inputs,
$y = (y_1, y_2, \ldots, y_K) \in \Re_{+}^{k}$ is the desirable outputs,
$b = (b_1, b_2, \ldots, b_N) \in \Re_{+}^{N}$ is the undesirable outputs,

and production possibility set $P(x)$ is denoted as:

$$P(x) = [(y, b) : (x, y, b) \in T] \tag{2}$$

The production possibility set contains the desirable and undesirable outputs that can be jointly produced for given inputs. It was assumed that this set is closed, bounded and convex and that $P(x)$ satisfies strong disposability (SD) of desirable outputs and inputs, weak disposability (WD) of undesirable outputs and desirable and undesirable outputs are null-joint. These assumptions are written as [1, 3, 10]:

$$(y, b) \in P(x)\ y' \leq y \text{ than } (y', b) \in P(x) \tag{3}$$

$$(y, b) \in P(x) \text{ and } x' \geq x \text{ than } (y, b) \in P(x') \tag{4}$$

$$(y, b) \in P(x) \text{ and } 0 \leq \theta \leq 1 \text{ than } (\theta y, \theta b) \in P(x) \tag{5}$$

$$(y, b) \in P(x) \text{ and } b = 0 \text{ than } y = 0 \tag{6}$$

SD of desirable outputs and inputs implies that if a desirable and undesirable output vector producible from a given amount of input then it is possible to produce either less desirable output (Eq. 3) or produce the same amount of desirable and undesirable outputs using more inputs (Eq. 4). WD (Eq. 5) implies that the opportunity cost of reducing undesirable outputs is that some desirable output must be foregone. Finally, desirable and undesirable outputs are null-joint (Eq. 6) if the only way to produce zero undesirable outputs is to also produce zero desirable ones.

Once the output set has been defined, to obtain the measure of technical efficiency the directional distance function was introduced that simultaneously reflects the potential expansion in outputs and the potential reduction in inputs [1, 3, 11]. The directional distance function on the output set $P(x)$ was defined as:

$$\overrightarrow{D_O}(x, y, b) = \sup\{\theta : ((1 + \theta)y, (1 - \theta)b) \in P(x)\} \tag{7}$$

The directional output distance function defined in Eq. (7) seeks to increase the desirable outputs and decrease undesirable outputs in a ratio of θ simultaneously, while keeping the inputs unchanged. Furthermore, to evaluate the technical inefficiency for each DMU, the directional output distance function in Eq. (7) can be implemented by a linear programming technique, similar to the DEA approach. Mathematical model with assumption of weak disposability in undesirable outputs was formulated as follows [1, 4, 5]:

$$\overrightarrow{D_W}(x, y, b) = \text{Max } \theta_O \tag{8}$$

$$\begin{aligned} s.t. \sum_{j=1}^{J} \lambda_j y_{jk} &\geq (1 + \theta_O) y_{Ok},\ k = 1, \ldots, K \\ \sum_{j=1}^{J} \lambda_j b_{ji} &= (1 - \theta_O) b_{Oi},\ i = 1, \ldots, N \\ \sum_{j=1}^{J} \lambda_j x_{jm} &\leq x_{Om},\ m = 1, \ldots, M \\ \sum_{j=1}^{J} \lambda_j &= 1, \lambda_j \geq 0,\ j = 1, \ldots, J \\ \theta &\in \Re \end{aligned}$$

Mathematical model with assumption of strong disposability in undesirable outputs was formulated as follows [1, 4, 5]:

$$\overrightarrow{D_S}(x, y, b) = \text{Max } \beta_O \tag{9}$$

$$\begin{aligned} s.t. \sum_{j=1}^{J} \lambda_j y_{jk} &\geq (1 + \beta_O) y_{Ok}, \ k = 1, \ldots, K \\ \sum_{j=1}^{J} \lambda_j b_{ji} &\geq (1 - \beta_O) b_{Oi}, \ i = 1, \ldots, N \\ \sum_{j=1}^{J} \lambda_j x_{jm} &\leq x_{Om}, \ m = 1, \ldots, M \\ \sum_{j=1}^{J} \lambda_j &= 1, \lambda_j \geq 0, \ j = 1, \ldots, J \\ \beta &\in \Re \end{aligned}$$

In Eqs. (8) and (9), $\overrightarrow{D_W}$ and $\overrightarrow{D_S}$ stand for distance functions with assumptions of, respectively, weak and strong disposability, x denotes inputs, y denotes outputs and λ represents intensity multiplier. The key difference between these two models is that the equality is imposed on the second constraint in (Eq. 8), while the counterpart in (Eq. 9) is set with inequality. Both θ_i and β_i measure the distances between the observations to efficient frontiers, namely the level of inefficiencies. Obviously, the higher θ_i (β_i)—the less efficient is the DMU$_i$, also DMU$_i$ is efficient if and only if θ_i (β_i) equals to zero.

Under the context of the BCC model [2, 10], an alternative method to deal with desirable and undesirable outputs in DEA was introduced using a linear transformation approach to treat undesirable outputs which then are incorporated into standard BCC DEA model as derived factors. For the purpose of preventing convexity relations a linear monotone decreasing transformation is used [10]:

$$\overline{b_{ji}} = -b_{ji} + \delta_j > 0 \tag{10}$$

where: δ_j is a proper translation value that makes $\overline{b_{ji}} > 0$.

Based upon the above linear transformation, the standard BCC DEA model can be modified.

3 Efficiency Evaluation of Citybus Microtrips

The exhaust emissions test under real road operating conditions was performed in the city of Poznan [8]. The test was carried out on the bus route number 67 in the pre-selected conditions to ensure the highest possible reflection of the actual average traffic conditions. The passenger count on a bus route number 67 was in line with the average passenger count in all the bus routes in the city of Poznan.

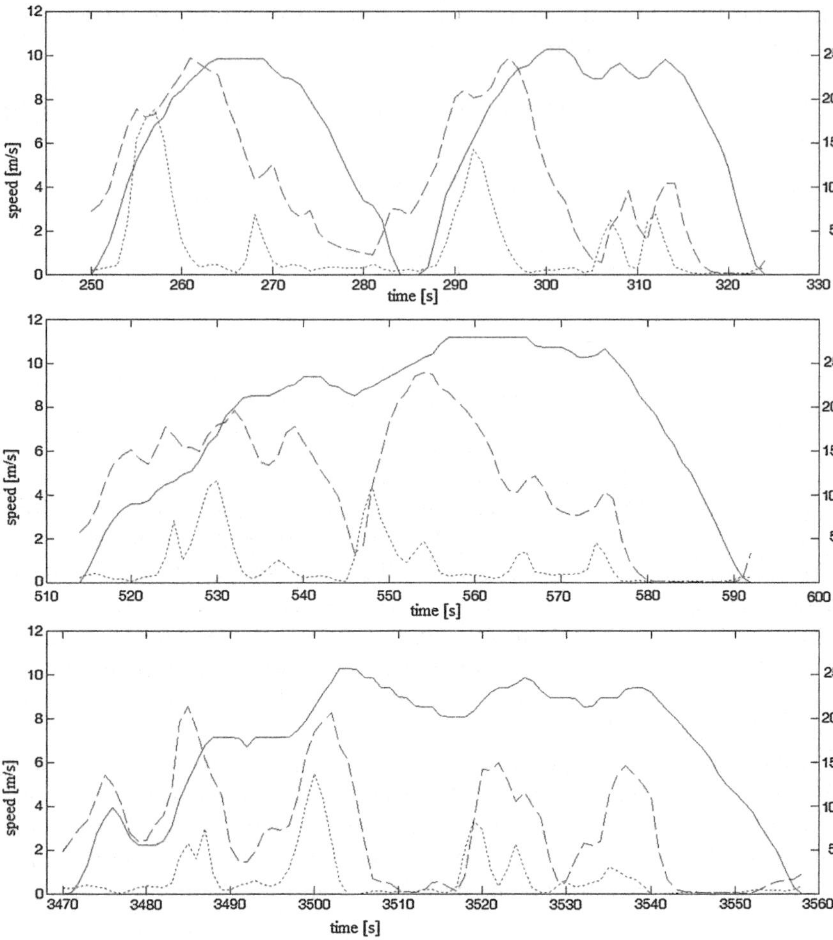

Fig. 1 Citybus speed [m/s] (*solid line*), emission of CO [mg/s] (*dotted line*) and emission of NO_x [mg/s] (*dash line*) measured during **a** microtrip number 4, **b** microtrip number 7, **c** microtrip number 28

Also, the traffic congestion was similar to that usually observed in the city of Poznan. The test was carried out between Os. Sobieskiego and Os. Debina, Poznan.

In order to measure the concentration of the individual emissions, a portable analyzer for emission testing SEMTECH DS by SENSORS Inc. was used. The analyzer allowed the measurement of the concentration of the individual emissions with a simultaneous measurement of mass flow rate of the exhaust gases.

In the tests the measurements of exhaust emission were performed with the resolution of 1 s and, for comparison, signals such as fuel consumption, engine speed, load, vehicle speed, engine temperature from an on-board diagnostic system were registered (CAN SAE J1939).

Table 1 Values of the primary factors

DMU	v_{mean} [m/s]	v_{max} [m/s]	P^{CO}_{mean} [mg/s]	P^{HC}_{mean} [mg/s]	$P^{NO_x}_{mean}$ [mg/s]	F_{mean} [gal/s]	t_{SA} [s]	t_{CD} [s]
1	6.98	12.53	29.42	0.105	115.26	0.0010	29	29
2	6.98	10.72	27.05	0.135	137.85	0.0011	24	27
3	7.08	10.72	25.57	0.113	100.13	0.0007	23	35
4	6.66	10.28	30.15	0.096	107.40	0.0009	36	39
5	6.38	10.28	19.25	0.085	96.65	0.0008	39	48
6	7.48	10.72	10.90	0.090	106.23	0.0008	21	40
7	7.65	11.17	20.52	0.088	114.01	0.0009	37	42
8	3.49	9.83	9.43	0.070	80.54	0.0005	85	59
9	6.88	9.83	26.97	0.090	102.25	0.0008	31	47
10	4.13	11.61	16.27	0.069	74.76	0.0005	90	51
11	5.70	9.83	13.47	0.078	94.38	0.0008	14	21
12	2.28	8.50	11.15	0.073	56.44	0.0003	123	62
13	2.78	9.39	10.37	0.087	71.34	0.0005	85	51
14	5.75	9.83	18.70	0.180	123.83	0.0010	38	54
15	2.42	9.39	14.11	0.087	79.43	0.0005	109	50
16	6.51	12.97	18.09	0.097	101.37	0.0008	47	70
17	4.76	8.50	14.65	0.095	70.08	0.0005	39	68
18	8.55	13.86	16.14	0.202	71.59	0.0006	32	69
19	2.64	9.39	11.77	0.135	71.24	0.0004	67	61
20	1.77	5.36	7.63	0.108	72.88	0.0003	111	62
21	3.97	8.50	17.39	0.105	86.99	0.0006	51	47
22	4.65	9.39	31.07	0.099	89.60	0.0007	32	34
23	3.47	10.72	17.40	0.072	86.05	0.0005	166	116
24	8.46	11.61	21.94	0.063	108.75	0.0009	38	58
25	7.91	12.08	22.42	0.055	113.44	0.0009	24	36
26	8.60	12.97	15.95	0.040	113.94	0.0008	68	118
27	6.73	12.08	15.07	0.048	145.87	0.0011	47	64
28	6.77	10.28	17.25	0.039	71.68	0.0005	31	58
29	5.62	10.28	32.82	0.051	107.94	0.0008	24	26
30	4.84	9.78	14.49	0.037	78.68	0.0006	28	31

The data registered during the bus route experiment comprises a total of 30 microtrips (DMUs). Exemplary microtrips (number 4, 7, and 28) were presented on the Fig. 1, where the speed [m/s] of the citybus, emission of CO [mg/s] and emission of NO_x [mg/s] were shown. As it may be seen, even the microtrips have similar duration their characteristics are quite different.

The collection of the primary factors characterizing DMU was created based on the data gathered during each microtrip:

- average speed of the citybus: v_{mean},
- maximal speed of the citybus: v_{max},
- average emission of carbon monoxide: P^{CO}_{mean},

- average emission of hydrocarbons: P^{HC}_{mean},
- average emission of mono-nitrogen oxides NO and NO_2: $P^{NO_x}_{mean}$,
- average fuel consumption: F_{mean},
- the number of seconds the bus was in "idle" or "acceleration" states: t_{SA},
- the number of seconds the bus was in "cruise" or "deceleration" state: t_{CD}.

The values of above factors are presented in the Table 1. Some additional factors characterizing DMUs, derived from the primary factors, were also used in the models:

- total fuel consumption: F_{sum},
- total emission of carbon monoxide: P^{CO}_{sum},
- total emission of hydrocarbons: P^{HC}_{sum},
- total emission of mono-nitrogen oxides NO and NO_2: $P^{NO_x}_{sum}$,
- transformed value of average emission of carbon monoxide in accordance with (Eq. 10), where proper translation value equals to 201 [mg/s]: Q^{CO}_{mean},
- transformed value of average emission of carbon monoxide in accordance with (Eq. 10), where proper translation value equals to 0.58 [mg/s]: Q^{HC}_{mean},
- transformed value of average emission of mono-nitrogen oxides NO and NO_2 in accordance with (Eq. 10), where proper translation value equals to 343 [mg/s]: $Q^{NO_x}_{mean}$.

The efficiency analysis of citybus microtrips (trips of the bus between two consecutive bus stops treated as decision making units—DMUs) was performed using five DEA-based models:

- Model A, where input = $\{F_{mean}\}$, output = $\{v_{mean}, v_{max}\}$,
- Model B, where input = $\{F_{mean}, t_{SA}\}$, output = $\{v_{mean}, v_{max}, t_{CD}\}$,
- Model C, where input = $\{F_{mean}\}$, output = $\{v_{mean}, v_{max}, Q^{CO}_{mean}, Q^{HC}_{mean}, Q^{NO_x}_{mean}\}$,
- Model D, where input = $\{F_{mean}\}$, desirable output = $\{v_{mean}, v_{max}\}$, undesirable output = $\{P^{CO}_{mean}, P^{HC}_{mean}, P^{NO_x}_{mean}\}$,
- Model E, where input = $\{F_{sum}\}$, desirable output = $\{v_{mean}, v_{max}\}$, undesirable output = $\{P^{CO}_{sum}, P^{HC}_{sum}, P^{NO_x}_{sum}\}$.

Several different variants of the DEA methods were applied to the models A, B, C, D and E. The first letter in the name of the variant denotes the applied distance function type (R—radial, S—slack based, D—directional function), the second letter indicates the orientation of the model (I—input oriented, O—output oriented), the third letter specify the type of the return to scale method (C—constant return to scale, V—variable return to scale). The name of the DEA variant may also contain the superscript letters "WD" indicating that the output factors related to the pollution emissions are set in the model to be weakly disposable.

Efficiency measures λ of DMUs (citybus microtrips) of Model A (variants: ROC, SOC, ROV and SOC), Model B (variants ROV, SOV, RIV and SIV) and Model C (ROC, SOC, ROV and SOV) are presented in Table 2. Efficiency measures λ and inefficiencies β and θ of DMUs (citybus microtrips) calculated for

Table 2 Efficiency measures λ of DMUs (citybus microtrips) of variants of Model A, Model B and Model C

DMU	Model A				Model B				Model C			
	ROC	SOC	ROV	SOV	ROV	SOV	RIV	SIV	ROC	SOC	ROV	SOV
1	0.550	0.530	0.903	0.857	0.950	0.693	0.897	0.786	0.572	0.400	0.959	0.920
2	0.438	0.427	0.813	0.794	0.887	0.756	0.805	0.750	0.477	0.321	0.930	0.867
3	0.676	0.653	0.824	0.799	0.938	0.876	0.958	0.956	0.737	0.585	0.950	0.909
4	0.547	0.533	0.775	0.759	0.778	0.662	0.722	0.698	0.611	0.453	0.925	0.880
5	0.559	0.557	0.745	0.743	0.746	0.696	0.722	0.715	0.645	0.498	0.967	0.902
6	0.669	0.628	0.871	0.821	1	1	1	1	0.732	0.560	1	1
7	0.623	0.590	0.889	0.847	0.894	0.722	0.738	0.734	0.674	0.489	0.966	0.930
8	0.752	0.590	0.794	0.622	0.811	0.690	0.827	0.691	0.755	0.692	1	1
9	0.578	0.542	0.799	0.753	0.812	0.746	0.786	0.786	0.642	0.476	0.941	0.888
10	0.942	0.765	0.985	0.801	0.986	0.790	0.983	0.849	0.942	0.808	1	1
11	0.553	0.541	0.709	0.687	1	1	1	1	0.650	0.517	1	1
12	1	1	1	1	1	1	1	1	1	1	1	1
13	0.795	0.563	0.838	0.635	0.838	0.677	0.887	0.772	0.795	0.745	1	1
14	0.432	0.424	0.709	0.690	0.715	0.696	0.693	0.674	0.488	0.354	0.953	0.807
15	0.743	0.482	0.796	0.529	0.797	0.588	0.807	0.662	0.743	0.676	0.979	0.767
16	0.699	0.646	0.935	0.839	0.942	0.825	0.746	0.736	0.699	0.538	0.990	0.933
17	0.783	0.756	0.896	0.826	1	1	1	1	0.958	0.891	0.993	0.939
18	1	1	1	1	1	1	1	1	1	1	1	1
19	0.872	0.598	0.909	0.721	1	1	1	1	0.873	0.794	0.997	0.900
20	0.618	0.465	0.683	0.645	1	1	1	1	0.981	0.804	1	1
21	0.550	0.493	0.613	0.528	0.619	0.552	0.738	0.718	0.628	0.545	0.963	0.770
22	0.559	0.514	0.677	0.602	0.677	0.561	0.827	0.799	0.614	0.511	0.929	0.814
23	0.779	0.581	0.835	0.603	1	1	1	1	0.780	0.661	0.976	0.816
24	0.666	0.610	0.984	0.907	0.989	0.852	0.813	0.753	0.711	0.501	0.997	0.970
25	0.633	0.614	0.920	0.897	1	1	1	1	0.691	0.500	0.982	0.968
26	0.735	0.708	1	1	1	1	1	1	0.789	0.588	1	1
27	0.452	0.436	0.871	0.827	0.876	0.792	0.640	0.588	0.490	0.331	0.999	0.929
28	0.874	0.846	0.939	0.869	1	1	1	1	1	1	1	1
29	0.517	0.495	0.741	0.696	0.851	0.689	0.817	0.774	0.586	0.442	0.983	0.872
30	0.742	0.683	0.758	0.702	1	1	1	1	0.840	0.747	1	1

Table 3 Efficiency measure λ and inefficiency measures β and θ of DMUs (citybus microtrips) of variants of Model D and Model E

DMU	Model D								Model E			
	DOC		DOV		DOC^{WD}		DOV^{WD}		DOV		DOV^{WD}	
	λ	β	λ	β	λ	θ	λ	θ	λ	β	λ	θ
1	0.902	0.109	0.903	0.108	0.921	0.086	1	0	1	0	1	0
2	0.883	0.132	0.889	0.125	0.905	0.105	1	0	0.922	0.085	0.922	0.084
3	0.891	0.123	0.897	0.114	0.939	0.064	1	0	1	0	1	0
4	0.896	0.116	0.906	0.104	0.925	0.081	1	0	0.887	0.127	1	0
5	0.903	0.108	0.913	0.095	0.913	0.095	0.918	0.089	0.866	0.154	0.867	0.154
6	1	0	1	0	1	0	1	0	1	0	1	0
7	0.907	0.102	0.913	0.096	0.919	0.088	0.919	0.088	0.961	0.041	0.974	0.027
8	1	0	1	0	1	0	1	0	0.773	0.294	1	0
9	0.900	0.111	0.910	0.099	0.926	0.080	0.973	0.028	0.902	0.109	0.991	0.009
10	1	0	1	0	1	0	1	0	0.894	0.119	1	0
11	0.904	0.106	0.938	0.066	0.910	0.099	0.939	0.065	1	0	1	0
12	1	0	1	0	1	0	1	0	0.603	0.657	1	0
13	0.959	0.043	0.968	0.033	0.959	0.043	0.968	0.033	0.704	0.420	0.892	0.121
14	0.870	0.150	0.875	0.143	0.880	0.136	0.880	0.136	0.825	0.212	0.880	0.137
15	0.911	0.097	0.923	0.083	0.966	0.035	0.967	0.034	0.662	0.510	0.898	0.114
16	0.924	0.082	0.962	0.039	0.924	0.082	0.985	0.015	0.990	0.010	0.990	0.010
17	0.895	0.118	0.926	0.080	1	0	1	0	0.776	0.288	0.875	0.144
18	1	0	1	0	1	0	1	0	1	0	1	0
19	0.916	0.092	0.918	0.089	0.935	0.069	1	0	0.683	0.464	1	0
20	0.862	0.160	1	0	1	0	1	0	0.444	1,252	1	0
21	0.883	0.133	0.897	0.115	0.894	0.118	0.900	0.111	0.699	0.431	0.722	0.385
22	0.890	0.124	0.901	0.110	1	0	1	0	0.808	0.238	1	0
23	0.938	0.066	0.940	0.063	1	0	1	0	0.767	0.303	1	0
24	0.944	0.059	0.965	0.036	0.960	0.042	1	0	1	0	1	0
25	0.948	0.054	0.954	0.048	0.972	0.029	1	0	1	0	1	0
26	1	0	1	0	1	0	1	0	1	0	1	0
27	0.964	0.037	0.981	0.019	1	0	1	0	0.974	0.027	1	0
28	1	0	1	0	1	0	1	0	0.957	0.045	1	0
29	0.940	0.063	0.959	0.043	1	0	1	0	0.995	0.005	1	0
30	0.991	0.009	1	0	1	0	1	0	1	0	1	0

Model D (variants: DOC, DOV, DOC^{WD} and DOV^{WD}) and Model E (variants DOV and DOV^{WD}) are presented in Table 3.

4 Conclusions

In this chapter diverse DEA-based methods were used to examine the impact of the undesirable outputs on efficiency and inefficiency of Citybus microtrips (DMUs). The undesirable outputs are those related to the pollution emission (CO, HC, NO_x).

Results show that application of directional distance function DEA model significantly influence the set of efficient microtrips comparing to the sets obtained by radial and slack based DEA models. Moreover, disposability feature (strong or weak) on undesirable outputs poses a significant difference in the final efficiency evaluation. The analysis conducted during the experiment indicated that two microtrips (number 12 and 18) are efficient in all tested models under all variants. The different results of DMU efficiency and inefficiency measures of Model D and Model E are the base to emphasis the importance of the process variable selection—this problem is currently under research.

Traditionally, the characteristics of microtrips are computed using the functional approach. The efficiency measure approach presented in this chapter is an innovative way enabling the assessment of the quality of multi-dimensional microtrips. Results of this chapter provide the reason to support continued research targeted to conversion recognized efficient frontier microtrip into tactical and operation activities of bus trips designers and bus trip drivers.

Acknowledgments The work was partially financed under the grant N N509 559240 from the National Science Centre in Cracow.

References

1. Chung YH, Färe R, Grosskopf Shawna (1997) Productivity and undesirable outputs. A directional distance function approach. J Environ Manage 51:229–240
2. Cooper WW, Seiford LM, Tone K (2007) Data envelopment analysis. A comprehensive text with models, applications, references and DEA-solver software [M]. New York: Springer Science + Business Media, LLC
3. Färe R, Grosskopf S (2004) Modeling undesirable factors in efficiency evaluation: comment. Eur J Oper Res 157:242–245
4. Färe R, Grosskopf S, Pasurka CA (2007) Environmental production functions and environmental directional distance functions. Energy 32:1055–1066
5. Lin ETJ, Lan LW (2009) Accounting for accidents in the measurement of transport inefficiency: a case of Taiwanese bus transit. Int J Environ Sustain Dev 8(3–4):365–385
6. Liu WB, Meng W, Li XX, Zhang DQ (2010) DEA models with undesirable inputs and outputs. Ann Oper Res 173:177–194

7. B. Starr McMullen, Noh DW (2007) Accounting for emissions in the measurement of transit agency efficiency: A directional distance function approach [J]. Transportation Research Part D 12, 1–9
8. Merkisz J, Pielecha J (2010) Emissions and fuel consumption during road test from diesel and hybrid buses under real road conditions. IEEE Vehicle Power and Propulsion Conference (VPPC'10), CD: 1–6
9. Rogers MM, Weber WL (2011) Evaluating CO_2 emissions and fatalities tradeoffs in truck transport. Int J Phys Distrib Logistics Manage 11(8):750–766
10. Seiford LM, Zhu J (2002) Modeling undesirable factors in efficiency evaluation. Eur J Oper Res 142:16–20
11. Lei Y, Jia S, Shi Q (2009) Research on Transportation-Related Emissions: Current Status and Future Directions. J Air Waste Manag Assoc 59:183–195
12. Zhou P, Ang BW, Poh KL (2008) Measuring environmental performance under different environmental DEA technologies. Energy Economics 30:1–14

Extended Charge Motion Design: CAE Based Prediction of Gasoline Engine Pre-ignition Risk

Jens Ewald, Matthias Budde, Bastian Morcinkowski, Rüdiger Beykirch, Adrien Brassat and Philipp Adomeit

Abstract A simulation model to predict pre-ignition phenomena in spark ignition engines is presented. The model uses a semi-detailed reaction scheme to calculate auto-ignition reactions of the in-cylinder charge at the end of and after the compression stroke. In order to reduce simulation time for the calculation of auto-ignition chemistry, a zonal chemistry approach is employed, which simplifies the description of the thermodynamic engine cylinder state. Auto-ignition of gasoline is modeled by a chemical reaction scheme of a primary reference fuel (PRF). Boundary conditions for the pre-ignition analysis are obtained from flow simulations of the in-cylinder charge motion and mixture formation as well as FE heat load analysis for accurate wall temperature determination. By means of this approach, physical mechanisms due to the influence of hot surfaces are studied. In order to investigate whether lubrication oil from the piston head has the potential to promote auto-ignition events, initial two-phase flow simulation studies are presented. Results of the chemistry predictions are compared to experimental measurements from the single cylinder engine test bench.

Keywords Combustion · Modeling · Abnormal combustion · Pre-ignition · Simulation

F2012-A09-009

J. Ewald · M. Budde · R. Beykirch · P. Adomeit
FEV GmbH, Aachen, Germany
e-mail: sengstock@fev.de

A. Brassat (✉)
FEV China Co., Ltd, Dalian, China

B. Morcinkowski
RWTH Aachen University, Aachen, Germany

SAE-China and FISITA (eds.), *Proceedings of the FISITA 2012 World Automotive Congress*, Lecture Notes in Electrical Engineering 190,
DOI: 10.1007/978-3-642-33750-5_42,

Nomenclature

BDC	Bottom dead center
°CA	Degree crank angle
CAE	Computer aided engineering
CFD	Computational dynamics
DI	Direct injection
FEA	Finite element analysis
HTC	Heat transfer coefficient
PRF	Primary reference fuel
SCE	Single cylinder engine
TDC	Top dead center

1 Introduction

The recent gasoline engine development trend to improve fuel economy aims to reduce engine displacement while simultaneously increasing the mean effective pressure by boosting. This so-called engine downsizing has brought up again a complex set of problems related to abnormal combustion. The theoretically possible gain in engine power output and efficiency at full load is limited by the onset of irregular combustion phenomena like pre-ignition or knock. Furthermore, if such events occur too frequently during engine operation or if the cylinder pressure magnitude during such events is too high, engine failure can be expected.

The demand to optimize engine efficiency therefore requires an engine design which promises to have a low propensity towards abnormal combustion. The assessment of this propensity requires appropriate combustion models already in an early stage of the engine development process. In order to accomplish this, all relevant influencing parameters must be identified and appropriately taken into account. A correct classification of abnormal combustion helps to identify all these parameters.

A widely accepted phenomenological classification of abnormal combustion is due to Heywood [1]. Heywood distinguishes between spark knock which can be controlled by spark timing and surface ignition. Among the latter phenomenon pre-ignition is categorized, an abnormal combustion event which occurs prior to spark timing.

The presently published state of research (for example, see the comprehensive discussion in) indicates [2] that there are a large variety of different causes for abnormal combustion events. It is not expected to find one common cause of abnormal combustion which is valid for all engine configurations. Especially the onset of engine knock is still difficult to predict by means of corresponding models (see also [3, 4]) without the knowledge of engine specific modeling parameters.

In this work we limit ourselves as a first step to predict the propensity of pre-ignition events. Here, we will account for the following influences:

- Compression ignition of the thermodynamic in-cylinder state charge and
- Surface ignition of the in-cylinder charge due to wall temperature hot-spots.

These two mechanisms are considered in this work as the two most important influence factors for the onset of pre-ignition. Using the tools presented below, their possible impact on pre-ignition phenomena can be quantitatively assessed. However, it needs to be kept in mind that there are further influencing factors as for example the occurrence of lube oil emissions into the cylinder charge mixture or—in case of a DI engine—the injection, evaporation and mixing behavior of a multicomponent fuel spray characterized by a composition of differently volatile fuel components. Also, alongside with the aforementioned mechanisms, lube oil dilution due to fuel spray impingement on the cylinder liner and the additive composition of the lube oil itself may have influence on the occurrence of stochastic pre-ignition events. The current status of research shows qualitatively such inter-relationships [2], but quantitative assessment is still difficult. Therefore, a fundamental investigation is presented below.

2 Computational Model

The simulation model used here has already been presented in [5]. The outline of the model is given below. Basis of the pre-ignition analysis is a 3D-CFD in-cylinder simulation including intake, compression, and—if required—fuel spray. The result of this simulation is the thermodynamic state of the cylinder. From here, data at a prescribed crank angle position, in this case shortly before the end of compression stroke, is taken as initial condition of the pre-ignition analysis. The in-cylinder mixture is then subdivided into multiple zones according to unburned temperature, residual gas fraction, and relative air/fuel ratio. The range of minimum and maximum values in the cylinder in terms of temperature, residual gas fraction, and relative air/fuel ratio, respectively is subdivided into a prescribed number of sections which in turn determine the boundary value limits of the zones. By means of this, each computational cell can be attributed to exactly one thermodynamic zone. Thereby the three dimensional space of the engine cylinder is converted into a three dimensional phase space. Now, the reaction chemistry does not need to be calculated in every computational CFD cell, but only for each zone. Depending on the number of zones employed, this drastically reduces the computational efforts.

Heat transfer from surfaces (piston crown, liner, chamber head, spark plug, valve bottoms) acting on the zones is also quantified. In order to obtain wall surface temperatures to be as realistic as possible, a full cycle CFD simulation over the whole four engine stokes including the simulation of the regular combustion progress is carried out, see also [5]. The pre-ignition tool employs the open-source software library Cantera [6] and it is described in detail also in [7].

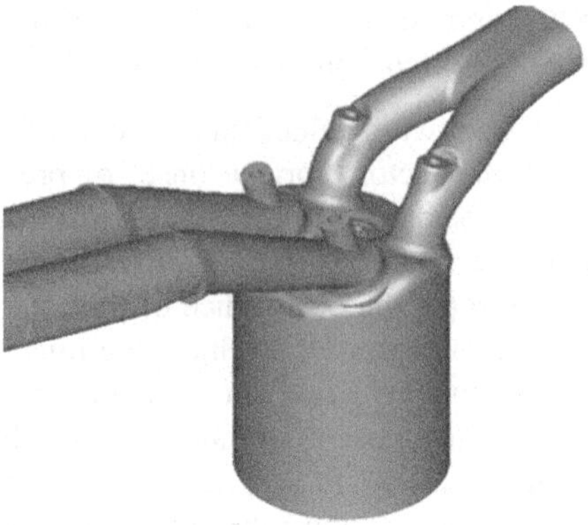

Fig. 1 Engine cylinder geometry as represented by CFD model

3 Engine Configuration and Simulation Setup

The engine analyzed is a single cylinder research engine (see Fig. 1) with the parameters given in Table 1. Fuel direct injection is accomplished by a multi-hole injection system, centrally mounted.

The engine operating point investigated is given in Table 2. Running the engine on this operating point, the pressure trace exhibited a large number of stochastic pre-ignition events. Only 29 individual engine cycles were measured on the engine test bench. Approximately 35 % of those cycles showed pre-ignition or mega-knock events. The other cycles were identified either as conventional knocking or as regular combustion events.

The CFD simulation of the engine cycle is carried out with the software STAR CD and the engine mesh tool ES-ICE. The in-cylinder region was meshed in CFD by approx. 0.9 million cells at piston bottom dead center, at top dead center 300,000 cells are still active. The pre-ignition chemistry analysis is initialized with the temperature and residual gas distribution at 700 °CA. An a priori analysis revealed that it was not required to calculate auto-ignition reactions prior to that crank angle position.

In the pre-ignition tool the RON 95 gasoline fuel was modeled as a PRF91 surrogate consisting of 9 % n-heptane and 91 % iso-octane employing a semi-detailed reaction mechanism by Andrae [8]. Due to the fact that the motored octane number (MON) of common commercial gasoline fuel blends is lower than the RON number, an average of both numbers was assumed for the iso-octane concentration of the PRF model fuel in the calculation.

Results for further operating points against the pre-ignition tool has been tested can be found in [5].

Table 1 Geometrical engine parameters

	Unit	Value
Bore	[mm]	75
Stroke	[mm]	82.5
No Valves/Cylinder	[–]	4
Compression Ratio	[–]	variable, 8.5 … 12

Table 2 Engine operating point

Engine Speed (1/min)	Engine IMEP (bar)	Fuel	Injection system	Air/Fuel ratio (–)	Compression Ratio (–)	Manifold pressure (kPa)	EVC (°CA)	IVO (°CA)	Spark Advance (°CAbTDC)
1500	30	RON 95	Multi-hole		10	319	1 mm@ 340° CA	1 mm @ 370° CA	−12

4 Simulation Results and Discussion

As a first step a four stroke CFD simulation including a prescribed premixed combustion progress was carried out. The combustion progress was obtained from the test bench measurements. By means of averaging, spatial distributions of temperature and heat transfer coefficient (HTC) are obtained. This approach ensures that the influence on the charge motion on the heat transfer in the wall boundary layer and the temporal and spatial development of the premixed turbulent flame is realistically taken into account.

For the analysis, a heat load analysis was carried out using the cycle averaged CFD heat transfer coefficient distribution assuming a total heat flux to the structure of 18.86 kW. The peak temperature at the chamber head was predicted to be 477.5 K, while on the piston crown upper rim the peak temperature was calculated to be 635 K. On the other hand, the highest temperature was assumed to be found at the bottom of the exhaust valve surface as 1040 K, see also Fig. 2.

With these wall boundary conditions, the CFD flow simulation was repeated for the intake and compression stroke. At 700 °CA, the thermodynamic state was categorized into chemistry zones, see Fig. 3. As can be seen, most chemistry zones have a temperature of slightly higher than 650 K, but stay below 670 K. The integration of gas phase auto-ignition chemistry was carried out using the pre-ignition tool from 700 to 780 °CA. In Fig. 4, a comparison of the experimental individual pressure traces from test bench (grey curves) to the prediction of the pre-ignition tool (green solid line) is carried out, showing that with the temperature prediction of the CFD code at 700 °CA, only a marginal increase in pressure due to heat release at 740 °CA can be seen. This simulation can be referred to as the base calculation. However, as can be seen from Fig. 5a, there are zones which auto-ignite. At 700 °CA, those zones have temperatures which are higher than 670 K. In fact, if one disregards the mass in each zone with respect to the heat

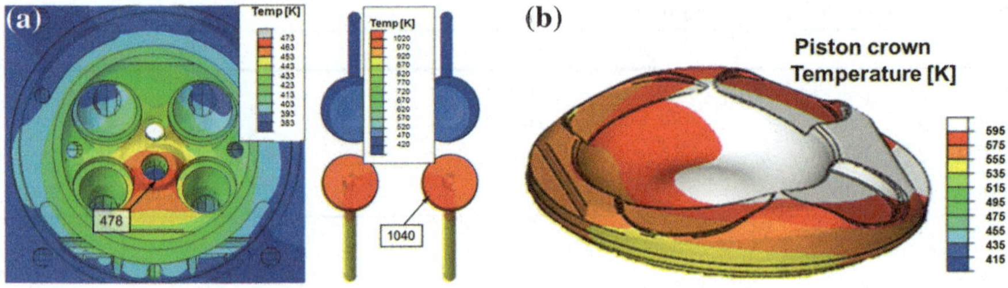

Fig. 2 Surface temperature distribution from FE head load analysis. (**a**) Cylinder head and valves (**b**) Piston crown [5]

Fig. 3 Individual pressure traces from test bench pressure indication and pressure trace results from the pre-ignition tool prediction with the thermodynamic cylinder distribution from CFD and with the initial temperature scaled from the CFD result [5]

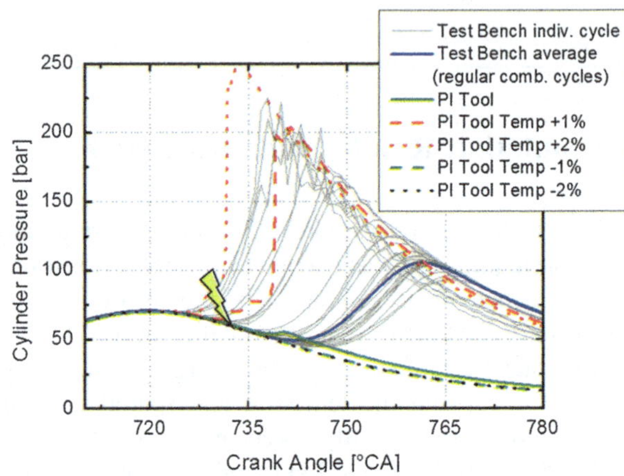

release rate, see Fig. 5b, most zones having a temperature of higher than 670 K show auto-ignition reactions. As the main portion of the cylinder mass has a temperature of slightly below that temperature level, pre-ignition/mega-knock is not predicted.

This observation is in line with the sensitivity analyses shown in Fig. 4: decreasing the temperature distribution in the pre-ignition tool at 700 °CA by 1 % (approximately 7 K) by simple scaling of the temperature initial condition, suppresses all significant heat release by auto-ignition while increasing the temperature by 1 or 2 % by scaling exhibit a pressure increase which is comparable to the pressure curves of the pre-igniting cycles from test bench.

In Fig. 6, the temperature distribution calculated by the chemistry in the pre-ignition tool is mapped back into the 3D space of the in-cylinder region. It can clearly be seen that there is a significant increase of temperature below the exhaust valves.

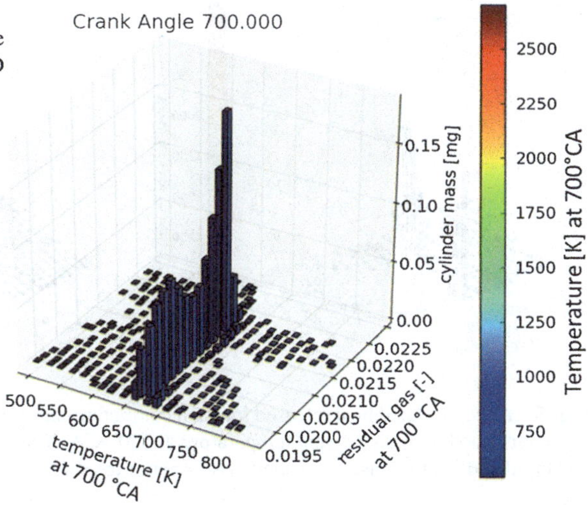

Fig. 4 Mass distribution of the chemistry zones taken from CFD simulation at 700 °CA ATDC

The sensitivity study shows that the pre-ignition tool is able to predict the distance between the cycle-averaged thermodynamic in-cylinder charge state and a different state which can cause pre-ignition/mega-knock events. However, the average in-cylinder temperature state as simulated by CFD does not exhibit abnormal combustion. As shown by the sensitivity analyses, only a marginal increase in average in-cylinder temperature level can lead to abnormal combustion.

5 Oil Droplet Study

In the previous section, the effect of thermodynamic conditions as pressure and temperature in the cylinder on the occurrence of pre-ignition events was studied. On the other hand it is also known that deposits on combustion chamber or piston surfaces and lube oil emissions are able to promote auto-ignition reactions [9]. Lube oil consists mainly of long-chain alkanes which have a small auto-ignition delay time. Furthermore, additives in the lube oil can potentially further promote auto-ignition, but the exact additive composition is not known in most cases and therefore the pertinent chemistry only insufficiently known.

Lube oil emissions into the cylinder can occur to shear-off from the piston ring pack on the top land. At the piston top dead center of gas exchange or combustion in the cylinder the piston has the highest acceleration downwards such that at these positions shear-off of lube oil has the highest likelihood. Lube oil can be diluted by fuel spray impinging on the cylinder liner which results in a reduction of the lube oil viscosity. Thereby, lube-oil emissions into the cylinder can be promoted.

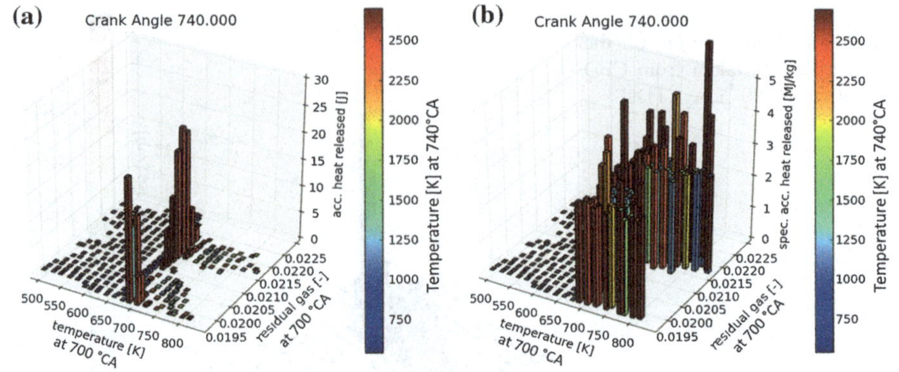

Fig. 5 **a** Accumulated heat released for each computational zone. The bars representing each zone are color coded with the current temperature at 740 °CA. **b** Specific accumulated heat release rate (accumulated heat release per zone/mass in zone) for same crank angle position 740 °CA [5]

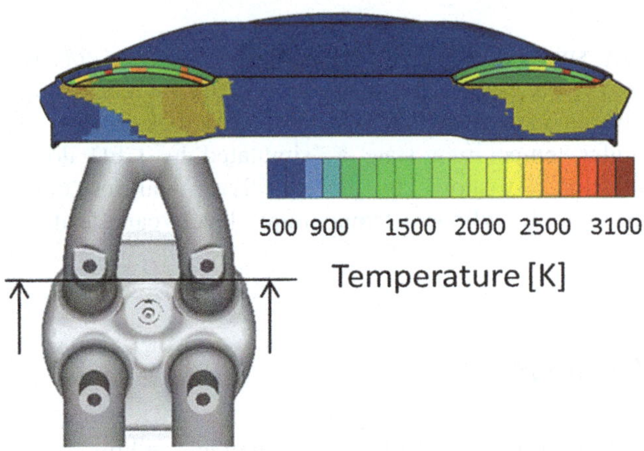

Fig. 6 Plot of temperature distribution in the cylinder as calculated from the pre-ignition tool at 735 °CA, where the temperature from each zone is mapped back into the engine geometry at 700 °CA. The in-cylinder region is cut vertically though the exhaust valves as indicated by the sketch below

The subsequently presented investigations aim to assess which conditions of sheared lube-oil in terms of shear-off time and oil droplet diameter are a candidate to promote auto-ignition in the unburnt. The investigations are based on the following hypotheses: Only if the lube oil evaporates significantly around top dead center position of combustion, the evaporated lube oil can form an oil/air mixture that can auto-ignite and spread a premixed flame front. If the droplet evaporates completely significantly before the end of the compression stroke, turbulent mixing can dilute the oil/air mixture such that auto-ignition may be prevented. Secondly, the lube oil droplets require a certain residence time in the cylinder such

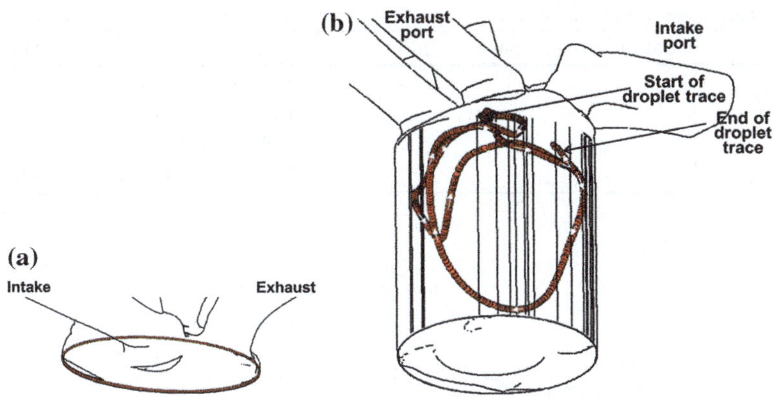

Fig. 7 **a** Initialization of oil droplets (*red points*) in the piston crevice at TDC. **b** Exemplary tracking of one droplet's path during intake and compression stroke

that they can spread from the piston top land deeply enough into the cylinder region to have an effect.

In order to assess the possibility of the aforementioned hypotheses, a two-phase flow simulation of lube oil distribution and evaporation in the cylinder is carried out. The goal of this simulation is to study the individual droplet paths from different starting point locations in above the top land until beyond the compression stroke.

The oil droplets are modeled as a two-component mixture of 56 mass % n-dodriacontane ($C_{32}H_{66}$) and 44 mass % 1,1,2,2-tetraphenylethane ($C_{26}H_{22}$). This surrogate oil provides a sufficient approximation for the boiling curve of SAE 5 W-40 oil [10]. 360 lube oil droplets are initialized circumferentially in the cylinder above the top land, see Fig. 7a. By means of this approach, it can be statistically checked by the one simulation where lube oil droplets originating from different top land positions will be transported to. The particle density of the introduced lube oil droplets is so low that interaction among lube oil droplets can be neglected.

Four different simulation cases will be carried out as parameter study. In Table 3 the parameter variations are given. The oil droplets are initialized in the CFD domain at the two top dead center piston positions of the four stroke cycle. The purpose of this parameter variation is to understand whether droplets sheared off immediately at the combustion TDC position have more potential impact on pre-ignition than at TDC of gas exchange.

The second parameter variation is with regards to the initial oil droplet diameter. Two different diameter sizes are considered. The aim of this variation is to understand which droplet diameter range after shear-off from the piston top is a candidate for prevailing until top dead center of combustion, thereby possibly promoting pre-ignition events.

Table 3 Parameter variation for lube oil initialization used in the oil droplet study

Case	Initial oil droplet diameter [μm]	Droplet initialization in CFD domain
OD1	11	TDC gas exchange
OD2	11	TDC combustion
OD3	1	TDC gas exchange
OD4	1	TDC combustion

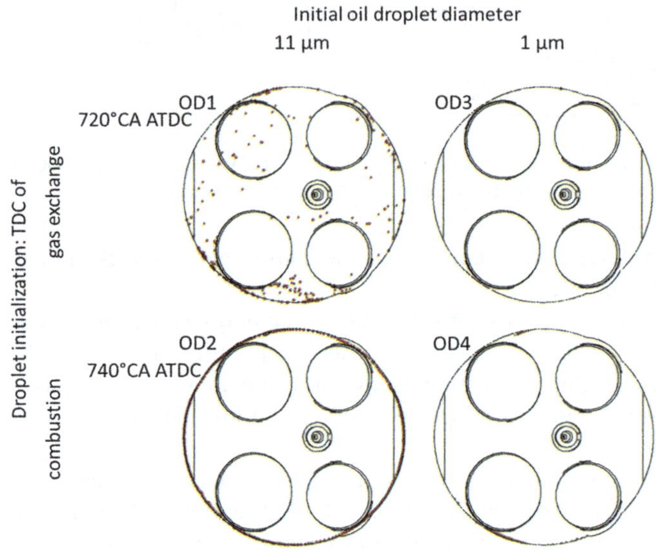

Fig. 8 Oil droplet simulation results for all simulation cases

In Fig. 7b, an exemplary path for one single droplet in case OD1 over the complete simulation range from TDC of gas exchange after 30 °CA ATDC of combustion is shown. As can be seen, the droplet follows a complex path. During the intake phase the droplet is sucked down with the piston movement. It does not rest on the top land position in the piston where it has been initialized but rather propagates initially in circumferential position along the liner. During the subsequent compression phase, the droplet propagates towards the middle of the cylinder.

In Fig. 8, plots of the oil droplet distribution at or after the compression stroke, respectively, for all simulation cases shown in Table 3 are given. For OD1, it can be seen that droplets at 720 °CA are still present and are spread throughout the entire cylinder region. Close to the inlet valves in the vicinity to the cylinder liner, however, an accumulation of liquid droplets can be seen. Simulation results for OD2, where the liquid droplets have been initialized at TDC of combustion, show 20 °CA later that the oil droplets still remain above the piston ring top land and did not propagate at that time towards the center of the cylinder. On the other hand, for

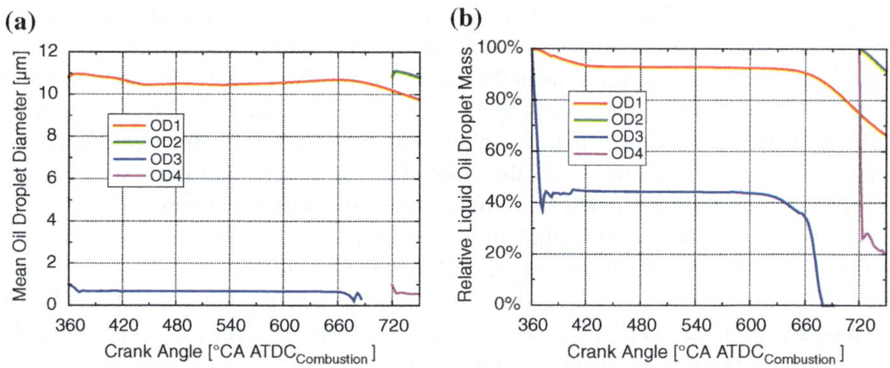

Fig. 9 **a** Development of mean lube oil diameter and **b** reduction of liquid lube oil mass over degree crank angle for all simulation cases

OD3 where the initial droplet diameter was much smaller than for OD1 and 2, no droplets are present anymore in the cylinder region at TDC of combustion. In Fig. 9b it can be seen that the rest of the liquid oil evaporates for this simulation case between 660 and 690 °CA, early before TDC of combustion. The results for case OD4 show that the small droplets, initialized in the computational domain at TDC of combustion evaporate for the largest part quickly after their initialization whereas the remaining oil droplets to not spread from piston top land into the cylinder region.

In Fig. 9 the temporal evolution of mean oil droplet diameter and relative liquid oil droplet mass can be seen. The plots mainly underline the previously made statements. The OD cases in which the oil droplets are sheared-off from the top land at TDC of combustion do not appear to be relevant for promoting pre-ignition events according to the previously stated hypothesis since they do not leave the piston top land area which is expected to be well cooled such that auto-ignition reactions can be expected to be inhibited. For the cases with droplet shear-off at TDC of gas exchange it is concluded that sheared droplets of 1 µm initial diameter are too small to be relevant for auto-ignition. Those droplets evaporate under the thermodynamic boundary conditions of the engine cycle early before TDC completely such that after that turbulent mixing is suspected to have enough time to dilute the oil concentration sufficiently. As a consequence, pre-ignition of evaporated oil/air spots is prevented.

6 Conclusions

A simulation methodology to predict pre-ignition due to compression ignition in gasoline engines has been presented and discussed. For this method, FEA to predict a realistic temperature distribution of the in-cylinder wall surface, CFD simulation to obtain the thermodynamic state and mixing in the cylinder, and a pre-ignition tool that calculates the heat release rate and pressure increase due to auto-ignition chemistry reactions in the gas phase has been used. For the validation of the tool, results for one operating points under high load conditions has been utilized. It has been shown that the stochastic pre-ignition engine cycle pressure increases can be predicted with this tool-chain if the temperature state in the cylinder is slightly augmented in comparison to the base CFD simulation.

In order to assess the potential influence of lube oil droplet to promote pre-ignition events, a basic parameter study has been carried out. It was shown that lube oil droplets that are sheared-off from the piston top land at TDC of gas exchange can prevail until TDC of combustion and spread over the cylinder region if the initially sheared-off droplet diameters have a sufficient diameter of the order of 10 μm. It is concluded that lube oil droplets have the potential to promote pre-ignition events under these conditions. Further work is required to better understand this possible mechanism leading to pre-ignition.

References

1. Heywood JB (1988) Internal combustion engine fundamentals. McGraw-Hill, New York
2. Rothenberger P, Zahdeh A, Anbarasu M, Göbel T, Schäfer J, Schmuck-Soldan S (2010) Experimental pre-ignition investigations of boosted spark ignition engines in combination with lightintensified high speed camera and CFD. In: 3. Tagung Ottomotorisches Klopfen—irreguläre Verbrennung, Berlin
3. Douaud A, Eyzat P (1978) Four-octane-number model for predicting the anti-knock behavior of fuels and engines. In: SAE Paper 780080
4. Franzke D (1981) Beitrag zu Ermittlung eines Klopfkriteriums der ottomotorischen Verbrennung und zur Vorausberechnung der Klopfgrenze, Dissertation, Technische Universität München
5. Ewald J et al. (2012) CAE tool chain for the prediction of pre-ignition risk in gasoline engines. In: COMODIA, Fukuoka, Japan
6. Goodwin D (2012) Cantera: an object-oriented software toolkit for chemical kinetics, thermodynamics, and transport processes [Online]. Available http://code.google.com/p/cantera/. Access on Jan 2012
7. Morcinkowski B, Budde M, Ewald J, Schernus C, Adomeit P (2010) Simulation of irregular combustion characteristics caused by chemical kinetics as part of engine port and combustion chamber layout. In: 3. Tagung Ottomotorisches Klopfen—irreguläre Verbrennung, Berlin
8. Andrae J (2009) Development of a detailed kinetic model for gasoline surrogate fuels. Fuel Bd 87(11–12):2013–2022
9. Dahnz C, Spicher HKMU, Schießl R, Magar M, Maas U (2012) Investigations on pre-ignition in highly supercharged SI engines. In: SAE Paper 2010-01-0355
10. Budde M, Jakob M, Ehrly M, Lamanna G (2011) Entwicklung einer Simulationsmethode zum Dieselkraftstoffeintrag in Motorenöl unter besonderer Berücksichtigung der Betriebsbedingungen im Regenerationsbetrieb, FVV final report, research project No. 990

Electric Water-Pump Development for Cooling Gasoline Engine

Wenxin Cai, Shenglin Xiong, Lizhi Fang and Shaoping Zha

Abstract This paper introduced the usage, energy saving characteristic and the development background of electric water-pump (EWP) for cooling system of gasoline engine. Based on a new type of self-developed gasoline engine, we work out the performance target of the EWP according to thermal balance experiment in the self-developed gasoline engine, and set the main design ideals and the structure plan. At the same time, the paper introduced the design method of impeller, motor and the controlling circuit board. During the design process, CAE tools was used to predict the performance of the EWP and the motor. The performance experiment of the trial products was carried out, it indicates that CAE's results get a better compliance with experiment, and the design target was obtained.

Keywords Vehicle · Engine · Cooling · Electric · Water-pump

1 Introduction

The sales amount of vehicle increased from 40,000,000 per year to 64,000,000 since 1980–2000. And the quantity of vehicle of the whole world increased from 400,000,000 to 800,000,000. Fast increased amount of the vehicle caused the global range of atmospheric pollution, the greenhouse effect, oil resource exhaustion and material resources reduction. These make human pay more

F2012-A09-011

W. Cai (✉) · S. Xiong · L. Fang · S. Zha
Powertrain Department, Technical Center, Dongfeng Motor Corporation, Shanghai, China
e-mail: xiongsl@dfmc.com.cn

SAE-China and FISITA (eds.), *Proceedings of the FISITA 2012 World Automotive Congress*, Lecture Notes in Electrical Engineering 190,
DOI: 10.1007/978-3-642-33750-5_43,

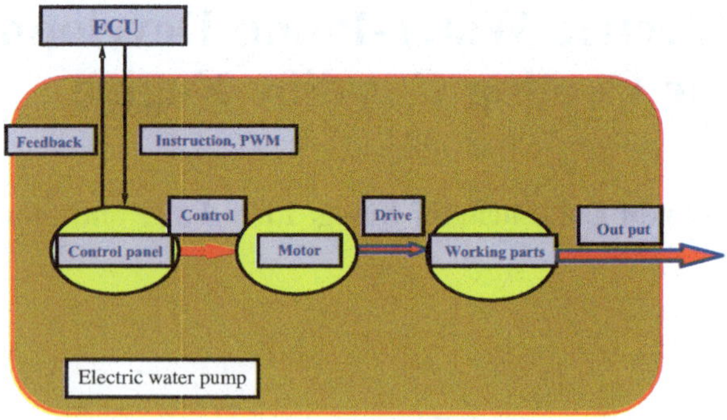

Fig. 1 Structure of the EWP

attention to the development of low pollution, high efficiency electric and hybrid vehicles. For electric and hybrid vehicles, EWP is an inevitable trend.

EWP can not only improve the efficiency of water-pump (WP) but also reduce the over-run loss of mechanical water-pump (MWP) because it doesn't need belt drive. Moreover, EWP can provide the engine with adequate amount of coolant at the right time, thus it also reduces the over-cooling loss, especially reduce warm-up time during the cold start process of the engine. For example there is a kind of hybrid vehicle abroad equipped with a EWP cooling engine in a hybrid vehicle, which improves the engine thermal efficiency about 2 %. Therefore, EWP used in engine cooling system is an effective method to improve fuel economy both in hybrid vehicle and traditional engines.

EWP is not only suitable for traditional engines, but also an indispensable key technology in current hybrid vehicles and the future electric vehicles. So it is necessary to give more attention and more resources on making the MWP electrified, especially to develop a serial EWP for all kinds of vehicles. And it is important to meet the international standards.

So far, there are few domestic companies carrying out research and development on EWP, and almost all are just in innovation phase. But it is about 10 years since the research and development were carried out in this area overseas, and mass production has become true.

Figure 1 shows the structures of the EWP, it is indicated that EWP is constituted by three main part, they are working parts of impeller, motor parts and the control circuit board. The structures of the EWP indicate the following issues and the key technology of the product:

- System design aspects: Analyze the requirement of the cooling system;
- Impeller dynamic aspects: Optimize the design and calculating of impeller and runner, including improving the efficiency of the impeller;
- Motor aspects: Design and development of the small and high efficiency brushless dc motor;
- Control board aspects: design of the PWM mode circuit broad to control digital motor, and the matching research to dc motor.

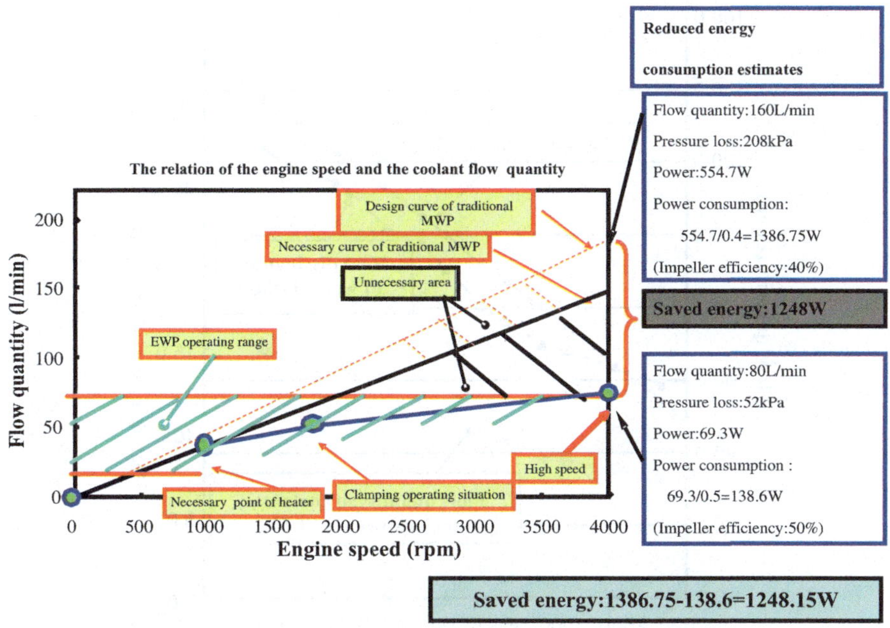

Fig. 2 The effect of fuel economy

2 Design and Development of EWP

Compared with the traditional mechanical water-pump (MWP), EWP has following advantages: acceleration of engine warm-up process, reduction of belt drive loss and providing required coolant flow for engine cooling system. EWP can make the engine work temperature stay at the desired range and reduce pump power consumption of over running. Figure 2 indicates the reason why a EWP reduces the fuel consumption. The horizontal ordinate represents the speed of engine (rpm), and the vertical ordinate represents the flow quantity of the coolant (l/min). The black curve is the necessary amount of the traditional MWP, the red curve is the target design flow quantity with considered margin and the blue curve is the needed flow quantity of an operating engine. It can be found that the area between the red curve and the blue curve is the energy can be saved by using a EWP, the energy-saving is about 1.25 kW when the engine speed is 4,000 rpm. Therefore, it is necessary to analysis the cooling requirement of the engine before developing the EWP in order to determine the performance of the EWP.

The design work of the EWP includes the following aspects: determination of required performance in cooling system, design of the impeller, the stator and rotor of the motor, the assembly layout and the driver control board, formulation the drive strategies and carrying out some related experiments. This article is to introduce the process of the EWP designing after analyzing the requirement based on one self-developed engine.

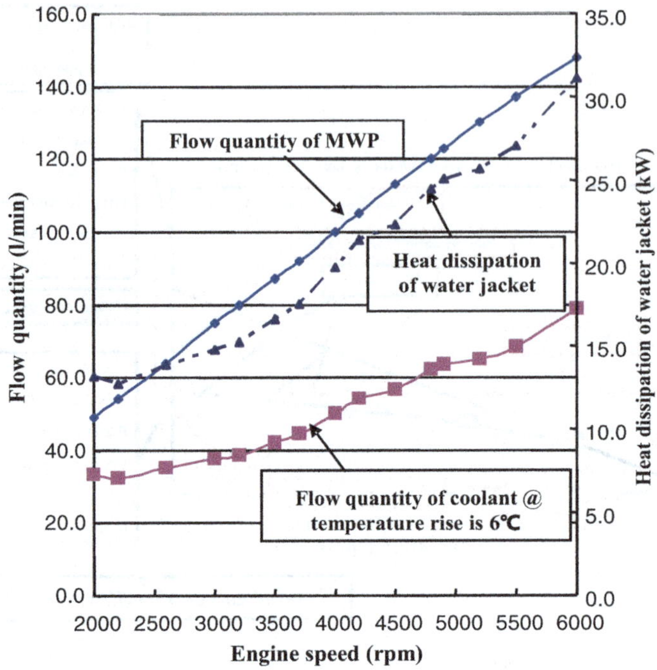

Fig. 3 Results of thermal balance experiment

2.1 Analysis of the Engine Cooling Requirement

First of all, a thermal balance experiment was carried out on an engine equipped with a MWP. Experiment ambient temperature was around 25 ± 3 deg. and the coolant temperature of the outlet was set to 90 ± 3 deg. Then started the experiment, measured the data such as the engine speed and power, coolant flow quantity and the temperature of the coolant both inlet and outlet of the engine on conditions of the engine external characteristics.

Figure 3 shows the results of the thermal balance experiment. The horizontal ordinate is the engine speed (rpm), and the left vertical ordinate is the coolant flow quantity (l/min), right vertical ordinate is the energy taken away by the coolant. It shows that as the engine speed increases the coolant flow quantity of the MWP increases accordingly and achieves the maximum value 150 l/min at the engine top speed 6,000 rpm. Because the temperature difference between the inlet and outlet, the energy dissipates via the coolant is close to 31 kW, but the coolant temperature value increased from inlet to outlet is only about 3 deg. Currently, this value should be more higher about 5 ± 3 deg [1] according to design target. The red curve in Fig. 3 indicate the calculated coolant flow quantity on condition that the coolant temperature increases 6 deg. As it shows in the figure, the maximum value of the coolant value is 80 l/min when the engine speed achieves 6,000 rpm.

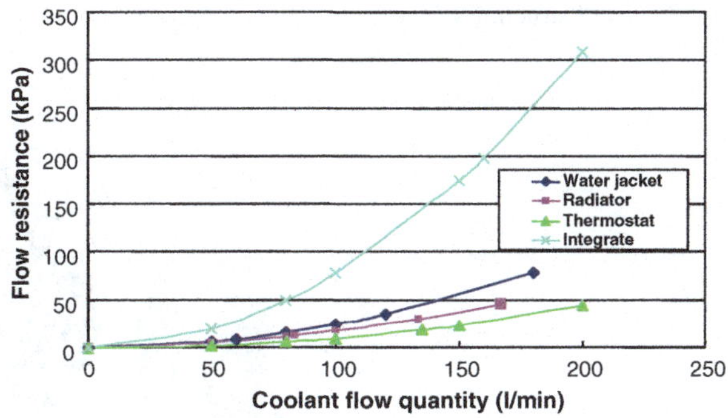

Fig. 4 Flow resistance characteristic of the engine cooling system

That means if the EWP can provide the engine with coolant no less than 80 l/min, which is about a half of the MWP, it is enough to meet the cooling requirement of the engine. From the descriptions above, the electric maximum flow of the WP can be set to 80 l/min. After figuring out the cooling requirement of the engine, it is necessary to hold a cooling system experiment of the whole system and match analysis of the flow resistance for determination of the pump head and the style of the EWP.

Figure 4 shows the flow resistance curves of the main components and synthesis flow resistance curve of the all components, the horizontal ordinate is the flow quantity (l/min), the vertical ordinate is the flow resistance (kPa). The blue curve in the diagram, the pink curve and the orange curve are respectively the flow resistance characteristic curve of the thermostat, the water jacket of the engine and the radiator. The black curve is the synthetic flow resistance curve of the three curves. From the chart, it is found that the flow resistance is about 40 kPa when the flow is 80 l/min, considering the flow resistance of the pipes and other parts which has not been taken into account, so it is required to increase the pump head to 65 kPa with an allowance.

Based on the experiment data above, we set the character of the EWP that is the water head is higher than 65 kPa when flow quantity is 80 l/min.

2.2 *Design of the Overall and the Impeller*

The design of the impeller and overall EWP were carried out for the cooling requirement. Firstly, the rough shape was confirmed according to the engine boundary. Figure 5 shows the structure of the EWP; we selected a close-type impeller in order to improve the efficiency, and a digital integration style for the control board and the motor.

The main performance parameters of the EWP and the impeller parameters are shown in Table 1.

Q is the flow quantity, n is motor speed, and H is EWP head. D is the outer diameter of impeller. The parameter defining process presents as follows. First

Fig. 5 Detail structure of the EWP

Table 1 EWP and impeller parameters

$Q/(l\ min^{-1})$	H/kPa	n/rpm	D/mm
80	65	4500	48

Table 2 Motor requirements(design input)

U/V	I/A	n/rpm	T/(Nm)	Pe/W	η/%
12	16.5	4500	0.34	158	80

select an appropriate diameter of the impeller according to the layout boundary, then calculate and determine the rated speed according to the pump head, target flow quantity and the diameter of the impeller, and at last, define the internal diameter, import and exit angle and design the impeller cap. High efficiency was taken into account during the process above [2]. Flow analyses and performance prediction were done for impeller structure, and the results meets the targets well [3].

2.3 Design of Motor and Control Board

As the requirement of impeller, the parameters such as the output power and the available torque of the motor were calculated according to Eqs. 1 and 2, and the motor efficiency achieved 80 %. The parameter requirement shows in Table 2.

$$Pe = \frac{\rho gHQ}{1000} \tag{1}$$

$$Pe = \frac{nT}{9550} \tag{2}$$

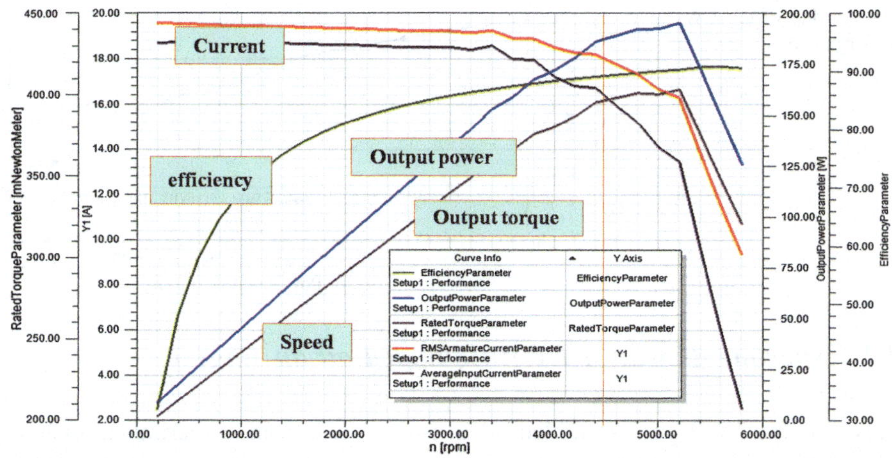

Fig. 6 Predict results of the motor

g is gravity acceleration and ρ is coolant density. U and I are the input voltage and current. T and Pe are the output torque and power, and η is motor efficiency.

It is continued to hold a layout design and designed the stator of the motor [4] and determined the profile dimensions within the scope of the installation space. Also, the motor character simulation was carried out according to the parameters. Figure 6 shows the results. In the figure, the horizontal ordinate is the rotational speed of the motor, and the ordinate shows the output of torque and power, the current and efficiency curve. When the motor rotational speed is 4500 rpm, the output torque is 0.39 Nm, which is larger than the target value 0.34 Nm, and similarly, the output power is 180 W, that is more than the design requirements 158 W.

The control board was displayed at the back of the pump body, some chip and components was welded on it, and a silica membrane was sticked on to the board for purpose of heat conduction.

3 Experiments

3.1 EWP Characteristic Experiments

A flow character experiment was carried out with a EWP manufactured with the above design, and Fig. 7 shows the results.

The blue curve in Fig. 7 is the performance of the EWP, and the pink point is the target spot which is 65 kPa @ 80 l/min. It is clear that the EWP performance is a little better than the target. So we continue to carry out the experiment with a EWP equipped on an engine.

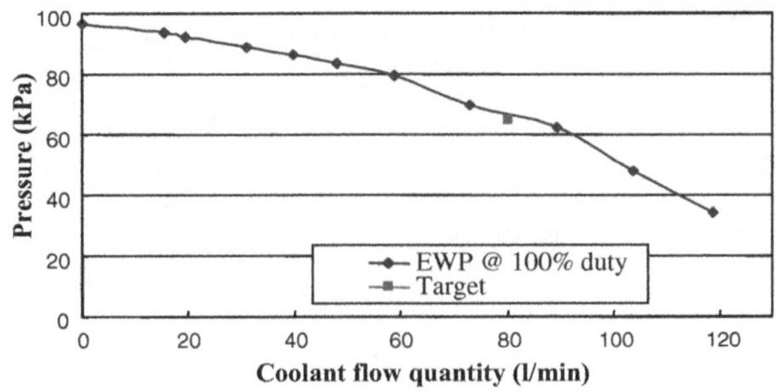

Fig. 7 Map of the EWP performance

3.2 Experiments with a EWP Equipped on an Engine

This time the EWP was assembled on an engine to do some experiments such as cooling performance and fuel economy. During the experiment, the EWP operated at a designed point, and the engine export water temperature was controlled around 90 ± 2 deg. The engine operated at external characteristics and part load conditions.

Figure 8 shows the temperature and coolant flow quantity curve of EWP compared with MWP at external characteristics. In the illustration, it is found that the flow quantity proportionally increases as the engine speed, approaching to 180 l/min at 6,000 rpm, while the EWP flow quantity is stable about 83 l/min. The EWP temperature curve indicates that the highest temperature rise is about 6.5 deg, but the MWP temperature rise is blow 5 deg. Figure 9 shows the engine torque and power curve of the two kinds of pump and Fig. 10 is the engine oil temperature comparison. The experiment results indicate that, the engine oil temperature and output torque and power are almost the same between the engines. From the above, we can draw a conclusion that the EWP can fully meets the cooling requirements of the engine, and has the same cooling effect with the MWP for the engine.

Figure 11 shows the fuel consumption rate comparison of the two kinds of WP on external characteristics condition, the fuel consumption rates of EWP is lower on all conditions, so a EWP will reduce the fuel consumption about 1.6 %. Figure 12 is the comparison of fuel consumption rate at specific part operating condition, and we can see that it is lower of EWP than MWP at all spots, by reducing 1.6–3.9 %. In order to verify the fuel economy of the EWP, it is decided to continue the experiment on vehicle.

3.3 Experiments of a EWP Equipped on a Vehicle

This experiment was carried out to verify the cooling effect and the fuel economy according to the NEDC cycle of the vehicle equipped with a EWP on a vehicle experiment bench.

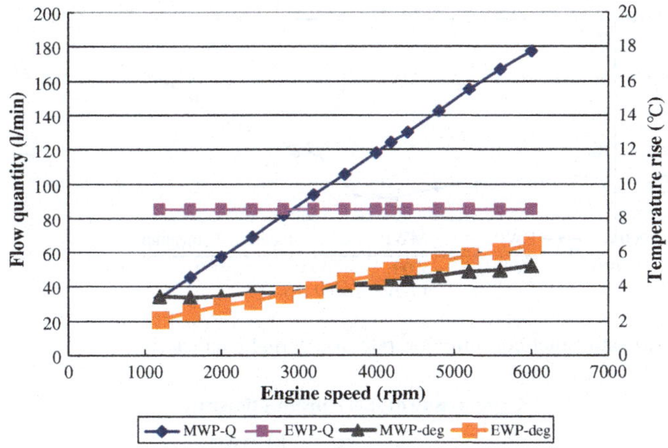

Fig. 8 The relationship of coolant temperature and the engine speed

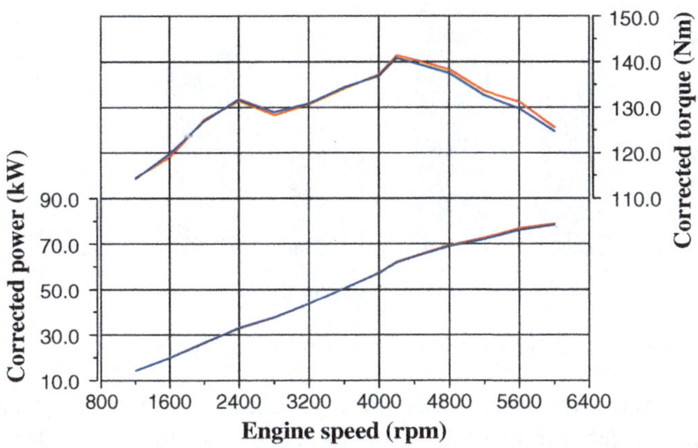

Fig. 9 The external characteristics of the engine equipped with a EWP

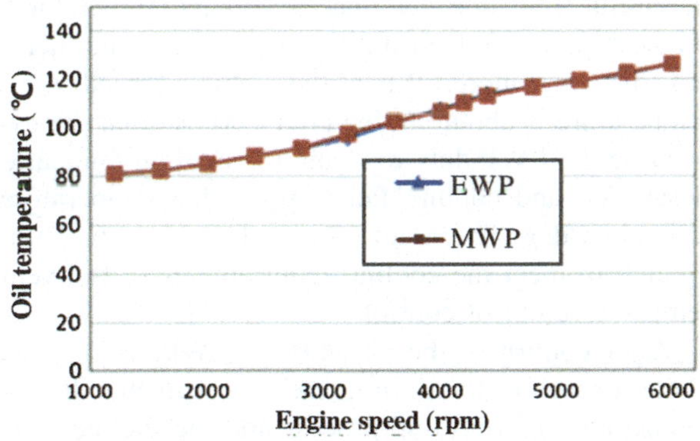

Fig. 10 Comparison of the oil temperature curves

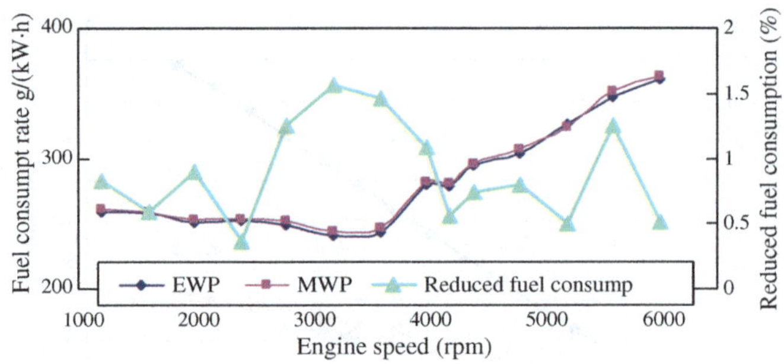

Fig. 11 Comparison of fuel consumption rate at external characteristics

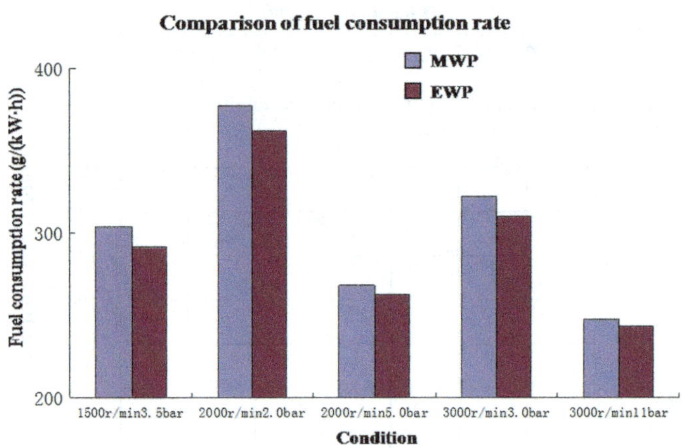

Fig. 12 Comparison of fuel consumption rate at part load conditions

First some trial experiment was done to confirm if the EWP can operate smoothly, and the laboratory room temperature was set to 25 ± 2 deg. When the engine was warmed-up, the vehicle was run at 120 and 60 km/h stably simulating the high speed condition and the urban conditions. During the experiment, data especially the coolant temperature was measured and recorded. Figure 13 shows the results, it is indicated that the coolant temperature rise as the time and achieves the peak point 85 deg till 200 s at 60 km/h, and then the coolant temperature does not rise any more. When the vehicle speed is 120 km/h, the peak coolant temperature is about 96 deg after 1300 s, then the temperature dropped to 93 deg because the EWP duty rose and the cooling fan started. And while the EWP duty descended and cooling fan stopped, the temperature rose again and fluctuating around 94 deg. The results above show that, the ability of the EWP is completely enough to meet the cooling requirements of the vehicle and will not lead to the temperature out of control.

After confirmed the capacity of EWP, we continued to carry out the NEDC cycle experiments, during the experiment, the rising curve of the coolant temperature of warm-up process and the fuel economy were measured.

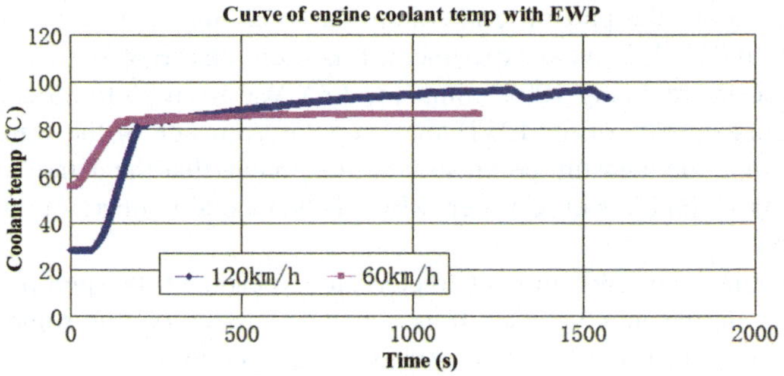

Fig. 13 The coolant temperature maps of the vehicle

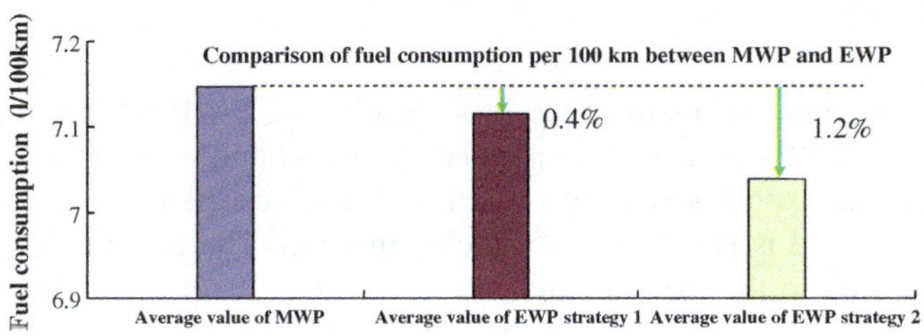

Fig. 14 Fuel consumption comparison of the two kinds of pump

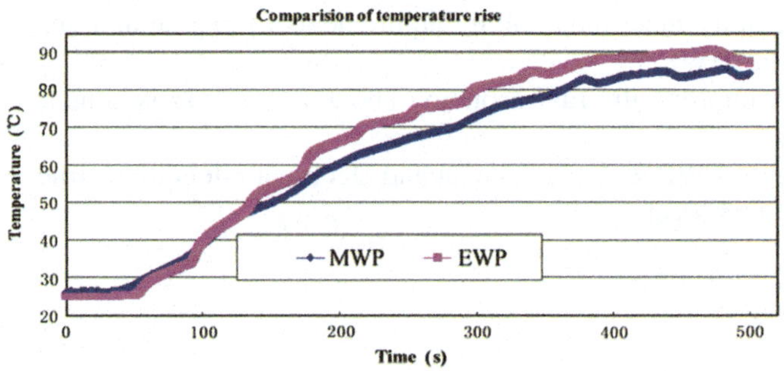

Fig. 15 Coolant temperature comparison of vehicle with EWP & MWP in NEDC cycle experiment

Figure 14 shows the fuel consumption rate of NEDC cycle both a EWP and a MWP equipped on a vehicle. During the experiment, the strategy of the EWP was optimized according to the coolant rise curve, such as set three kinds of duties according to different temperature and made the EWP operate intermittently during warm-up process in order to meets the different requirements of the different operating conditions. Afterwards, we get two typical strategies by summarizing in the experiment. Compared to MWP, strategy 2 can decrease the fuel consumption about 1.2 %, so strategy 2 was choose as the final strategy.

Figure 15 shows the part of temperature curve comparison between EWP and MWP during the NEDC cycle experiment. It is found that EWP can accelerate the warm-up process, reducing 100 s compared to MWP from 25 to 85 deg. Because the intermittent operate of the EWP, the temperature inside of the water jacket is even higher than the measure point, so it is imaginable that the actual temperature rise of the engine block is even faster, which is beneficial to decrease the emission of the engine.

After finishing the EWP flow characteristic experiment, the pump experiment equipped on engine and vehicle. It is scheduled to carry out some reliability experiment, optimization of cooling strategy and structure.

4 Conclusion

This article presents the main process of developing EWP for one independent engine, and the EWP experiments equipped on engine and vehicle were completed successfully. The experiment shows that EWP matches the cooling requirements of vehicle well, and is beneficial to vehicle emission. The conclusion presents as follows according to the experiments:

a. It is feasible route to design the EWP based on thermal balance experiment. And integrated structure EWP has many advantages such as simple structure and easy installation.
b. EWP can fully meet the cooling requirements of engine and vehicle, and can replace the MWP.
c. EWP can improve the fuel economy about 1.6 %, and is beneficial to engine emission.
d. This kind of EWP with mechanical and electrical integration structure achieves the advanced level.

References

1. Yuan Zhaocheng (2008) Engine design. China Machine Press, Beijing
2. Guan Xingfan (1995) Modern technical manual of pump. Aerospace Press, Beijing
3. LI L, Liu Y (2006) Flow superposition design method and CFD simulation for the impeller of water pump. J Beijing Univ Aeronaut Astronautics 36(6):705–707
4. Qin Zenghuang (2004) Electric Engineering. Higher Education Press, Beijing

Two Types of Variable Displacement Oil Pump Development

Dan OuYang, Wenxin Cai, Lei He and Yaojun Li

Abstract Regulations in fuel consumption and emission become more and more strict, it is required to give more research on finding methods of saving energy and reducing emissions, also, reducing energy loss of engine attachments is an effective method. Oil pump play a key role in engine lubrication system, it provides power to make oil flow for lubrication and cooling of moving parts in engine. Due to discharge effect of the pressure relief valve in oil pump, thus, traditional oil pump loses a lot of energy as engine works in a high speed period. On the contrary, variable displacement oil pump has a certain effect on energy-saving by reducing oil pump displacement directly when an engine is running in a high speed period. This article gives research on the structural characters and energy-saving effect on both rotor type and vane type of variable displacement oil pump.

Keywords Oil pump · Gasoline engine · Fuel consumption · Rotor pump · Vane pump

1 Introduction

As a key part in engine lubrication, Oil pump provides enough oil to ensure the lubrication and cooling of the moving parts such as crankshaft bearing and pistons, and also it provides pressure for hydraulic-pressure parts. The volume of oil

F2012-A09-014

D. OuYang (✉) · W. Cai · L. He
Powertrain Engineering, Technical Center, Dongfeng Motor Corporation, Wuhan, China
e-mail: ouyangdan@dfmc.com.cn

Y. Li
Technique Department, Dongfeng Motor Pump Corporation Limited, Wuhan, China

SAE-China and FISITA (eds.), *Proceedings of the FISITA 2012 World Automotive Congress*, Lecture Notes in Electrical Engineering 190,
DOI: 10.1007/978-3-642-33750-5_44,

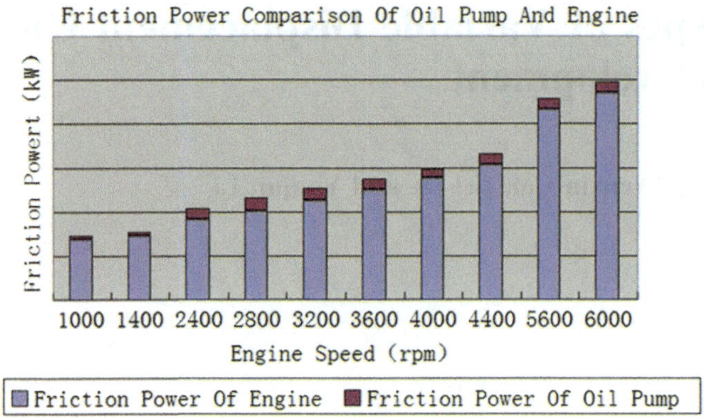

Fig. 1 Compariation of friction power in a developed engine

pumped by a traditional pump will increase as the engine's speed rises. In order to meet requirement of an engine's performance and durability, oil pump is designed to ensure the engine's lubrication requirement for engine idling condition. Thus, it causes an extra ability to pump oil in engine high-speed condition, moreover, the extra pump ability will cause a loss of engine power. The automobile companies all over the world are now keeping on researching to find out a way to reduce the displacement when an engine works on a high speed. The loss of power reduces significantly when the displacement of an oil pump decreases. Then the fuel consumption comes down so that the aim to save energy and reduce emission is reached.

Figure 1 shows the contrast data of the friction power between a traditional oil pump and an engine under motoring condition.

As Fig. 1 shows, the friction power of an oil pump takes 8 % of the friction power of an engine in average. A traditional oil pump usually loses a lot of extra energy by releasing excess lubricating oil through a pressure relief valve due to an excess capacity of pumping when it works in a high speed. A variable displacement oil pump reduces the loss of energy by decreasing the capacity for pumping oil when it runs in a high speed.

The effect of energy saving is described in Fig. 1, the engine speed is shown on the abscissa and oil pressure on the ordinate.

In Fig. 2, the blue color represents the designed pressure curve of a traditional oil pump which shows that it can meets the required lubricating effect on parts like VVT, camshafts and crankshaft bearing when the engine works in a low speed. However, when engine works in a high speed, the pumping capacity is beyond what an engine needs so that it causes an excess lubrication capacity. A variable displacement pump can adjust the pumping capacity based on the engine's actual need, as represented by dotted lines in Fig. 2. Therefore, variable displacement oil pump have energy-saving effect about 1 % of total power [1].

The structure of a variable displacement pump may be diverse. It can be a rotor type or a vane. And it also can be driven by crankshaft directly or by a chain gear.

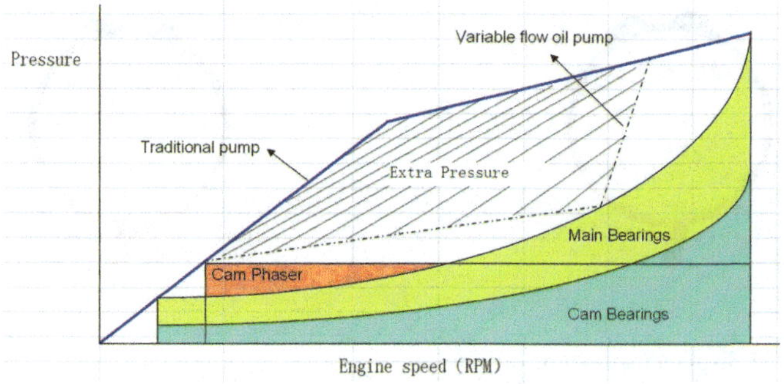

Fig. 2 The effect of energy saving on variable displacement oil pump

Fig. 3 Rotor oil pump

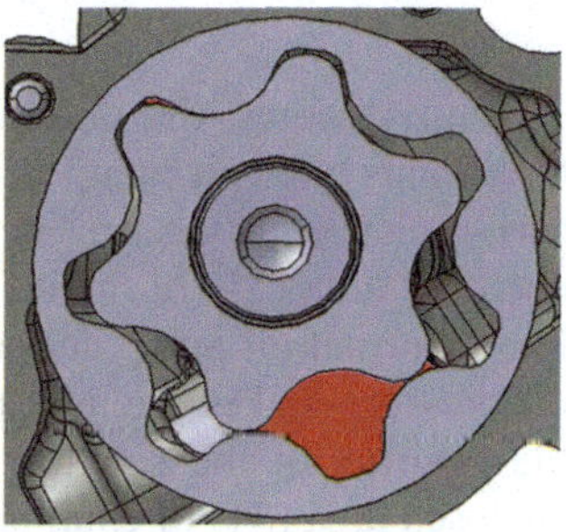

The pilot valve may be mechanical controlled or solenoid controlled. So it can match with various types of engines. The effects of fuel saving and cost saving are different between various types of variable displacement pumps. This article mainly studies the rotor type and vane type variable displacement pump.

2 The Structure of the Oil Pump

2.1 The Simple Calculation of Rotor Pump Displacement

In the process of rotor spinning, the space between the inner and outer rotors forms an enclosed space, the enclosed space will become vacuum from small to big and the oil is absorbed when it is in Oil absorption cavity. When the enclosed space is near the oil discharge, it will turn from big to small and the oil is pressured and discharged. And when the rotor spins a circle, the process of oil absorption and discharge in an inner-outer rotor couple will happens once. So the calculation of the rotor oil pump displacement can be as followers:

The red enclosed area as shown in Fig. 3 is the biggest enclosed space of lubrication oil pump. It can be supposed the area of the red areas is S1, the thickness of the rotor is B, the number of teeth of the inner rotor is n, therefore the displacement Q of the oil pump, is Q = B*S1*n*10^(−6) [*litres/rev*].

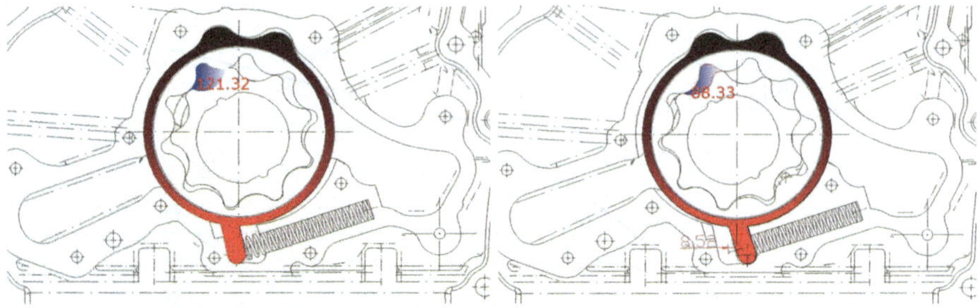

Fig. 4 Variable displacement rotor oil pump

2.2 The Calculation of Variable Displacement Rotor Pump

As Fig. 4 shows, the difference between variable displacement rotor pump and traditional rotor oil pump is adjusting ring. Under the Pre-tightening force of the spring, the adjusting ring is in the position of the biggest displacement, and the enclosed area (S1) of the oil pump rotor is also the biggest, thus the pump works like a normal oil pump at that moment. When discharge pressure is increased, the oil pressure makes the adjusting ring moving right. When the adjusting ring is in the position of the smallest displacement as shown in right side of Fig. 4, the enclosed area (S1) of coupling rotor is the smallest and it causes low displacement of the oil pump. According to the equation of the displacement Q = B*S1*n*10^(−6) [litres/rev], when oil pressure value exceeds a constant value, the displacement of the oil pump is also decreased.

The above rotor-type variable displacement oil pump works by adjusting the relative phase of the inner-outer rotors and changing the maximum sealed volume of them, at the same time, the displacement of the oil pump is adjusted too. To achieve the goal of variable displacement, it is necessary to add an adjustment-ring with tooth designed in lateral. As the outer rotor driven by the adjustment-ring is rotated at an angle, the relative phase of the inner-outer rotors is changed, but the eccentricity of them doesn't change, which ensures both the meshing working normally.

2.3 Van Pump Displacement Calculation

Figure 5 is a calculations diagram for traditional displacement of vane type oil pump. Parameters of stator and rotor are indicated D and d in diameter, B in width. Angle between the two blades is β. Number of blades is Z. Eccentricity distance of stator and rotor is e. When the rotor turn around, the sealed volume variation of two adjacent blades is $V_1 - V_2$. Predicted AB and CD as arcs whose center is O_1:

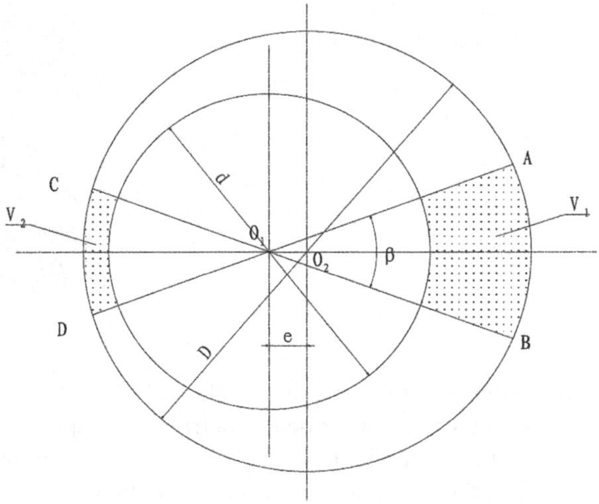

Fig. 5 Fixed displacement vane oil pump

$$V_1 = \frac{1}{2}[(\frac{D}{2} + e)^2 - (\frac{d}{2})^2]\beta B$$

$$V_2 = \frac{1}{2}[(\frac{D}{2} - e)^2 - (\frac{d}{2})^2]\beta B$$

Therefore, the displacement of the traditional displacement vane pump is:

$$V = (V_1 - V_2)Z = 2\pi DBe$$

From the above calculation, it can be concluded that the oil pump displacement is proportional to Eccentricity distance [2].

2.4 Displacement Calculation of Variable Displacement Vane Oil Pump

It is deducted above that vane pump displacement formula is $V = (V_1 - V_2) Z = 2\pi DBe$. From the formula, you can see the displacement of variable displacement van oil pump (Q) is proportional to the rotor eccentricity distance (e), and variable displacement van oil pump change the oil pump displacement by adjusting the eccentricity distance of rotor and stator.

Figure 6 is the theorctical pump of variable displacement vane oil pump. In initial state, the adjustment ring is pushed to the maximum eccentricity position by spring preload, at this time the oil pump displacement is the largest. When the oil pressure is increased, the oil push the adjustment ring to compress spring, the rotor eccentricity is reduced, and then the oil pump displacement is decreased.

It can be found from the above, variable displacement vane oil pump has a larger displacement adjustment range than rotor type. Rotor pump adjust the displacement by changing the inner-outer rotors relative phase, and the maximum

Fig. 6 Theoretical pump of variable displacement vane oil pump

sealed volume, which require a relatively large rotation angle to achieve a large range of pump displacement control. The space restriction leads to a small adjusting range. Variable displacement vane oil pump regulate the displacement directly. In theory, it allows the eccentricity to be zero (i.e. the rotor and stator are concentric), and at this time the oil pump displacement is zero.

Vane pump has a stable oil pressure and low noise, but requires a high blade precision and radial clearance, so the efficiency is low. Rotor pump is simple in structure, low in manufacturing cost, but big in the friction loss.

3 Oil Consumption of Engine

Before determining the pump displacement, the engine demand needs to be investigated. Firstly, the following demand should be completed: (1) the main oil gallery pressure; (2) branch oil pressure requirements; (3) cylinder head oil gallery pressure requirements; (4) main bearing oil film thickness; (5) connecting rod bearing oil film thickness; (6) oil pump map. Input the above parameters to the simulation model, adjust the flow rate of oil pump at each speed point, and iteratively compute the oil flow requirements of engine at each speed point.

From above Fig. 7, you can find the flow rate of conventional oil pump exceeds the engine demand. The variable displacement oil pump is trying to make the oil flow rate closer to the engine requirement.

4 Sample of Variable Displacement Oil Pump

Trial-produce of oil pump sample generally refers to produce one or two theoretical sample, verifying the feasibility of the part design principle, as well as the rationality of the design tolerance. Theoretical pump is only used for the part test, and cannot be installed on the engine. Therefore, the theoretical pump shape is always simple and easy to machine.

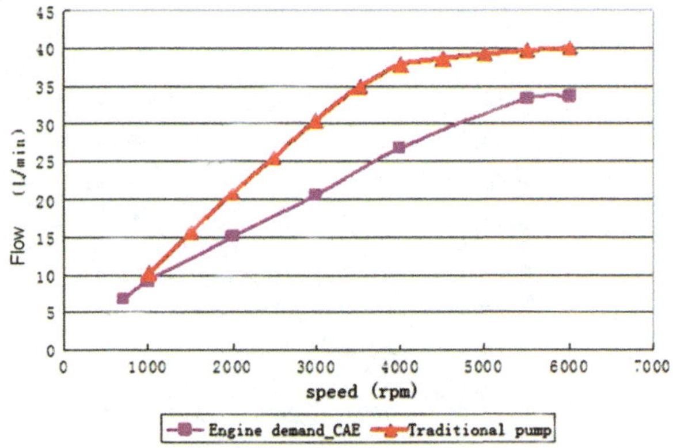

Fig. 7 Engine requirement of oil flow quantity

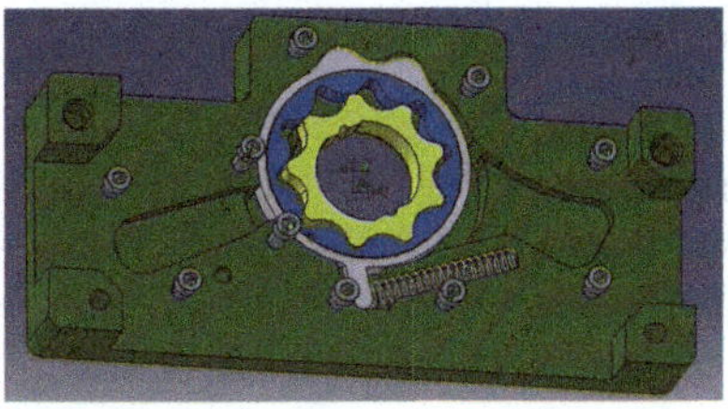

Fig. 8 Theoretical sample of variable displacement oil pump for rotor type

Figure 8 is the theoretical sample of variable displacement oil pump for rotor type. Compared with conventional oil pump, there is an additional adjustment ring. When the pressure of oil is high, it forces the adjustment ring to rotate, which makes the outer rotor turns certain angle. Meanwhile, the relative phase of inner-outer rotors changes, the maximum sealed volume decreased, and then the displacement of oil pump decreased. Through a series of part test to measure oil pressure and flow, we can determine whether the pump meet the requirement.

In Fig. 9, the abscissa is the speed of the engine, the vertical is the oil pressure of the main oil gallery. The red line is the pressure curve of the conventional oil pump, the blue line is the pressure curve of the variable displacement oil pump, and the pink line is the oil pressure requirement of the engine. We can find that the oil pressure of both oil pumps can meet the needs of engine. But, when the speed of engine is high, the oil pressure of the conventional oil pump exceeds the engine requirement, while, the oil pressure of the variable displacement oil pump is nearly the same as the engine requirement. It is obvious that the variable displacement oil pump can save power of engine.

Figure 10 is a variable displacement oil pump for vane type. It consists of the following components: drive shaft, rotor, stator, blade, spring, and ring. The rotor, stator as well as two adjacent blades make some separate spaces. When the rotor rotates, and the separate space comes to the location of the suction chamber, the separate space changes from small to large, which creates a vacuum, and the oil is inhaled. When the rotor comes to the outlet the oil pump, the separate space changes from large to small, and the oil is forced out. When the oil pressure

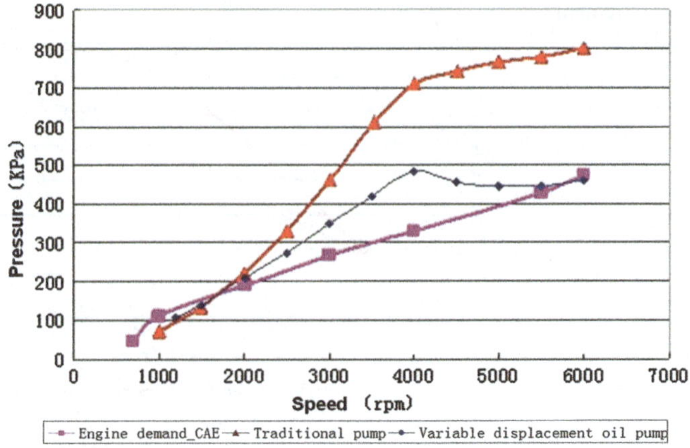

Fig. 9 Oil pump pressure character curve

Fig. 10 Variable displacement oil pump for vane type

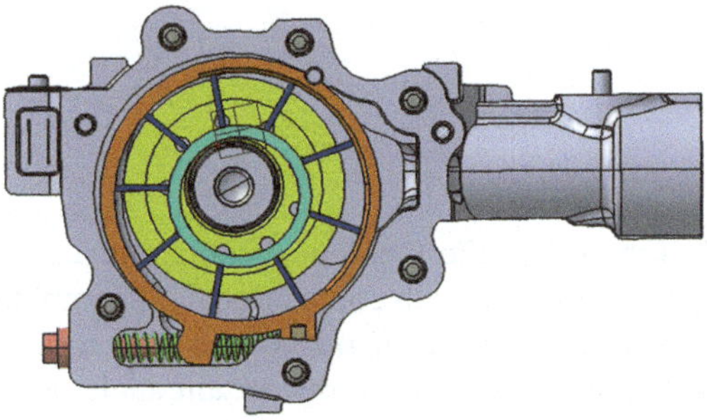

becomes high, it forces the stator to rotate a certain angle, the eccentricity of the stator and rotor becomes smaller, and the displacement of the oil pump can be calculated as: $V = (V_1 - V_2)Z = 2\pi DBe$. The displacement of the oil pump decreases as the eccentricity of the stator and rotor becomes smaller [3].

Variable Displacement Oil Pump for Vane Type is still in machining, there is no experimental data.

5 Variable Displacement Oil Pump Test

According to the structures mentioned above and the CAE calculation of oil pump displacement, we designed two samples. The following is sample test:

We tested the cooling system and lubrication system characteristics and energy-saving effect as engine equipped with variable displacement oil pump for rotor type. We kept the engine's water temperature at 90°, did not control oil temperature, and tested the engine full load speed characteristics. The variable

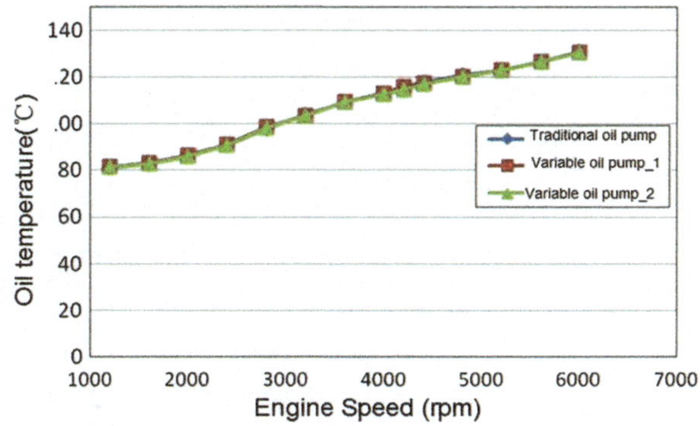

Fig. 11 Oil temperature curve of engine equipped with variable displacement oil pump

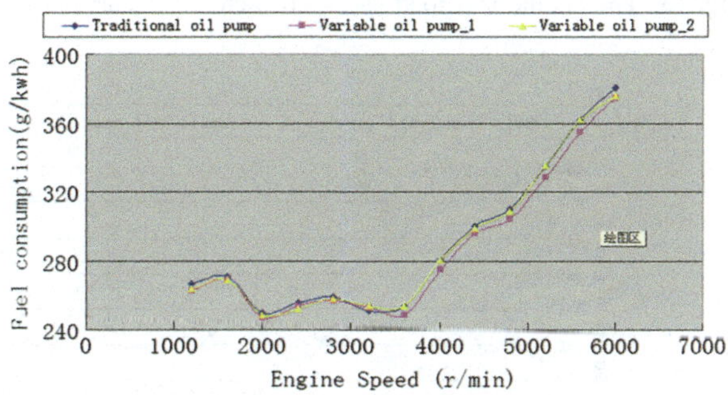

Fig. 12 The engine full load speed characteristics curves

displacement oil pump was tested two times in total, each test had four groups, averaged them as the test results.

Figure 11 compares the temperature of engine equipped with variable displacement oil pump and traditional oil pump. The abscissa is the engine speed, the vertical is the oil temperature. The test results show that the temperature trend of traditional oil pump and variable displacement oil pump is basically the same: oil temperature increases with the speed, and temperature value is basically the same. It can be concluded that the variable displacement oil pump has the same function as traditional pump, it can meet the lubrication and cooling requirements of engine. Figure 12 is the engine full load speed characteristics fuel consumption comparison as engine equipped with traditional oil pump and variable displacement oil pump. Both of them have the same fuel consumption trend: fuel consumption increases with the speed. Engine equipped with variable displacement oil pump has certain advantages, it has fuel consumption 0–10 g lower than the traditional pump at each speed, and 4 g lower at most of the rotational speed. The fuel saving effect is considerable.

Figure 13 is the comparison of the rate of fuel consumption under part load speed characteristics at the speed of 1600 rpm, as the load increase, the fuel consumption rate is descending. Compared with the traditional oil pump, the engine's fuel consumption rate is a little lower at different load when it equipped the variable displacement oil pump.

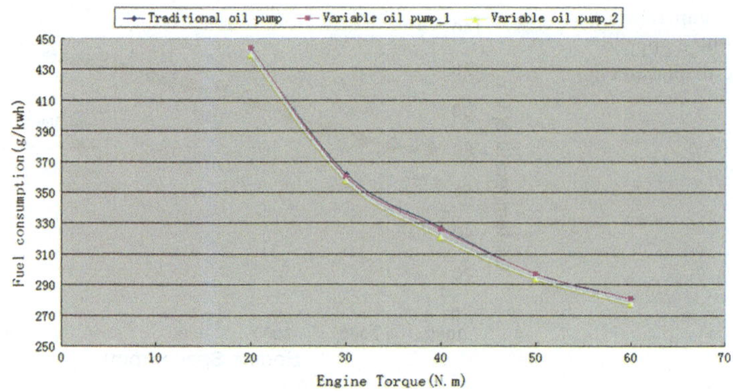

Fig. 13 The engine part load speed characteristics curves at 1600 rpm

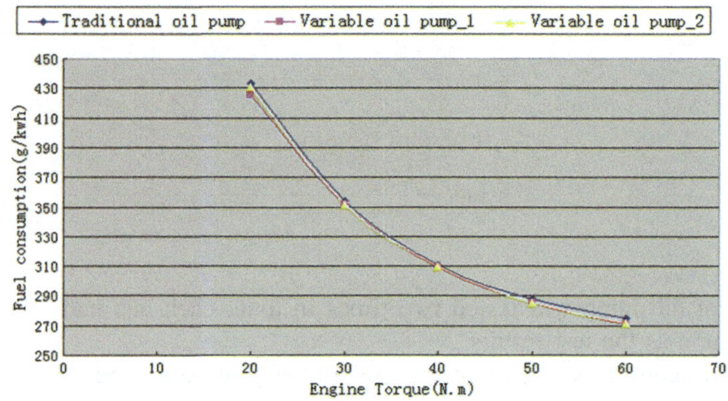

Fig. 14 The engine part load speed characteristics curves at 2000 rpm

Figure 14 is the comparison of the rate of fuel consumption under part load speed characteristics at the speed of 2,000 rpm. We also can find that the engine's fuel consumption rate is a little lower at different load when it equipped the variable displacement oil pump. It can improve the engine's fuel economy.

Figure 15 is the compare of the engine's fuel consumption rate when the engine equipped two different types of oil pump, part load speed characteristics at the speed of 3,200 rpm. The engine's fuel consumption rate is reduce as the load is increase. And the engine is at the same fuel consumption rate as it equipped the two types of oil pump. So the variable displacement oil pump is no advantage at the fuel consumption rate. Then we found that the volumetric efficiency of the variable displacement oil pump is only 45 %, it is much low than the traditional oil pump's 90 %. The low volumetric efficiency lead a lot of energy consumption. We need to do some thing to improve the oil pump's volumetric efficiency in the next stage.

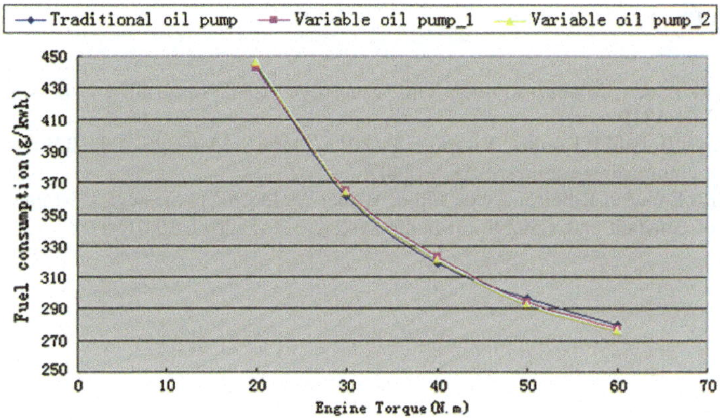

Fig. 15 The engine part load speed characteristics curves at 3200 rpm

The vane type of variable displacement oil pump's sample is been machining, so we haven't the test data. The vane type of variable displacement oil pump and the rotor type of variable displacement oil pump have the same principle of the fuel-efficient. Both of them can reduce the oil pump's capacity when the engine is at the high speed. The rotor type of variable displacement oil pump change the rotor's phase to reduce the Oil pump's capacity, and the vane type of variable displacement oil pump change the eccentricity. It can be predicted that the vane type of variable displacement oil pump also has a certain effect on oil-saving, and as a result of the smaller friction. Its oil saving effect will be even more significant.

6 Summary

Two types of the variable displacement pump can meet the demand of the engine, and have a certain degree of fuel-saving effect. But the displacement adjusting range of rotor type of variable displacement pump is small, and the friction is bigger, so the energy saving effect is smaller. Vane type of variable displacement pump realizes oil pump displacement regulation by changing the eccentricity of rotor, and has a wide adjusting range, but the control in tolerance size on Vane type of variable displacement pump is more stringent, processing is more difficult. The saving principle of all types of Variable displacement pump is the same. The main difference is the control mechanism which is to realize the variable displacement. There has been an electromagnetic valve controlled pump, displacement can be adjusted in real time to the oil pump, which can be more precisely matching engine demand and the fuel-saving effect will be more substantial.

References

1. Heiko Neukirchner, Mathias kramer, Tilo Ohnesorge (2002) Controlled vane type oil pump. SAE 2002-01-1319
2. Haas T, Esch E, Fahl P Kreuter, Kissinger F (2011) Optimized design of the lubrication system of modern combustion engines. SAE 912407
3. Joao Meira, Eduardo Ribeiro, Ayres Filho, Weber Melo (2011) Strategies for energy savings with use of constant and variable oil pump systems. SAE 2011-36-0150

Considerations on Influencing Factors of Carbon Deposit in Gasoline Direct Injection Engine

Changhoon Oh

Abstract This paper describes investigation of the carbon deposit formation phenomenon in Gasoline Direct Injection. A 2.4 L GDI engine and a 2.0 L port fuel injection engine, which were produced by the same manufacturer, were tested on an EC-type engine dynamometer. Non-additized RON 92 regular gasoline was used as the test fuel. The test cycle was based on the ASTM D6201 method, but in the experimental conditions, the average engine load and load-changing frequencies were different. After engine operation, collected carbon deposits were performed on Thermo Gravimetric Analysis (TGA) to investigate the primary source of the carbon deposits. Our investigation revealed that the weight of the carbon deposits on the intake valve and combustion chamber had an inverse-proportional relationship with the average engine load on the GDI engine. In addition, the shapes of carbon deposits that formed on the combustion chamber varied as a function of the engine load. The TGA results also showed that the intake valve deposits were largely affected by lubricants, while the carbon deposits in the combustion chamber were mainly generated from gasoline.

Keywords Gasoline direct injection · Intake valve deposit · Combustion chamber deposit · Engine load · Thermo gravimetric analysis

F2012-A09-015

C. Oh (✉)
GS Caltex, Seoul, Republic of Korea
e-mail: hoonjet@hanmail.net

SAE-China and FISITA (eds.), *Proceedings of the FISITA 2012 World Automotive Congress*, Lecture Notes in Electrical Engineering 190,
DOI: 10.1007/978-3-642-33750-5_45, © Springer-Verlag Berlin Heidelberg 2013

1 Introduction

In a gasoline engine, carbonaceous deposits are accumulated while the engine is operating. Deposits on the fuel injector, intake valves and combustion chamber can cause many problems for vehicle performance [1]. Intake valve deposits impede intake air flow, thus deteriorating engine performance and fuel efficiency. Combustion chamber deposits can be the reason of abnormal combustion, knocking. Generally intake valve deposits and fuel injector deposits are generally considered a more important problem, because combustion chamber deposits can be spontaneously oxygenated during high-load engine operation.

In conventional a Port Fuel Injection (PFI) engine, numerous studies on carbon deposit formation have been performed [2, 3], and now the problem related to intake valve deposit is easily solved by detergent fuel additives. Detergent is mixed with gasoline in the refinery, and it removes carbon deposits when fuel spray touches the intake valve. Many detergents have been invented and nowadays polymer (polyisobuthylene-amine) detergents, which have a highly reacted component on its tail, are widely used.

However, in a Gasoline Direct Injection (GDI) engine, a concept introduced in the late 1990 s, the situation is different. Although several that dealt with carbon deposit formation in GDI engines have been done, most of them focused only on carbon deposits on injector tips [4–6]. In a GDI engine, it is more difficult to remove carbon deposits on intake valves or combustion chambers than it is in a PFI engine, because unlike in a conventional PFI engine, there is no way to wash out carbon deposits in a GDI engine. In a GDI engine, fuel spray is directly injected inside the combustion chamber so it does not touch the intake valve. As a consequence, current fuel additives are no longer effective in GDI engines, so a new way to remove carbon deposits is needed. Carlisle et al. revealed that the amount of aromatic content in gasoline has a significant effect on deposit formation on intake valves and combustion chambers [7], and Parshinejad and Biggs showed that lubricants might be an important factor for intake valve deposits and combustion chamber deposits in GDI engines [8]. The purpose of this research was to investigate carbon deposit formation on intake valves and in combustion chambers in terms of engine operation conditions and to reveal the main source of carbon deposits in GDI engines.

2 Experimental Setup

Figure 1 shows a schematic diagram of the experimental apparatus. The engine RPM and the throttle (the relation between torque and throttle was calibrated before test) were controlled with an EC dynamometer. A 2.4 L GDI engine was used without modification. For the PFI engine, a 2.0 L PFI engine, which was produced by the same manufacturer as the 2.4 L GDI engine and shares some

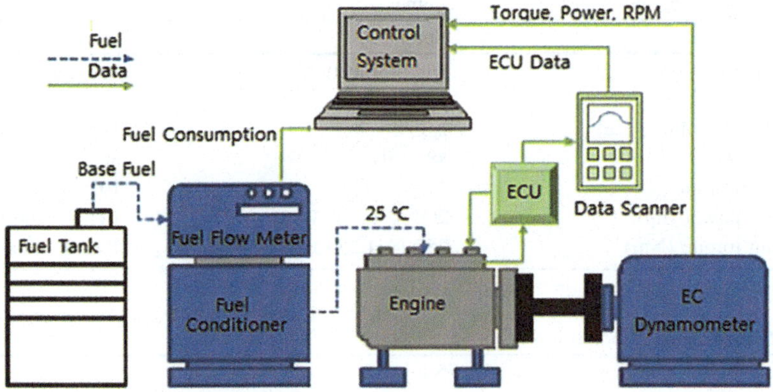

Fig. 1 Schematic diagram for experimental set-up

engine components, was chosen. Specifications of test engines are given in Table 1. As a test fuel, non-additized regular gasoline used in South Korea was chosen.

Table 2 and Fig. 2 show the test conditions for the experiment. All the experimental conditions were based on the ASTM D6201 test method, but the engine load and load-changing frequencies varied. The low-load conditions (#1, #2) were set to the lowest engine load that can stabilize engine operation, and the high-load conditions (#3, #4) were chosen as four times of throttle openness(almost maximum engine load that can perform long-time engine operation without damaging engine and dynamometer system) in a low-load condition. One cycle was composed of 24 min in low-frequency conditions (#1, #3), and in high-frequency conditions (#2, #4), one cycle was 6 min. However, in all cases the total operation time was 75 h. In the medium condition (#5), the average values of low - load and high-load conditions were chosen. For mixed-load conditions (#6), the average engine load and load-changing frequencies were the same as #5, but load gap between the 2,000 RPM operation and 2,800 RPM operation was expanded.

After the test, the amount of Intake Valve Deposit (IVD) was weighed without removing it from the intake valve. The thickness of the Combustion Chamber Deposit (CCD) in the piston top and the cylinder head were measured, and afterwards all the CCD was collected and weighed. The thickness measurement points are given in Fig. 3. The number and position of the measurement points followed the ASTM D6201 method, but some modifications were made in the piston top due to the different piston shape of the GDI engine.

Table 1 Technical specification for test engine

	Engine A	Engine B
Fuel injection	Direct injection	Port fuel injection
Engine layout	In-Line 4	
Displacement volume (cc)	2,359	1,997
Bore × stroke	88 × 97	86 × 86
Compression ratio	11.3:1	10.5:1
Maximum power (hp)	200 @ 6,300 RPM	143 @ 6,000 RPM
Maximum torque (Nm)	252 @ 4,250 RPM	190 @ 4,000 RPM

Table 2 Experimental condition

Operation time	75 h	
Cycle composition	2,000 RPM (1/3) + 2,800 RPM (2/3)	
Fuel temperature (°C)	25 ± 5 °C	
	Low	High
Throttle openness (%)	2,000 RPM: 5 2,800 RPM: 7.5	2,000 RPM: 20 2,800 RPM: 30
Cycle length (min)	24 (8 + 16)	6 (2 + 4)

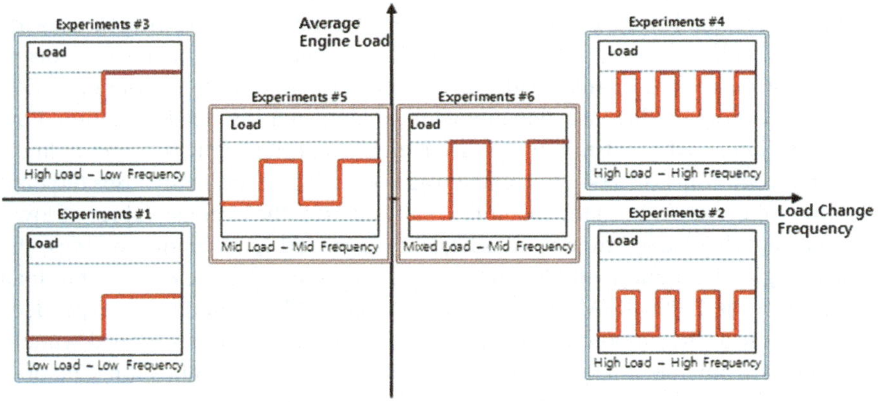

Fig. 2 Schematic figure of each experimental condition

3 Result

3.1 Investigation on Carbon Deposit Formation in GDI Engine

The test results for IVDs and CCDs is given in Fig. 4a, b. In case of injector tip deposits, the temperature of the injector tip should be higher than T90 of the gasoline fuel to accelerate deposit formation [6]; thus, more carbon deposits

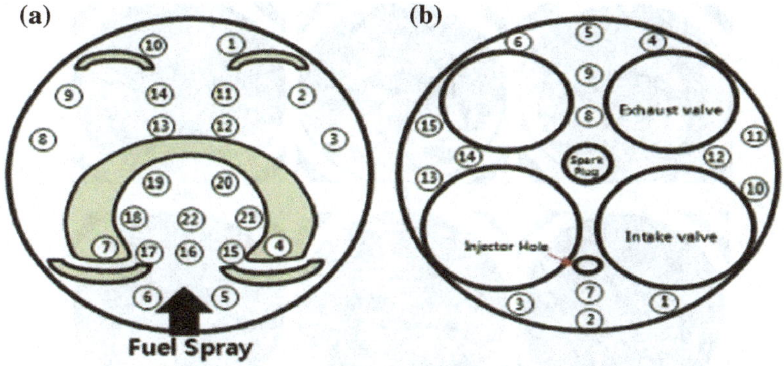

Fig. 3 CCD thickness measurement position. **a** Piston Top. **b** Cylinder head

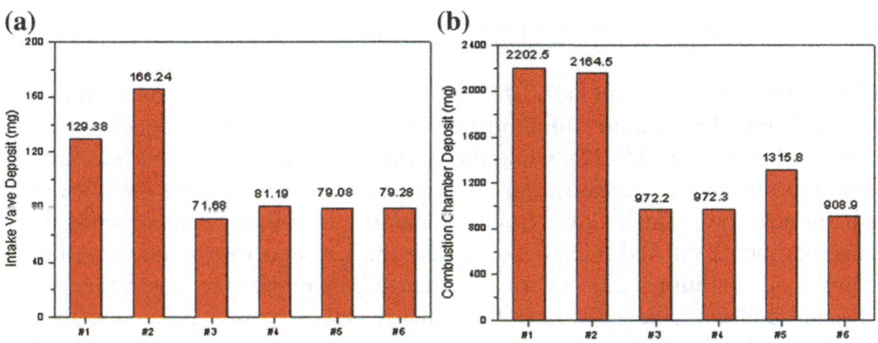

Fig. 4 Weight of intake valve deposit (**a**) and combustion chamber deposit (**b**) for each condition

formed during medium or high load operation. However, in this case, the weight of carbon deposits is inversely proportional to engine load for both IVDs and CCDs. On the other hand, load-changing frequency was not as significant as the average engine load. We suspect that under high-load conditions, a relatively higher flame temperature burned out carbon deposits, and thus a lower amount of carbon deposits accumulated.

In #5 and #6, which have the same average load, the amount IVD was almost the same, but the weight of the CCDs in #6 was only 70 % of the level of #5. The weight of the CCDs in experiment #6 was similar to high-load conditions. This similarity is shown in Fig. 5 in the photos of the piston top after the test. As clearly seen in the pictures, in low-load conditions and mid-load conditions (#1, #2 and #5), the CCDs on the piston top formed a fan shape. However, in contrast to this in high-load conditions (#3, #4) the CCD formed a donut ("O") shape, and #6 also followed the same pattern.

The change in deposit thickness as a function of the measurement point is plotted in Fig. 6, which shows that the average carbon deposit thickness in #1, #2

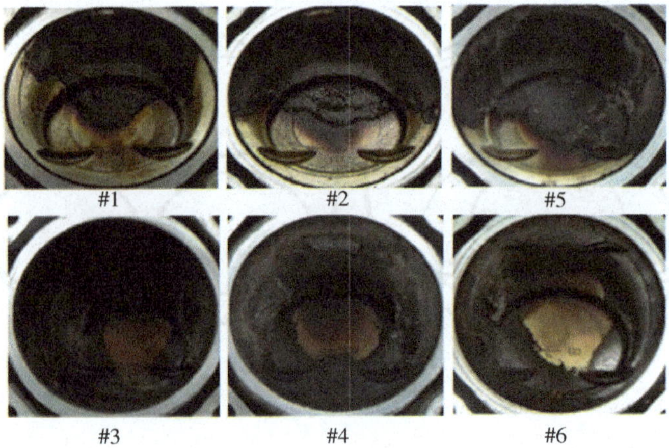

Fig. 5 Pictures of piston top (#1 cylinder) for each experimental condition

and #5 is much thicker than in #3, #4 and #7. In Fig. 6a, it is also observed that in high and mixed loads, there are almost no carbon deposits on the concave part of the piston bowl (point 15–22), while the thickness of carbon deposits was around 300 μm in the other three cases. For all cases, carbon deposits were the thickest at the point near the exhaust valves (points 2 and 9). For cylinder head deposits, low-load experiments showed thick deposits near the fuel injector (points 4–6), but in the mid-load conditions, this pattern was changed. In all the other cases except two low-load conditions, deposits near the exhaust valves (points 1–3) were thicker than those of other points. These differences seem to be caused by the fuel injection timing and amount. In a GDI engine, fuel spray is injected directly into the cylinder chamber during the compression stroke, while a PFI engine uses a pre-mixed fuel mixture during the intake stroke. Thus, the location where the fuel spray contacts the piston top and the size of the fuel spray differ as a function of the engine load.

These experimental results show that CCD formed during the low-load operation could be oxidized during high-load operation mode; thus the CCD formation pattern was similar to that in the high-load operation. It also suggests that the CCD formation pattern in a GDI engine could be different according to the mounted position or nozzle shape of the fuel injector, injection timing or type of engine (either spray guided or wall-guided). However, to reveal more precisely how the fuel spray affects CCD formation, further research on optical engines is needed.

Therefore, maintaining a high enough temperature in the combustion chamber is needed to inhibit formation of carbon deposits. In addition, it is already known that the surface temperature of metal is an important factor in carbon deposit formation [9]. Thus, problems relating to carbon deposits could be more serious for owners of hybrid cars because hybrid cars usually maintain a low engine load due to the aid of a battery and for city commuters who do not have many opportunities to accelerate their cars.

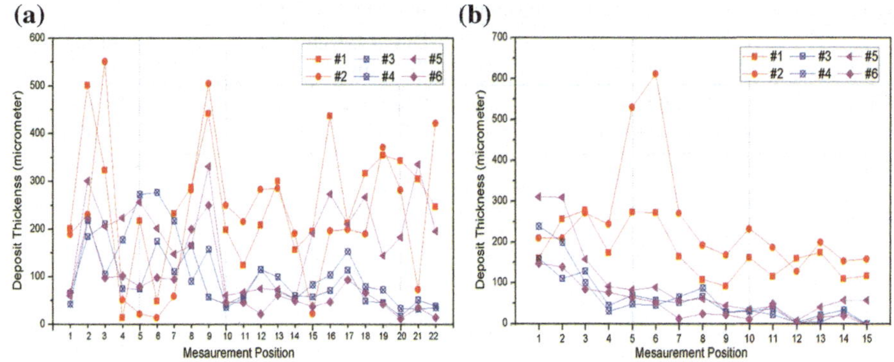

Fig. 6 CCD Thickness for each experimental condition. **a** Piston top. **b** Cylinder head

Engine torque and fuel efficiency were measured at the start and end of the test, but no distinguishable difference was observed across all the experimental conditions. There was also no significant change in the data related to fuel injection from ECU, but for all the experimental conditions, the exhaust temperature increased by around 10–20 °C, which could be the insulation effect of combustion chamber deposits [10]. Weak knocking was detected only in experimental condition #4.

3.2 Comparison Between Source of Carbon Deposit Formation on GDI and PFI Engine

To investigate the source of carbon deposits, we made a comparison between the PFI engine and the GDI engine. For the PFI engine, the ASTM D6201 test method was used. However, the test conditions for engine torque were modified due to the difference between an official Ford 2.3 L engine and the test engine. For the GDI engine, the experiment conditions of #6 (the experiment duration was increased to 100 h to equate with the ASTM D6201) were chosen because its experimental conditions are most similar to the ASTM D6201.

Thus, the source of carbon deposits can be investigated by analyzing amount of metallic compound in carbon deposits. Amount of metallic compounds can give information on source of because gasoline does not contain metal elements and only lubricants fuel additive contain metallic compounds. Table 3 shows the average weight of the carbon deposits and Fig. 7 shows the TGA results(Nitrogen purging with flow rate of 100 ml/min) for carbon deposits from the PFI engine and the GDI engine. In the TGA test with carbon compounds, the residue left over 450 °C can be considered as metallic compounds. As the table and figure show, intake deposits from the GDI engine contained more metallic compounds than

Table 3 Intake valve deposit and combustion chamber deposit for each engine

		Direct injection	Port injection
IVD	Weight	79.28 mg	129.65 mg
CCD	Weight	945.8 mg	555.1 mg
	Avg. thickness	123.64 μm	125.00 μm

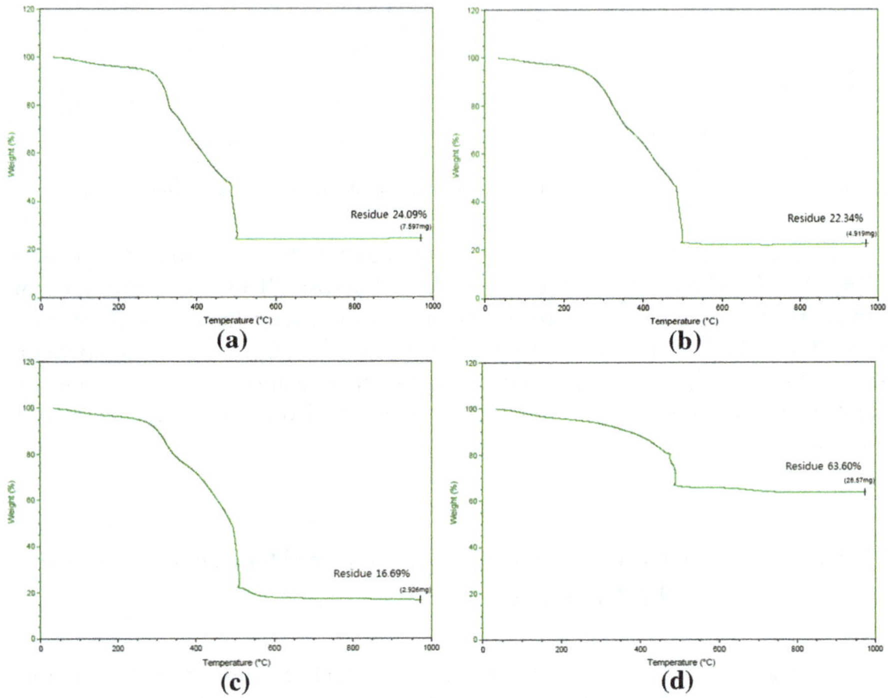

Fig. 7 Test result for each carbon deposit. **a** GDI engine: IVD Result (Residue 24.09 %). **b** GDI engine: CCD Result (Residue: 22.34 %). **c** PFI engine: IVD Result (Residue: 16.69 %). **d** PFI engine: CCD Result (Residue: 63.60 %)

those of the PFI engine. However, the weight of carbon deposits from the GDI engine was only around 60 % level when compared to that of the PFI engine.

As a consequence, Fig. 7 suggests that lubricants played a more important role in formation of intake valve deposits in the GDI engine than in the PFI engine because intake valve deposits from the GDI engine contained more metallic compounds than the PFI engine. The GDI engine has its fuel injectors inside the combustion chamber so there is no contact with the fuel spray; thus, liquid gasoline is a less important source of intake valve deposits. On the other hand, CCDs from the GDI engine had a much lower amount of metallic compounds than the PFI engine, and the absolute amount of CCDs in the GDI engine was much higher than that in the PFI engine. We speculate that unburned fuel or large droplets from the fuel injector that were stuck to the combustion chamber contributed to the formation of carbon deposits. Thus, complete combustion of fuel will be essential to remove combustion chamber deposits in the GDI engine.

4 Conclusion

In this research, carbon deposit formation in a GDI engine was investigated. A 2.4 L GDI engine was used to experiment with various engine loads and load-changing frequencies. Carbon deposits from a GDI engine and a PFI engine were collected, and TGA was performed to reveal the main source of deposit formation. The following conclusions were drawn from the experiments.

1. A GDI engine generates carbon deposits on its intake valve as a PFI engine does although fuel spray does not make contact with the intake valves. However, the absolute amount of carbon deposit was much smaller in the GDI engine.
2. The weight of carbon deposits formed on the combustion chamber was much greater in the GDI engine than in the PFI engine. In addition to this, the shape of the carbon deposits was different as a function of engine. It seems that a difference in fuel injection quantity and timing caused these phenomena.
3. For both the intake valve and combustion chamber, more carbon deposits were generated when the engine load was low. However, the engine load-changing frequency was not an important factor in carbon deposit formation.
4. The TGA results showed that lubricant oil could be an important source of intake valve deposits in a GDI engine, while combustion chamber deposits mainly come from unburned fuel.

References

1. Ramadhas A, Singh V, Subramanian M, Acharya G et al (2011) Impact of fuel additives on intake valve deposits, combustion chamber deposits and emissions
2. Richardson CB, Gyorog DA, Beard LK (1989) A laboratory test for fuel injector deposit studies, SAE Technical Paper, 892116
3. Hilden DL (1988) The relationship of gasoline diolefin content to deposits in multiport fuel injectors
4. Ashida T, Takei Y, Hosi H (2001) Effects of fuel properties on SIDI fuel injector deposit, SAE Technical Paper, 2001-01-3694
5. DuMont RJ, Evans JA, Feist DP, Studzinski WM, Cushing TJ (2009) Test and control of fuel injector deposits in direct injected spark ignition vehicles, SAE Technical Paper, 2009-01-2641
6. Von Bacho PS, Sofianek JK, Galate-Fox JM, McMahon CJ (2009) Engine test for accelerated fuel deposit formation on injectors used in gasoline direct injection engine, SAE Technical Paper, 2009-01-1496

7. Carlisle HW, Frew RW, Mills JR, Aradi AA et al (2001) The effect of fuel composition and additive content on injector deposits and performance of an air-assisted direct injection spark ignition research engine, SAE Technical Paper 2001-01-2030
8. Parsinejad F, Biggs W (2011) Direct injection spark ignition engine deposit analysis : combustion chamber and intake valve deposits, SAE Technical Paper, 2011-01-2110
9. Guthrie PW (2001) A review of fuel, intake and combustion system deposit issues relevant to 4-stroke gasoline direct fuel injection engines, SAE Technical Paper, 2001-01-1202
10. Orhan A, Semih E (2004) Carbon deposit formation from thermal stressing of petroleum fuels. Preprints Papers Am Chem Soc Div Fuel Chem 49(2):764–766

Air Intake Modules with Integrated Indirect Charge Air Coolers

Juergen Stehlig, Rene Dingelstadt, Johann Ehrmanntraut, Rolf Mueller and James Taylor

Abstract Gasoline downsizing and turbocharging has become a key contributor to respect future legislation regarding CO_2 emissions and engine dynamics. In turbocharged engines, increasing charge pressure will cause rising charge air temperatures. Increased charge air temperatures also increase the thermal strain on charge air cooler and air intake module, and have a negative effect on combustion. As a result, the engine's response characteristics and torque output will deteriorate, accompanied by increased fuel consumption. By cooling the charge air, combustion is improved significantly, and the tendency to knocking decreases. At full load, the spark can be advanced, and combustion can be more efficient. Due to the higher mass flow, the turbocharger will also respond more readily. At lower engine speeds, this will lead to improved dynamics and torque. At high speed and load, the cooled charge air will also reduce the requirements for overfueling and thus fuel consumption. We have investigated the effects of charge air temperature on combustion in a 1.2 L three-cylinder gasoline engine. Our results show improvements in consumption, torque and dynamics. At increasing specific and absolute engine performance, even higher potentials can be expected.

F2012-A09-022

J. Stehlig (✉)
MAHLE Filtersysteme GmbH, Germany
e-mail: juergen.stehlig@mahle.com

R. Dingelstadt
MAHLE International GmbH, Germany

J. Ehrmanntraut · R. Mueller
Behr GmbH & Co. KG, Stuttgart-Feuerbach, Germany

J. Taylor
MAHLE Powertrain Ltd, Northampton, UK

SAE-China and FISITA (eds.), *Proceedings of the FISITA 2012 World Automotive Congress*, Lecture Notes in Electrical Engineering 190,
DOI: 10.1007/978-3-642-33750-5_46, © Springer-Verlag Berlin Heidelberg 2013

Keywords Charge air cooling · Turbocharging · Downsizing · CO_2 reduction · Reduction of fuel consumption

1 Introduction

In the foreseeable future, the internal combustion engine will continue to dominate in most vehicles. Therefore, its development has to comply with future emission standards. In addition to reducing emission, maintaining driving pleasure, dynamics and comfort continue to be the main target.

One way of combining these requirements is to increase specific engine output while reducing capacity. One of the main factors in achieving this goal is engine charging, however, rising charge air temperatures can be problematic. To solve this problem, MAHLE and Behr have developed a new charge air cooling system which is capable of dealing with future development steps in combustion engines. This two-stage charge air cooling system is called "cascaded charge air cooling". It is a significant further development of integrated indirect charge air cooling (i^2CAC).

Expertise in various areas is required in order to fully understand and successfully develop the charge air cooling system. MAHLE and Behr are ideal partners, providing core competence in cooling systems, air management and the interaction of engine systems. In addition, our cooperation combines all necessary development tools and appliances.

2 Influence of Charge Air Temperature

The effectiveness of gasoline engines depends, among others, on the combustion phasing. For physical reasons, the optimum of effectiveness is achieved when 50 % of the fuel is burnt at 8–10° CA after TDC. In order to prevent knocking at higher loads, the spark must be retarded, which exceeds the optimum (50 % mass fraction burned point) and causes a decrease in the engine's effectiveness. Exhaust temperatures will rise to critical values within the engine's range of operation.

In order to prevent damages to parts of the exhaust system and the turbine, additional fuel is usually injected for cooling. High degrees of pressure charging and higher compression ratios, as well as low octane rating, will compound these issues.

Charge air temperature reduction immediately influences the tendency to knocking as well as the requirements for overfueling, expanding the effective operating range of the engine.

Configuration	-	Inline-3
Capacity	cm^3	1200
Bore	mm	83
Stroke	mm	73.9
Compression Ratio	-	9.3:1
Turbocharger	-	BMTS 2011
Peak Power	kW	120 / 5000 – 6000
Specific Power	kW/l	100
Peak Torque	Nm	286 / 1600 – 3500
Specific Torque	Nm/l	238
ECU	-	MAHLE Flexible ECU (MFE)
Calibration	-	AFT_NCOD_2011-04-07A
Injector	-	Bosch Multi Hole (7 hole)
Spark Plug	-	Bosch M10
Engine Weight	Kg	125
Fuel	RON	98
Oil	-	0W40 Mobil 1
Coolant	-	50/50 G12+

Fig. 1 1.2 L three-cylinder MAHLE downsizing engine

3 Engine Test Results

3.1 Test Setup

A highly pressure charged 1.2 L three-cylinder gasoline engine has been used by MAHLE and Behr to examine the new cascaded charge air cooling system. Our main criteria for choosing this engine was its high specific performance, with 30 bar BMEP and 100 kW/L as well as high pressure charging requirements (Fig. 1), which offered ideal conditions to observe the effects of changes in the charge air temperature.

The engine testing was focused on fuel consumption, engine dynamics as well as performance and torque characteristics. Steady state measurement points were in the areas of knock limitation and overfueling (Fig. 2). In addition, dynamic characteristics were assessed with a load step at 2,000 rpm. Charge air temperatures inside the plenum after exiting the cooler were varied between 30 and 70 °C.

3.2 Full Load at Low Speed

Lowering the charge air temperature at full load and 1,400 rpm while maintaining the spark angle at the knock limit results in increased torque (Fig. 3). This effect can be explained by the higher density of the charge air, which results in increased engine air mass and resultant turbocharger feedback loop. Despite the increased torque, specific consumption remains at a constant level.

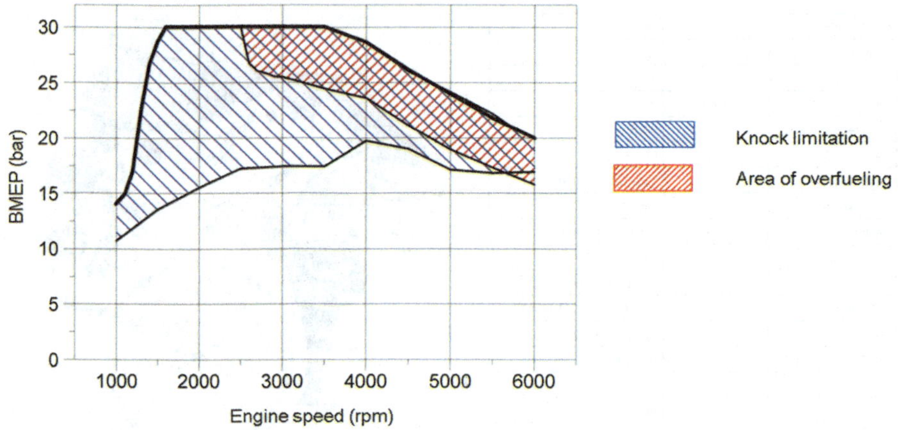

Fig. 2 Torque/speed curve for turbo-charged gasoline engine

Fig. 3 Run-up line torque (1,400 rpm, full load)

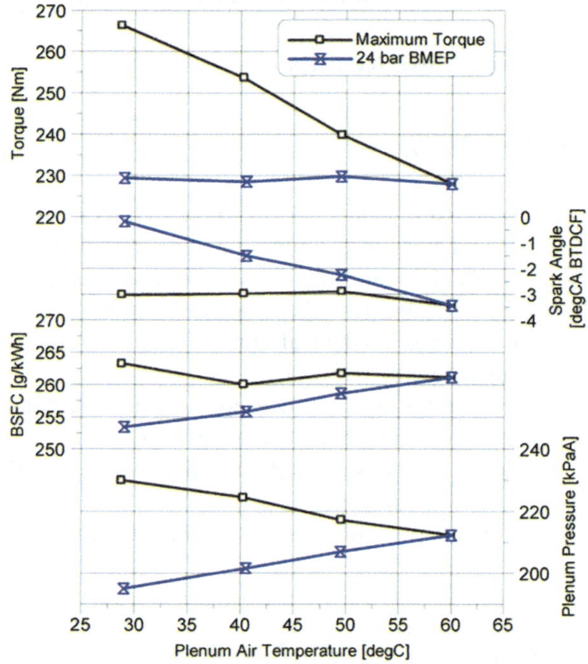

The lowered charge air temperature may also be used to reduce specific consumption at fixed torque. This is achieved by advancing the spark at the knock limit towards optimized effectiveness. Therefore, less fuel is required to maintain the torque level at this operating point. For this, injection pulse width is reduced and the quantity of air is adjusted by opening the waste gate. As a result, reduced pressure charging further lowers the charge air temperature.

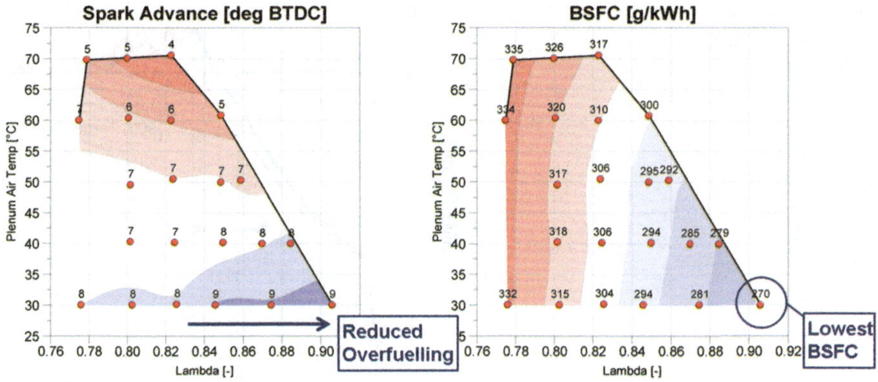

Fig. 4 Plenum temperature and Lambda maps at high load (5,000 rpm, 24 bar BMEP)

3.3 Full Load at High Speed

At high speed and load, exhaust temperatures may frequently reach critical values. Cooler charge air temperatures allow combustion phasing to be advanced, reducing end of combustion and exhaust temperature. As a result, less fuel needs to be injected for additional cooling, thus lowering consumption. Measurements at an operating point of 5,000 rpm and 24 bar BMEP have shown that cooler charge air and advanced spark have a positive effect on consumption in this case as well. The greatest advantage, at this operating point, is the reduction of overfueling (Fig. 4).

3.4 Transient Response

Due to the lower charge air temperature, air mass flow is increased, resulting in improved response and dynamics of the exhaust turbocharger. Measurements at an operating point of 2,000 rpm have shown advantages in the transient response. In Fig. 5, the charge air temperature is lowered from 60 to 40 °C, reducing the time to achieve 90 % of the target torque by 12 %.

4 Charge Air Cooling Systems

Charge air cooling systems currently in the market can be separated into direct (air/air) and indirect (air/coolant) charge air cooling. Currently, direct charge air cooling is predominant in gasoline as well as diesel engines, with a global market share of more than 80 %.

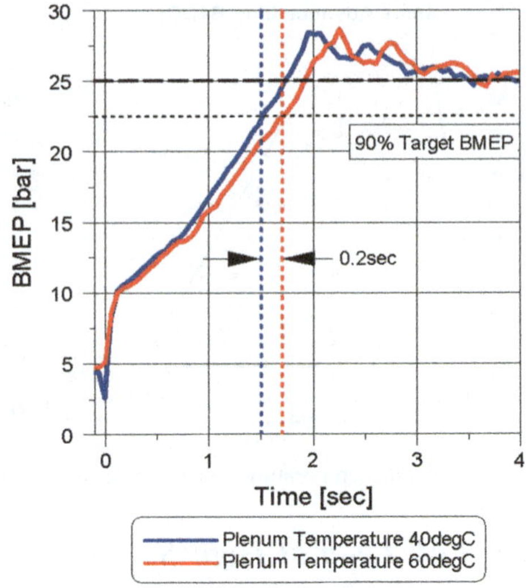

Fig. 5 Load step between 4 and 25 bar BMEP (2,000 rpm)

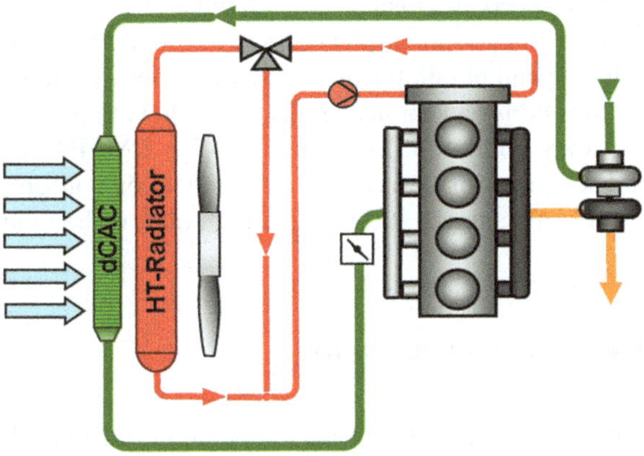

Fig. 6 Direct charge air cooling concept

4.1 Direct Charge Air Cooling

On direct charge air cooling concepts (Fig. 6), the charge air warmed by the compressor is cooled directly in the vehicle front end by an air/air cooler. In order to minimize pressure loss, the charge air cooler must have a suitable flow cross-section on the charge air side. Therefore, the available space is the main factor limiting the efficiency of direct charge air cooling. In common designs, a direct charge air cooler can achieve charge air exit temperatures of approximately 20–30 K above ambient temperature. The main advantages of this concept are low weight, complexity and cost.

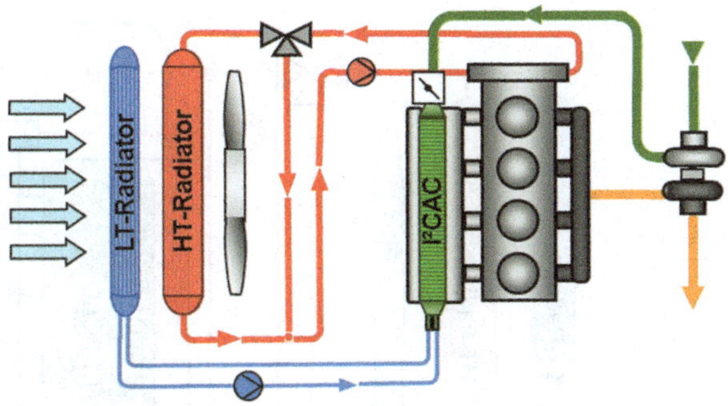

Fig. 7 Integrated indirect charge air cooling concept

4.2 Indirect Charge Air Cooling

Indirect charge air cooling utilises a liquid coolant to reduce the temperature. This coolant circulates in a separate low-temperature circuit. Heat is released to the surroundings via a low-temperature radiator in the vehicle's front end. Using a liquid coolant permits smaller flow cross-sections, allowing the low-temperature radiator to be designed as a full-face part. Compared to the direct charge air cooler, this solution has significant package advantages in the front end. The system's performance increases by approximately 10 K while at the same time pressure loss is reduced.

Therefore, indirect charge air cooling fulfills higher requirements for engine performance and response [1, 2].

4.3 Integrated Indirect Charge Air Cooling

One way of further exhausting the potential of indirect charge air cooling is to integrate the charge air cooler into the intake manifold [3] (Fig. 7). The most obvious advantage of this solution is a further reduction of space requirement.

In addition to package advantages, the common housing will also reduce cost. Also, it was possible to verify that this is the ideal part design to reduce pressure loss (by up to 80 %). It is therefore the basis for the cascaded system described below.

5 Cascaded Charge Air Cooling

Increasing specific performance has great potential to further increase the efficiency of combustion engines. In order to be able to demonstrate this, it was necessary to increase pressure charging, thus raising the requirements for cooling the charge air. With increased performance requirements for the engine, the physical limits of indirect charge air cooling are reached sooner. These physical

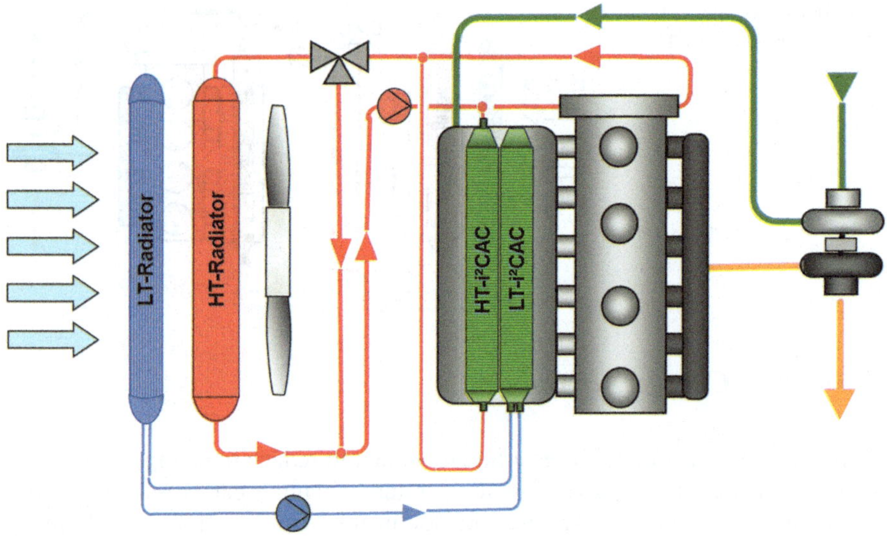

Fig. 8 Integrated cascaded charge air cooling concept

limits are set by the maximum heat rejection of the low-temperature radiator in the front end, and they cannot be pushed further by design changes of existing charge air coolers. Increasing cooling performance by enlarging the low-temperature radiator in the front end is not feasible due to the unfavorable relation between required additional space and additional performance.

The only remaining solution to further increase the cooling performance is to lower the temperature level in the low-temperature coolant. In the self-contained system of indirect charge air cooling, this temperature level depends exclusively in the quantity of heat transferred and, depending on for example the driving situation and weather, the flow rate of cooling air.

Basically, there are two ways of influencing the temperature level. The first variant, which is theoretical, is to increase the frontal area of the low-temperature radiator, thereby increasing the mass flow of cooling air. Practically, this is very difficult due to package conditions and would need power for the fan.

The second option is to insert a pre-cooling stage in front of the existing charge air cooler. With this solution, the temperature of the charge air is reduced before entering the system (Fig. 8). MAHLE and Behr have represented this pre-cooling on the test rig with the existing engine cooling system. In this test, charge air temperature enters the first charge air cooler at 200 °C. The first cooler is supplied with coolant at 90 °C exiting from the engine block. With this arrangement, we can demonstrate permanent average cooling of the charge air by 80 K down to 120 °C. Entrance temperature at the original cooler is now significantly lower, so that the temperature level in the coolant can be reduced by approximately 10 K. This difference can be transferred immediately to the charge air exit temperature (20 K compared to direct charge air cooling).

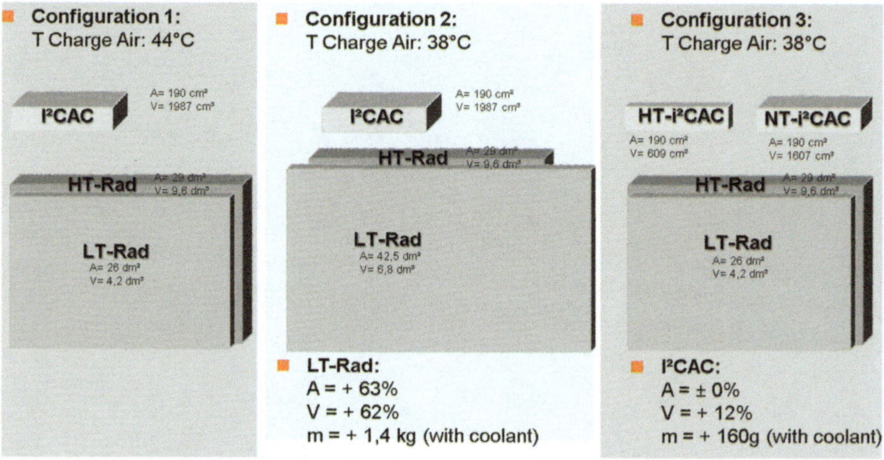

Fig. 9 Benefits of integrated cascaded charge air cooling concept

Due to the higher average heat flow density of typically 2–4 kW/dm^2 of the main radiator, compared to approximately 0.5–1 kW/dm^2 of the low-temperature radiator, this temperature difference can be reached efficiently. Comparatively few measures must be taken in the high-temperature cooling system. If an attempt was made to achieve the same level of charge air temperature exclusively with the low-temperature cooling circuit, the low-temperature radiator would require unrealistic dimensions (Fig. 9).

6 Summary

Integrated cascaded charge air cooling enables further steps in downsizing as well as higher levels of pressure charging. Present disadvantages of higher degrees of pressure charging (such as retarded spark for knock suppression and increased overfueling) are reduced by cascaded charge air cooling.

In addition, it offers the opportunity to choose a higher compression ratio, resulting in further advantages in fuel consumption in load ranges relevant to the legislative cycle.

References

1. Zuck B (2012) Downsizing—Auswirkungen auf das Wärmemanagement Proceedings. Wärmemanagement des Kraftfahrzeugs VIII expert-Verlag
2. Eilemann A, Stehlig J (2011) Indirekte Ladeluftkühlung—Ein Wegbereiter zur CO_2^- Reduzierung OEM Supplier
3. Szengel R et al (2007) Der TSI-Motor mit 90 kW MTZ 07-08 2007

The manufacturer's authorised representative in the EU is Springer Nature Customer Service Centre GmbH, Europaplatz 3, 69115 Heidelberg, Germany. If you have any concerns regarding our products, please contact ProductSafety@springernature.com

Printed and bound by CPI Group (UK) Ltd, Croydon, CR0 4YY
07/01/2026
02030926-0010